Electrode Materials in Energy Storage Technologies

Electrode Materials in Energy Storage Technologies

Applications in Lithium-, Sodium-, Potassium-, Sulfur- and Zinc-Based Rechargeable Batteries

Edited by Liqiang Xu

WILEY-VCH

Editor

Prof. Liqiang Xu
School of Chemistry and Chemical Engineering
Shandong University
Jinan
Shandong, 250100
China

And

School of Chemistry and Chemical Engineering
Ningxia University
Yinchuan
Ningxia, 750021
China

Cover Image: Courtesy of Liqiang Xu

Library of Congress Card No.: applied for

British Library Cataloguing-in-Publication Data
A catalogue record for this book is available from the British Library.

Bibliographic information published by the Deutsche Nationalbibliothek
The Deutsche Nationalbibliothek lists this publication in the Deutsche Nationalbibliografie; detailed bibliographic data are available on the Internet at <http://dnb.d-nb.de>.

The manufacturer's authorized representative according to the EU General Product Safety Regulation is Wiley-VCH GmbH, Boschstr. 12, 69469 Weinheim, Germany,
e-mail: Product_Safety@wiley.com.

Print ISBN: 978-3-527-35358-3
ePDF ISBN: 978-3-527-84576-7
ePub ISBN: 978-3-527-84578-1
oBook ISBN: 978-3-527-84577-4

Typesetting Straive, Chennai, India

Contents

About the Editor

Professor Liqiang Xu is a prominent researcher in fields of novel energy storage and synthetic chemistry. He has been recognized as Distinguished Young and Middle-Aged Scholar at Shandong University, High-Level Talent of Shandong Province, a Senior Member of the Chinese Chemical Society, a Fellow of the Royal Society of Chemistry, and a Changjiang Scholar Distinguished Professor of the Ministry of Education, P. R. China.

He as the first/corresponding author has published over 140 papers in top journals such as *Chemical Society Reviews*, *Journal of the American Chemical Society*, *Angewandte Chemie International Edition*, *Advanced Materials*, and so on.

He has significantly advanced high-performance electrodes, tackling many key challenges in the development of next-generation energy storage systems.

Through his comprehensive creative research efforts, high-quality teaching, and active journal editorial board member activities, Professor Xu continues to push the frontiers of electrochemical energy storage, contributing to the future development of sustainable and high-performance battery technologies.

Preface

Welcome to our book named *Electrode Materials in Energy Storage Technologies: Applications in Lithium-, Sodium-, Potassium-, Sulfur-, and Zinc-Based Rechargeable Batteries*. This book is all about the exciting world of rechargeable battery technology and the amazing materials that make it all possible.

We're living in a time where finding better ways to store energy is super important. With climate change and the need for renewable energy, creating better batteries is a hot topic. In this book, we dive into the latest and greatest research on different electrode materials. From the well-known lithium-ion batteries to the newer promising sodium-, potassium-, sulfur-, and zinc-based rechargeable batteries, we cover it all.

Each chapter is packed with information about how these batteries work, how to design and improve the critical electrode materials, *in situ* detections, the structure and property relationship investigations, and the challenges we face in making them even better. It's a real team effort, with scientists, engineers, and industry experts all working together to push the boundaries of what's possible.

Whether you're a researcher, teacher, student, or just someone curious about rechargeable batteries, we hope this book provides you a deeper understanding and sparks new ideas. We're grateful to all the contributors who helped make this book a reality. Our goal is to inspire and drive the development of next-generation energy storage technologies that will help create a sustainable future.

Thanks for joining us on this journey. We can't wait for you to dive in and explore the fascinating world of electrode materials and energy storage.

Happy reading!

Prof. Liqiang Xu
Shandong University
and
Ningxia University

1

Research Developments on Lithium-Ion Batteries

Lishan Yang, Da Xiong, Yangfan Li, Xiang Wang, Boyao Gan, and Xinkang Li

National and Local Joint Engineering Laboratory for New Petrochemical Materials and Fine Utilization of Resources, Hunan Normal University, Changsha 410081, PR China

Outline

The endurance of electric vehicles and the energy storage capacity of power grids demand higher performance from energy storage devices. Since their commercialization in 1991 by Sony, LIBs have rapidly gained traction in both industry and scientific research due to their high output voltage, high energy density [1], and long cycle stability [2, 3]. LIBs are primarily composed of cathode materials [4], anode materials [5], separators, and electrolytes [6]. Over the past three decades, the energy density of LIBs has increased nearly fourfold, solidifying their role in modern technology. The 2019 Nobel Prize in Chemistry was awarded to three pioneering scientists in LIBs research, underscoring their significant contributions to technological advancement [7].

The United States initially led the development of LIBs, followed by Japan and South Korea. Although China entered the LIB industry later, it surpassed Japan and South Korea in lithium-ion car battery patents by 2023 [8, 9]. The LIBs industry value chain includes raw materials, intermediate products, batteries, system integration, and battery recycling, with varying profit margins and revenues across these segments. Battery and new energy vehicle manufacturers dominate the supply chain, but raw material prices significantly impact battery costs. The trend toward vertical integration in the LIBs industry blurs traditional boundaries between upstream, midstream, and downstream sectors, with companies increasingly diversifying their operations.

Declining raw material costs (e.g., lithium carbonate) further enhance LIBs affordability, positioning them for steady growth until 2030. [10].

1.1 Polyanion Cathodes

Polyanion-type cathodes offer several advantages, including low cost, high specific capacity, excellent cycle performance, and enhanced safety [11]. Currently, the main

Electrode Materials in Energy Storage Technologies: Applications in Lithium-, Sodium-, Potassium-, Sulfur- and Zinc-Based Rechargeable Batteries, First Edition. Edited by Liqiang Xu.

Table 1.1 $LiFePO_4$ and $LiFe_{1-X}Mn_XPO_4$ cathode materials production information.

Cathodes	Synthetic technologies	Major producers
$LiFePO_4$	Spray-drying method Solid-phase method Hydrothermal method	Hunan Yuneng New Energy Battery Material, Dynanonic, Fulin Precision Machining
$LiFe_{1-X}Mn_XPO_4$	Spray-drying method Solid-phase synthesis	Ronbay Technology, Lithitech, HCM

commercial polyanion-type cathode materials include $LiFePO_4$ and $LiFe_{1-X}Mn_XPO_4$. Common production methods and manufacturers are summarized in Table 1.1.

1.1.1 Lithium Iron Phosphate

In 1997, Goodenough and colleagues first proposed that $LiFePO_4$ could be used as a cathode material for LIBs [12], has a theoretical specific capacity of 170 mAh g^{-1}. During the charge–discharge process, $LiFePO_4$ and $FePO_4$ are mutually converted. The structure is shown in Figure 1.1a. $LiFePO_4$ belongs to the orthorhombic crystal system and is classified in the Pnma space group. The oxygen atoms create a hexagonal structure, with lithium and iron atoms situated at the center, forming regular FeO_6 and LiO_6 octahedra. A strong covalent bond is formed between P and O, which stabilizes the oxygen atom and prevents its release due to oxidation during charging. The de/intercalation of Li^+ during cycling does not cause a significant change in the volume of $LiFePO_4$, which gives $LiFePO_4$ material strong thermodynamic stability. The P-O tetrahedral structure, interspersed within the Fe-O framework, impedes the deintercalation and intercalation of Li^+ ions during the cycling process. This leads to a low Li^+ diffusion rate and reduced electronic conductivity in $LiFePO_4$ batteries [13]. Under low-temperature conditions, the Li^+ diffusion rate of the material will be further reduced, the charge transfer impedance will be significantly increased, and the battery capacity will be significantly attenuated.

In view of the above problems of $LiFePO_4$, the modification methods are mainly carried out in three aspects [14]: (i) Surface coating: The conductivity between $LiFePO_4$ particles was improved by constructing a stable C coating layer. The conductive coating formed on the surface of the material can effectively control the size of secondary particles, reduce the transmission distance of Li^+ in the material, and improve electrochemical performance. (ii) Ion doping: There are three types of doping: lithium-site doping, iron-site doping, and lithium-iron mixed doping. The ion radius of lithium-site doping is small, such as Mg, Al, and Na. The iron-site-doped ions are mainly Mg, Cu, V and Ti. The doping and function of each site are detailed in Table 1.2. (iii) Morphology control: Nanocrystallization can effectively improve the diffusion rate of Li^+. The preferred crystal orientation was optimized, and the crystal structure of $LiFePO_4$ was changed.

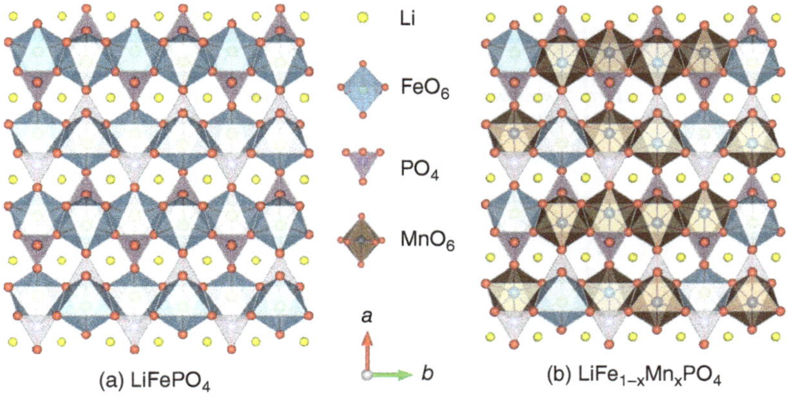

Figure 1.1 The crystal structure simulations of (a) $LiFePO_4$ and (b) $LiFe_{1-X}Mn_XPO_4$.

Table 1.2 The functions of elemental doping on $LiFePO_4$.

Doping sites	Elements	The effects of elemental doping
Li	Mg	Promotes the migration of Li^+
	Ti	Improves conductivity and enhances rate ability
Fe	Cu, V, Ti	Enhances rate ability
	Zn	Increases the transmission space for Li^+ and enhances rate ability
Li and Fe	La, Mg	Improve low-temperature performance
	Mo	Improve conductivity and enhance rate ability

Source: [11]. Electrochemical Energy Reviews 2021/John Wiley & Sons.

1.1.2 Lithium Manganese Iron Phosphate

The structure of $LiFe_{1-X}Mn_XPO_4$ is analogous to that of $LiFePO_4$, as illustrated in Figure 1.1b. $LiFe_{1-X}Mn_XPO_4$ features a voltage platform of 4.1 V for Mn^{2+}/Mn^{3+} and a voltage platform of 3.4 V for Fe^{2+}/Fe^{3+}; the solid solution of iron phosphate and manganese phosphate is formed when all lithium ions are removed [15, 16]. The addition of Mn atoms not only enhances the performance of $LiFePO_4$ materials but also brings new challenges. The difference in the proportion of manganese and iron introduced in the material will significantly cause differences in electrochemical performance: When the content of manganese is too low, the platform voltage is not significantly increased, and the energy density of the material is not significantly increased. When the manganese content is excessively high, it can induce the Jahn–Teller effect, causing lattice distortion. The dissolution of manganese reacts with the electrolyte, which negatively impacts the cycle stability and capacity retention rate [17–19]. The modification is carried out from three aspects [20]: ion doping, surface coating, and morphology control to improve the intrinsic conductivity, rate

Table 1.3 The performance of various $LiFe_{1-X}Mn_XPO_4$ cathode materials.

Cathode materials	Potential window/ V (vs. Li^+/Li)	Discharge capacity at 0.1C ($mAh\,g^{-1}$)
$LiFe_{0.5}Mn_{0.5}PO_4$	2.7–4.5	142 (0.05C)
$LiFe_{0.6}Mn_{0.4}PO_4$	2.0–4.5	160.2
$LiFe_{0.7}Mn_{0.3}PO_4$	2.5–4.5	162
$LiFe_{0.9}Mn_{0.1}PO_4$	2.5–4.6	145

Source: [21]/John Wiley & Sons.

performance, and phase transition mechanism. The performance of $LiFe_{1-X}Mn_XPO_4$ with common ratios is shown in Table 1.3.

With the rapid growth of the new energy automotive industry and the demand for base station energy storage, significant progress has been made, and the demand for low-cost, high-performance cathodes will continue to rise. Lithium iron phosphate will continue to dominate, and lithium manganese iron phosphate is expected to be more widely used in high-energy-density batteries.

1.2 Layered Oxide Cathodes

Layered oxide cathode materials are pivotal in LIBs due to their high energy density and excellent electrochemical performance [22]. Generally, layered oxide cathode materials include the earliest commercialized lithium cobalt oxide ($LiCoO_2$), high-energy-density nickel cobalt manganese oxides ($LiNi_{1-x-y}Co_xMnyO_2$, NCM), and the current research focus, lithium-rich phase materials ($xLi_2MnO_3{\cdot}yLiTMO_2$, LRM). Common production processes and manufacturers and 2023 annual production are listed in Table 1.4.

Discovered by Professor J.B. Goodenough in 1980, $LiCoO_2$ has a theoretical capacity of 274 $mAh\,g^{-1}$ and remains a primary cathode material for portable electronics [23]. $LiCoO_2$ features an α-$NaFeO_2$-type layered structure, belonging to the hexagonal crystal system, the R-3M space group. The thermodynamically stable O3 phase of $LiCoO_2$ consists of edge-sharing LiO_6 and CoO_6 octahedra. As Li^+ is extracted, the O3 phase undergoes three phase transitions: the first transition occurs at a Li^+ extraction level of 0.07–0.25, resulting in an H1–H2 transition; the second transition occurs at about 0.5 extraction, involving a disordered-to-ordered transition of Li^+ and a structural change from hexagonal to monoclinic; and the third transition occurs at 0.6–0.9 extraction, with the voltage rising to about 4.5 V, resulting in an O3–O1 transition, with the intermediate H1-3 phase causing significant lattice (c-axis) and volume contraction. Higher voltages can extract more lithium ions, releasing more energy, but high-voltage charging may lead to rapid capacity, efficiency, and cycle life decay due to surface degradation, destructive phase transitions, and uneven reactions [24]. Doping and surface coating methods are widely used

Table 1.4 Production informations of layered oxide cathode materials.

Cathodes	Synthesis technologies	Producers
$LiCoO_2$	Solid-phase synthesis Coprecipitation method Sol–gel method	RiseSun MGL, BSBM, Ronbay Tech., Xiamen Tungsten
NCM	Coprecipitation method Solid-phase synthesis	Ronbay Tech., Tianjin B&M, Minmetals New Energy, LG Chem
LRM	Coprecipitation method Solid-phase synthesis	Easpring, Ronbay Tech., BASF Shanshan, Ningxia Hanyao

for material modification. Coating materials include $LiAlO_2$, Al_2O_3, and lithium-ion conductors like Li_2ZrO_3; doping elements include Zr, Al, Ti, Nb, and multielement co-doping [25].

For example, for $LiCoO_2$ materials, the theoretical capacity calculation formula is as follows:

$$C = \frac{nF}{M} \tag{1.1}$$

C is the theoretical capacity in mAh g^{-1}, n is the number of moles of electrons transferred per mole of silicon (for $LiCoO_2$, $n = 1$), F is the Faraday constant (96 485 C mol^{-1}), and M is the molar mass of silicon (98 g mol^{-1}). For $LiCoO_2$, this results in a theoretical capacity of approximately 274 mAh g^{-1} [26].

Ternary materials (NCM) partially replace Co in $LiCoO_2$ with Ni and Mn to reduce costs and improve electrochemical performance through synergistic effects among the elements [27]. Different transition metals have unique physical properties: Ni increases cathode capacity, Co enhances charge–discharge kinetics, and Mn improves structural stability during cycling. Various NCM cathode materials are derived based on the ratio of Ni, Co, and Mn, such as 424, 333, 523, and 811. $LiNi_{1\text{-}x\text{-}y}Co_xMnyO_2$ also has an α-$NaFeO_2$-type layered structure (R-3m space group), with its crystal structure shown in Figure 1.2. Li^+ and transition metals alternately occupy the 3a (0 0 0) and 3b (1 1 1/2) sites, with O^{2-} located at the 6c (0 0 z) site.

In industrial production, the coprecipitation method is the standard synthesis method for polycrystalline ternary cathode materials. Increasing the synthesis temperature promotes grain growth, resulting in single-crystal particles with better stability and electrochemical performance. The theoretical specific capacity of ternary materials is about 275 mAh g^{-1}. Different compositions of ternary materials have different discharge-specific capacities in the voltage range of 2.7–4.3 V, usually

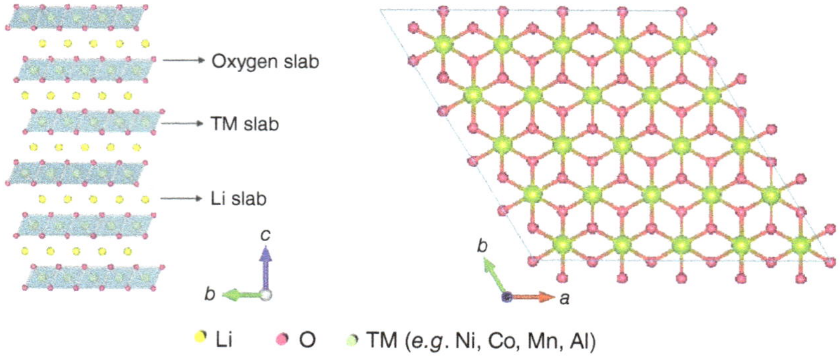

Figure 1.2 The crystal structure simulation of $LiNi_{1-x-y}Co_xMn_yO_2$.

increasing with higher Ni content. As the demand for longer driving ranges and lower battery costs in the electric vehicle market increases, the Ni content in ternary materials continuously increases, making nickel-rich ternary materials an important development direction [28]. However, in ternary LIBs, the cathode material undergoes phase transitions from the original hexagonal phase (H1) to two other hexagonal phases (H2 and H3) during lithium-ion extraction. The failure modes of ternary LIBs are complex behaviors: (i) Due to the similar ionic radii of Ni^{2+} (0.69 Å) and Li^+ (0.76 Å) and their low migration barriers, increasing the Ni content in the material leads to Li–Ni site exchange at the 3b lattice sites, resulting in Li^+/Ni^{2+} cation mixing. This cation mixing obstructs the Li^+ transport channels, leading to low initial efficiency and poor long-term cycling performance of the battery [29]. (ii) The extraction of Li^+ ions necessitates charge compensation, resulting in the formation of highly oxidative Ni^{4+} species. These species facilitate the decomposition of the electrolyte on the cathode surface, consuming active lithium ions in the system and compromising the thermal stability of the material [30]. (iii) H^+ species in the electrolyte attack the surface of the cathode material, causing the dissolution of transition metal ions from the electrode surface. This induces heterogeneous interfacial chemical reactions, increasing interfacial impedance. Additionally, the dissolved transition metal ions may migrate to the anode surface and deposit on the solid electrolyte interphase (SEI) layers [31]. To address these issues, ion doping, morphology control, and concentration gradient methods are employed to enhance the electrochemical performance of ternary materials [32, 33]. Table 1.5 lists the effects of different element doping, and Table 1.6 shows the types and main functions of commonly used coating materials.

When the Co position in the layered structure also contains Li, the material is referred to as lithium-rich cathode material $x\text{Li}_2\text{MO}_3{\cdot}(1-x)\text{LiM}_2$ (M = Ni, Co, Mn). This material typically consists of a composite of Li_2MnO_3 and $LiMO_2$-layered structures in varying proportions, with a theoretical capacity of up to 280 mAh g^{-1}. $LiMO_2$ belongs to the hexagonal crystal system with the same α-$NaFeO_2$ configuration (R-3m space group) as $LiCoO_2$, while Li_2MnO_3 has a superstructure ordering of Li and Mn atoms, belonging to the monoclinic crystal system.

Table 1.5 The effects of elemental doping on NCM cathodes.

Elements	Main effects
Mg	Cation ordering and "pillar" effect
Al	Relieving phase transition and refining particles
Zr	Cation ordering and "pillar" effect
Ti	Cation ordering and "pillar" effect
B	Modulating microstructure
W	Decreasing the surface passivation layer
Na	Improving Li^+ kinetics and mitigating transition metals migration
K	Improving Li^+ kinetics and mitigating transition metals migration
F	Enhancing the oxygen framework stability
S	Enhancing the oxygen framework stability
Nb	Spheroidization of primary particles

Source: [32]/John Wiley & Sons.

Table 1.6 Various coating materials and main effects.

Coating layers		Main effects
Metal oxides	Al_2O_3 TiO_2 CeO_2	Initiative coating protective layers
Phosphates	$AlPO_4$ $Cu_3(PO_4)_2$	Improvement of cyclic and thermal stability
Fluorides	AlF_3 LiF	Prevention of corrosion by hydrofluoric acid
Lithium transition metal oxides	Li_2ZrO_3 $Li_4Ti_5O_{12}$ $LiAlO_2$ $Li_{1.3}Al_{0.3}Ti_{1.7}(PO_4)_3$	Protective layer with excellent lithium-ion conductivity
Carbon materials	PPy CNT	Improvement of structural stability and electronic/Li^+ conductivity

Source: [33]. Oxygen loss in layered oxide cathodes/John Wiley & Sons.

Lithium-rich phase materials have a high lithium/transition metal molar ratio and higher discharge capacity. However, during charging, the oxygen anions in lithium-rich phase materials partially oxidize to O^- radicals or oxygen gas and escape from the lattice, reacting with the electrolyte [34]. During cycling, transition metals enter the Li layer, causing the material to transition to a spinel structure, lowering the discharge plateau [35]. Surface modification, bulk doping, and electrolyte design methods can alleviate lattice oxygen loss and metal ion migration in lithium-rich manganese-based materials during cycling, improving cycle performance and initial coulombic efficiency [36].

1.3 Spinel Structure Cathodes

Lithium manganese oxide ($LiMn_2O_4$) and high-voltage lithium nickel manganese oxide (LNMO) are both spinel-type cathode materials utilized in LIBs [37]. They are characterized by their high energy density, excellent cycle stability, and prolonged service life, rendering them extensively applied within the battery industry [38]. According to the statistics of Huajing Industrial Research Institute, the output of the lithium manganate industry in 2021 has been greatly improved compared with 2020, with an output of 87 400 tons, an increase of 45.9%. The import volume was 305.84 tons. The year-on-year growth was 42.05%, and the export volume was 468.96 tons, a decrease of 8.88%. Common production processes and major manufacturers are delineated in Table 1.7 for reference.

The first lithium manganate cathode material with three-dimensional lithium-ion channels was developed by Hunter in 1981 [39]. In LIBs, charge and discharge with $LiMn_2O_4$ as the cathode are shown as follows:

Cathode:

$$LiMn_2O_4 \underset{\text{discharge}}{\overset{\text{charge}}{\rightleftharpoons}} Li_{1-\lambda}Mn_2O_4 + \lambda Li^+$$

During the charging process, 8a lithium ions traverse the 8a–16c–8a channel to dissociate from the three-dimensional lattice, while Mn^{3+} undergoes electron loss and oxidation to become Mn^{4+}. Upon discharging, lithium ions are incorporated into the 8a position under the influence of electrostatic force, while Mn^{4+} acquires electrons and is reduced to Mn^{3+}, ultimately leading to a structural transformation into $LiMn_2O_4$. Manganese dioxide was regarded as one of the most promising cathode materials for LIBs due to its theoretical specific capacity of 148 mAh g^{-1} and a discharge voltage plateau at 4.15 V. However, the Jahn–Teller effect gives rise to significant capacity degradation issues such as material deterioration and electrolyte

Table 1.7 $LiMn_2O_4$ and $LiNi_{0.5}Mn_{1.5}O_4$ cathode materials production information.

Cathodes	Synthetic technologies	Major producers
$LiMn_2O_4$	Solid-phase synthesis Microwave synthesis Hydrothermal synthesis	DX Energy, Boshi High-Tech., Xiangtan Electrochemical, Xinxiang Hongli, South Manganese
$LiNi_{0.5}Mn_{1.5}O_4$	Solid-phase synthesis	Boshi High-Tech., BSBM

decomposition [40]. The most effective solution to this problem involves partially substituting Mn with other metal ions (such as Li, Mg, Al, Ti, Cr, Ni, and Co) and modifying the surface of the materials [41, 42].

High-pressure lithium nickel manganate $LiNi_{0.5}Mn_{1.5}O_4$ was first reported by Blasse in 1964 as specific $LiNi_{0.5}Mn_{1.5}O_4$ along with a series of mixed metal spinel oxides [43]. Lithium nickel manganate crystals have two different structures: one is a disordered structure, which is also known as the spinel type and expressed as D-LNMO, in which nickel and manganese ions are disordered at the 16d site of the lattice. The other is the olivine structure, represented as O-LNMO, in which nickel and manganese ions are orderly distributed in the lattice, occupying the 4a and 12d sites, respectively. Its reversible capacity is 146.7 mAh g^{-1}, similar to that of lithium manganate; in the battery charging process, lithium ions will be removed from a specific point through a three-dimensional channel; Ni^{2+} and Ni^{3+} will be oxidized into Ni^{4+}; and $LiNi_{0.5}Mn_{1.5}O_4$ will eventually be converted into $Ni_{0.5}Mn_{1.5}O_4$; and the discharge process is opposite to the charging process. The high-voltage lithium nickel manganate discharge voltage can reach 4.7 V, mainly because of its unique chemical composition and electrochemical properties; nickel provides a higher voltage; and manganese helps to inhibit the excessive oxidation of nickel. However, when it is operated under high-voltage conditions during the cycle, side reactions occur between its surface and electrolyte, resulting in a short cycle life due to rapid capacity decline [44]. Doped metal ions and surface coatings can be used to reduce side reactions [45, 46].

Lithium manganate is an important precursor for the synthesis of other LIBs cathode materials, such as lithium-rich materials, multicomponent materials, and solid lithium manganese oxides. Also, the solid-state battery made of high-pressure lithium nickel manganate material as a positive electrode has higher energy density, higher safety, and lower cost advantages than ordinary lithium ions. It is believed that under this development trend, lithium manganate and high-pressure lithium nickel manganate will be more widely used.

1.4 Anode Materials

Anode materials are crucial for the electrochemical performance of LIBs. Pure lithium metal is ideal due to its high capacity and lightweight, but safety concerns with lithium dendrites limit their use. Currently, anode materials are categorized into carbon and non-carbon types. Carbon materials include graphite, soft carbon, and hard carbon. Non-carbon materials encompass silicon-based materials, lithium titanate, and tin-based materials [47–49].

The ideal anode material for LIBs should possess low chemical potential, high coulombic efficiency, and good electrical conductivity, stability, and compatibility; should be abundant; and should have low cost. Currently, graphite carbon is the most widely used anode material, offering the best overall performance to meet these criteria. In battery production, anode materials account for about 10% of the total material cost. Research is focused on developing low-cost, high-capacity anode materials. Silicon-based materials are promising for next-generation high-energy-density LIBs due to their high theoretical capacity, environmental friendliness, and abundant reserves. The specific details of common lithium anode materials are shown in Table 1.8 [56, 57, 59].

For example, for silicon-based materials, the theoretical capacity calculation formula is as follows:

$$C = \frac{n\mathrm{F}}{\mathrm{M}} \tag{1.2}$$

C is the theoretical capacity in $\mathrm{mAh\,g^{-1}}$, n is the number of moles of electrons transferred per mole of silicon (for silicon, $n = 4.4$), F is the Faraday constant ($96\,485\ \mathrm{C\,mol^{-1}}$), and M is the molar mass of silicon ($28.0855\ \mathrm{g\,mol^{-1}}$). For silicon, this results in a theoretical capacity of approximately $4200\ \mathrm{mAh\,g^{-1}}$ [60].

As new energy vehicles and energy storage markets rapidly develop, the need for high-performance anode materials will continue to increase. While graphite anode materials will remain prevalent, emerging alternatives like silicon-based and SiO_x materials are anticipated to see broader application in high-energy-density batteries. Advancements in technology and process enhancements will be crucial for the progress of anode materials.

1.5 Cell Technology

The cell technology of energy storage batteries refers to the unit technology used to store and release electrical energy in energy storage batteries, mainly involving cell materials and cell integration technology. The cell technology is a key factor in determining the output power density, cost, and service life of energy storage batteries [61, 62].

1.5.1 Battery Cells

With the rapid increase in demand in the energy storage market, various types of energy storage cells have been developed, as shown in Figure 1.3, and have been widely used. Coin cells have the advantages of small size, safe and convenient use, and wide applicability in various electronic products. The full-cell button battery assembled from lithium iron phosphate/graphite and NCM811/silicon can achieve the specific energy of 100–160 and $300–350\ \mathrm{Wh\,kg^{-1}}$ and the specific power of 160–230 and $300–350\ \mathrm{Wh\,kg^{-1}}$, respectively [63]. The specific energy (E) and specific power (P) are calculated as follows:

$$E = \frac{k \cdot \Delta U \cdot C_p}{M} \tag{1.3}$$

k is the mass fraction of the active material, that is, the sum of the cathode active material mass m_p and the anode active material mass m_n divided by the total cell

Table 1.8 Main parameters of various anode materials for lithium-ion batteries.

Anode	Lithium storage mechanism	Storage potential/V (vs. Li^+/Li)	Specific capacity/($mAh\,g^{-1}$)	Initial coulombic efficiency/%	Major producers	Ref.
Graphite	Intercalation mechanism	~ 0.1	372	85–90	BTR, BSBM, etc.	[50]
Hard carbon	Adsorption–intercalation mechanism, intercalation–adsorption mechanism, adsorption–intercalation–adsorption mechanism, etc.	0–1.5	200–600	~ 70	Kureha Corporation, SGL Carbon, etc.	[51–53]
Soft carbon	Adsorption–intercalation mechanism, etc.	0–1.2	> 350	~ 50	BSBM, BTR, etc.	[54]
Si	Alloying reaction mechanism	< 0.1	650–4200	65–93	Shin-Etsu Chemical Co., Ltd., SUMCO Corporation, etc.	[55]
Si/C composite anode	Alloying reaction and intercalation mechanism	0.01–1.5	700–3500	47–95	Group14 Technologies, Showa Denko, etc.	[56, 57]
SiO_x	Conversion reaction and alloying reaction mechanism	0.01–0.5	785–2875.2	44–82.2	Mitsubishi Chemical, Toppan Printing, etc.	[58]

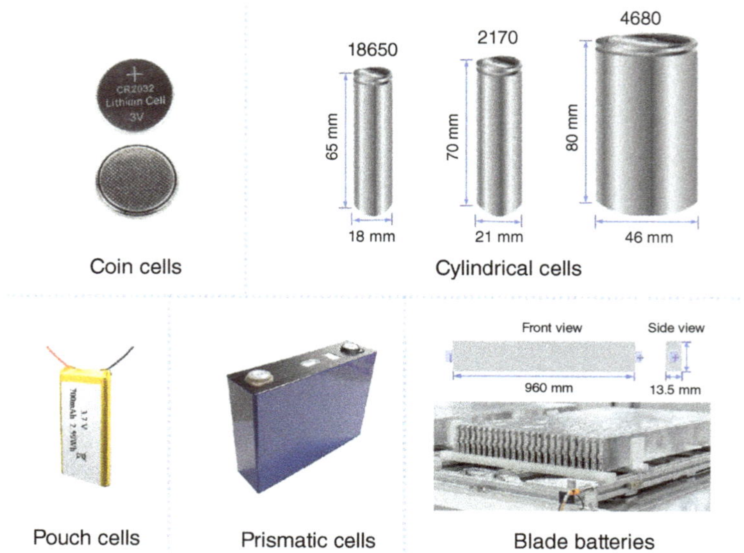

Figure 1.3 The pictures of commercial coin cells, cylindrical cells, pouch cells, prismatic cells, and blade batteries.

mass. ΔU is the average voltage difference between the cathode and anode. C_p is the gram capacity of the cathode active material. M is the total battery mass.

$$P = \frac{E}{\sqrt{t}} \tag{1.4}$$

E is the specific energy. t is the discharge time.

Additionally, the cylindrical cells have the characteristics of large capacity, long cycle life, and wide operating ambient temperature, which have gained a firm foothold in the power market with the unified model and standardized production. Japan's Sony company first designed a standard LIB, the 18 650 cylindrical cells (18 mm diameter and 65 mm axial length) [64], but the energy density is lesser and the charging speed is slower. Compared to the 18 650 battery, the energy density of the whole package composed of the 2170 battery (21 mm diameter and 70 mm axial length) can be further improved, and the charging speed can be faster. Later, Tesla's 4680 battery (46 mm diameter and 80 mm axial length) only changed in terms of cell design, while the levelized cost of electricity (LCOE) of the cell was reduced by about 14% compared with the 2170 battery, and the power of a single cell was increased to 5.48 times. Many other battery types, such as pouch cells, have the advantage of high energy density and lightweight but require additional protection against battery damage and thermal runaway. Prismatic cells are currently the most widely used cells in the international field, which have advantages such as high strength, small internal resistance, and high space utilization; however, it is difficult to unify the standard production process, and the heat dissipation during operation is poor.

The blade battery is a single battery with a long and thin blade-like shape and adopts cell-to-pack (CTP) module-free integration technology [64]. The cell structure adopts the design of the battery pack instead of the traditional shell structure; meanwhile, the beam and cell of the battery are directly applied by the blade battery; and finally, the top and bottom are connected by aluminum plates.

BYD launched the lithium iron phosphate blade batteries in 2020, which increase the energy density by 50%, reduce the manufacturing cost by 30% without changing the battery system, and can withstand safety level tests such as collision, high temperature, and puncture, which greatly expands the application field of LIBs.

The blade battery is a single battery with a long and thin blade-like shape and adopts CTP module-free integration technology. Due to the change in the battery structure, the design of the battery pack cancels the shell structure of the traditional battery. The blade battery acts as the beam and cell of the battery and then adopts the design of honeycomb aluminum plates, with two high-strength aluminum plates pasted on the upper and lower sides. The blade batteries are arranged within this structure.

1.5.2 Cell Integration Technology

In the early days, major battery manufacturers mainly used cell-to-module (CTM) cell integration [65], which is a mode that integrates battery cells directly into modules, and the general configuration method is "cell–module–pack loading," but the space utilization rate of the module configuration mode is only 40%, and with the increase of energy density requirements of power trams, it is gradually replaced by technologies such as CTP and cell to chassis (CTC). CTP integration technology integrates the battery cells directly into the battery pack, eliminating the need for standardized module linking, also known as "module-less battery" [66]. From the perspective of product performance, compared to conventional battery packs, the CTP method has increased the volume utilization rate by 15–20% and the production efficiency by 50%, and the energy density can reach more than 200 Wh/kg, while effectively reducing costs. As a result, CTP has become a mainstream battery cell integration technology [67]. Products with typical CTP technology include BYD's blade batteries, CATL's Kirin batteries, and Svolt Energy Technology Co., Ltd. (SVOLT's) short-knife batteries. The CTC first proposed by Tesla is considered to be the key core technology in the next stage of new energy vehicles. CTC technology directly integrates the battery cell into the vehicle chassis [68]. It reduces the loss of space caused by the connection between the car body and the battery cover, thus improving space utilization.

1.6 Electrolyte

The electrolyte is a vital component in every electrochemical device [69], which plays a pivotal role in promoting the migration of ions between the anode and cathode, and is vividly called the "blood" of the battery. The electrolyte not only plays an important role in regulating the performance of the electrode and the electrolyte interface but also directly impacts the specific capacity, internal resistance, charge and discharge performance rates, operating temperature range, and storage characteristics of the battery [68]. To date, the mainstream LIBs electrolytes are often composed of solvents, lithium salts, and functional additives.

The primary constituent of the electrolyte is an organic solvent, which can dissolve lithium salts and deliver a carrier for lithium ions. An ideal LIBs electrolyte requires organic solvents to meet a variety of conditions [70]: high dielectric constant to ensure sufficient dissolution of lithium salts; low viscosity; high electrochemical and chemical stability; low melting point, high boiling point, and high flash point; and non-toxic, harmless, and cost-effective. Currently, the commonly used organic solvents are mainly carbonates and organic ethers, as shown in Table 1.9. In addition, a single solvent is difficult to meet the requirements of batteries in extreme environments, so mixed solvents are often used to combine multiple excellent properties.

The application of lithium salt in LIBs needs to meet the following characteristics [71]: high solubility in organic solvents, easy dissociation to release Li^+ ions, and high conductivity. It exhibits excellent antioxidant stability and remains unaffected by electrochemical and thermodynamic reactions with organic solvents, electrode materials, and battery components. Moreover, it is environmentally friendly and boasts straightforward preparation, purification, and industrial scalability. Commonly used lithium salts are shown in Table 1.10.

Electrolyte additives refer to specific chemicals mixed into the electrolyte. In general, the number of additives is not more than 20% by weight or volume percentage. Compared with conventional electrolytes, including only solvents and lithium salts, electrolytes containing different additives have a huge impact on battery performance, which can adjust some physical and chemical properties of the battery and significantly enhance the electrochemical performance [73]. The common types of electrolyte addition and their effects are shown in Table 1.11.

Table 1.9 Physical properties of common organic solvents.

Solvents	Dielectric constant	Boiling point /°C	Melting point /°C
Ethylene Carbonate (EC)	89.78	248	36.4
Propylene Carbonate (PC)	64.9	242	−48.8
Dimethyl Carbonate (DMC)	3.11	91	4.6
Diethyl Carbonate (DEC)	2.81	126	−74.3
1,2-Dimethoxyethane (DME)	5.5	82.5	−58
Ethyl Methyl Carbonate (EMC)	2.96	110	−53
Ethyl Acetate (EA)	6.02	77	−84
Tetrahydrofuran (THF)	7.58	66	−108.5
Dipropyleneglycol Dimethyl Ether (DMM)	2.7	41	−105

Source: [71]/John Wiley & Sons.

Table 1.10 Common lithium salts and their advantages and disadvantages.

Salts	Advantages	Disadvantages
$LiPF_6$	Soluble, high ionic conductivity, and capable of forming a stable passivation film on the current collector surface	Poor thermal stability and prone to decomposition reactions
$LiBF_4$	Wide operating temperature range and good stability at high temperatures	Low ionic conductivity
LiBOB	High conductivity, a wide electrochemical window, and excellent thermal stability contribute to the formation of the SEI film	Low solubility
LiDFOB	Good film-forming and low-temperature performance, with higher solubility in carbonate solvents and higher electrolyte conductivity	Expensive
LiTFSI	Enhanced solubility and conductivity, along with a high thermal decomposition temperature and hydrolysis resistance	Corrosive to aluminum foil
LiFSI	High conductivity	Corrosive to aluminum foil
$LiPO_2F_2$	Excellent low- and high-temperature performance	Low solubility

Source: [72]/John Wiley & Sons.

Table 1.11 Common additives and their functions.

Additives	Functions
Fluoroethylene Carbonate (FEC)	SEI-forming additive
Biphenyl	Overcharge protection additive
Triphenyl Phosphate/Tributyl Phosphate	Flame-retardant additive
Graphene	Conductive additive

Source: [73]/John Wiley & Sons.

1.7 Binders

During the assembling of the electrodes, binders are polymer compounds which can adhere the cathode or anode active substances to the current collectors. The binder mainly has the following functions: first, the binder acts as a dispersant or thickening agent to improve the uniformity of electrode components. Second, the binder can bind the conductive agent, active substance, and fluid collector to maintain the integrity of the electrode structure. Finally, the wettability of the electrolyte is improved to promote the conveyance of Li^+ at the electrode–electrolyte interface [74, 75].

The desirable binder should have the following characteristics: (i) the binder should not have an electrochemical reaction with the electrolyte or other substances. (ii) The binder should have good dispersion, and its dispersion effect is related to surface charge density, main-chain flexibility, and electrostatic repulsion effect. (ii) The binder must have excellent mechanical traits, such as elasticity, hardness, and adhesion strength. These properties are related to the molecular weight and functional group of the binder. (iv) The binder also needs to have good electrical conductivity, which can be improved by gas-phase doping, electrochemical doping, solvent doping, and ion-exchange doping. (v) The binder should be able to maintain stability under a certain ambient temperature and humidity to avoid affecting the battery performance due to changes in temperature and humidity [76, 77]. Commonly used binders are listed in Table 1.12. However, there is currently no binder that fully meets all of the above requirements. Therefore, when selecting binders, the selection should be optimized according to the specific application environment and needs. By taking these factors into account, the binder that is most suitable for the specific application can be selected, thereby raising the performance and life of the battery.

With the incessant progress of LIBs technology, more and more new binders have begun to enter people's vision, such as polyimide (PI) [78, 79]. These new binders have significant potential to improve battery performance, extend battery life, and enhance environmental adaptability. In the future, the custom design of binders will become an important direction, and it will be developed according to factors such as the morphology, state, and functional groups of the material surface to satiate the demands of high-energy-density batteries.

1.8 Outlook

In the modern era, information and transportation technologies are increasingly dependent on rechargeable batteries, creating a pressing need for higher energy density. For over 15 years, LIBs have been the primary power source for portable electronic devices, achieving high maturity and reliability. However, the quest for greater energy density remains a focal point for global research communities and intensive industry research and development (R&D) efforts. This necessitates the development of new, advanced rechargeable LIBs and a comprehensive understanding of their working mechanisms during charge and discharge cycles.

Traditional LIBs utilize liquid electrolytes that typically contain highly volatile and flammable organic solvents, which increase the risk of fire and explosion and limit their ability to withstand high operating voltages. Additionally, liquid electrolytes react with lithium metal to form an unstable SEI layer, leading to the growth of harmful lithium dendrites. In contrast, solid-state LIBs employ solid electrolytes, significantly enhancing safety and reducing the risk of fire and explosion. Solid-state batteries offer higher energy density, longer cycle life, and faster charging speeds, thereby improving the user experience. With the development of solid electrolytes such as oxides, sulfides, and polymers, solid-state LIBs have

Table 1.12 Common binders and their advantages and disadvantages.

Types	Applications	Advantages	Disadvantages
Polyvinylidene Fluoride (PVDF)	$LiMO_2$ (M = Co, Ni, Mn, Al), $LiFePO_4$(LFP), $LiMn_2O_4$, $LiMn_xNi_yCo_zO_2$, Si, Sn, TiO_2, SnO_2, $NaMnO_2$, $NaV(PO_4)_2F_3$, MoS_2, MnO_2/MWCNT, MnO_2/graphene, etc.	Good corrosion resistance, good heat resistance, and good processability	Low adhesion strength, high cost, and poor electrical conductivity
Carboxymethyl Cellulose (CMC)	$LiMO_2$, $LiFePO_4$, $LiNi_{0.4}Mn_{1.6}O_4$, $LiMn_2O_4$, Si, $Li_4Ti_5O_{12}$, SnO_2, Sn, TiO_2, LI-S, P, Sb, oxides, $NaMnO_2$, $NaV(PO_4)_2F_3$, PFPT, etc.	High adhesion, can improve battery capacity and cycle life, and has good thermal stability and electrochemical characteristics	Large brittleness, poor elasticity, and easy moisture absorption
Styrene-Butadiene Rubber (SBR)	$LiMO_2$, $LiFePO_4$, TiO_2, Li-S, carbonaceous materials, Si, etc.	The high bonding force	High cost
Sodium Alginate (Alg-Na)	$LiMO_2$, carbonaceous materials, Si, MoS_2, metal oxides, etc.	Environmental protection, low cost, and good adhesion and mechanical properties	Improvements are needed to adapt to different material systems
Polytetrafluoroethylene (PTFE)	Li-O_2, $LiFePO_4$, carbonaceous materials, MnO_2/graphene, PANI/CNT/graphene, etc.	Good corrosion resistance, good heat resistance, and good aging resistance	High cost, high thermal expansion rate, and low mechanical strength
Polyacrylic Acid (PAA)	$LiFePO_4$, $LiMn_2O_4$, carbonaceous materials, Si, Li-S, P, Sn, oxides, etc.	The low expansion coefficient and high thermal diffusion coefficient can ensure the structural integrity of the silicon-based electrode during the cycle	Great brittleness

Source: Refs. [78, 79].

broad application prospects in renewable energy storage, consumer electronics, and electric transportation and are expected to lead a new revolution in battery technology. Although aqueous LIBs currently face the challenge of lower energy density, their safety, low cost, and environmental friendliness make them an emerging research hotspot.

In the exploration of battery working mechanisms, in-situ characterization techniques provide multi-angle real-time data analysis by monitoring operating batteries with high temporal resolution. For example, in-situ X-ray diffraction (XRD), Raman spectroscopy, and scanning transmission electron microscopy (STEM) can be used to observe phase changes and lattice parameters of electrodes or electrode–electrolyte interfaces in real time during charge–discharge cycles, providing critical insights and data support for studying battery working and failure mechanisms. To achieve higher battery energy density, it is essential to promote innovation in cell structure and platformization of battery systems. This involves seeking the optimal cell structure by integrating factors such as lightweight, safety, and rate performance and aligning with typical structures like blade batteries, large cylindrical batteries, and large pouch batteries, while accelerating the standardization of cell sizes.

With the widespread application of computers in materials science, theoretical simulations can help researchers understand the microstructure and electrochemical reaction mechanisms of battery materials, guiding material design and performance optimization. Additionally, artificial intelligence technologies, such as machine learning and deep learning, can be employed for big data analysis and prediction, accelerating the discovery of new materials and optimization of battery performance. Finally, combining theoretical simulations and artificial intelligence methods can enable predictions and management of the entire battery lifecycle, enhancing battery efficiency and lifespan.

References

1 Wang, Q., Lu, T., Xiao, Y. et al. (2023). Leap of Li metal anodes from coin cells to pouch cells: challenges and progress. *Electrochemical Energy Reviews* 6 (1): 22.

2 Manthiram, A. (2020). A reflection on lithium-ion battery cathode chemistry. *Nature Communications* 11 (1): 1550.

3 Liu, T.; Yu, L.; Liu, J. et al. (2024). Ultrastable cathodes enabled by compositional and structural dual-gradient design. *Nature Energy* 9: 1252–1263.

4 Winter, M., Barnett, B., and Xu, K. (2018). Before Li ion batteries. *Chemical Reviews* 118 (23): 11433–11456.

5 Xiao, J., Shi, F., Glossmann, T. et al. (2023). From laboratory innovations to materials manufacturing for lithium-based batteries. *Nature Energy* 8 (4): 329–339.

6 Chen, Y., Qin, Q., Zhao, L. et al. (2024). Analysis of China's patent landscape for new energy storage technologies. *Energy Storage Science and Technology*. 13 (06): 2089–2098.

7 Xie, J. and Lu, Y.-C. (2024). Designing nonflammable liquid electrolytes for safe Li-ion batteries. *Advanced Materials* 2312451.

8 Deng, Z., Chen, S., Yang, K. et al. (2024). Tailoring interfacial structures to regulate carrier transport in solid-state batteries. *Advanced Materials* 2407923.

9 Orangi, S., Manjong, N.B., Clos, D.P. et al. (2023). Trajectories for lithium-ion battery cost production: can metal prices hamper the deployment of lithium-ion batteries. *Batteries & Supercaps* 6 (12): e202300346.

10 Clément, R.J., Lun, Z., and Ceder, G. (2020). Cation-disordered rocksalt transition metal oxides and oxyfluorides for high energy lithium-ion cathodes. *Energy & Environmental Science* 13 (2): 345–373.

11 Liu, Y., Li, W., and Xia, Y.Y. (2021). Recent progress in polyanionic anode materials for Li (Na)-Ion batteries. *Electrochemical Energy Reviews* 4 (3): 447–472.

12 Padhi, A.K., Nanjundaswamy, K.S., and Goodenough, J.B. (1997). Phospho-olivines as positive-electrode materials for rechargeable lithium batteries. *Journal of the Electrochemical Society* 144 (4): 1188.

13 Li, B., Xiao, J., Zhu, X. et al. (2024). Enabling high-performance lithium iron phosphate cathodes through an interconnected carbon network for practical and high-energy lithium-ion batteries. *Journal of Colloid and Interface Science* 653: 942–948.

14 Zhao, T., Mahandra, H., Marthi, R. et al. (2024). An overview on the life cycle of lithium iron phosphate: synthesis, modification, application, and recycling. *Chemical Engineering Journal* 485: 149923.

15 Nekahi, A., MR, A.K., Li, X. et al. (2024). Sustainable $LiFePO_4$ and $LiMn_xFe_{1-x}PO_4$ (x=0.1–1) cathode materials for lithium-ion batteries: A systematic review from mine to chassis. *Materials Science and Engineering: R: Reports* 159: 100797.

16 Manthiram, A. (2020). A reflection on lithium-ion battery cathode chemistry. *Nature Communications* 11 (1): 1550.

17 Deng, Y., Yang, C., Zou, K. et al. (2017). Recent Advances of Mn-Rich $LiFe_{1-y}Mn_yPO_4$ ($0.5 \leq y < 1.0$) Cathode materials for high energy density lithium ion batteries. *Advanced Energy Materials* 7 (13): 1601958.

18 Li, S., Zhang, H., Liu, Y. et al. (2024). Comprehensive understanding of structure transition in $LiMn_yFe_{1-y}PO_4$ during delithiation/lithiation. *Advanced Functional Materials* 34 (4): 2310057.

19 Yang, R., Chang, L., Luo, S. et al. (2024). A critical revelation of lithium ferromanganese phosphate (LMFP) performance in a Mn-rich cathode for Li-ion batteries using Fe equivalents to occupy a Mn site. *Journal of Materials Chemistry C* 12 (14): 4961–4976.

20 Nasajpour-Esfahani, N., Garmestani, H., Bagheritabar, M. et al. (2024). Comprehensive review of lithium-ion battery materials and development challenges. *Renewable and Sustainable Energy Reviews* 203: 114783.

21 Sun, K., Luo, S.-H., Du, N. et al. (2024). Research progress of lithium manganese iron phosphate cathode materials: from preparation to modification. *Electroanalysis* e202400120.

22 Manthiram, A.J.N. (2020). A reflection on lithium-ion battery cathode chemistry. *Nature Communications* 11 (1): 1550.

23 Mizushima, K., Jones, P.C., Wiseman, P.J., and Goodenough, J.B. (1981). Li_xCoO_2 ($0< x\leqslant 1$): A new cathode material for batteries of high energy density. *Solid State* 3: 171–174.

24 Lyu, Y., Wu, X., Wang, K. et al. (2021). An overview on the advances of $LiCoO_2$ cathodes for lithium-ion batteries. *Advanced Energy Materials* 11 (2): 2000982.

25 Konar, R., Maiti, S., Shpigel, N., and Aurbach, D. (2023). Reviewing failure mechanisms and modification strategies in stabilizing high-voltage $LiCoO_2$ cathodes beyond 4.55 V. *Energy Storage Materials* 103001.

26 Lyu, Y.C., Wu, X., Wang, K. et al. (2021). An Overview on the Advances of $LiCoO_2$ Cathodes for Lithium-Ion Batteries. *Advanced Energy Materials* 11 (2).

27 de Biasi, L., Schwarz, B., Brezesinski, T. et al. (2019). Chemical, structural, and electronic aspects of formation and degradation behavior on different length scales of Ni-rich NCM and Li-rich HE-NCM cathode materials in Li-ion batteries. *Advanced Materials* 31 (26): 1900985.

28 Li, W., Erickson, E.M., and Manthiram, A.J.N.E. (2020). High-nickel layered oxide cathodes for lithium-based automotive batteries. *Nature Energy* 5 (1): 26–34.

29 Manthiram, A., Song, B., and Li, W. (2017). A perspective on nickel-rich layered oxide cathodes for lithium-ion batteries. *Energy Storage Materials* 6: 125–139.

30 Sim, R. and Manthiram, A. (2024). Factors influencing gas evolution from high-nickel layered oxide cathodes in lithium-based batteries. *Advanced Energy Materials* 14 (8): 2303985.

31 Deng, Z., Liu, Y., Wang, L. et al. (2023). Challenges of thermal stability of high-energy layered oxide cathode materials for lithium-ion batteries: a review. *Materials Today* 236–261.

32 Liu, C., Cui, Z., and Manthiram, A. (2024). Tuning dopant distribution for stabilizing the surface of high-nickel layered oxide cathodes for lithium-ion batteries. *Advanced Energy Materials* 14 (3): 2302722.

33 Zhang, H., Liu, H., Piper, L.F. et al. (2022). Oxygen loss in layered oxide cathodes for Li-ion batteries: mechanisms, effects, and mitigation. *Chemical Reviews* 122 (6): 5641–5681.

34 Zhao, S., Yan, K., Zhang, J. et al. (2021). Reaction mechanisms of layered lithium-rich cathode materials for high-energy lithium-ion batteries. *Angewandte Chemie International Edition* 60 (5): 2208–2220.

35 Zhao, S., Guo, Z., Yan, K. et al. (2021). Towards high-energy-density lithium-ion batteries: strategies for developing high-capacity lithium-rich cathode materials. *Energy Storage Materials* 34: 716–734.

36 Kim, S., Cho, W., Zhang, X. et al. (2016). A stable lithium-rich surface structure for lithium-rich layered cathode materials. *Nature Communications* 7 (1): 1–8.

37 Li, W., Lee, S., and Manthiram, A. (2020). High-nickel NMA: a cobalt-free alternative to NMC and NCA cathodes for lithium-ion batteries. *Advanced Materials* 32 (33): 2002718.

38 Song, J., Wang, H., Zuo, Y. et al. (2023). Building better full manganese-based cathode materials for next-generation lithium-ion batteries. *Electrochemical Energy Reviews* 6 (1): 20.

39 Hunter, J.C. (1981). Preparation of a new crystal form of manganese dioxide: λ-MnO_2. *Journal of Solid State Chemistry* 39: 142–147.

40 Zuo, C., Hu, Z., Qi, R. et al. (2020). Double the capacity of manganese spinel for lithium-ion storage by suppression of cooperative Jahn–teller distortion. *Advanced Energy Materials* 10 (34): 2000363.

41 Yang, S., Yan, P., Bao, W. et al. (2023). Surface magnesium substitution at spinel lithium manganate 8a tetrahedral sites for suppressed manganese dissolution and enhanced cycle stability. *ACS Energy Letters* 8: 4278–4286.

42 Chen, C., Feng, J., Li, J. et al. (2022). Functional fiber materials to smart fiber devices. *Chemical Reviews* 123: 613–662.

43 Blasse, G. (1966). Ferromagnetism and ferrimagnetism of oxygen spinels containing tetravalent manganese. *Journal of Physics and Chemistry of Solids.* 27: 383–389.

44 Lee, S., Su, L., Mesnier, A. et al. (2023). Cracking vs surface reactivity in high-nickel cathodes for lithium-ion batteries. *Joule* 7: 2430–2444.

45 Thackeray, M.M. and Amine, K. (2021). Layered Li–Ni–Mn–Co oxide cathodes. *Nature Energy* 6: 933–933.

46 Dopilka, A., Larson, J.M., Cha, H., and Kostecki, R. (2024). Synchrotron near-field infrared nanospectroscopy and nanoimaging of lithium fluoride in solid electrolyte interphases in Li-ion battery anodes. *ACS Nano* 18 (23): 15270–15283.

47 Khan, M., Yan, S., Ali, M. et al. (2024). Innovative solutions for high-performance silicon anodes in lithium-ion batteries: overcoming challenges and real-world applications. *Nano-Micro Letters* 16 (1): 179.

48 Feng, Y., Zhou, L., Ma, H. et al. (2022). Challenges and advances in wide-temperature rechargeable lithium batteries. *Energy & Environmental Science* 15 (5): 1711.

49 Shi, Q.T., Zhou, J.H., Ullah, S. et al. (2021). A review of recent developments in Si/C composite materials for Li-ion batteries. *Energy Storage Materials.* 34: 735–754.

50 Schomburg, F., Heidrich, B., Wennemar, S. et al. (2024). Lithium-ion battery cell formation: status and future directions towards a knowledge-based process design. *Energy & Environmental Science* 17: 2686–2733.

51 Dai, F. and Cai, M. (2022). Best practices in lithium battery cell preparation and evaluation. *Communications Materials* 3: 64.

52 Spielbauer, M., Soellner, J., Berg, P. et al. (2022). Experimental investigation of the impact of mechanical deformation on aging, safety and electrical behavior of 18650 lithium-ion battery cells. *Journal of Energy Storage* 55: 105564.

53 Shen, K., Sun, J., Zheng, Y.J. et al. (2022). A comprehensive analysis and experimental investigation for the thermal management of cell-to-pack battery system. *Applied Thermal Engineering* 211: 118422.

54 Dasari, S.M., Srivastav, P., Shaw, R. et al. (2013). Optimization of cell to module conversion loss by reducing the resistive losses. *Renewable Energy* 50: 82–85.

55 Wang, H.B., Wang, S.Y., Feng, X.N. et al. (2021). An experimental study on the thermal characteristics of the cell-to-pack system. *Energy.* 227: 120338.

56 Man, Q.Y., An, Y.L., Liu, C.K. et al. (2023). Interfacial design of silicon/carbon anodes for rechargeable batteries: a review. *Journal of Energy Chemistry* 76: 576–600.

57 Ge, M.Z., Cao, C.Y., Biesold, G.M. et al. (2021). Recent advances in silicon-based electrodes: from fundamental research toward practical applications. *Advanced Materials* 33 (16).

58 Jin, C.Y., Sun, Y.D., Yao, J. et al. (2022). No thermal runaway propagation optimization design of battery arrangement for cell-to-chassis technology. *ETransportation* 14: 100199.

59 Yao, F., Guenes, F., Huy Quang, T. et al. (2012). Diffusion mechanism of lithium ion through basal plane of layered graphene. *Journal of the American Chemical Society* 134 (20): 8646–8654.

60 Wang, M.S., Wang, G.L., Wang, S. et al. (2019). In situ catalytic growth 3D multi-layers graphene sheets coated nano-silicon anode for high performance lithium-ion batteries. *Chemical Engineering Journal* 356: 895–903.

61 Stevens, D.A. and Dahn, J.R. (2000). High capacity anode materials for rechargeable sodium-ion batteries. *Journal of the Electrochemical Society* 147 (4): 1271.

62 Stevens, D.A. and Dahn, J.R. (2000). An in situ small-angle X-ray scattering study of sodium insertion into a nanoporous carbon anode material within an operating electrochemical cell. *Journal of the Electrochemical Society* 147 (12): 4428.

63 Schomburg, F., Heidrich, B., Wennermar, S. et al. (2024). Lithium-ion battery cell formation: status and future directions towards a knowledge-based process design. *Energy & Environmental Science* 17 (8): 2686–2733.

64 Bai, P., He, Y., Zou, X. et al. (2018). Elucidation of the sodium-storage mechanism in hard carbons. *Advanced Energy Materials* 8 (15): 1703217.

65 Wang, D., Zhou, J., Li, Z. et al. (2019). Uniformly expanded interlayer distance to enhance the rate performance of soft carbon for lithium-ion batteries. *Ionics* 25 (4): 1531–1539.

66 Sun, L., Liu, Y.X., Wu, J. et al. (2022). A review on recent advances for boosting initial coulombic efficiency of silicon anodic lithium ion batteries. *Small* 18 (5): 2102894.

67 Zhang, M., Liang, N., Hao, D. et al. (2023). Recent advances of SiO_x-based anodes for sustainable lithium-ion batteries. *Nano Research Energy* 2: e9120077.

68 Xu, K. (2021). Li-ion battery electrolytes. *Nature Energy* 6: 763.

69 Ye, C., Tu, W., Yin, L. et al. (2018). Converting detrimental HF in electrolytes into a highly fluorinated interphase on cathodes. *Journal of Materials Chemistry A* 6 (36): 17642.

70 Xu, K. (2004). Nonaqueous liquid electrolytes for lithium-based rechargeable batteries. *Chemical Review* 104 (10): 4303–4418.

71 Zhang, Y., Lu, Y., Jin, J. et al. (2024). Electrolyte design for lithium-ion batteries for extreme temperature applications. *Advanced Materials* 36 (13): 2308484.

72 Liu, Y.K., Zhao, C.Z., Zhang, X.Q. et al. (2023). Research progresses of liquid electrolytes in lithium-ion batteries. *Small* 19: 2205315.

73 Zhu, Y.X., Ge, M.Y., Ma, F.C. et al. (2024). Multifunctional electrolyte additives for better metal batteries. *Advanced Functional Materials* 34: 2301964.

74 Zou, F. and Manthiram, A. (2020). A review of the design of advanced binders for high-performance batteries. *Advanced Energy Materials* 10 (45): 2002508.

75 Chen, H., Ling, M., and Hencz, L. (2018). Exploring chemical, mechanical, and electrical functionalities of binders for advanced energy-storage devices. *Chemical Reviews* 118 (18): 8936–8982.

76 Dou, W., Zheng, M., and Zhang, W. (2023). Review on the binders for sustainable high-energy-density lithium ion batteries: status, solutions, and prospects. *Advanced Functional Materials* 33 (45): 2305161.

77 Lux, S.F., Schappacher, F., and Balducci, A. (2010). Low cost, environmentally benign binders for lithium-ion batteries. *Journal of the Electrochemical Society* 157 (3): A320.

78 Shi, Y., Zhou, X., and Yu, G. (2017). Material and structural design of novel binder systems for high-energy, high-power lithium-ion batteries. *Accounts of Chemical Research* 50 (11): 2642–2652.

79 Fransson, L., Eriksson, T., and Edström, K. (2001). Influence of carbon black and binder on Li-ion batteries. *Journal of Power Sources* 101 (1): 1–9.

2

Evaluation of Sodium-Ion Battery: Fast-Charging Next Generation

Muhammad Mamoor and Liqiang Xu

Key Laboratory of Colloid and Interface Chemistry (Ministry of Education), School of Chemistry and Chemical Engineering, Shandong University, Jinan 250100, China

2.1 Motivation for Exploring Na-Ion Batteries

Sodium-ion batteries (SIBs) have captured considerable interest from researchers and scientists as a compelling alternative to lithium-ion batteries for energy storage. With their unique advantages, SIBs have the potential to significantly advance energy storage systems and address several pressing challenges. One primary motivation for investigating SIBs lies in the abundance of sodium. Unlike lithium, which is relatively scarce and concentrated in specific regions, sodium is widely available in seawater, rocks, and various minerals [1]. This abundance positions SIBs as a more sustainable and accessible energy storage option. Affordability is another driving factor for exploring SIBs. Lithium-ion batteries (LIBs) rely on costly materials like cobalt and nickel, which contribute to their high production costs. In contrast, SIBs utilize more affordable materials, making them a cost-effective alternative [2]. This cost advantage could accelerate the adoption of energy storage solutions, particularly in developing regions and large-scale applications. Safety considerations also show a pivotal act in the interest surrounding SIBs. LIBs are known for their flammability and the hazard of thermal runaway [3]. In comparison, SIBs demonstrate improved thermal stability due to the larger ionic radius of sodium ions. This improved safety profile makes SIBs particularly appealing for applications where safety is crucial, such as in electric vehicles and grid-scale energy storage [4]. A short summary of LIB and SIB properties is provided in Table 2.1.

The scalability and compatibility of SIBs are additional factors driving exploration in this area. Many existing LIB manufacturing facilities can be repurposed or adapted to produce SIBs with minimal modifications. This compatibility streamlines the transition process for battery manufacturers and facilitates rapid scaling of production, potentially expediting commercialization [5]. Although SIBs currently lag behind LIBs in terms of energy density, substantial advancements have been achieved in improving their performance. Researchers are actively developing advanced electrode materials and electrolyte formulations to enhance the capacity,

Electrode Materials in Energy Storage Technologies: Applications in Lithium-, Sodium-, Potassium-, Sulfur- and Zinc-Based Rechargeable Batteries, First Edition. Edited by Liqiang Xu.

Table 2.1 Comparison between the properties of LIB and SIB.

Parameters	Li–ion battery	Na-ion battery
Cost	High	Low
Energy density	High	Moderate/high
Safety	Low	High
Materials	Scarce	Earth abundant
Voltage	−3.04 V vs. Li^+/Li	−2.71 V vs. Na^+/Na
Ionic size	0.76 Å	1.02 Å
Relative atomic mass	6.94	−2.71
Cycling stability	High	High
Efficiency	High (>90%)	High (>90%)
Temperature range	−25 to 40 °C	−40 to 60 °C
Capacity (mAh g^{-1})	3829	1160
Price (US$ per ton)	5800	250–300
Coordination preference	Octahedral and tetrahedral	Octahedral and prismatic
Desolvation energy in PC (kJ mol^{-1})	218	157.3
Remarks	Transportation restriction at discharge rate	Less mature technology; easy transportation

cycling stability, and overall performance of SIBs [6]. With further advancements, SIBs could become a viable option for diverse applications, including stationary energy storage, portable electronics, and integration with renewable energy sources. Exploring SIBs also contributes to diversifying the energy storage landscape. Dependence on lithium-ion technology poses the risk of resource depletion and supply chain constraints [7]. By expanding research and development efforts in SIBs, we can foster a more diversified and resilient energy storage ecosystem, reducing reliance on a single technology.

Moreover, SIBs present a compelling prospect for bolstering environmental sustainability. The extraction and processing of lithium have significant environmental consequences, such as water pollution and habitat disruption. Sodium, being more abundant and widely distributed, offers a potentially greener alternative. Embracing SIBs can help minimize the environmental footprint associated with energy storage technologies and advance the journey toward a more sustainable future [8]. Moreover, SIBs hold promise for grid-scale energy storage. With the growing reliance on renewable energy sources such as solar and wind, there is a pressing need for efficient and cost-effective storage solutions. SIBs can effectively store excess renewable energy during periods of low demand and discharge it when needed. This capability supports grid stability, reduces dependence on fossil fuels, and enhances the overall reliability of the energy system. Investing in research and development of SIBs provides opportunities for technological innovation and economic growth

[9]. Countries that support SIB development can nurture a thriving battery industry, create jobs, and stimulate economic development. Furthermore, the knowledge and expertise gained from SIB research can be transferred to other areas of energy storage, leading to advancements across related fields. However, it is important to acknowledge that challenges remain in the establishment of SIBs [10]. Ongoing research and collaboration among scientists, engineers, and industry experts are essential to address these challenges and fully realize the potential of the SIB technology.

2.2 Fundamental of SIBs

SIBs have garnered substantial interest as a promising technology for energy storage. Understanding the comprehensive fundamentals of SIBs involves exploring key components, working principles, and important characteristics that contribute to their potential applications and performance. The electrodes are crucial components of SIBs, consisting of an anode and a cathode. During the charging process, sodium ions are inserted into the anode material, which acts as a host for sodium storage [11]. Conversely, in discharge, the cathode receives the sodium ions. Researchers are exploring a range of electrode materials, including carbon-based substances, transition metal oxides, and polyanionic compounds, to improve the electrochemical performance of SIBs.

The electrolyte is crucial for enabling the displacement of sodium ions between the anode and cathode. It can be an organic solvent with a sodium salt or a solid-state electrolyte. The choice of electrolyte influences important factors such as ionic conductivity, stability, and battery safety [12] (comprehensively discussed in Section 2.4). Scientists are working on developing electrolytes with improved conductivity, enhanced electrochemical stability, and superior safety features to optimize the performance of SIBs. SIBs operate based on the reversible process of intercalating and deintercalating sodium ions within the host materials of the electrodes. Sodium ions travel from the cathode to the anode for storage when the charging process is taking place. Conversely, during discharge, these ions are released from the anode and move back to the cathode [13]. This intercalation process directly impacts the capacity and overall performance of SIBs. Ongoing research aims to explore new materials with enhanced sodium-ion intercalation properties to foster energy density and secure the cycling stability of SIBs. Energy density, a key attribute of batteries, indicates the amount of energy stored per unit volume or mass [14]. However, researchers are actively working on improving the energy density of SIBs by investigating larger capacity electrode materials and innovative approaches. Research into advanced electrode materials, including transition metal oxides and polyanionic compounds, aims to boost the storage capacity and improve the energy density of SIBs [15].

Another critical aspect is the cycle life and stability of SIBs. Cycle life refers to the ability of a battery to endure repeated charge and discharge cycles without significant degradation in performance. Researchers are focusing on developing electrode materials and electrolyte formulations that minimize side reactions, structural deterioration, and electrolyte decomposition to extend the cycle life of SIBs. Through advanced characterization techniques and computational modeling, scientists aim to understand the degradation mechanisms and design more durable electrode materials for long-lasting SIBs [16], ensuring the safety of energy storage systems is a top priority. While SIBs demonstrate improved safety compared to LIBs, challenges such as dendrite formation and electrolyte decomposition still need to be addressed. Researchers are exploring innovative solutions, including the incorporation of electrolyte additives and protective coatings, to mitigate dendrite growth and strengthen the stability of SIBs, thus making them safer for various applications. For the widespread adoption of SIBs, scalable and cost-effective manufacturing processes are crucial [17]. Efforts are striving to develop production methods that can be seamlessly integrated into existing battery manufacturing infrastructure. Leveraging knowledge and experience gained from lithium-ion battery production, they aim to streamline the manufacturing of SIBs and achieve efficient and cost-effective large-scale production [18].

SIBs hold promise for a range of applications. Their cost-effectiveness, safety benefits, and the widespread availability of sodium make them especially well suited for applications such as grid-scale energy storage, electric vehicles, portable electronics, and renewable energy integration. Grid-scale SIBs have the potential to store surplus energy from renewable sources and discharge it as required, contributing to a stable and dependable power supply [19]. Furthermore, SIBs can power electric vehicles, offering a cost-effective and environmentally friendly alternative to traditional combustion engines. In conclusion, comprehending the comprehensive fundamentals of SIBs involves investigating electrode materials, electrolyte composition, intercalation processes, energy density, cycle life, safety considerations, manufacturing, and applications. Continued research and development are essential to address existing challenges and fully realize the potential of SIBs as a competitive alternative to LIBs across diverse applications. Ongoing advancements in the SIB technology can revolutionize energy storage and contribute to a sustainable future. The widespread adoption of LIBs in consumer electronics and electric vehicles has created robust supply chains, established extensive manufacturing infrastructure, and built significant consumer familiarity [20]. Overcoming these barriers and achieving widespread adoption of SIBs will require continued research, technological advancements, and supportive policies. SIBs offer scalability, temperature tolerance, flexibility in electrolyte selection, and synergy with renewable energy. However, they also present challenges concerning lower power density, limited cycle life, reduced energy efficiency, slower charging speeds, and limited market penetration. Current research and development efforts are dedicated to overcoming these challenges and fully realizing the potential of SIBs as a practical energy storage solution. With further advancements, SIBs have the potential to transform the energy storage landscape, facilitating the sustainable and efficient use of renewable energy sources and powering various applications in a cost-effective and environmentally friendly manner [21].

2.3 Cathode Materials: Strategies for Improvement

In this section, famous and potential cathode materials will be the center of discussion, including layered oxides, polyanions, Prussian blue, and organic compounds. Ongoing research advancements aim to predict exceptional cycling performance, profound stability, and mitigating capacity fade. As Na^+ ions possess larger ionic sizes, improvements in conductivity are still a key challenge that is under struggle through structural optimization, doping, and some modifications.

2.3.1 Layered Oxides

Cathodes play a vital role in SIBs, with various materials being investigated to optimize performance. These include polyanionic compounds [22], Prussian blue analogs [23], layered oxides [24], and organic-based materials [25]. Transition metal-based layered oxides (Na_xTMO_2, where TM represents transition metals like Fe, Mn, Ni, Co) are particularly attractive because of their high specific capacity, relatively simple preparation, and similarity to successful $LiTMO_2$ in LIBs [26]. However, Na_xTMO_2 materials encounter several challenges attributed to the large Na^+ ion radius, including multiphase evolution, sluggish Na^+ ion diffusion, poor air stability, and limited cycling lifespan. These issues hinder SIB commercialization, necessitating various optimization strategies [27]. To enhance electrochemical characteristics, it is crucial to understand degradation mechanisms and implement strategies to address negative impacts on air stability, specific capacitance, and cycling performance, especially at high voltages. In recent decades, numerous notions and design strategies have been employed to foster the performance of SIBs by optimizing the structures of layered oxide cathodes. It investigates the structural characteristics of layered oxides and their inherent limitations, providing a summary of the failure mechanisms associated with air exposure and cathode–electrolyte interfaces (CEIs). To tackle issues such as irreversible multiphase transitions, low energy density, and inadequate air stability, various strategies can be applied, either individually or in combination [28]. These strategies are designed to address phase transition issues, boost air stability, and increase the energy density of layered oxide cathodes.

2.3.1.1 Challenges – Structure Characteristic and Failure Mechanism

This passage discusses three major challenges faced by layered oxide cathode materials in SIBs:

1) Irreversible phase transition:
   - O3-phase materials can transition to the P3 phase when the charging process is taking place.
   - Further irreversible transitions occur at higher voltages.
   - P2-type materials can transform into O2 type during charging.
   - These transitions lead to reduced Na^+ ion diffusion kinetics, mechanical stress, lattice distortion, and crystal cracking.

- This leads to reduced specific capacity, diminished rate performance, lower Coulombic efficiency, decreased energy density, and compromised cycle stability.

2) Low-energy density:
   - Sodium is denser and has a higher standard electrode potential compared to lithium.
   - The larger radius of the Na^+ ion results in slower diffusion rates and less efficient insertion and extraction processes.
   - SIBs exhibit lower energy density compared to LIBs.
   - Increasing the working potential window can improve energy density but may cause severe irreversible phase transitions.
3) Poor air stability:
   - Layered oxide cathodes and other sodium-based materials are particularly prone to degradation when exposed to air.
   - Hygroscopic degradation occurs in humid air, leading to structural changes and particle destruction.
   - H^+/Na^+ proton exchange with water removes Na^+ ions from interlayers, increasing transition metal valency and expanding interlayer spacing.
   - CO_2 reacts with Na_xTMO_2, entering TMO_2 layers as CO_3^{2-} species.
   - Electron-insulating alkalis (NaOH, Na_2CO_3) form on the material surface.
   - These reactions cause denaturation of PVDF binder, particle agglomeration, and slurry gelation during electrode preparation.
   - Aluminum foil can be compromised by surface alkali residues, leading to diminished cell performance.

These challenges significantly impact the practical feasibility and performance of SIBs, necessitating the development of strategies to address these issues for improved battery technology. The structure consists of edge-sharing TMO_6 octahedrons forming $(TMO_2)_n$ layers [29]. Alkali metal ions occupy octahedral (O), tetrahedral (T), or prismatic (P) sites between these layers [30]. Oxygen atoms in layered oxides can occupy A, B, or C sites. The stacking sequence of these sites determines the structure type:

1) ABAB: T1-type structure
2) ABBA: P2-type structure
3) ABBCCA: P3-type structure
4) ABCABC: O3-type structure

The numbers 1, 2, and 3 refer to the count of transition metal layers within the unit cell. Na_xTMO_2 with O- and P-type structures is extensively researched for SIB cathodes as illustrated in Figure 2.1. Komaba narrated in his research that Claude Delmas introduced Rouxel plots to distinguish stacking structures based on the Na content, ionic bonding, and covalent bonding strength [32]. Nevertheless, this approach encounters limitations when dealing with ions that have similar Pauling electronegativities or in multicomponent systems where ions possess identical chemical valences but differ in other aspects. These structural features

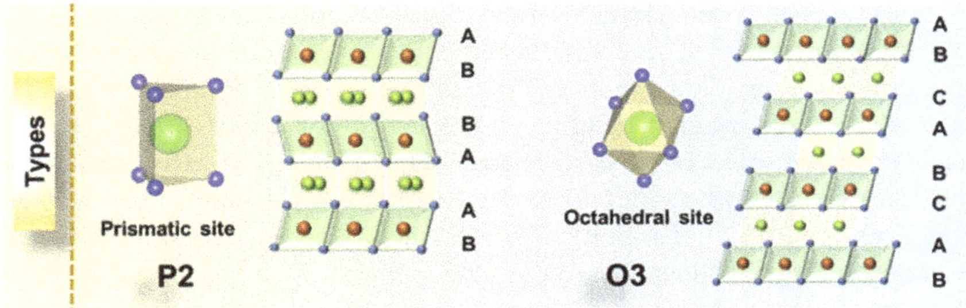

Figure 2.1 Various layered oxide cathode materials and their phase transitions. Source: [31]/with permission of John Wiley and Sons.

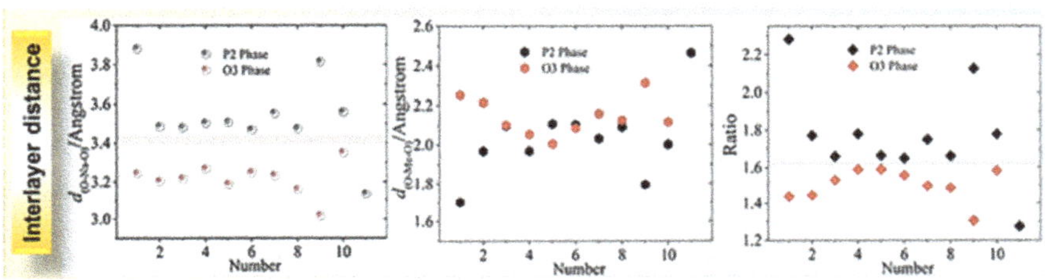

Figure 2.2 Interlayer distances: the perpendicular spacing of oxygen sheets with Na ions, $d_{(O-Na-O)}$ (left), transition metal spacing, $d_{(O-TM-O)}$ (middle), and the ratio between $d_{(O-Na-O)}$ and $d_{(O-TM-O)}$ (right). Source: [33]/with permission of John Wiley and Sons.

play a crucial role in determining the electrochemical performance of layered oxide cathodes in SIBs and offer a fundamental basis for comprehending their behavior in battery applications.

Zhao et al. identified a relationship between interlayer distance and P/O-type structures in layered oxides, as illustrated in Figure 2.2 [33]. The interlayer spacing is influenced by electrostatic interactions between NaO_2 and TMO_2 layers. Higher sodium content increases electrostatic attraction, reducing $d_{(O-Na-O)}$ and favoring the O3 phase. Conversely, the P2 phase is dominated by electrostatic repulsion, resulting in larger $d_{(O-Na-O)}$. The key difference between P2 and O3 structures lies in the ratio of interlayer distances $d_{(O-Na-O)}$ to $d_{(O-TM-O)}$. A critical ratio of 1.62 was identified: ratios above 1.62 typically form P2 phases, while those below form O3 phases. This discovery sheds light on the core concepts behind the genesis of O3 and P2 configurations. Nevertheless, structural characterization remains essential to verify the specific stacking structure, as the mere ratio does not suffice to predict or guide the formation of these configurations.

2.3.1.2 Honeycomb-Ordered Superlattice Structure: Crystal Parameter Modulation

The incorporation of two distinct transition metals in $Na_x[M1,M2]O_2$ layered oxides has significantly enhanced their appeal, thanks to the dual potential benefits this combination provides. The differences between the transition metals M1 and M2 within the layers play a crucial role in cationic ordering, which in turn influences phase transitions and overall structural stability. A pronounced variation in M—O bond energy and ionic radii generally encourages an ordered arrangement, whereas smaller differences result in a more disordered structure [34]. Furthermore, variations in cationic ordering or disordering are linked to intricate phase transitions and unstable air stability, leading to particle cracking and changes in structural stress. These issues adversely affect the charge–discharge profile and reduce electrochemical performance [35]. To design high-performance cathodes for SIBs, it is essential to develop a highly ordered layered oxide framework. The honeycomb-ordered superlattice structure stands out in this regard, as it significantly improves structural integrity, thermal stability, and phase stability, making it a promising choice for SIB cathodes [36]. In a typical honeycomb-ordered superlattice structure, such as $Na_3[M1,M2]O_6$, the ratio of M1 to M2 is maintained at 2 : 1. M1 is consistently a divalent transition metal, whereas M2 is generally a heavier metal with variable valence, and these two metals exhibit significantly different ionic radii [37]. In the "ab" plane, this atomic arrangement results in a honeycomb lattice, where each $M2O_6$ octahedron is interspersed among six $M1O_6$ octahedra, with edges shared between them [38]. Yuan et al.'s (2014) research on the honeycomb-ordered superlattice structure marked a significant development in this field [39]. Their study focused on the honeycomb-layered O3-$Na_3Ni_2SbO_6$, characterized by an unconventional morphology. In this structure, a SbO_6 octahedron is surrounded by six NiO_6 octahedra, which share edges, creating a distinct superlattice configuration. Sodium ions are accommodated within the Ni_2SbO_6 layers, as illustrated in Figure 2.3.

During charge and discharge cycles, the $Na_3Ni_2SbO_6$ electrode undergoes a reversible phase transformation, transitioning partially from the O3 to the P3 phase. This behavior aligns with a two-phase sodium-ion extraction mechanism, as illustrated in Figure 2.4a. As charging progresses, the structure transitions entirely to a P3 layered configuration, indicative of a Na-deficient phase. Sodium ions migrate out from the O3-type structure, leading to an increase in the Na interlayer spacing. This expansion results from the growing electrostatic repulsion between the MO_2 slabs, continuing until the voltage reaches 3.67 V. As charging progresses to 3.7 V, the structure undergoes a transition from P3 to O1. At the full charge of 4 V, the $NaNi_2SbO_6$ phase is fully realized. This stage marks the complete return

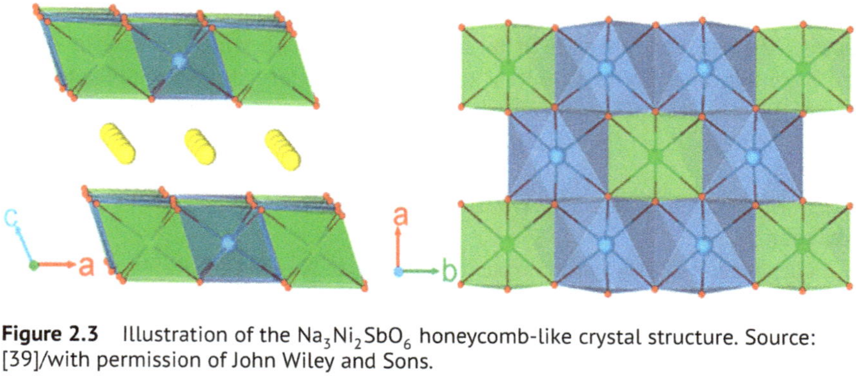

Figure 2.3 Illustration of the $Na_3Ni_2SbO_6$ honeycomb-like crystal structure. Source: [39]/with permission of John Wiley and Sons.

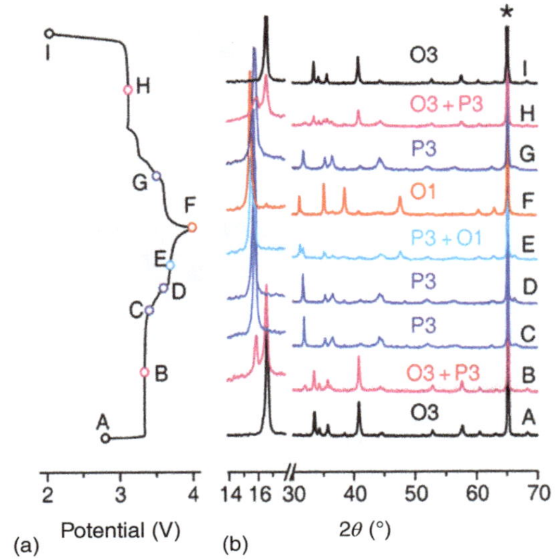

Figure 2.4 (a) Charge–discharge profiles of $Na_3Ni_2SbO_6$ and (b) *ex-situ* XRD analysis. Source: [39]/with permission of John Wiley and Sons.

from the O1-$NaNi_2SbO_6$ state (fully charged) back to the original O3-$Na_3Ni_2SbO_6$ state (discharged), illustrating the reversibility of the phase transformations throughout the electrochemical cycle. Charge–discharge profiles of $Na_3Ni_2SbO_6$ with *ex situ* X-ray diffraction (XRD) are also provided in Figure 2.4b at manipulated corresponding voltages.

2.3.2 Recent Progress of Phosphate-Based Polyanion

In this part, the discussion will be on the recent performance-based advancement of phosphate-based polyanion. Phosphate-based polyanion cathodes fall into four primary categories: phosphates, oxyphosphates, pyrophosphates, and mixed phosphate compounds. The phosphate category is further divided into NASICON-type and $NaMPO_4$ structures. Research on these materials is unevenly distributed, with NASICON-type phosphates receiving the most attention, as illustrated in Figure 2.5. PO_4 polyhedra hold a key position in these materials by strengthening the crystal structure and providing a robust 3D framework for phosphate compounds. The covalent bonding of oxygen atoms within PO_4 polyhedra contributes to the superior cycling and rate performance of these materials [40]. This structural characteristic is a key factor in their effectiveness as cathode materials.

2.3.2.1 Tuning Voltage and Capacity of NASICON

In the pursuit of superior NASICON materials, researchers are exploring innovative strategies to enhance cycling and electrochemical performance. Nanostructure tailoring aims to optimize ionic conductivity and diffusion pathways by controlling particle size and shape. Compositional tuning through careful doping seeks to improve sodium-ion mobility and structural integrity. Interfacial engineering, including core–shell designs and gradient doping, shows promise in balancing

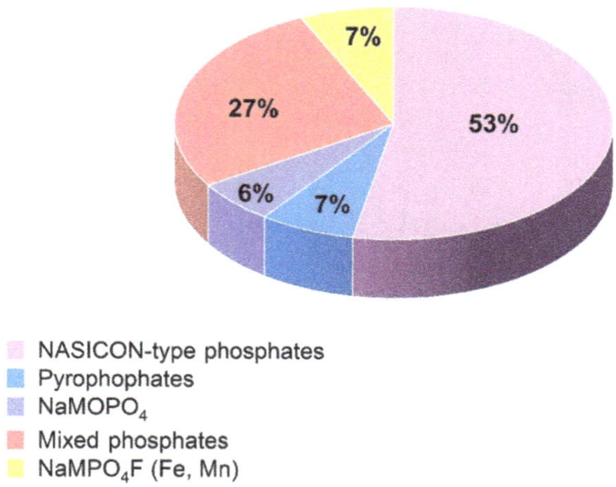

Figure 2.5 Phosphate-based polyanions.

conductivity with stability. Advanced synthesis methods, such as sol–gel combined with spark plasma sintering, are being refined to produce high-quality, dense ceramics. Composite approaches integrating NASICON with polymers or carbon nanomaterials aim to achieve synergistic improvements in properties. Surface modification techniques, like atomic layer deposition and fluorination, are being investigated to enhance stability and suppress unwanted reactions. Defect engineering, through controlled oxygen vacancies and grain boundary optimization, offers another avenue for improving ionic conductivity. These multifaceted approaches hold the potential for significant advancements in the SIB technology and related energy storage applications. Li et al. introduced a medium-entropy cathode with a NASICON structure, designated as $Na_{3.5}V_{0.5}Mn_{0.5}Fe_{0.5}Ti_{0.5}(PO_4)_3$ (Me-NVMP) [41]. Figure 2.6a reveals that Me-NVMP features irregular polyhedral particles, with dimensions varying between 750 nm and 2.75 μm. Figure 2.6b,c illustrates the initial four charge–discharge cycles at 0.1 C, along with cyclic voltammetry (CV) curves recorded at $0.2\,mV\,s^{-1}$ within the voltage range of 1.8–4.4 V (vs. Na^+/Na). The Me-NVMP electrode exhibits two distinct plateaus: one extending from 2 to 2.5 V and another from 3.3 to 3.6 V. These features are indicative of the redox reactions involving the Mn^{2+}/Mn^{3+} couple. The Me-NVMP electrode demonstrates specific capacities of $165.8\,mAh\,g^{-1}$ at 0.1 C, $159\,mAh\,g^{-1}$ at 0.5 C, $157\,mAh\,g^{-1}$ at 1 C, $150\,mAh\,g^{-1}$ at 2 C, $144\,mAh\,g^{-1}$ at 5 C, $132\,mAh\,g^{-1}$ at 10 C, $121\,mAh\,g^{-1}$ at 20 C, $98\,mAh\,g^{-1}$ at 50 C, and $69\,mAh\,g^{-1}$ at 100 C, as illustrated in Figure 2.6d. Li et al. engineered $Na_{2.5}VTi_{0.5}Al_{0.5}(PO_4)_3$ as a cathode material, utilizing a combination of $V^{2+}/V^{3+}/V^{4+}/V^{5+}$ and Ti^{3+}/Ti^{4+} redox couples to enhance its electrochemical performance [42]. The improved specific capacities of $Na_{2.5}VTi_{0.5}Cr_{0.5}(PO_4)_3$ relative to $Na_2VTi(PO_4)_3$ can be traced to the activation of a voltage plateau at 4.0 V, as well as enhanced voltage plateaus related to the V^{3+}/V^{4+} and V^{2+}/V^{3+} redox couples. Figure 2.6d illustrates the electrochemical cycling performance of these electrodes. At a rate of 10 C, $Na_{2.5}VTi_{0.5}Al_{0.5}(PO_4)_3$ achieves an initial specific capacity of $96.7\,mAh\,g^{-1}$, retaining an exceptional capacity after 1000 cycles. Furthermore,

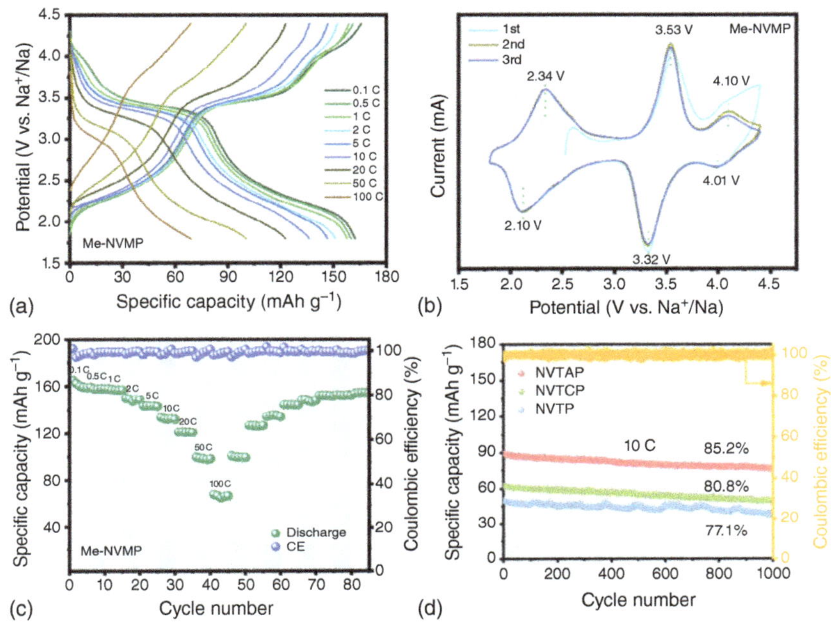

Figure 2.6 (a) Electrochemical characteristics of the Me-NVMP ($1\,C = 171\,mAh\,g^{-1}$) cathode, (b) CV curves of Me-NVMP, (c) rate performance of the Me-NVMP cathode. Source: Reproduced with the permission of [41]/John Wiley & Sons, and (d) cycling performance at 10 C. Source: Reproduced with the permission of [42]/John Wiley & Sons.

$Na_{2.5}VTi_{0.5}Cr_{0.5}(PO_4)_3$ demonstrates superior capacity retention, maintaining and surpassing the performance of $Na_2VTi(PO_4)_3$.

According to Gu et al., the stable phase transitions observed during electrochemical processes have sparked significant interest in vanadium-based polyanions for SIBs, with a notable focus on $Na_3V_2(PO_4)_3$ (NVP) [43]. The electron exchange within the $V^{3+/4+}$ redox couple in NVP is typically facilitated by the minimal crystal structure changes occurring during battery operation. Despite its potential, utilizing the advantageous $V^{4+/5+}$ redox couple remains challenging because of the significant energy barrier, which results in lower energy efficiency. This study demonstrates how the multilevel redox within the NVP framework (MLNP) shifts the reaction mechanism to a homeostatic solid solution process, effectively overcoming the high-energy barrier of $V^{4+/5+}$ at elevated voltages through successive redox couples of transition metals such as vanadium (V), iron (Fe), titanium (Ti), and chromium (Cr). Wei's group carried out density functional theory (DFT) under first-principles calculations to thoroughly investigate the $Na_3Cr_2(PO_4)_3$, and calculations identified phase transition ($R\bar{3}c \leftrightarrow P\bar{3}c_1 \leftrightarrow R\bar{3}c$) during the removal of sodium contents [44]. In this regard, Prof. Zeng's group unveiled different ordering sites for Na ions targeting to enhance ionic conduction [45]. Figure 2.7a demonstrates the schematic understanding of NASICON, illustrating the influence of bottleneck size for ionic conduction. To examine this phenomenon, the energy portfolio elucidates the diffusion of

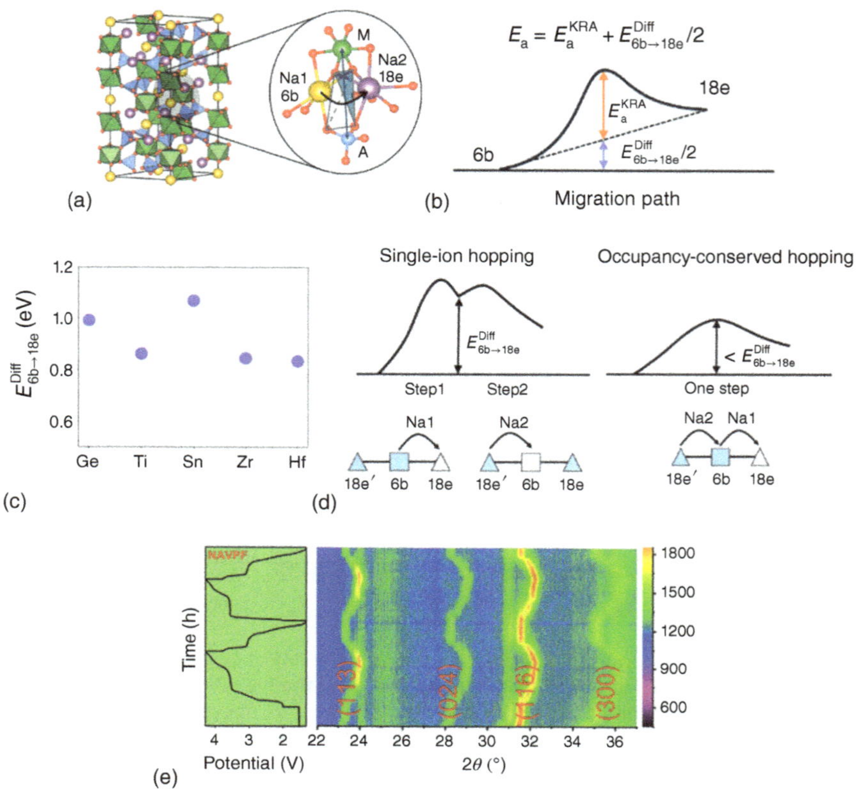

Figure 2.7 (a) Illustration of the NASICON structure with $R\overline{3}c$ symmetry, (b) demonstration of Na diffusion between 6b and 18e sites, (c) calculation of site energy difference for various metals, and (d) mechanism of Na-ion migration through two different hopping processes. Source: Reproduced with the permission of [45]/John Wiley & Sons. (e) *In situ* XRD patterns of NAVPF cathodes over the initial four cycles within the voltage range of 1.5–4.6 V along with sodium storage mechanism. Source: Reproduced with the permission of [46]/John Wiley & Sons.

Na ions in NASICON along ordered Na sites as defined in Figure 2.7b. There are two sites: lower (6b) and upper sites (18e), dedicating the difference in energy to order the tendency of Na ions. It attributes the fact that a larger difference in energy between these two sites will demonstrate the stronger ordering at lower energy sites (6b). The associated activation barrier of Na diffusion from the 6b site to the 18e site comprises two parts: one part belongs to the kinetically resolved activation (KRA) barrier, and the other part is site energy difference. As a result, a higher activation barrier is contributed due to the large energy site difference. Figure 2.7c illustrates the calculated values of the energy difference between two sites for different metals. For further insight, two different diffusion mechanisms in NASICON are discussed: (i) single ion hopping (SIH) and (ii) occupancy-conserved hopping (OCH) (see Figure 2.7d). The SIH process follows two subsequent hopping steps in which Na1 first jumps from the 6b site to the nearby 18e site; afterward, Na2 jumps from 18e′ toward the adjacent 6b site. On the other side, the OCH process demonstrates that Na1 hops from 6b toward the 18e site; then immediately, Na2 follows through hopping from the adjacent 18e′ site toward the initial 6b site from Na1 departed at the start.

Zhu et al. introduced an innovative NASICON-type $Na_{3.5}MnCr_{0.5}Ti_{0.5}(PO_4)_3$ cathode, synthesized through a combination of electrospinning and stepwise sintering techniques [46]. The structural evolution of $Na_3Al_{2/3}V_{4/3}(PO_4)_{2.9}F_{0.3}$ (NAVPF) during electrochemical cycling was investigated using *in situ* XRD analysis as displayed in Figure 2.6j, and was recorded over the first four cycles between 1.5 and 4.6 V. Examination of the diffraction patterns revealed a consistent shift in peak positions, particularly for the (113), (024), (116), and (300) reflections. These peaks exhibited a systematic migration toward higher angles during the charging process, followed by a return to their initial positions upon discharge. This behavior is indicative of a highly reversible structural transformation within the NAVPF material. The periodic nature of the peak shifts suggests that NAVPF undergoes predominantly solid solution reactions throughout the charge–discharge cycle.

2.3.3 Optimizing Composition and Morphology: Prussian Blue Analog

On account of increased theoretical capacity, low cost, and open framework structure, Prussian blue analogs (PBAs) are highly compelling materials for SIBs. The stable framework of PBAs ensures that structural stability and integrity are maintained during sodiation and desodiation processes. Primarily, vacancies in the $[Fe(CN)_6]$ framework lead to a reduction in both the number of active redox pairs and the unit mass. Moreover, Fe atoms bonded to just one $[Fe(CN)_6]$ unit may undergo changes in their valence states, a behavior frequently observed in nonstoichiometric semiconductors. Liu et al. reported that in PBA(Cu)□Fe, a small amount of FeHS is replaced by Cu dopant in the structure (Figure 2.8) [47]. The presence of Fe vacancies leads to a minor contraction in the lattice structure, as evidenced by the reduced lattice parameters of PBA□Fe and PBA(Cu)□Fe compared to PBA. Professor Xu's group has recently targeted the fast fading in capacity and also addressed the sluggish kinetic reactions through the strategy based on synergistic bimetallic reaction centers. In the findings, the FeVO-PBA (FV) framework demonstrated exceptional performance, achieving an impressive rate capability

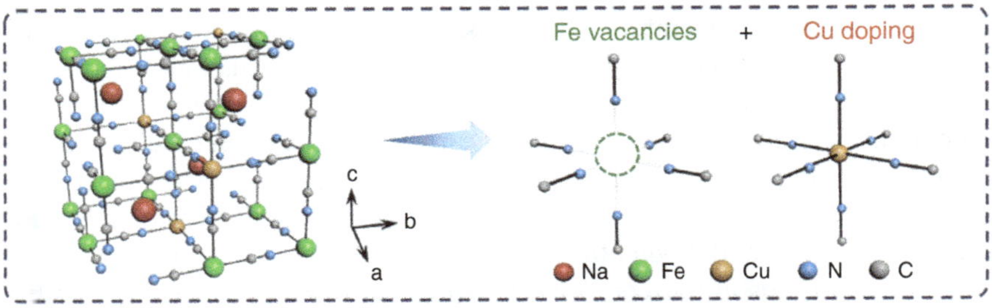

Figure 2.8 Structural diagram of PBA(Cu)□Fe, illustrating the arrangement of copper and iron within the framework. Source: Reproduced with the permission of [47]/John Wiley & Sons.

with capacities of 148.9 mAh/g at 0.5 C and 56.1 mAh/g at 100 C. It also exhibited outstanding cycling durability, maintaining stability over 30,000 cycles, alongside a high energy density of 259.7 Wh/kg in full-cell configurations. Additionally, it operated effectively across a broad temperature range, from −60 °C to 80 °C [48].

In another study by Professor Xu's group, the relationship between structural regulation and electrochemical performance was explored to develop efficient sodium-ion battery (SIB) materials. Seven Cobalt-based PBAs (CoM-PBAs) were synthesized using a low-energy co-precipitation method, incorporating various 3d transition metals (M = V, Mn, Fe, Co, Ni, Cu, Zn). DFT simulations and experimental results revealed that an optimal p-band center of cyanide linkages (CN) facilitates Na^+ intercalation and diffusion, while the d-band center of metal cations correlates with electrode stability. Among the tested materials, CoV-PBA exhibited the best performance, with a high specific capacity of 565 mAh/g at 0.1 A/g, 97.7% capacity retention over 15,000 cycles, and a rate capability of 174.7 mAh/g at 30 A/g [49].

Additionally, Liu et al. demonstrated the use of vacancy-regulated PBAs, synthesized via precipitation conversion, as cathodes in SIBs. This approach achieved impressive energy densities exceeding 500 Wh kg^{-1} [50]. *Ex-situ* XRD analysis is conducted to examine the alterations in the crystal structure of the Eg-$NaFeCl_2$ sample (see Figure 2.9a). The crystal structure remains cubic throughout most of the potential range. At 3.8 V, a partial tetragonal phase appears, and at the end of charge, the structure comprises a blend of both cubic and tetragonal phases. During charging, Eg-$NaFeCl_2$ crystalline phase transitions from a cubic to a blend phase of both cubic and tetragonal, and the other way around during discharge (Figure 2.9b). The performance of the samples is assessed in half-cell configurations, focusing on their rate capabilities. The Eg-$NaFeCl_2$ sample demonstrates an exceptional rate performance and maintains a high specific capacity, aligning well with earlier findings. The Eg-$NaFeCl_2$ sample exhibits specific capacity under lower current density, resulting in an exceptional energy density exceeding 500 Wh kg^{-1} (Figure 2.9c,d).

2.3.4 Exploring Sustainable Alternatives: Organic Cathodes

Organic batteries present several benefits, including the use of widely available and cost-effective materials, as well as rapid reaction kinetics. The organic batteries are positioned as a compelling alternative to traditional technologies, particularly excelling in ultralow-temperature conditions. In organic molecules, the reduction of unsaturated bonds such as C=C, C=N, and C=O often leads to the formation of radical intermediates. This process generates a negative charge on nitrogen or oxygen atoms and an unpaired electron on carbon atoms. Despite their potential, these organic intermediates exhibit instability and can frequently dissolve in electrolyte, resulting in considerable capacity degradation. Nanostructured organic electrodes offer substantial potential for advancing SIBs. Their improved electron and ion transport capabilities, coupled with a greater exposure of redox-active sites, have the potential to enhance both the actual capacity and rate performance of these batteries. Despite the progress in nanostructured organic electrodes, finding simple and affordable methods for their fabrication remains a significant challenge and is

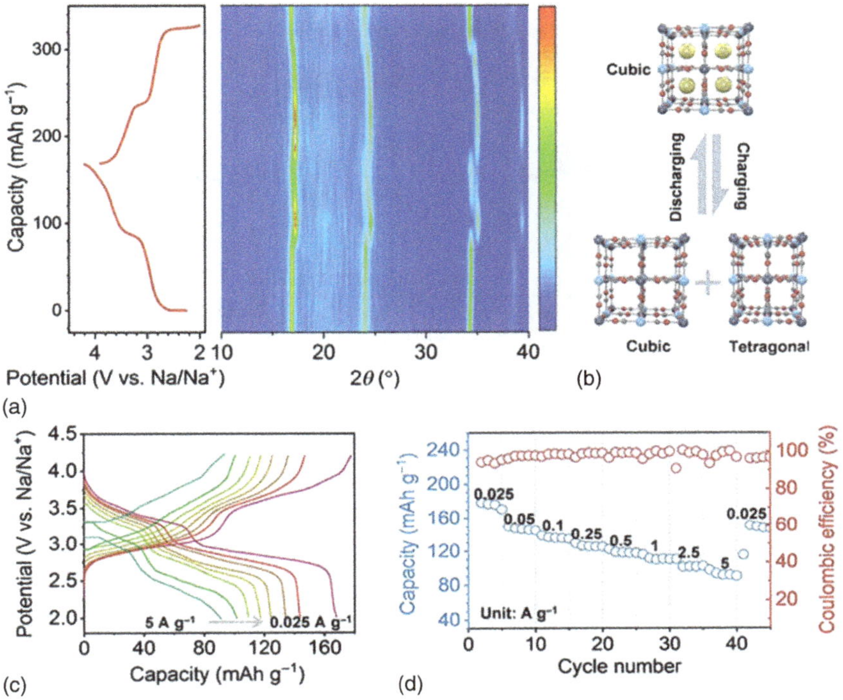

Figure 2.9 *Ex situ* XRD and normal electrochemical measurement of the Eg-$NaFeCl_2$ sample. (a) *Ex situ* XRD measurement, (b) schematic diagram of the crystal structure change, (c) charge–discharge curves from 5 to 0.025 $A\,g^{-1}$, and (d) rate performance of half batteries. Source: Reproduced with the permission of [50]/John Wiley & Sons.

Figure 2.10 Molecular design strategy of NDI-Ona. Source: [51] permission from/American Chemical Society.

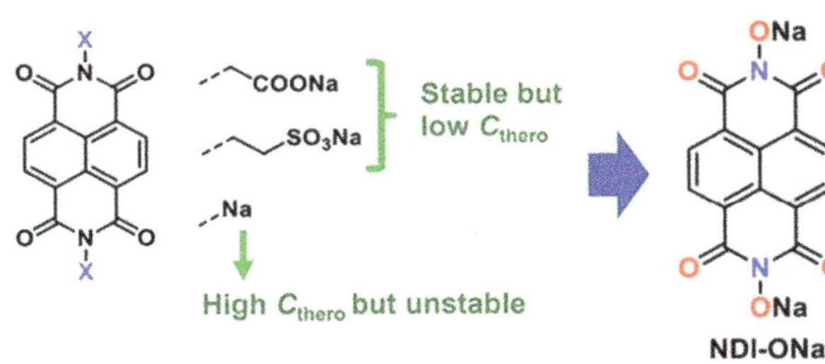

still relatively uncommon. Xing et al. investigated the creation of high-performance sodium–organic batteries by self-assembling nanostructured cathodes derived from naphthalene diimide [51]. Three methods for salinizing aromatic imides involve the use of sodium carboxylate, sodium disulfonate, and sodium imidate (Figure 2.10). The initial two approaches are effective in generating stable ionic compounds. Nonetheless, the design, which theoretically maximizes capacity, faces challenges due to its susceptibility to hydrolysis from trace moisture. This highlights the need for improved approaches to develop stable sodium host materials that can achieve high specific capacity in sodium–organic batteries.

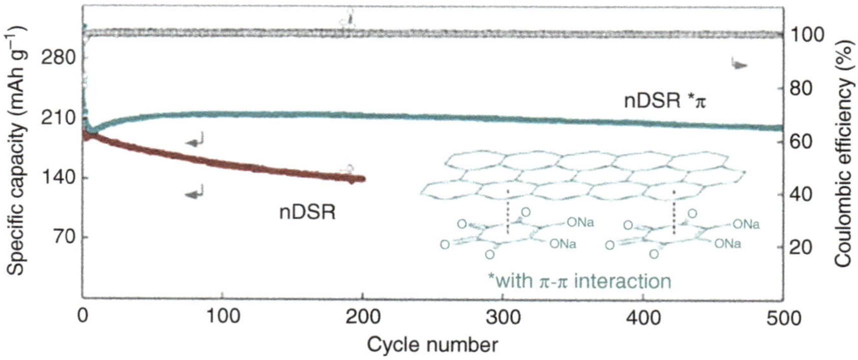

Figure 2.11 Cycling stability of nDSR and nDSR*π at ambient temperature. Source: [52]/American Chemical Society.

Xu et al. demonstrated remarkable durability in organic-sodium batteries at extremely low temperatures by employing disodium rhodizonate organic electrodes combined with single-layer graphene oxide [52]. This combination stabilizes multiphase conversions and prevents the consumption of organic radical intermediates. The cycling performance of nDSR is adversely affected by the dissolution of intermediates, leading to swift capacity degradation at 30 °C as illustrated in Figure 2.11. This rapid loss is attributed to the "shuttle effect" caused by soluble species. To stabilize the molecular structure of nDSR, single-layer graphene oxide (GO) was introduced, enhancing the cycling stability of the electrode. The findings suggest that nDSR*π holds promise for practical use in ultralow-temperature applications. In half-cell setups, however, a pure Na metal anode faces challenges in maintaining long-term cycling performance at such temperatures, primarily due to slow Na^+ ion diffusion and the risk of dendrite formation. A proposed setup of nDSR*π||PBA is used to ensure the complete redox reaction of the organic electrode. This configuration aimed to address the limitations seen with Na metal anodes in ultralow temperatures to prevent capacity decay. At −40 °C, the nDSR*π||PBA setup demonstrated impressive cycling performance, starting with an initial capacity of 192 mAh g^{-1} at 100 mA g^{-1}.

2.4 Performance Optimization Strategies of Different Electrolytes for Fast Charging

In this part, a comprehensive discussion will be carried out to demonstrate the unique features of efficient electrolytes utilized for fast charging in SIBs. Performance optimization strategies for fast charging in Na-ion batteries focus on improving electrolyte properties and electrode–electrolyte interactions. Electrolyte desolvation strategies aim to reduce the energy barrier for Na^+ ion transport by using solvents with lower binding energies to Na^+ ions or incorporating additives that weaken Na^+ ion–solvent interactions. Optimizing the electrode–electrolyte

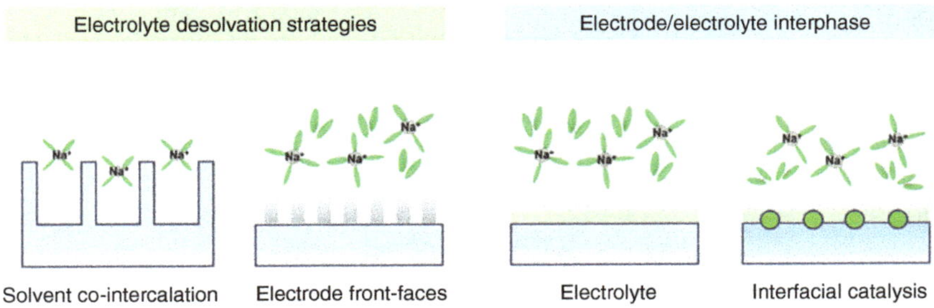

Figure 2.12 Electrolyte desolvation strategies and electrode–electrolyte interphase. Source: [53]. Reproduced with the permission of/John Wiley & Sons.

interphase, particularly the solid electrolyte interface (SEI), is crucial for enhancing fast-charging capabilities. This involves developing electrolyte formulations that form stable and thin SEI layers, allowing rapid Na^+ ion transport while protecting against further electrolyte decomposition. Other approaches include using highly concentrated electrolytes to promote faster desolvation, exploring co-intercalation mechanisms with specific solvents, and designing nanostructured electrode surfaces to facilitate ion transport (Figure 2.12). Additionally, research is focused on developing novel electrolyte systems, such as ionic liquids or deep eutectic solvents, which can offer unique solvation properties and potentially lower desolvation energies. The use of electrolyte additives that can modify the SEI composition and structure is another promising strategy to enhance fast-charging performance. Furthermore, tailoring the porosity and surface chemistry of electrode materials can synergistically work with electrolyte optimization to improve Na^+ ion diffusion kinetics at the interface. Balancing these strategies to achieve high ionic conductivity, stable interfaces, and efficient desolvation is key to enabling fast charging in Na-ion batteries.

2.4.1 Electrolyte Desolvation Strategies

The kinetics of Na^+ ion transport at the electrode–electrolyte interface during rapid charging are heavily influenced by the solvation structure of the electrolyte bulk. Critical factors involve the desolvation of Na^+ ions and the breakdown of solvents within the solvation sheath, forming SEI. The desolvation of Na^+ ions is the most energy-intensive stage, posing a significant barrier to the rapid charging of anode materials. To attain high performance, it is inevitable to design electrolytes and electrode material surfaces to reduce the activation energy of desolvation. Na^+ ion–solvent co-intercalation reactions, which occur without the desolvation process, show promise for rapid sodium storage. Studies have demonstrated the effectiveness of Na^+ ion–solvent co-intercalation in achieving high-rate performance. For instance, Jache and Adelhelm demonstrated the utilization of an anode material named graphite in SIBs, with Na^+ ions co-intercalating with diglyme (DGM) within the graphite structure [54]. Kim et al. discovered to employ $NaPF_6$ in a diglyme (DGM) electrolyte, exhibiting in a graphite anode that demonstrated

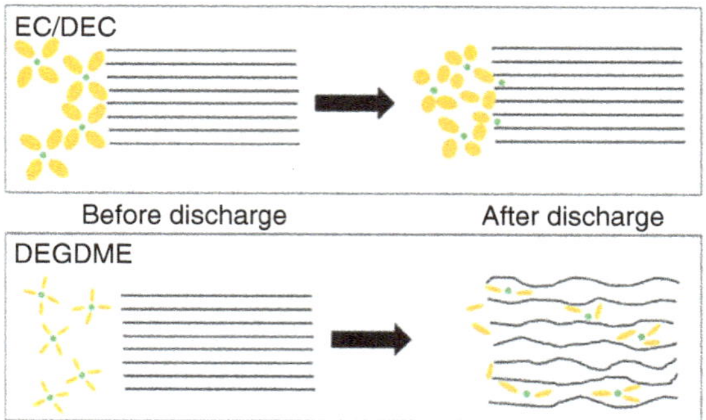

Figure 2.13 Proposed schematic illustrating sodium storage in natural graphite with different electrolyte systems. Source: [55]. Reproduced with permission of/John Wiley & sons.

a capacity of 100 mAh g^{-1} at 5.0 A g^{-1} [55]. The influence of various electrolyte solvent species on redox potential and SEI formation was examined, resulting in a crystalline few-layered graphene anode achieving approximately 100 mAh g^{-1} with 30.0 A g^{-1} (Figure 2.13). Additionally, highly concentrated electrolytes, due to weak solvent co-intercalation, can exhibit fast desolvation processes, reducing the desolvation energy barrier. However, high viscosity and limited ionic conductivity pose challenges for fast-charging performance under high-concentration electrolyte conditions.

To enhance solvation behavior and battery performance, another effective approach involves varying the electrolyte concentration to adjust the number of solvent molecules, alongside the use of multiple solvents. At elevated voltages, the risk of side reactions increases due to the breakdown of free anions and solvents at the electrode–electrolyte interface. Increasing the electrolyte concentration alters the solvation structure surrounding Na^+ ions. When solvent molecules are insufficient, cations and anions tend to form ion pairs, resulting in the clustering of anions. Reducing free solvent molecules broadens the potential window by minimizing solvent decomposition at the interface. In the desolvation process of Na^+ ions, using highly concentrated electrolytes enhances Na^+ ion diffusion by reducing solvent co-intercalation. In these electrolytes, the SEI primarily comprises decomposition products of anions, differing from the composition typically found in conventional electrolytes. Although an extended voltage range and more rapid desolvation are advantageous, highly concentrated electrolytes have limitations such as high cost and reduced ionic conductivity due to their high viscosity. An alternative approach involves using ultralow-concentration electrolytes (Figure 2.14a) [56]. To address issues of low initial Coulombic efficiency (ICE) and high redox potential in solvent co-intercalation, a co-adsorptive mechanism has been proposed. This mechanism involves reversibly storing solvated Na^+ ions in the ribboned grain boundaries and mesopores of microcrystalline graphite fibers. These fibers, prepared from natural cotton, exhibit high graphitization, multiple graphene layers, and small pore sizes. This structure benefits from a low energy barrier of the co-adsorptive mechanism, achieving fast-charging performance within 4 minutes, a high ICE of 92.5%, and enhanced cycling stability (Figure 2.14b) [57].

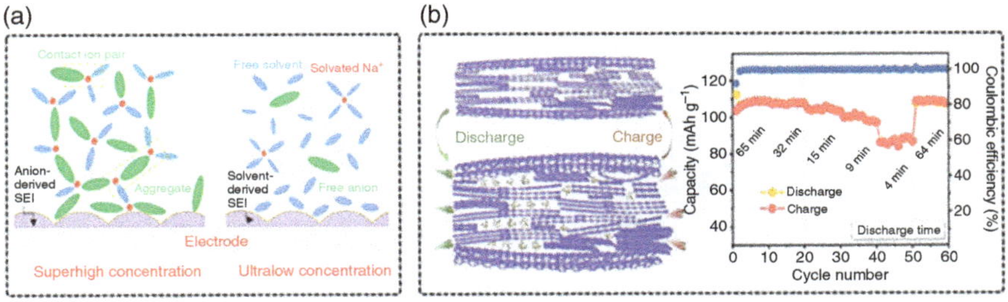

Figure 2.14 (a) Models of solvation and interfaces for superhigh and ultralow-concentration electrolytes. Source: [56]/John Wiley & sons/CC BY 4.0. (b) Schematic illustrating the volume expansion of microcrystalline graphite fiber (MCGF) during charge and discharge, with corresponding rate performance and discharge duration across different current densities. Source: [57]. Reproduced with the permission of/John Wiley & Sons.

2.4.2 Electrode–Electrolyte Interphase

The Na^+ ion desolvation process underscores the importance of the solvation structure and the reduction decomposition of the solvation sheath in SEI formation, crucial for Na^+ ion transport kinetics and fast-charging performance. Designing electrolyte and electrode materials systematically is vital for rapid Na^+ ion transport within the SEI. Ether electrolytes, with higher lowest unoccupied molecular orbital (LUMO) compared to carbonate electrolytes, resist excessive decomposition and produce a high-quality SEI, enhancing charge transport kinetics and rate performance for anodes of SIBs. Additionally, ether-based electrolytes benefit alloy-type anodes due to a thin, elastic, and ionically conductive SEI structure, enhancing ion transport and structural stability. When the electrochemical potential of the anode (μ_A) drops below the LUMO level of the solid-state electrolytes (SSEs), both the solid–electrolyte interface (SEI) and the CEI are formed, or if the electrochemical potential of the cathode (μ_C) goes beyond the highest occupied molecular orbital (HOMO) level of the SSEs (Figure 2.15). Ideally, the SEI and CEI act as barriers, preventing further anode reduction and cathode oxidation; this setup facilitates the reversible electrochemical reactions between the electrode and electrolyte interfaces. Nevertheless, continuous reactions often occur if the electrochemical potentials exceed the stability limits of the electrolytes. To ensure thermodynamic stability, the electrochemical potentials μ_A and μ_C must be placed within the stability window of the electrolyte. This positioning limits the open-circuit voltage (V_{oc}) of the battery cell. A protective SEI layer contributes to kinetic stability, facilitating a higher V_{oc} provided that the gap between eV_{oc} and the band gap (E_g) is not excessively large.

$$eV_{oc} = \mu_A - \mu_C \leq E_g \tag{2.1}$$

Electrons travel from the anode through an external circuit to the cathode, and during discharge, cations migrate into the electrolyte to maintain charge balance, a process that reverses when charging. The energy density of the battery is governed by

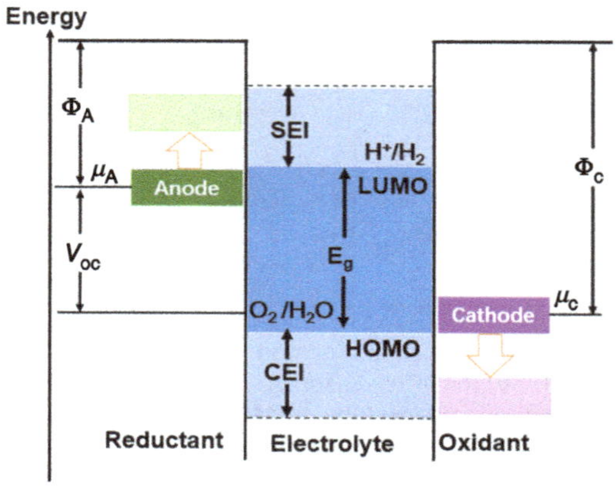

Figure 2.15 Diagram illustrates the energy levels in an aqueous electrolyte under open-circuit conditions. Φ_A and Φ_C represent the work functions of the anode and the cathode, respectively. E_g represents the allowable energy window of the electrolyte for maintaining thermodynamic stability. To ensure kinetic stability, the formation of a solid–electrolyte interface (SEI) layer is necessary when the electrochemical potentials μ_A exceed the lowest unoccupied molecular orbital (LUMO) and/or μ_C falls below the highest occupied molecular orbital (HOMO).

the product of Λ and V_{oc}, where Λ represents the amount of reversible charge transfer capacity per unit weight (Ah g^{-1}). The value of Λ diminishes with the charging or discharging rate, due to the slower ionic motion within the battery compared to the electronic current. In aqueous electrolytes, high H^+ ion conductivity can be achieved in liquid water. However, the band gap of around 1.3 eV restricts the open-circuit voltage (V_{oc}). To achieve higher open-circuit voltages (V_{oc}) and increased energy density, research is focused on nonaqueous electrolytes with larger band gaps (E_g), leading to lithium-ion battery development. The limiting electrochemical potential (μ_C) of the cathode in SIBs is determined by the HOMO levels of the salt.

In bismuth (Bi) anodes, Lee et al. discovered that the small adsorption energy between DME molecules and Bi nanoparticles (NP) forms a 3D-networked nanostructure in DME solvent, enabling rapid ion diffusion and transfer, reaching an impressive capacity of 325 mAh g^{-1} even at 150 C (Figure 2.16a) [58]. In liquid environments, metal nanoparticles (NPs) often cluster together to minimize surface energy when they come into proximity [59]. In battery systems, this tendency varies with the electrolyte solvent due to the interaction between polarized solvent molecules and the NPs. In general, metal nanoparticles carry a positive charge [60], which draws the negatively charged parts of solvent molecules toward them, resulting in the formation of a solvation layer. A thicker solvation layer makes coalescence more difficult. To investigate this, the effects of various electrolyte solvents, including DME, EC, and DEC, on the behavior and aggregation of Bi nanoparticles were examined through computational analysis. By positioning these solvent molecules 0.3 nm from the Bi NP surface, their interactions were observed. The results revealed that DME stayed away from the Bi surface, whereas EC and DEC were found to adhere to it, as depicted in Figure 2.16a. Adsorption energies

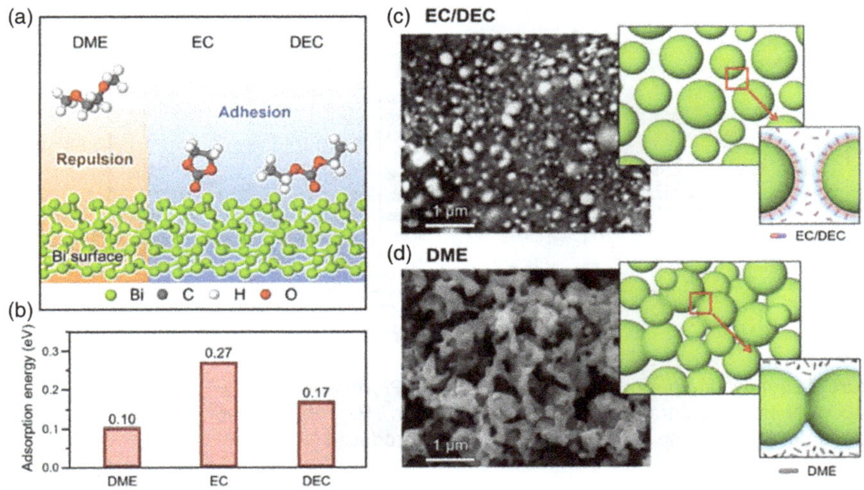

Figure 2.16 (a) DFT calculations to simulate the interactions between the Bi surface and EC, DEC, and DME molecules, (b) calculated adsorption energies for EC, DEC, and DME molecules interacting with the Bi surface, (c) Bi nanoparticles (NPs) extracted from the 20×-cycled, inset solvation layer on the Bi NPs, and (d) formation of porous Bi nanostructures from the coalescence of fragmented Bi NPs in DME solvent, with insets illustrating the DME molecules' distribution around the Bi NPs. Source: Reproduced with permission from [58]/John Wiley & Sons.

further quantified these tendencies (Figure 2.16b). DME had low adsorption energy (0.10 eV $atom^{-1}$), whereas EC and DEC had higher adsorption energies (0.27 and 0.17 eV $atom^{-1}$, respectively). This indicates that EC and DEC are more inclined to create a solvation layer, as illustrated in the inset schematic of Figure 2.16c. The positively charged outer surface of this layer can attract additional ions or molecules, creating a liquid–solid interface akin to an "electrical double layer." This interaction repels adjacent Bi NPs, increasing their separation and preventing coalescence (Figure 2.16c). In contrast, the low adsorption energy of DME means it does not adhere to Bi NP surfaces, allowing the NPs to potentially form a 3D networked structure over time (inset schematic of Figure 2.16d).

Designing electrolyte composition to regulate the electrode–electrolyte interphase can also improve anode rate performance. Wang et al. proposed a weakly solvating electrolyte with tetrahydrofuran solvent and $NaPF_6$ salt, resulting in lower ion desolvation energy and homogeneous SEI, significantly enhancing the rate performance of commercial hard carbon (HC) anodes (Figure 2.17) [61]. High-concentration and localized high-concentration electrolytes also impact SEI formation, facilitating ion transport. The influence of electrode material on electrode–electrolyte interphase formation is gaining attention. In a tetrahydrofuran (THF)-based electrolyte, characterized by its weak solvation environment, sodium ions (Na^+) undergo swift desolvation. This process results in the creation of a consistent SEI that is enriched with NaF and organic compounds on the surface of the commercial HC anode. This enhancement stabilizes an interface and promotes

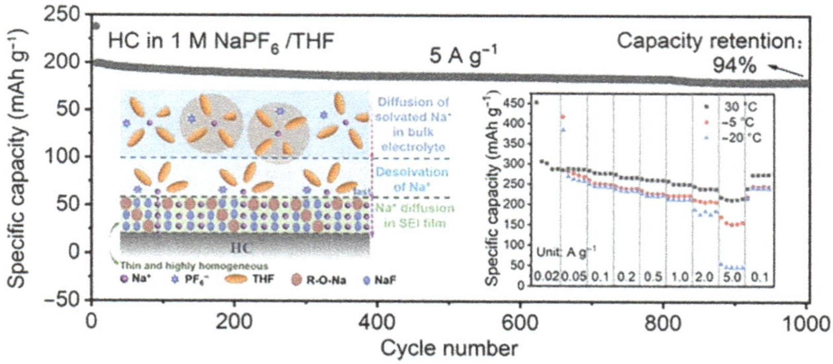

Figure 2.17 Cycling performance of HC in 1 M $NaPF_6$ dissolved in THF at $5\,A\,g^{-1}$. Insets include schematic diagrams of Na^+ ion migration in THF (left) and rate performance over a wide temperature range on right. Source: [61]. Reproduced with permission of/John Wiley & sons.

rapid sodium-ion (Na^+) migration within the SEI, leading to improved high-rate performance, extended durability, and excellent low-temperature operation for the HC anode. At −20 °C, the electrode possesses a substantial capacity of 74% of its room-temperature value – while operating at $2\,A\,g^{-1}$, maintaining 90% of its initial capacity after 1000 cycles.

Song et al. emphasized the undeniable importance of the SEI in extending battery lifespan, especially for high-capacity anodes that experience substantial electrolyte decomposition resulting from particle fragmentation [62]. They describe the creation of a highly effective ultrathin SEI, measuring just 3–4 nm, achieved through a Cu^+ ion-optimized double electrical layer (EDL). The distinctive double electrical layer (EDL) created with a SO_3CF_3–Cu complex deposited on CuS in a $NaSO_3CF_3$/diglyme electrolyte ensures remarkable durability, allowing the electrode to sustain even after 7000 cycles (Figure 2.18). In a comparable approach for the CuS anode, the use of Cu^+-tailored double electrical layers was found to minimize electrolyte depletion and form an exceptionally thin CuF_2-rich SEI. This configuration effectively reduces activation energy and enhances the conductivity of Na^+ ions within the SEI.

For carbon-based anode materials, the reversible C=O bond was uniformly implanted on the HC surface via a surface–functionality–reconstruction strategy, promoting the selective reduction of $NaPF_6$ salts to create a consistent, layered, and inorganic-rich SEI (Figure 2.19a). HC doped with single Zn atoms was produced through high-temperature calcination of a precursor made from 2,4-diaminophenol formaldehyde resin and zinc acetate. The rapid decomposition of $NaPF_6$ was facilitated. This development enhanced interface sodium-ion storage kinetics, resulting in an impressive rate capacity of $140\,mAh\,g^{-1}$. B-doped Bi-interconnected nanoparticles catalyzed electrolyte decomposition, creating a NaF-dominant SEI that significantly improves sodium-ion diffusion and mechanical stability; this approach achieves exceptional performance, delivering $203\,mAh\,g^{-1}$ at $80.0\,A\,g^{-1}$ (Figure 2.19b). *Ex situ* XPS, TOF-SIMs, and HRTEM are used to study SEI

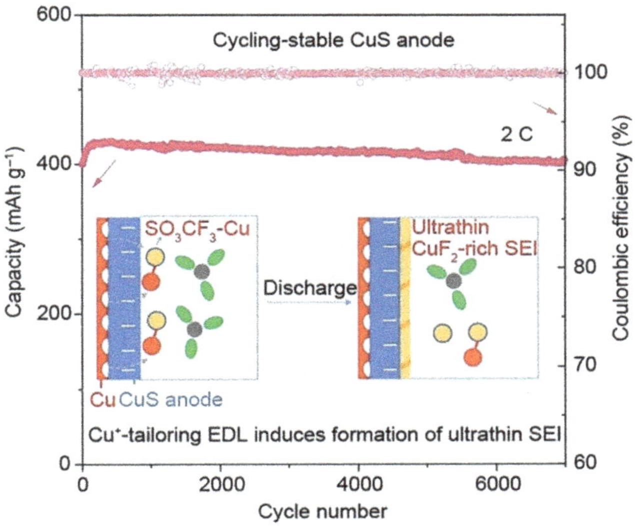

Figure 2.18 Long-term cycling stability and relative Coulombic efficiency of CuFG at 2 C. The inset illustrates the Cu^{+}-enhanced EDL, created by micro-dissolution of the Cu current collector, which promotes the development of an ultrathin, CuF_2-rich SEI on the CuS anode. Source: [62]. Reproduced with permission of/John Wiley & sons.

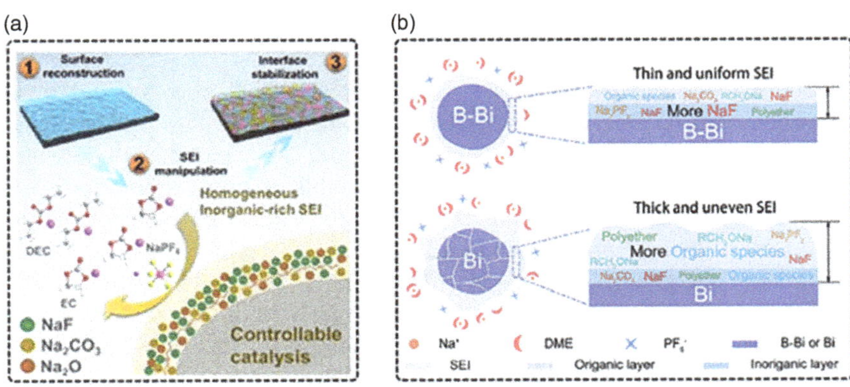

Figure 2.19 (a) Controllable SEI design on HC via surface reconstruction featuring abundant C=O (carbonyl) bonds. Source: [63]. Reproduced with permission of/John Wiley & sons. (b) Schematic showing the SEI thickness and composition formed by pure Bi and B-doped Bi-interconnected nanoparticles. Source: [64]. Reproduced with permission of/John Wiley & sons.

structure and composition, favoring higher inorganic components (such as NaF) for fast-charging performance. DFT calculations reveal the interface catalytic mechanism of SEI formation, providing insights into adsorption energy, binding energy, radical distribution, and HOMO/LUMO energy levels. This interfacial catalytic mechanism offers broad research opportunities for optimizing SEI and enhancing Na^+ ion transport. However, effective SEI regulation is only a prerequisite for fast-charging performance; achieving high ICE, reaction reversibility, and structural stability is essential for practical fast-charging anode materials.

2.4.3 Organic Electrolyte

Electrolytes are composed of solvents, solute salts, and additives. Ether-based solvents, known for their low dielectric constants and susceptibility to oxidation, are less favorable compared to ester-based solvents. The latter are preferred in NIFCs due to their superior ionic conductivity, enhanced oxidation resistance, and higher dielectric constants. Often, mixed solvents combine the benefits and drawbacks of multiple solvents, making it crucial to control their proportions for optimal performance. Common sodium salts with large anions and weak anion–cation associations include inorganic ($NaClO_4$ [65], $NaPF_6$) [66] and organic salts ($NaSO_3CF_3$ [67], $Na[(FSO_2)_2N]$ [68], $Na[(CF_3SO_2)_2N]$) [69]. Additives help address weaknesses like unstable SEI films and flammability, by forming passivation layers, reducing electrolyte flammability, and preventing overcharge. Additives are a key focus in electrolyte research, and the interplay of solvents, salts, and additives determines the electrolyte's properties, as shown in Figure 2.20a. Unlike half-cells, NIFCs use non-sodium anodes. In fuel-cell systems, it is crucial to assess the compatibility between the electrolyte and both electrodes simultaneously. Various strategies for electrolyte tailoring have been developed, including composition control [71], concentration adjustment [72], new recipe design [73], and functional modifications [74], as illustrated in Figure 2.20b.

Ether-based electrolytes are prone to oxidation at high voltages and often incompatible with cathodes, yet they are valued for forming thin SEI films and their compatibility with graphite anodes. In 2018, TMO-based batteries with diglyme electrolytes showed performance comparable or superior to ester-based systems. The GPC anode facilitated reversible intercalation and de-intercalation processes by allowing both diglyme and Na^+ ions to co-intercalate within its structure. A $Na_{0.66}Ni_{0.26}Zn_{0.07}Mn_{0.67}O_2$//1 M $NaPF_6$ in diglyme//GPC device demonstrated stable cycling, while $Na_3V_2(PO_4)_2F_3$//HC with the same electrolyte showed high initial irreversibility and capacity fading, emphasizing the need for system-specific electrolyte designs. Sadan et al. introduced a DME electrolyte for $Na_3V_2(PO_4)/_3$, achieving an excellent capacity at high current density (500 C) in a $Na_3V_2(PO_4)_3$//Sn full cell. Wu's team formulated a tailored ether-based electrolyte (NDD-3), consisting of 3.04 M $NaPF_6$ dissolved in a solvent blend of diethylene glycol dimethyl ether (DEGDME) and 1,3-dioxolane (DOL) at a 10 : 1 volumetric ratio [75]. The electrolyte demonstrated high oxidative stability and good compatibility with anodes, aiding in the formation of a dense inorganic interphase. The graphite//NDD-3//$Na_3V_2(PO_4)_2O_2F$ system exhibited exceptional rate performance under extended cycling durability, maintaining stability over 1300 cycles with a loss of approximately 0.0077% per cycle. This performance highlights the viability of Na^+ ion–solvent co-intercalation within graphite anodes. The NIFC cell, incorporating HC and $Na_3V_2(PO_4)_2O_2F$ with the NDD-3 electrolyte,

(a)

(b)

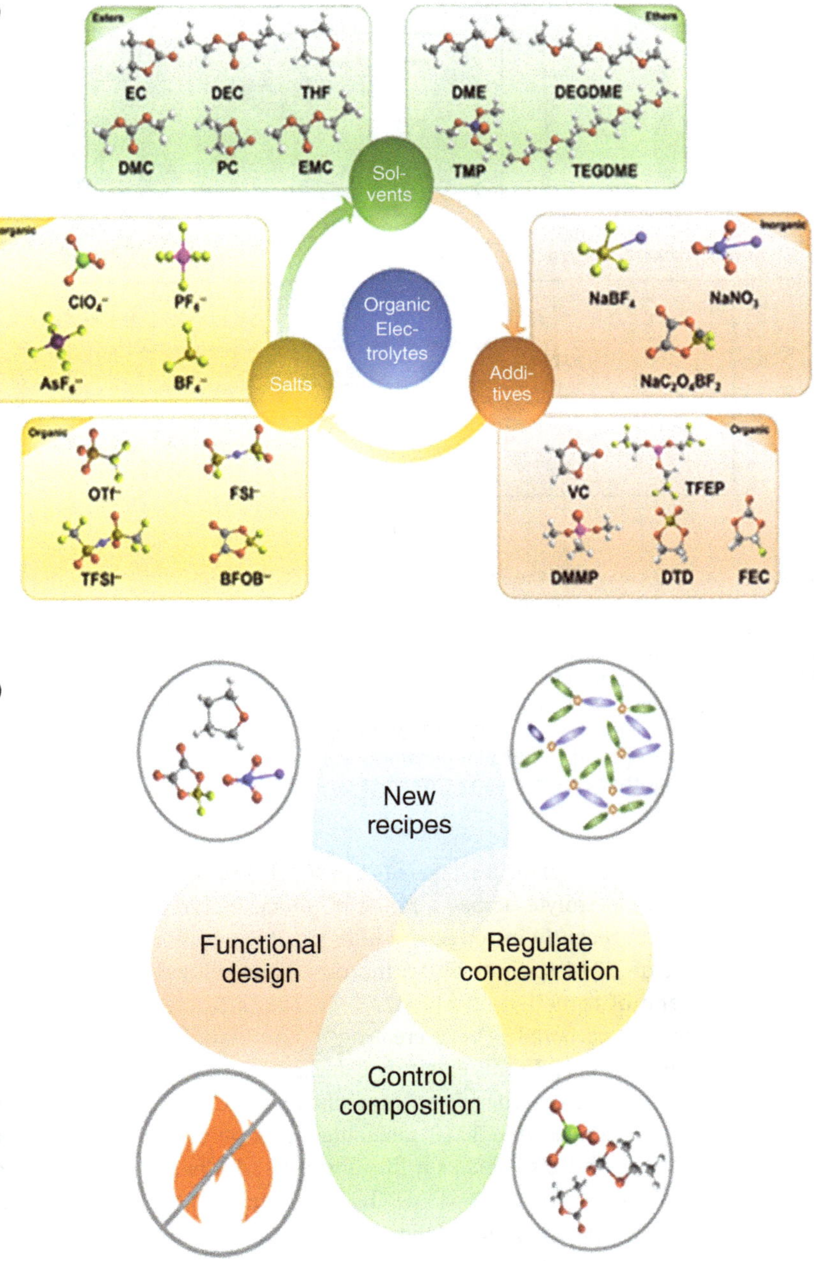

Figure 2.20 (a) Different formulations of the electrolyte and (b) enhanced approaches for liquid electrolyte optimization. Source: [70]/John Wiley & Sons/CC BY 4.0.

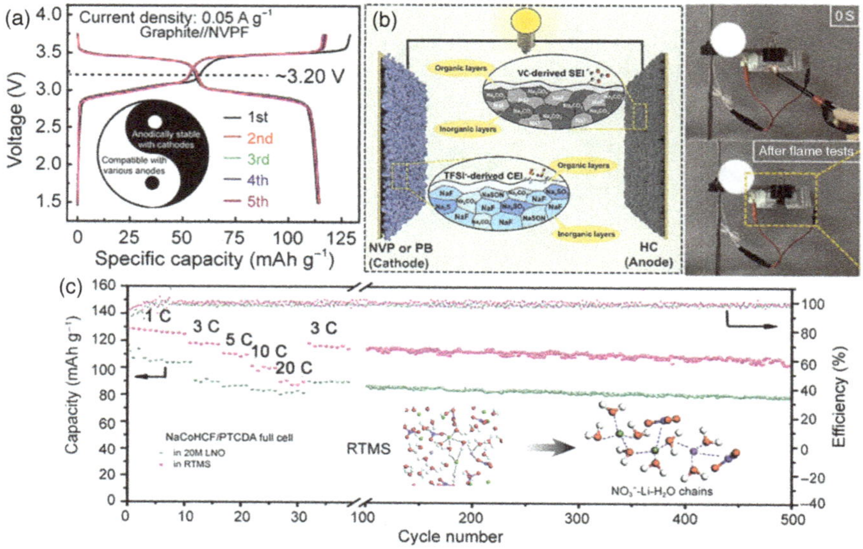

Figure 2.21 (a) GCD curves for the initial five cycles at 0.05 $A\,g^{-1}$ in NVPF//HC full cells, with an inset schematic highlighting the benefits of NDD-3 over other ether-based electrolytes. Source: [75]. Reproduced with permission of/John Wiley & Sons. (b) Diagram illustrating the formation of the CEI and SEI on the cathode and anode in a 1.2 M NaTFSI-TMP/BTFE/VC electrolyte (left), along with flammability tests conducted on the pouch cell (right). Source: Reproduced with permission from [76]/John Wiley & Sons. (c) Performance metrics for the NaCoHCF//PTCDA full cell, including rate and cycling stability, with an inset showing molecular dynamics simulations of RTMS aqueous solutions. Source: Reproduced with the permission of Wang et al. [77]. Copyright © 2021 Wiley-VCH GmbH.

delivered an impressive average voltage of 3.72 V. This showcases the versatile potential of NDD-3 electrolyte across a range of anode materials (Figure 2.21a). Chou's research team recently introduced a highly stable and nonflammable electrolyte, consisting of 1.2 M sodium bis(trifluoromethanesulfonyl)imide (NaTFSI) dissolved in a blend of trimethyl phosphate (TMP), bis(2,2,2-trifluoroethyl) ether (BTFE), and vinylene carbonate (VC), creating a formulation denoted as 1.2 M NaTFSI-TMP/BTFE/VC [76]. In this electrolyte design, BTFE with its high fluorine content spread out concentrated salt–solvent clusters. VC, on the other hand, plays a dual role: it reduces free TMP molecules disrupted by BTFE, restoring the solvated structure, and also acts as a film-forming additive. This formulation enabled the successful operation of $Na_3V_2(PO_4)_3$//HC and PB//HC full pouch cells (Figure 2.21b) to achieve good cycling stability over 120 cycles with specific capacity retention exceeding 85%. Han's and Fang's group jointly developed a cost-effective room-temperature hydrated molten salt (RTMS) electrolyte system composed of $LiNO_3 \cdot 3H_2O$ and $NaNO_3$, providing an expansive electrochemical stability range of 3.1 V [77]. With this electrolyte, a full cell was constructed featuring a cobalt hexacyanoferrate (NaCoHCF) cathode paired with a perylene-3,4,9,10-tetracarboxylic dianhydride (PTCDA) anode. This NIFC demonstrated excellent rate performance and maintained 94.7% capacity retention over 500 cycles as seen in Figure 2.21c. The remarkable electrochemical performance can be attributed to the minimal free water content and the improved ion mobility, stemming from the strong interaction between Na^+ ions and NO_3^- groups that facilitates efficient ion release. The summary of other electrolytes at wide temperature range is provided in Table 2.2.

2.5 Electrolyte Additives

Modifying electrode materials is vital for improving electrochemical performance in sodium batteries, but the electrolyte formulation is equally crucial, embodying the adage "four ounces can move a thousand pounds" [79]. As the "lifeblood" of the battery, the electrolyte enables rapid ion movement and forms protective films on electrode–electrolyte interfaces, reducing degradation and extending cycle life [80]. Though used in small quantities, electrolyte additives are essential, comprising natural or synthetic organic and inorganic compounds that enhance ionic conductivity, rate performance, and other specific functions [81]. With the growth of the new energy industry and advancements in portable electronics, power batteries are pushing toward higher voltage and energy density [82], presenting new challenges for electrolyte additives [83]. Armand's group has led pioneering research, introducing a range of additives optimized for SIBs at room temperature [84]. Concurrently, Lin's group reviewed progress in nonaqueous sodium-ion electrolyte additives, focusing on their role in enhancing battery resilience under harsh conditions [85]. Additionally, Cui's team explored the ion transport mechanism and solvation structure of sodium-ion electrolytes, showcasing significant advancements in the field [86].

2.5.1 Specific Groups of Additives

In liquid electrolytes, every component – from solvents and salts to additives – plays a vital role in forming durable interface films on cathodes and anodes, providing essential protection [84]. These elements collectively determine the electrochemical and thermal stability of the electrolyte. Salts and solvents present in smaller quantities are often referred to as additives. Researchers carefully select these additives based on their compatibility with appropriate oxidation and reduction potentials, ensuring they decompose in a predetermined sequence. This strategic selection forms more stable and ionically conductive solid–liquid interface films, crucial for expanding the operational temperature range and enhancing the safety profile of the electrolyte [86]. As summarized in Figure 2.22, additives serve a dual function: they act as a "reinforcement army," strengthening the interface film and extending the battery's electrochemical lifespan, and as "support forces," providing additional benefits to the electrolyte and enhancing the operational safety of commercial batteries. Metaphorically, if a SIB is likened to a flower, then the cathode and anode materials represent the soil, the solvent, and electrolyte salt are the essential nutrients, and the additives are the butterflies in pollination, crucial for ensuring continuous and efficient battery operation.

Table 2.2 Summary of electrolyte-wide temperature for SIBs.

System	Electrolyte	Improved strategies	Temperature (°C)	Cycling stability/cycles/ current density ($mA\,h\,g^{-1}$)/–/($mA\,g^{-1}$)	Conductivity ($mS\,cm^{-1}$)
Liquid electrolyte	2 M $NaClO_4$-0.3DMSO	Disruption of HB network	−50–45	68/100th/133 90/100th/665	0.11
	H_2O/FA(3 : 7)	Disruption of HB network	−50	80/8000th/600	1.75
	1 M $NaNO_3$ in glycerol and water (2 : 1)	Disruption of HB network	−10	40/350th/100	—
	NaTFSI:H_2O:AND (mol ratio, 1 : 1 : 2.5)	Disruption of HB network	−20	90/100th/-	3.4 (RT)
	3.86 M $CaCl_2$ + 1 M $NaClO^4$ in water(m: mol kg^{-1})	Disruption of HB network	−30	74.5/1000th/150 23/6000th/1500	7.13
	3.5 M $Mg(ClO_4)_2$ + 0.5 M $NaClO_4$ in water	Disruption of HB network	−60	23.2/10000th/1064	4.60
	25 M NaFSI + 10 M NaFTFSI in water	Disruption of HB network	−10	45/500th/26.6	11.8(30 °C)
	1 M Na_2SO_4-SiO_2 hydrogel electrolyte	Disruption of HB network	−30	61.8/500th/130	0.07
	0.5 M $NaPF_6$ in diglyme	Optimization of solvation structures	−20 to −60	98.02/1000th/128.3	—
	0.8 M $NaPF_6$ in THF/PC/PFPN	Optimization of solvation structures	−40	71.5/100th/0.2C	2.3

	1 M NaFSI in EC/PC/DEC-6%ES	Optimization of solvation structures	−40	74/200th/0.1C	0.75
	0.3 M $NaClO_4$ in EC/PC-FEC	Optimization of solvation structures	−25	94.9/1000th/1C	101.6/1C
	2 M $NaNO_3$ in different sugar solutions	Disruption of HB network	−50–80	—	8.536
	$NaClO_4$-H_2 O-urea-DMF	Optimization of solvation structures	−50–50	—	13.85 (50 °C)
	0.8 M NaOTf and 0.2 M $NaBF_4$ in diglyme	Increase in conductivity	−40	50/80th/110	0.8
	1 M NaODFB in DME	Increase in conductivity	55	249.9/150/100	11.61
	1 M $NaPF_6$ in EC/PC-3% PS + 0.5% NaDOFB + 1% SN + 3% VC	Increase in conductivity	55	110/60th/12.8	—
	1 M $NaBF_4$ in (BMP − TFSI)-based ILs	Increase in conductivity	75	90/100th/-	4.8
	1 M NaFSA in Py13-FSA	Increase in conductivity	60	193/65th/50	11.9
	1 M NaFSI in Pyr1,3FSI/Pyr1,4FSI	Increase in conductivity	60	193/65th/50 130/80th/50	-
	PFSA-Na membranes	Increase in conductivity	−10	40/160th/100	0.0488
Solid electrolyte					
	0.04 M $NaClO_4$ in PAN:PC(1 : 4)	Increase in conductivity	−20	65/300th/–	1.8 (RT)
	PVDF-HFP-based electrolyte	Increase in conductivity	−30 to 130	72.3/100th/0.05 C	0.153 (−30 °C)
	MTP92	Increase in conductivity	−10 to 60	75/400th/2 C 60/1500th/2 C	1.0 (−20 °C)

(Continued)

Table 2.2 (Continued)

System	Electrolyte	Improved strategies	Temperature (°C)	Cycling stability/cycles/ current density $(mA\,h\,g^{-1})/-/(mA\,g^{-1})$	Conductivity $(mS\,cm^{-1})$
	Cross-linked PMMA	Increase in conductivity	70	95/100th/600	21
	METEPP-MMA-TFMA gpe	Increase in conductivity	60	120/100th/1000	6.29 (RT)
	MATEMP-MMA-TFMA gpe	Increase in conductivity	60	91.5/200th/–	5.13 (RT)
	SPDPA	Increase in conductivity	60	93.5/200th/270	4.96
	PMMA-b-PPEGMA	Increase in conductivity	60	89.9/80th/–	0.13
	$Na_3\ Zr_2\ Si_2\ PO_{12}$	Increase in conductivity	200	80/26th/–	0.19
	$Na_2\ Ti_2\ Si_1\ P_2\ O_{12}$	Element doping	300	—	17
	OIPC/$NaNTf_2$ mixtures	Composite materials	40	—	$10^{-1.8}$
	5 wt% Al_2O_3-PVDF-HFP	Composite materials	70	93.3/600th/500(RT)	1.3
	$Na_3\ NH_2\ B_{12}\ H_{12}$	Composite materials	80	102/100th/23.9	0.1
	$Na_{3.1}\ Zr_{1.95}\ Mg_{0.05}\ (SiO_4)_2(PO_4)/Na_2\ B_4\ O_7{\cdot}10H_2O$	Composite materials	60	30/62th/0.588	—

Source: [78]/John Wiley & Sons.

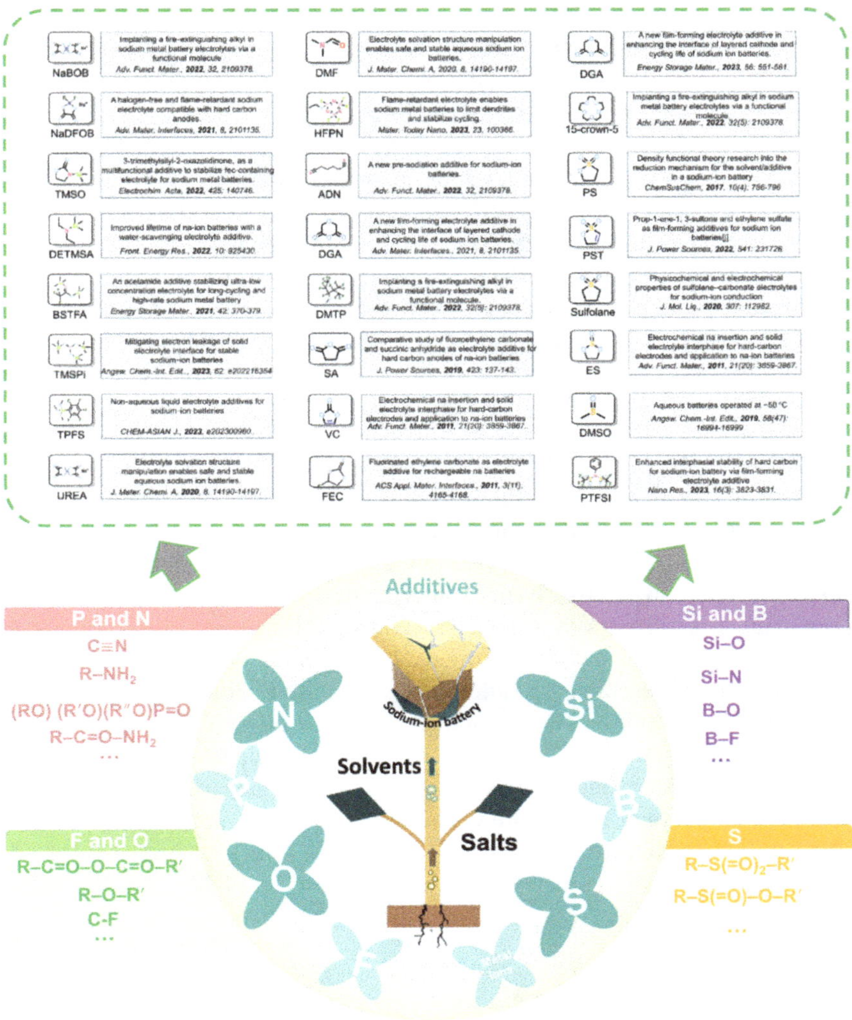

Figure 2.22 Metaphorical relationship diagram of sodium battery and additives. Source: [87]. Reproduced with permission of/John Wiley & Sons.

i) Fluoric (C—F) and Oxygen-Containing Groups (R—C=O—O—C=O—R′, R—O—R′)

The individual effects of simple groups are well known; for example, sodium ions (Na^+) interact weaker with fluorine than with oxygen atoms. While interactions in mixtures can be complex, distinct influences from unique additives can still be observed. In electrolyte chemistry, the interaction between Na^+ ions and oxygenated functional groups, such as hydroxyl and carboxylate, is similar to that with lithium ions. These groups, with their lone pairs and high electronegativity, reduce the

Na^+ ion–propylene carbonate (PC) interaction and aid in Na^+ ion desolvation. During discharge, metal ions interact with oxygen in ester groups, weakening the ion–solvent interaction due to oxygen's electron-donating nature. Leveraging SEI components with proven performance from previous studies is an innovative approach. For example, additives with anhydride groups, like succinic anhydride (SA) and diglycolic anhydride (DGA), decompose to produce more Na_2CO_3/sodium alkyl carbonates on the electrode surface compared to ester group additives. This decomposition mechanism hinders lattice oxygen diffusion in the cathode [88, 89]. Fluorine-containing functional groups also contribute to forming a NaF-enriched SEI following the oxidative decomposition of additives. This NaF-enriched SEI enhances interface stability and guides the design of artificial molecules (like $C_{60}(CF_3)_6$) or new fluorine-containing additives [90]. An optimal fluoride complex content in the SEI/CEI is beneficial.

ii) **Sulfonyl (R—S(=O)2—R′) and Sulfoxide (R—S(=O)—R′) groups**

The sulfur–oxygen chemical bond uniquely boosts the interface stability formed on graphite electrodes. Compounds similar to 1,3,2-dioxathiolane 2,2-dioxide (DTD), such as 1,3-propane sultone (PS) and prop-1-ene-1,3-sultone (PST), exhibit comparable inhibition of ion–solvent co-intercalation in graphite [91]. When reaching the cathode, molecules of PST and DTD with sulfonyl groups decompose to yield RSO_3—, $ROSO_2$—, and SO_{32}— anions. These anions complex with transition metal ions, forming insoluble transition metal sulfates or sulfites, thus preventing transition metal dissolution. The exploration of such additives is driven by the need to overcome the limitations of PC-based solvents, such as capacity loss and reduced cycle stability [92]. Sulfoxide additives offer versatility as they can dissolve in most solvents, including water mixed with various molar fractions, enabling the creation of low-freezing-point electrolytes (Figure 2.23a,b) [93].

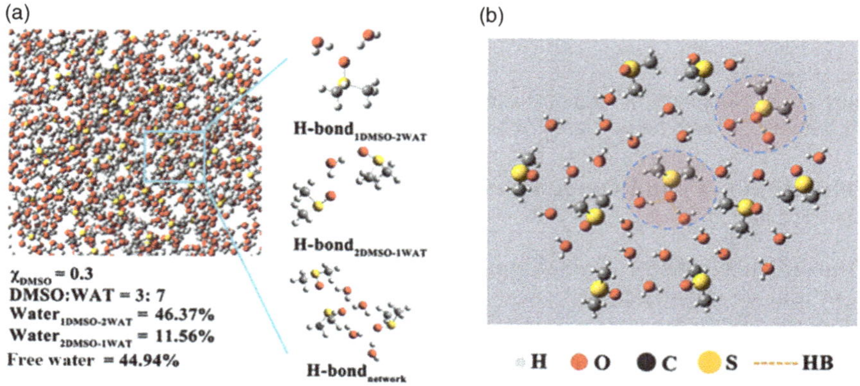

Figure 2.23 (a) MD simulations of analyzing the conformation of the system (the ratio of dimethyl sulfoxide [DMSO] is 0.3) and (b) local structure of the system from MD simulations. Source: [93]. Reproduced with permission of/John Wiley & sons.

iii) **Phosphate Ester ($(RO)_nP(O)(OH)_{3-n}$) and Nitrogenous Group (R—NH_2, R—C=O—NH_2, C≡N)**

Phosphorus and nitrogen, both from the same periodic group, exhibit numerous similar properties, including flame-retardant effects in their compounds. Fluorophosphate, with abundant P—O and P—F bonds, enhances the SEI layer's conductivity on the HC electrode surface and lowers the overpotential in graphite electrodes during charge–discharge cycles [94, 95]. Difluorophosphate salts, derived from the breakdown of hexafluorophosphate MPF_6 (M = Li, Na, or K) during hydrolysis, notably improve ionic conductivity due to their fluorine-rich and covalently bonded phosphorous structure. Nitrile-based additives are also gaining attention. Nitriles have a less favorable reduction potential compared to carbonate solvents, aiding in the formation of a low-impedance SEI film on cathode surfaces. They can react with residual acids and water to form amides, which remove impurities and stabilize the electrolyte. The C≡N group in nitrile complexes with high oxidation state Na^+ ion promotes SEI formation, especially in ether solvents [96]. Adiponitrile (ADN), with its symmetrical chain structure and dual C≡N bonds, reacts with Na^+ to form NaF and NaCN, effective components of the CEI film, balancing charge and enhancing performance across a wide temperature range [97]. Additionally, amide and amino groups have high water solubility, offering potential applications in aqueous sodium-ion electrolyte systems.

iv) **Boron Group (B—O, B—F) and Silyl Group (Si—N, Si—O)**

Borate, known for its affordability, is widely used in flame-retardant electrolyte formulations. Bis(oxalato)borate (BOB—) salt is particularly compatible with phosphate electrolytes, helping to prevent the degradation of TMP and triethyl phosphate (TEP). The radial distribution function shows that the double-bonded oxygens on the BOB— anion are primary coordination sites for Na^+ ions, which improves the electrolyte's conductive efficiency as free Na^+ tends to coordinate with the double-bonded oxygen on TEP. The B—F bond has a bond energy of 613 kJ mol^{-1}, higher than that of H—F (565 kJ mol^{-1}) and P—F (490 kJ mol^{-1}), indicating its greater stability [98]. In carbonate-based electrolytes, HF corrosion is a major factor in the degradation of interfacial layers, and borate additives enhance battery cycle performance by reducing HF corrosion [99]. The Si—O bond, with its long length, high bond dissociation energy, and exceptional stability at high temperatures, is significant. The Si—N functional group sequesters OH— from H_2O and F— from HF, forming stable products with N atoms, which bind to PF_5 and impede the hydrolysis of additives. Additives like *N,N*-Diethyl-1,1,1-trimethylsilanamine (DETMSA), *N,O*-bis(trimethylsilyl)trifluoroacetamide (BSTFA), and 3-(trimethylsilyl) oxazolidin-2-one (TMSO) not only boost ionic conductivity but also stabilize the interfacial layer [100, 101].

v) **Inorganic Salt Additives**

Inorganic salts are often used as additives in SIBs to introduce anions that help form protective interfacial films and to incorporate metal cations that modify

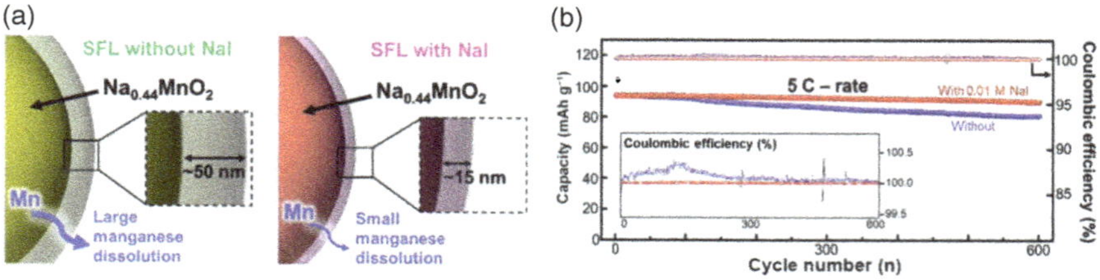

Figure 2.24 (a) Diagram illustrating the impact of the NaI additive on the system, highlighting its role in enhancing performance and stability during electrochemical processes, and (b) galvanostatic cycling performance of Na/1 M $NaClO_4$ in EC(4)/DEC(6)/$Na_{0.44}MnO_2$ cells, comparing results with and without the addition of 0.01 M NaI electrolyte additive. Source: [104]. Reproduced with permission of/John Wiley & sons.

the electrode interface, thereby enhancing battery performance. Common inorganic sodium salts, such as sodium difluoro-oxalate borate (NaODFB), sodium bis(trifluoromethanesulfonyl)imide (NaTFSI), sodium trifluoromethanesulfonate ($NaCF_3SO_3$), sodium carbonate (Na_2CO_3), sodium bis(fluoro-sulfonyl)imide (NaFSI), and sodium fluoride (NaF), are widely reported for use in SIBs [71, 102]. Moeez and colleagues utilized Na_2CO_3 and NaF as functional additives directly dissolved in an electrolyte to artificially ensure stability and thick CEI layer on the $Na_{0.67}Fe_{0.5}Mn_{0.5}O_2$ (NFMO) cathode before electrochemical cycling [103]. This approach inhibits the solubilization of transition metals within the electrolyte and prevents capacity degradation by protecting the interfacial layer, thereby improving cycling performance. Similarly from Figure 2.24a, Chae et al. outlined NaI electrolyte additive, demonstrating that iodide species react with manganese ions to create MnI_2 and MnI_3, which precipitate on the surface of NMO, creating a thin, protective surface film that reduces manganese dissolution, leading to enhanced cycling performance and an increase in capacity retention from 86% to 96% after 600 cycles (Figure 2.24b) [104]. Xie et al. addressed the parasitic effects caused by residual coordinated water during the charge–discharge process of PBAs by adding $AlCl_3$, which formed an $Al_2O_3{\cdot}3H_2O$ layer that depletes residual water and shields the electrode [105]. Shao et al. synthesized $Na_2CoFe(CN)_6$ (CoHCF) as a cathode material for aqueous SIBs and introduced an electrolyte with 1 wt% $CoSO_4$ as an additive in 1 m Na_2SO_4; the addition of Co^{2+} ions helped stabilize the framework structure of CoHCF and prevent dissolution during cycling, significantly enhancing cycling stability [106]. Moreover, adding 0.05 m $RbPF_6$ or 0.05 m $CsPF_6$ and 2 wt% FEC to the $NaPF_6$ baseline electrolyte in EC:PC (4 : 1, vol%) helped form highly conductive and stable passivation layers on HC, resulting in lower cell resistance [107]. Incorporating Rb^+ or Cs^+ ions into Na/HC cells enhanced both the specific capacity and the stability, while also diminishing polarization effects. This modification led to more consistent capacity retention over time.

iv) **Film-Forming Additives**

Besides the aforementioned additives, other commonly used types in SIBs include halogen salts, ionic liquids, anhydrides, anthraquinones, and boron compounds.

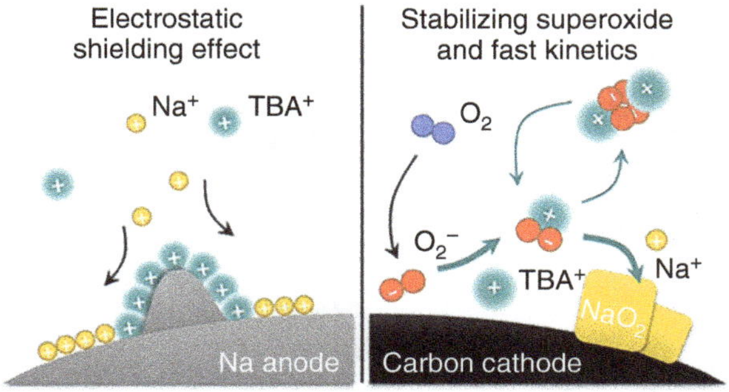

Figure 2.25 Schematic diagram of the electrostatic shielding effect of TBA^+ additive. Source: [112]. Reproduced with permission of/John Wiley & Sons.

Halogen salt additives, in particular, are famous to diminish the energy barrier at the interface during Na^+ charge transfer. Akkisetty et al. reported that using a halogen-rich additive, 2-chloro-1,1,2-trifluoroethyl difluoromethyl ether (enflurane), forms a thinner and more stable SEI layer containing sodium chloride and fluorinated salts. The inclusion of 10 v/v% enflurane in PC improved the reversible capacity and cycling performance of HC half-cells over 100 cycles compared to an additive-free electrolyte. Ionic liquids as electrolyte additives in SIBs can control side reactions at the electrode surface and enhance Na^+ ion mobility within the passivation layer. Manohar et al. [108] showcased that incorporating 5 wt% [C_3mpyr][NTf_2] into the organic electrolyte (1 M $NaClO_4$ in PC), combined with a $Na_3V_2(PO_4)_3$/C cathode and Na anode, led to an impressive discharge capacity of 107 mAh g^{-1} and a capacity retention of approximately 99.7%. This performance notably exceeded that of the electrolyte without the additive. Park et al. first introduced the *in situ* formation of electroconductive polymers for SIBs on $Na_3V_2(PO_4)_2F_3$ (NVPF) [109]. A layer of poly(1-aminoanthraquinone) (PAAQ) was deposited onto the NVPF, creating an efficient electron conduction pathway that enhances electron transfer at the cathode. TEM images of NVPF particles after a pretreatment cycle with 0.06 wt% AAQ show the particles covered by a thin PAAQ layer. Consequently, the Na//NVPF cell incorporating 0.06 wt% AAQ exhibited an enhanced initial discharge capacity of 99.3 mAh g^{-1}, outperforming cells that lacked the AAQ additive.

vii. **Other Additives Containing Specific Elements**

Extra metal ions help mitigate sodium dendrite formation through electrostatic shielding and redox reactions [110, 111]. As illustrated in Figure 2.25, tetrabutylammonium (TBA^+), with its larger ionic volume, adsorbs onto the electrode surface. This electrostatic shielding effect directs Na^+ ions to deposit uniformly on the sodium metal surface, reducing dendrite formation. Different additives offer various advantages. When using co-additives, it is crucial to consider the "stacking effect," which involves determining the oxidation or reduction sequence of different additives based on their redox properties [113]. The stability of the initially formed SEI layer is essential for stable electrochemical cycles. For example, while Li_2S_8 and $LiNO_3$ exhibit synergistic effects, the combination of Na_2S_6 and $NaNO_3$ has detrimental effects [114]. There are many other additives which are presented in Table 2.3.

Table 2.3 Summary of electrochemical performance of batteries with different additives and corresponding blank electrolytes.

Additive	Formulation of electrolyte	Cell type	Voltage range (V)	Working temperature (°C)	Rate or current density ($mA\,g^{-1}$)	Capacity ($mAh\,g^{-1}$) Capacity retention cycling cycles
FEC	1 M $NaClO_4$ + 10 vol% FEC in	SPAN‖Na	0.5–2.8	25	100 $mA\,g^{-1}$	75 270.2% 100 cycles
Blank	1 M $NaClO_4$ in TEGDME	SPAN‖Na	0.5–2.8	25	100 $mA\,g^{-1}$	193 13.6% 100 cycles
NaBF4	0.9 M $NaPF_6$ + 0.1 M $NaBF_4$ in G2	$Na[Cu_{1/9}Ni_{2/9}Fe_{1/3}Mn_{1/3}]O_2$ ‖Na	2–4	25	0.5 C (1 C = 127 $mA\,g^{-1}$)	100 82% 200 cycles
SA and FEC	1 M $NaClO_4$ + 5 vol% FEC + 3 wt% SA in PC	$Na_{0.6}Li_{0.15}Ni_{0.15}Mn_{0.55}Cu_{0.15}O_2$ ‖Na	2–4.5	30	0.2 C (1 C = 120 $mA\,g^{-1}$)	99.4 89.9% 150 cycles
FEC	1 M $NaClO_4$ + 5 vol% FEC in PC	$Na_{0.6}Li_{0.15}Ni_{0.15}Mn_{0.55}Cu_{0.15}O_2$ ‖Na	2–4.5	30	0.2 C (1 C = 120 $mA\,g^{-1}$)	64.9 54.6% 150 cycles
SA	1 M $NaPF_6$ + 0.5 wt% SA in EC/DEC (1/1, v/v)	HC‖Na	0–2	25	0.5 C	80 47% 30 cycles
FEC	1 M $NaPF_6$ + 0.5 wt% FEC in EC/DEC (1/1, v/v)	HC‖Na	0–2	25	0.5 C	231.2 92.5% 3210 cycles
FEC and PS	1 M $NaPF_6$ + 2 vol% FEC + 2% PS in EC/DMC (1 : 1, v/v)	$Na(Ni_{0.4}Mn_{0.4}Cu_{0.1}Ti_{0.1})_{0.999}La_{0.001}O_2$ ‖HC	1.2–4.4	25	0.1 C (1 C = 150 $mA\,g^{-1}$)	125 82% 50 cycles
FEC and DTD	1 M $NaPF_6$ + 2 vol% FEC+ 2 vol% DTD in EC/DMC (1 : 1, v/v)	$Na(Ni_{0.4}Mn_{0.4}Cu_{0.1}Ti_{0.1})_{0.999}La_{0.001}O_2$ ‖HC	1.2–4.4	25	0.1 C (1 C = 150 $mA\,g^{-1}$)	120 79.4% 50 cycles

FEC and PES	1 M $NaPF_6$ + 2 vol% FEC+ 0.2 vol% PES in EC/DMC (1 : 1, v/v)	$Na(Ni_{0.4}\ Mn_{0.4}\ Cu_{0.1}\ Ti_{0.1})_{0.999}\ La_{0.001}\ O_2$ ||HC	1.2–4.4	25	0.1 C (1 C = 150 mA g^{-1})	55 40.4% 50 cycles
$C_{60}\ (CF_3)_6$	1 M $NaPF_6$ + 0.1 vol% $C_{60}\ (CF_3)_6$ in DME	$NaV_2(PO_4)_3$@C||Na	2.7–3.7	25	10 C (1 C = 118 mA g^{-1})	108 98% 1000 cycles
Blank	1 M $NaPF_6$ in DME	$NaV_2(PO_4)_3$@C||Na	2.7–3.7	25	10 C (1 C = 118 mA g^{-1})	75 64% 1000 cycles
FEC	1 M $NaPF_6$ + 3 vol% FEC in EC:DMC(1 : 1, v/v)	HC||Na	0.003–2	25	0.1 C (1 C = 372 mA g^{-1})	230 94% 130 cycles
DMCF	1 M $NaPF_6$ + 1.5 vol% fluorinated dimethyl carbonate(DMCF) in EC:DMC (1 : 1, v/v)	HC||Na	0.003–2	25	0.1 C	96 48% 10 cycles
FEC	1 M NaTFSI+3 vol% FEC in EC:DMC(1 : 1, v/v)	HC||Na	0.003–2	25	0.1 C	190 95% 110 cycles
9F	1 M NaOTf+50 mm 9F in G2	FeS_2||Na	0.6–2.6	25	80 mA g^{-1}	400 33% 300 cycles
Blank	1 M NaOTf in G2	FeS_2||Na	0.6–2.6	25	80 mA g^{-1}	200 22% 140 cycles
15-C-5	1 M $NaClO_4$ + 2 wt% 15-C-5 in EC:DEC(1 : 1, v/v)	HC||Na	0.001–0.5	25	50 mA g^{-1}	251 95.69% 300 cycles
Blank	1 M $NaClO_4$ in EC:DEC(1 : 1, v/v)	HC||Na	0.001–2.5	25	50 mA g^{-1}	210.80 86.82% 300 cycles

(Continued)

Table 2.3 (Continued)

Additive	Formulation of electrolyte	Cell type	Voltage range (V)	Working temperature (°C)	Rate or current density ($mA\,g^{-1}$)	Capacity ($mAh\,g^{-1}$) Capacity retention cycling cycles
PS, NaDOFB, SN, and VC	1 M $NaPF_6$ in EC/PC + 3 vol% PS + 0.5 vol% NaDOFB+1 vol% SN + 3 vol% VC	$Na_3\,V_2\,(PO_4)_2\,F_3$ ‖HC	2–4.3	55	0.1 C (1 C = 128 $mA\,g^{-1}$)	108 99% 60 cycles
Blank	1 M $NaPF_6$ in EC-PC	$Na_3\,V_2\,(PO_4)_2\,F_3$ ‖HC	2–4.3	55	0.1 C (1 C = 128 $mA\,g^{-1}$)	75 77% 60 cycles
$Na_2\,CO_3$	1 M $NaClO_4$ + 2 wt% $Na_2\,CO_3$ in PC:FEC (98 : 2 by wt%)	$Na_{0.67}\,Fe_{0.5}\,Mn_{0.5}\,O_2$ ‖Na	1.5–4.3	25	0.5 C	60 36.8% 500 cycles
NaF	1 M $NaClO_4$ + 2 wt% NaF in PC: FEC (98 : 2 by wt%)	$Na_{0.67}\,Fe_{0.5}\,Mn_{0.5}\,O_2$ ‖Na	1.5–4.3	25	0.5 C	58 33.6% 500 cycles
Blank	1 M $NaClO_4$ in PC: FEC (98 : 2 by wt%)	$Na_{0.67}\,Fe_{0.5}\,Mn_{0.5}\,O_2$ ‖Na	1.5–4.3	25	0.5 C	30 18.6% 500 cycles
$CoSO_4$	1 M $Na_2\,SO_4$ + 1 wt% $CoSO_4$ in H_2O	$Na_2CoFe(CN)_6$‖C	0–1	25	2 C (1 C = 120 $mA\,g^{-1}$)	115 99.9% 100 cycles
SDS (Sodium dodecyl sulfate)	0.5 M $Na_2\,SO_4$ + 1 wt% SDS in H_2O	$NaTi_2\,(PO_4)_3$ ‖C	0–1	25	500 $mA\,g^{-1}$	84.8 81.2% 100 cycles
$SnCl_2$	50 mm $SnCl_2$ in 1 m $NaClO_4$ in EC:PC (1 : 1, v/v)	$NaV_2\,(PO_4)_3$ ‖Na	2.5–4	25	1 C	100 87% 250 cycles

Blank	1.0 M $NaClO_4$ in EC:PC (1 : 1, v/v)	$NaV_2(PO_4)_3$ ǀǀNa	2.5–4	25	1 C	80 75% 250 cycles
SbF_3	4 M NaFSI+1 vol% SbF_3 in DME	$NaV_2(PO_4)_3$ ǀǀNa	2–4	25	2 C	90.9 82.6% 1400 cycles
Blank	4 M NaFSI in DME	$NaV_2(PO_4)_3$ ǀǀNa	2–4	25	2 C	80 80% 1000 cycles
SbF_3	2 M NaFSI in DME + FEPE (1 : 1, molar ratio) + 1 wt% SbF_3	$NaV_2(PO_4)_3$ ǀǀNa	2–4	25	2 C	106.4 95% 1000 cycles
Blank	2 M NaFSI in DME + FEPE (1 : 1, molar ratio)	$NaV_2(PO_4)_3$ ǀǀNa	2–4	25	2 C	98 87% 1000 cycles
$EtAlCl_2$ and [EMIm]FSI	NaCl-buffered chloroaluminate electrolyte +1 wt% EtAlCl and 4 wt% [EMIm]FSI	$NaV_2(PO_4)_3$@CǀǀNa	2.5–3.8	25	150 mA g^{-1}	90 96% 460 cycles
Blank	1 M $NaClO_4$ + 5 wt% FEC in EC/DEC (1 : 1, v/v)	$NaV_2(PO_4)_3$@CǀǀNa	2.5–3.8	25	150 mA g^{-1}	80 79% 450 cycles
FEC and TMSO	1 M $NaPF_6$ + 5 wt% FEC + 0.5 wt% TMSO in EC/DEC (1 : 1, v/v)	$NaV_2(PO_4)_3$ ǀǀNa	2–4	25	2 C(1 C = 117 mA g^{-1})	103 93.4% 1400 cycles
FEC	1 M $NaPF_6$ + 5 wt% FEC in EC/DEC (1 : 1, v/v)	$NaV_2(PO_4)_3$ ǀǀNa	2–4	25	2 C (1 C = 117 mA g^{-1})	60 63.6% 1400 cycles

(Continued)

Table 2.3 (Continued)

Additive	Formulation of electrolyte	Cell type	Voltage range (V)	Working temperature (°C)	Rate or current density ($mA\,g^{-1}$)	Capacity ($mAh\,g^{-1}$) Capacity retention cycling cycles
DMSO	2 M $NaClO_4$ + 0.3 M DMSO in H_2O	$NaTi_2(PO_4)_3$@C‖C	0–1.6	−50	0.5 C	68 99% 100 cycles
$CaCl_2$	3.86 M $CaCl_2$ + 1 M $NaClO_4$ in H_2O	$Na_2CoFe(CN)_6$‖C	0–1.6	−30	1 C	64.6 86.7% 1000 cycles
HFPN	1 M $NaPF_6$ in EC:DEC:FEC:HFPN (1 : 1 : 0.05 : 0.05, v/v/v/v)	$NaV_2(PO_4)_3$ ‖Na	2.6–3.8	25	2 C	108 99% 800 cycles
Blank	1 M $NaPF_6$ in EC:DEC:FEC(1 : 1 : 0.05, v/v/v)	$NaV_2(PO_4)_3$ ‖Na	2.6–3.8	25	2 C	100 91% 800 cycles
BP (biphenyl)	1 M $NaPF_6$ in EC/DEC with 3 vol% BP	$Na_{0.44}MnO_2$ ‖Na	2–4	25	50 $mA\,g^{-1}$	106 98% 500 cycles
TAC	1 M $NaClO_4$ + 0.1 M TAC EC/DMC (1/1,v/v)	$NaV_2(PO_4)_3$ ‖Na	0–2.5	25	50 $mA\,g^{-1}$	92 96.2% 176 cycles
NaDFP	1 M $NaPF_6$ + 1 wt% NaDFP in EC/DMC (1/1,v/v)	HC‖Na	0.005–1.5	25	100 $mA\,g^{-1}$	180 76.3% 500 cycles
ADN	1 M $NaPF_6$ in EC/PC/DEC/FEC	$Na_{0.76}Ni_{0.3}Fe_{0.4}Mn_{0.3}O_2$ ‖HC	1.5–3.9	25	0.2 C (1 C = 115 $mA\,g^{-1}$)	75 75% 200 cycles

VC, NaODFB	1 M $NaPF_6$ + 3 vol% and TMSPi VC + 0.5 vol% NaODFB+1 vol%	$Na_3V_2(PO_4)_2F_3$ ‖HC	2–4.3	55	0.1 C (1 C = 128 mA g^{-1})	105 99% 60 cycles
Blank	1 M $NaPF_6$ in EC/DMC (v/v,1/1)	$Na_3V_2(PO_4)_2F_3$ ‖HC	2–4.3	55	0.1 C (1 C = 128 mA g^{-1})	30 99% 60 cycles
VC	1 M $NaBF_4$ + 3 vol% VC in TEP	$Na_{3.2}V_{1.8}Zn_{0.2}(PO_4)_3$ ‖HC	2–4.1	25	0.5 C	64.3 80.4% 200 cycle
Blank	1 M $NaBF_4$ in TEP	$Na_{3.2}V_{1.8}Zn_{0.2}(PO_4)_3$ ‖HC	2–4.1	25	0.5 C	20 28% 30 cycles
$NaBF_4$	0.9 M $NaPF_6$ and 0.1 M $NaBF_4$ in G2	$Na[Cu_{1/9}Ni_{2/9}Fe_{1/3}Mn_{1/3}]O_2$ ‖Al	0–3.8	25	0.5 C (1 C = 127 mA g^{-1})	88 84% 260 cycles
PTFSI	1 M $NaPF_6$ + 3 wt% PTFSI in EC/DEC (1/1, v/v)	HC‖Na	0.01–2.5	25	0.5 C (1 C = 250 mA g^{-1})	150 68% 500 cycles
Blank	1 M $NaPF_6$ in EC/DEC (1/1,v/v)	HC‖Na	0.01–2.5	25	0.5 C (1 C = 250 mA g^{-1})	100 50% 300 cycles

Source: [87]/John Wiley & Sons.

2.6 Performance Optimization Strategies for Fast-Charging Anode Materials

For SIBs, positive electrode materials are classified into five main types according to their sodium storage capabilities: carbon-based substances, intercalation transition metal oxides, conversion transition metal oxides, transition metal sulfides/selenides/phosphides, and alloy or phosphorous compounds (Figure 2.26a). Carbon anodes, known for their low cost, are crucial in commercial SIBs but have various drawbacks. For instance, HC anodes offer high capacities, low voltage, and minimal volume expansion, but their disordered structure leads to structural defects, resulting in poor ICE and poor structural stability. Additionally, the diffusion-controlled process with slow kinetics at low voltage can cause severe sodium dendrite formation, a major obstacle to fast-charging performance. Soft carbons, characterized by their relatively ordered carbon layer structure and

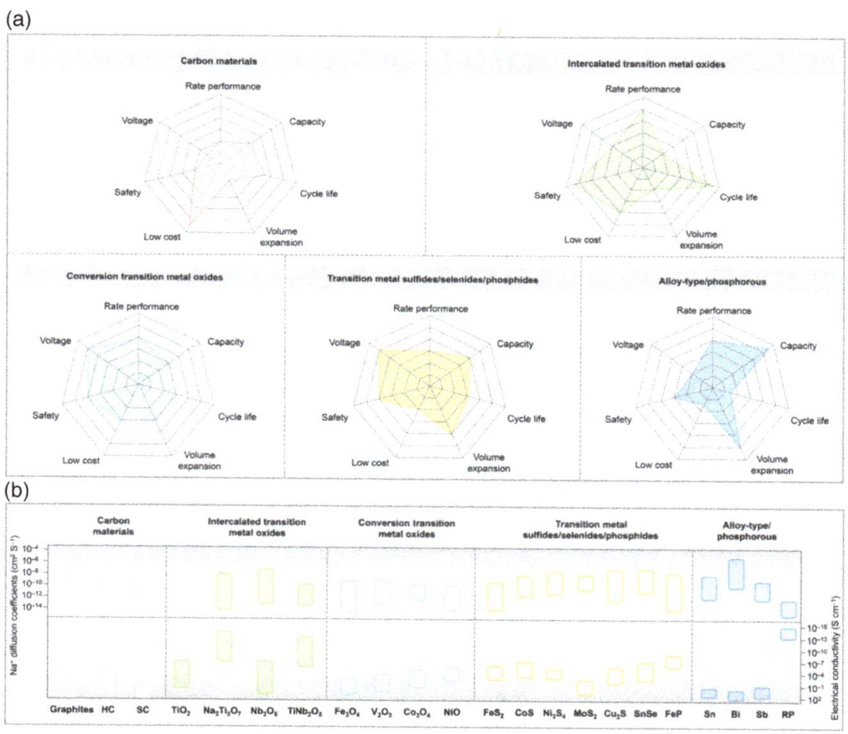

Figure 2.26 (a) Radar plots showcasing a comparison of discharge voltage, rate performance, theoretical capacity, cycle life, safety, cost-effectiveness, and volume expansion among various fast-charging anode materials, including carbon-based materials, intercalation transition metal oxides, conversion-type transition metal oxides, and transition metal sulfides/selenides/phosphides, as well as alloy-type and phosphorous anodes; (b) summary of Na^+ diffusion coefficients and electrical conductivity for these fast-charging anode materials, compiled from the literature. Source: [53]. Reproduced with permission of/John Wiley & Sons.

advanced graphitization, demonstrate excellent electrical conductivity, which is promising for fast sodium storage performance (Figure 2.26b). However, their small interlayer distances are not conducive to Na^+ insertion and extraction, and the limited active sites lead to worse sodium storage capacity. These challenges remain significant for improving the fast-charging performance of carbon anodes. Intercalated transition metal oxides offer suitable voltage platforms and excellent structural stability, making them safe and cost-effective candidates for fast-charging anodes. However, their low theoretical capacities and ionic conductivity hinder high-rate capacities [115, 116] (Figure 2.26b). Transition metal oxides, sulfides, selenides, and phosphides, with their large theoretical capacities, are promising for fast-charging anodes [117, 118]. However, their high discharge voltage leads to low-energy density for the full cell, and their cost and synthesis challenges further limit their application. Additionally, significant volume changes and low ionic/electronic conductivity led to poor cycling stability and reversibility during the (de)sodiation process. Alloy-type and phosphorous materials offer extremely high theoretical capacities and low discharge voltage but face structural stability issues due to severe volume changes during (de)sodiation [119, 120]. Nevertheless, alloys like bismuth (Bi), antimony (Sb), and tin (Sn), with high electrical conductivity and ionic diffusion coefficients, are expected to deliver excellent high-rate capacities for fast charging [121]. Thus, effective multiscale structural design is crucial to enhance ion/electron transfer capabilities, structural stability, and cycling life and rate capability, key aspects for fast-charging applications. The following section will thoroughly demonstrate the surface modification impact on anode materials amid fast charging.

2.6.1 Porous Engineering

While the energybarrier for solvated ions in the electrolyte is relatively low, the nonunit transfer number of Na^+ ion leads to a concentration gradient. This limitation reduces the depth of the Na^+ ion reaction in the electrode, significantly decreasing the utilization of active materials, especially during fast charging [122–124]. To enable rapid electrolyte diffusion in the electrode under fast-charging conditions, it is essential to analyze the effective Na^+ ion diffusion coefficient, which depends on the intrinsic Na^+ ion diffusion coefficient (D_0), the electrode's porosity ε, and the tortuosity τ of the electrode's pores [125]. Research has focused on optimizing electrode architectures to reduce τ and increase ε [126]. For example, an interpenetrated architecture of carbon nanofibers and vertically aligned graphene sheets, with *in situ* growth of molybdenum disulfide MoS_2 nanoflakes, was developed to enhance electrolyte penetration and facilitate fast Na^+ ion transfer [127]. These MoS_2 nanomaterials have numerous pores and a high specific surface area of $310\ m^2\ g^{-1}$, achieving capacity of $456\ mAh\ g^{-1}$. Therefore, developing strategies to reduce electrode tortuosity and enhance porosity is crucial for improving the fast-charging performance of anode materials in SIBs and warrants further attention. Designing a porous structure at the active material level can significantly improve reaction depth and active material utilization, thereby enhancing rate performance as shown in Figure 2.27.

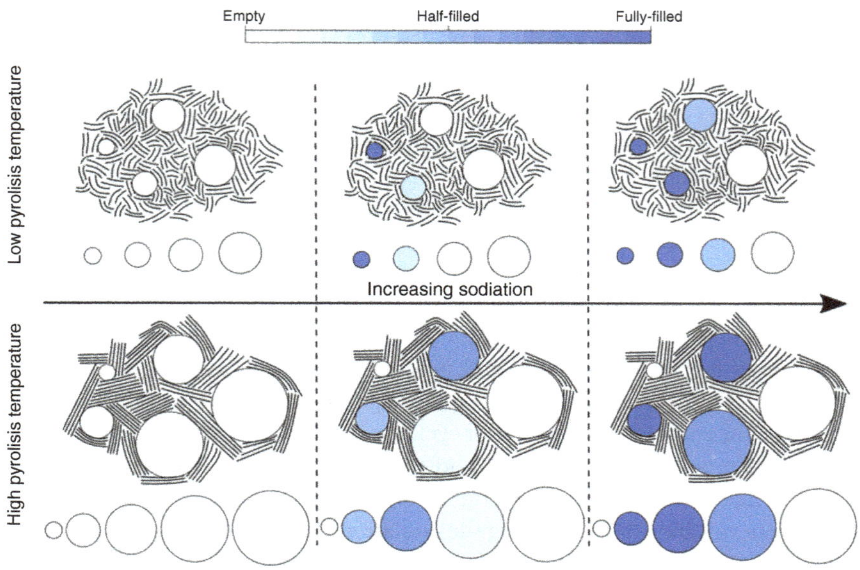

Figure 2.27 Schematic depiction of the microstructure and sodium-ion storage in closed pores of hard carbons (HCs) produced at different pyrolysis temperatures. The circles symbolize closed pores, and the black lines represent graphene layers. The blue gradient shows the level of sodium-ion filling within the pores, ranging from white for empty pores to dark blue for fully filled pores. Source: [128]. Reproduced with permission of/John Wiley & Sons.

Professor Xu team developed a novel porous heterojunction consisting of Ni-doped $FeSe_2/Fe_3Se_4$ encapsulated in Se-doped carbon through the selenization of metal–organic framework (MOF) precursors [129]. This innovative design features a high density of active sites and enhances both ion and electron mobility. The incorporation of Ni increases lattice defects and active sites, while the Se—C bonds and the carbon matrix maintain the structural stability of the composite, preventing aggregation of selenide nanoparticles during the charge and discharge cycles. The NF11S/C material showcases exceptional performance, achieving a capacity of 208.8 mAh g^{-1} after 2000 cycles at 8 A g^{-1} for SIBs. In a separate investigation led by Xu's team, they developed porous microspheres composed of CoS_2/Co_4S_3 encased in nitrogen-doped carbon (CoS_2/Co_4S_3@NC) using a straightforward solvothermal approach followed by annealing. These microspheres have fused as highly effective anode materials for SIBs, showcasing notable rate performance with capacities of 650 and 217 mAh g^{-1}. Furthermore, they demonstrate remarkable longevity, maintaining capacities after 1500 cycles and 3000 cycles [130]. Mesoporous HC materials, synthesized through various methods, have shown excellent rate performance in SIBs because of their open pore structure, which provides substantial specific surface area and numerous active sites. This results in sodium storage behavior dominated by a non-diffusion-controlled mechanism with excellent reaction kinetics [131]. However, despite their outstanding rate performance, mesoporous HC materials frequently exhibit reduced ICE and relatively high discharge voltage,

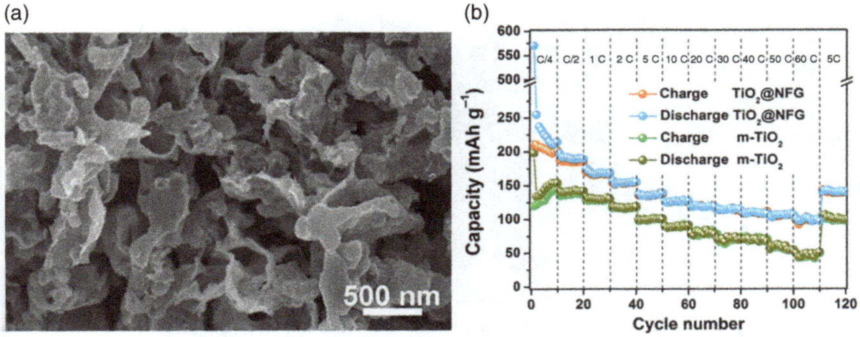

Figure 2.28 (a) Morphology and structure of the TiO_2@NFG hybrid as shown in FESEM images, and (b) electrochemical performance of the TiO_2@NFG electrode demonstrating cycling rate. Source: Reproduced with permission from [131]/John Wiley & Sons.

which may not be ideal for improving the energy density of the full cell. Moreover, various mesoscale TiO_2 structures have been synthesized using soft-templating or hard-templating methods to boost the rate performance of intercalated transition metal oxides [132]. As expected, the TiO_2@NFG hybrid electrode, which combines nitrogen-doped few-layered graphene with TiO_2, showcased remarkable durability and impressive rate performance, as illustrated in Figure 2.28a [131]. Additionally, upon returning the current density to 5 C following the high-rate testing, the electrode restored a reversible capacity of 140 mAh g^{-1}, which was sustained for over 3000 cycles without any loss and with a nearly 100% Coulombic efficiency (Figure 2.28b).

Mesoporous metal oxides like mesoporous titanium niobium oxide ($Ti_{0.88}Nb_{0.88}O_{4-x}$@C), with a surface area of 47 m^2 g^{-1} and a pore size of 28.4 nm, have been created using a confined acid–base pair self-assembly method to boost rate performance. These materials maintain a reversible capacity of 103 mAh g^{-1} at 50 C, outperforming nonporous samples in rate performance (Figure 2.29) [133]. Moreover, the design of multilevel pore or core–shell structures in transition metal oxides/sulfides/stannides/phosphides and alloy-type/phosphorous materials has been effective [134].

2.6.2 Nanostructuring

D hierarchically structured micro/nanomaterials (3D-HMNs) have been recognized as highly effective choices for energy storage applications, particularly in rechargeable batteries. These materials boast unique properties including unique porosities, extensive accessible spaces, remarkable capacity for volume accommodation capabilities, advantageous mass, and electron transport characteristics [135]. Their synthesis involves a range of strategies based on "Materials-Design-Properties" fundamentals, such as template approaches, electrospinning techniques, and 3D fabrication technologies. These approaches enable the construction of 3D structures

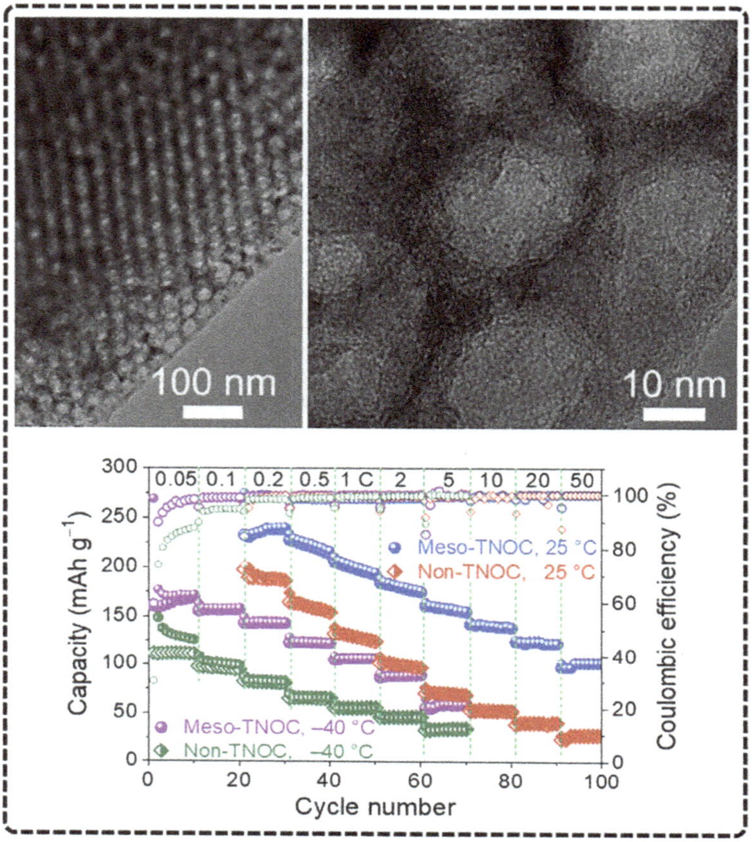

Figure 2.29 TEM images of the mesoporous TNOC below are rate performance comparisons between mesoporous and nonporous TNOC at 25 and –40 °C. Source: Reproduced with permission from [133]/John Wiley & Sons.

with optimized functions for battery applications. The field has experienced substantial progress in 3D hierarchically designed materials, which can take different forms and feature either uniform or irregular porous architectures. These structural characteristics are pivotal in energy applications. 3D-HMNs are particularly noteworthy as they combine the intrinsic attributes of each element, allowing for the attachment of large-mass active materials on conductive scaffolds. This results in extensive surface areas, highly porous architectures, and superior electrical conductivity. Additionally, these structures serve as efficient matrices for charge transport throughout thick electrodes, meeting the demands for effective mass transport and exchange. The 3D network design not only increases the availability of active sites but also offers sufficient room to handle volume changes that occur during battery use. Consequently, the application of 3D-HMNs in sodium-based batteries could address several challenges and lead to improved electrochemical performance.

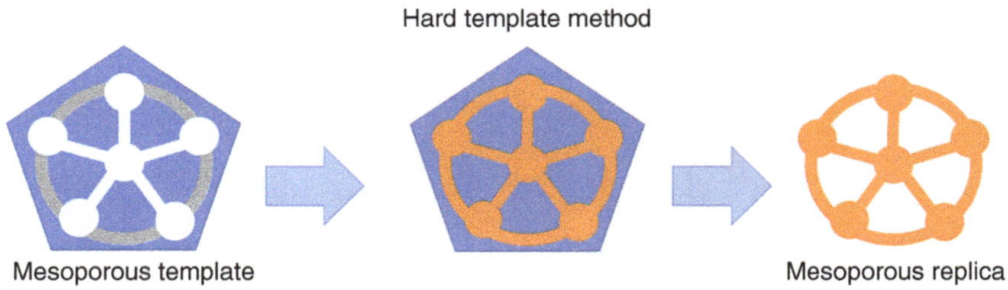

Figure 2.30 Schematic illustration of hard template method.

2.6.2.1 Templated-Assisted Strategies

i) **Hard Template:** Hard templates, particularly silica and salt, are effective for synthesizing 3D hierarchical micro/nanoarchitecture (3D-HMN) materials. A schematic illustration is provided in Figure 2.30. It is developed through a method using oleylamine to create 3D-ordered macroporous (3DOM) carbon with trapped quantum dots of transition metal sulfides [136]. After template removal, 3DOM transition metal quantum dots encapsulated in nitrogen-doped carbon (3DOM TMs-QDs@NC) with well-defined porous architectures, averaging around 150 nm in pore size, were successfully synthesized (Figure 2.32a,b). Similar structures were observed in 3D-inverse opal (IO) FeS-QDs@NC (Figure 2.32c,d). Salt templates (NaCl, Na_2CO_3, Na_2SiO_3) offer easier elimination through water-washing or high-temperature treatment [137]. A customizable approach is devised for 3D-HMN porous routes using NaCl salts, resulting in a conductive hollow pathway within an S-N-PCN matrix featuring pore sizes between 1 and 2 μm (Figure 2.32g,h). This hierarchical structure enhances electric/ion contact and reaction dynamics [138]. Other hard templates like MOFs, metal oxides, PMMA, and metal foams can also guide the creation of 3D linked framework architectures. For instance, hollow cubic N-doped carbon (hC-NC) was produced by utilizing Fe_2O_3 as a templating agent (Figure 2.32c,d) [139].

ii) **Soft Template:** An alternative method for producing 3D-HMN materials involves the use of soft templates, such as surfactants and block copolymers. Figure 2.31 depicts the schematic illustration of the soft template method.

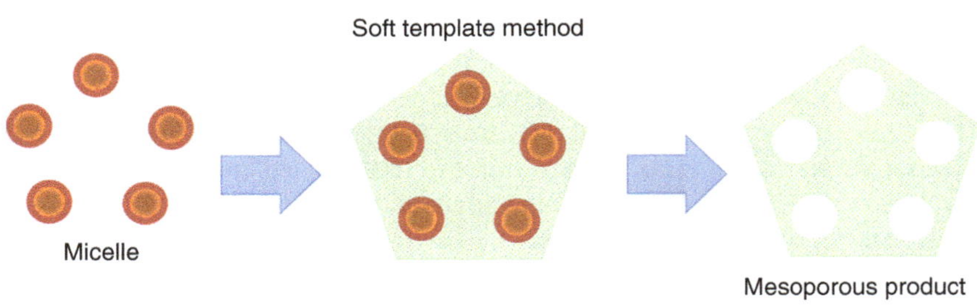

Figure 2.31 Schematic illustration of the soft template method.

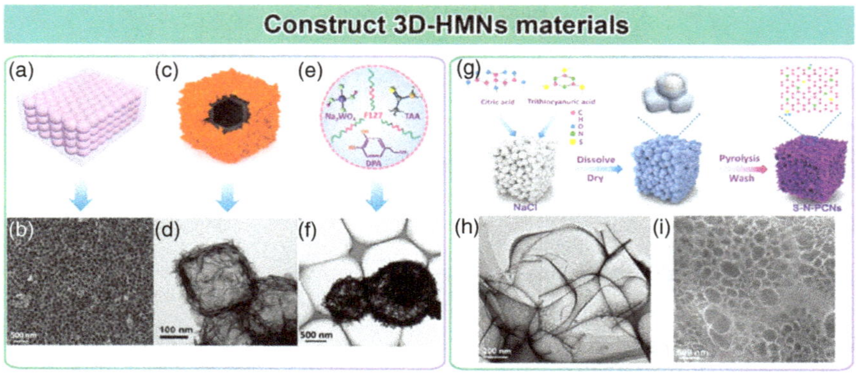

Figure 2.32 (a) Schematic of the silica template and (b) SEM images of 3DOM Co_9S_8-QDs@NC following template removal. Source: Reproduced with permission from [136]/John Wiley & Sons. (c) Schematic image of MoSSeNSs@NC/HC-NC and (d) SEM images of 3D MoSSeNSs@NC/hC-NC after the removal of the Fe_2O_3 hard template. Source: Reproduced with permission from [139]/John Wiley & Sons. (e) Schematic of the F127 soft template and (f) 3D HWS_2@NC. Source: Reproduced with permission from [141]/John Wiley & Sons. (g) Schematic of the NaCl hard template and (h) TEM images of S–N–PCNs following washing. Source: Reproduced with permission from [138]/John Wiley & Sons. (i) SEM images showing the 3D macroporous carbon (Se/N-3DMpC) created with a SeO_2 source template. Source: Reproduced with permission from [142]/Springer Nature.

These structure-directing molecules can form various organizational structures and offer precise regulation over the shape, structure, and dimensions of the end material [140]. The method entails controlling the sol–gel transition as precursors convert into oxides, with the soft templates being readily eliminated through solvent washing or pyrolysis. An example of this approach is shown in Figure 2.32e,f, which illustrates the synthesis of 3D hierarchical WS_2 nanoflowers (S-WS_2@NC) using pluronic F127 block copolymer as the template [141]. This method results in a distinctly different structure compared to the spherical WS_2 particles obtained without the F127 template. The use of soft templates allows for precise control over the final material's characteristics, making it a valuable technique in the synthesis of 3D-HMNs for various applications.

iii) **Self-Sacrificing Source Template:** This strategy creates 3D networks using synthetic materials that serve as both template and source, without introducing a third phase. For example, selenium dioxide (SeO_2) acts as a self-sacrificing template and selenium source to form 3D macroporous frameworks [143]. A straightforward method has been developed to produce Se- and N-codoped 3D macroporous carbon (Se/N-3DMpC) from SeO_2 powder, yielding a material with a substantial specific surface area of 711.86 $m^2\ g^{-1}$ and a macroporous architecture (Figure 2.32i). Tellurium (Te) nanotubes were also used as a source template to produce $MoS_{1.5}Te_{0.5}$@C nanowires-in-nanotube structures [144]. Similarly, P_2O_5 served as a desiccant in the fabrication of phosphorus-doped porous carbon nanosheets. Although template-based techniques allow for adjustable pore sizes, they often require expensive and time-consuming

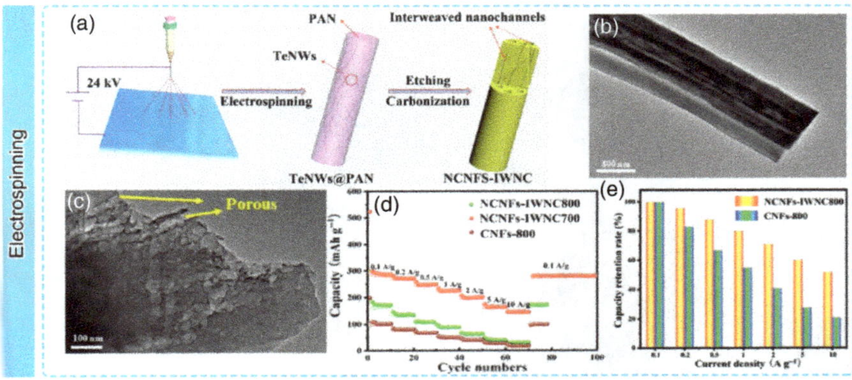

Figure 2.33 (a) Schematic representation of the synthesis process for NCNFs-IWNC, (b,c) TEM images of NCNFs-IWNC800, and (d) comparison of rate capabilities and (e) capacity retention rates between NCNFs-IWNC800 and other SIB electrodes. Source: Reproduced with permission from [145]/John Wiley & Sons.

processes for template extraction. Challenges in synthesis may arise due to potential mismatches between the external materials used and the internal templates employed. These challenges highlight the need for more efficient 3D-HMN fabrication methods [142].

iv) **Electrospinning Methods:** Electrospinning is a cost-effective, versatile technique for fabricating 3D nano/microfibers and porous networks. It uses electrohydrodynamic principles to stretch a liquid droplet under an electric field, forming a Taylor cone. The charged polymer solution jet undergoes stable linear motion followed by turbulent whipping, resulting in fine fibers deposited as a 3D fibrous internetwork. Common precursors include polymers (polyacrylonitrile (PAN), Polyvinylidene Fluoride (PVDF), Polyvinylpyrrolidone (PVP), Polyvinyl Alcohol (PVA), inorganic components (metal oxides, salts), and organic/inorganic composites. The development of electrospun porous fibrous networks has a profound effect on the conductivity and durability of materials used in energy storage. For SIB applications, a novel NCNFs-IWNC was created by integrating tellurium nanowires (TeNWs) into PAN electrospun nanofibers and then removing the TeNWs through etching (Figure 2.33a) [145]. Transmission electron microscopy (TEM) images showcased the intricate network of nanochannels and a higher density of active sites within the fibrous matrix (Figure 2.33b,c). Additionally, it exhibited remarkable cycling stability, retaining 51% of its capacity even after a 100-fold escalation in current density (Figure 2.33d,e).

v) **3D Printing Technologies:** 3D printing, specifically Direct Ink Writing (DIW), is a cutting-edge approach for constructing intricate 3D structures through additive manufacturing [146]. This extrusion-based method allows for the assembly of 3D-structured electrodes with robust mechanical strength and intricate architectures [147]. The process involves robotically extruding a continuous ink filament

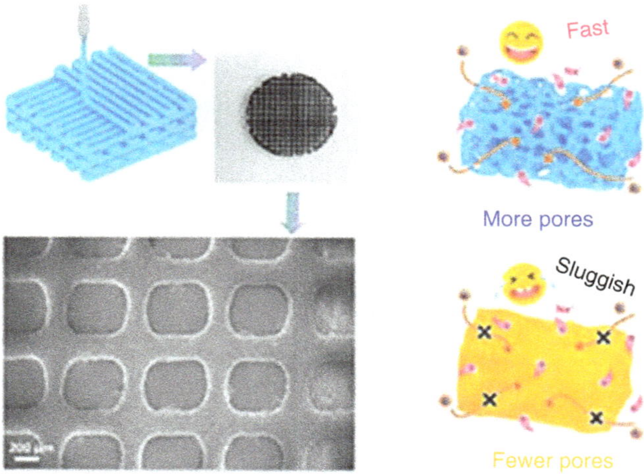

Figure 2.34 3D printing illustration. Source: Reproduced with permission from [149]/John Wiley & Sons.

from a micro-nozzle in layers, exhibiting shear-thinning behavior. The deposited inks quickly retain their shape due to a drying-induced shift from pseudoplastic to dilatant behavior [148]. A crucial aspect of this technique is the ink formulation, which can be customized to meet specific requirements. For instance, when fabricating a 3D-printed electrode, both ZnV_2O_4 NFs@N-PC and AC inks exhibit pseudoplastic behavior, with high apparent viscosities ranging from 10^3 to 10^4 Pa s at low shear rates [149]. This allows for smooth flow-through nozzles and quick solidification; this produces a 3D structure with sequentially layered lines (~200 μm wide) and numerous pores measuring 400–500 μm (see Figure 2.34). A different instance is the 3D-printed Cu_2S@ZnS/C, where intersecting filaments and expansive macropores form a comprehensive transport system, promoting rapid electron transfer [150].

SIBs are emerging as viable substitutes for LIBs because of the plentiful availability of sodium and comparable chemical properties. SIBs function based on a "rocking-chair" principle: amid charging, Na^+ ions migrate from the cathode to the anode via the electrolyte, while electrons flow externally to sustain electrical balance; this reverses during discharge. However, the large Na^+ ionic radius hampers performance. To improve SIBs, the printable NVPF and NTP anode and cathode inks were thoroughly mixed with high-performance conductive additives such as CNTs and EG (or just EG), along with aqueous polymer binders (Figure 2.35a) [151]. To illustrate their printability, a planar interdigitated NIMB with two printed layers (referred to as NIMB/EC-2L NVPF||NTP) was effectively fabricated by extruding the NVPF/EC and NTP/EC inks onto a PET substrate (Figure 2.35b). SEM image of printed NVPF/EC microelectrode is shown in Figure 2.35c. To explore the non-Newtonian behavior of these printable inks more deeply, a modeled extrusion test was performed by altering shear rates. As demonstrated in the modeled extrusion process (Figure 2.35d), the viscosities of all electrode inks

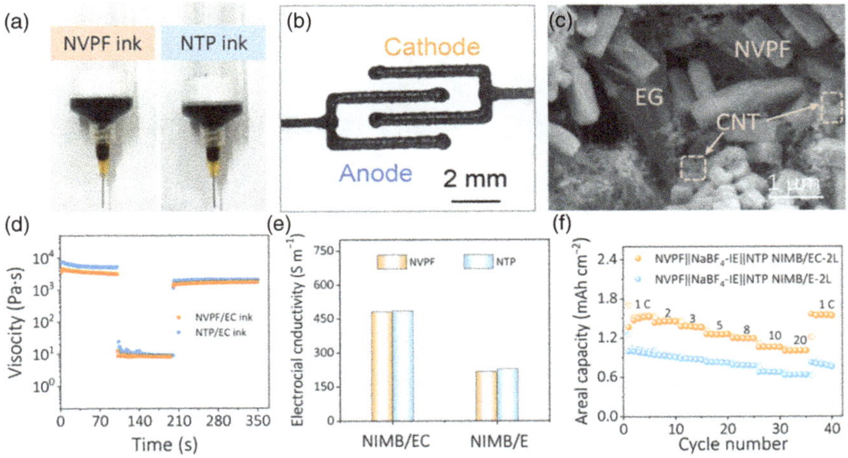

Figure 2.35 Rheological properties of the inks and morphology of the printed microelectrodes, (a) images of NVPF and NTP inks in syringes, (b) photograph of the printed NVPF||NTP NIMB/EC-2L, (c) SEM images of the printed NVPF/EC microelectrode, (d) viscosity of anode and cathode inks at shear rates of 0.1 and 100 s^{-1}, (e) electrical conductivity of the printed NVPF and NTP microelectrodes, and (f) rate capability measured from 1 to 20 C. Source: Reproduced with permission from [151]/John Wiley & sons.

dropped instantly when the shear rate was raised from 0.1 to 100 s^{-1}, and high viscosities were quickly reinstated when the shear rate was reverted to its original level. The 2D EG nanosheets and 1D CNTs formed a connected hierarchical 3D conductive framework that tightly encased the active electrode materials, leading to an electrical conductivity of about 480 S m^{-1} (Figure 2.35e). Remarkably, at a high rate of 20 C, the NVPF||NaBF4-IE||NTP NIMB/EC-2L achieved an areal capacity of 1.0 mAh cm^{-2}, which was comparable to, and twice as high as, the capacity of NVPF||NaBF4-IE||NTP NIMB/E-2L at both 1 and 20 C (Figure 2.35f).

2.6.3 Conductive Coating and Modification

The strategy as shown in Figure 2.36 combines particle-level encapsulation with long-range flexible carbon nanotube (CNT) networks in two steps: (i) encapsulating single particles with carbon layers and CNT networks, and (ii) depositing Si on the porous carbon skeleton, proceeding by carbon coating encapsulation. The integration of CNTs and carbon coatings prevents CNT-induced damage to the SEI and carbon layers and ensures efficient percolation. Long CNTs reduce compressive stress, minimize damage, and allow for smooth Na^+ diffusion because of the low energy barrier of the SEI. This structural design is scalable from micro- to macro-levels. Carbon nanotubes and coatings preserve Si anode conductivity and structural integrity at 1D and 2D scales, respectively, and can be applied to other electrodes [153]. The conductive network enhances electrode conductivity and reduces polarization resistance, improving electrochemical performance. An effective conductive agent should provide a stable mesoscale architecture for rapid

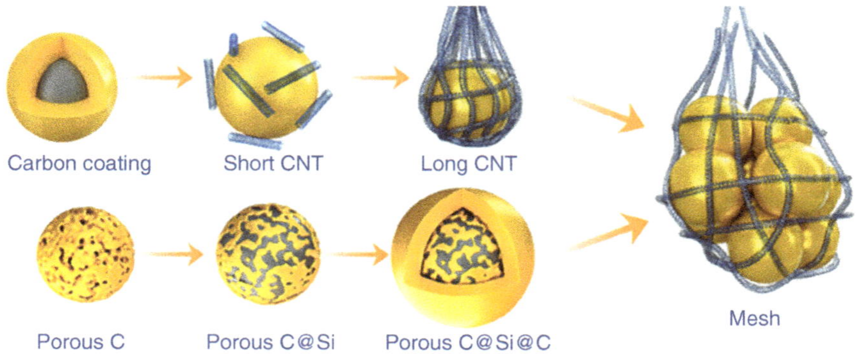

Figure 2.36 Strategy of carbon coating. Source: [152]. Reproduced with permission of/John Wiley & Sons.

electron-ion transfer and form a network that maintains mechanical stability, ensuring full active material capacity utilization and rapid electron transfer at high currents.

HC materials featuring abundant closed pore structures and nano-sized soft carbon coatings are promising anodes for SIBs. However, achieving precise control over closed pore structures and coating thicknesses to optimize performance at low-voltage platforms remains challenging. Sun et al. [154] developed PCHC-10, which has numerous closed pores of 0.45 nm and a nano-sized soft carbon coating, through the chemical crosslinking of preoxidized phenolic resin and pitch to create ester-based bonds (Figure 2.37a). It also maintained 91.4% capacity retention over 100 cycles, as shown in Figure 2.37b. Detailed electrochemical analyses and theoretical calculations revealed that the 0.45 nm pore size significantly enhances Na^+ ion storage at low-voltage platforms (Figure 2.37c).

2.6.4 Managing Volume Expansion: Alloy-Based Anodes

Antimony (Sb), with its high theoretical capacity, is a promising anode material for SIBs. Despite the larger radius of Na^+ ion (1.02 Å) potentially leading to greater capacity loss in Sb anodes due to significant volumetric changes during cycling, Sb anodes have unexpectedly demonstrated better electrochemical and structural stability. The reasons behind this performance anomaly are not fully understood. The stability is attributed to the presence of polycrystalline and amorphous phases, which enhance resistance to mechanical damage and contribute to improved cycling stability. The sodiation process primarily follows an alloying pathway. Initially, pristine Sb single crystals transform into amorphous Na_xSb and ultra-fine metallic NaSb and Na_3Sb nanoparticles (Figure 2.38a). The amorphous and polycrystalline phases are crucial for maintaining cycling stability. The amorphous phase, with its disordered structure, helps distribute stress more evenly in various directions, as bonds can form randomly rather than following fixed orientations as in crystalline phases. This reduces stress anisotropy and improves structural stability. Additionally, the amorphous phase helps prevent large cracks and maintains

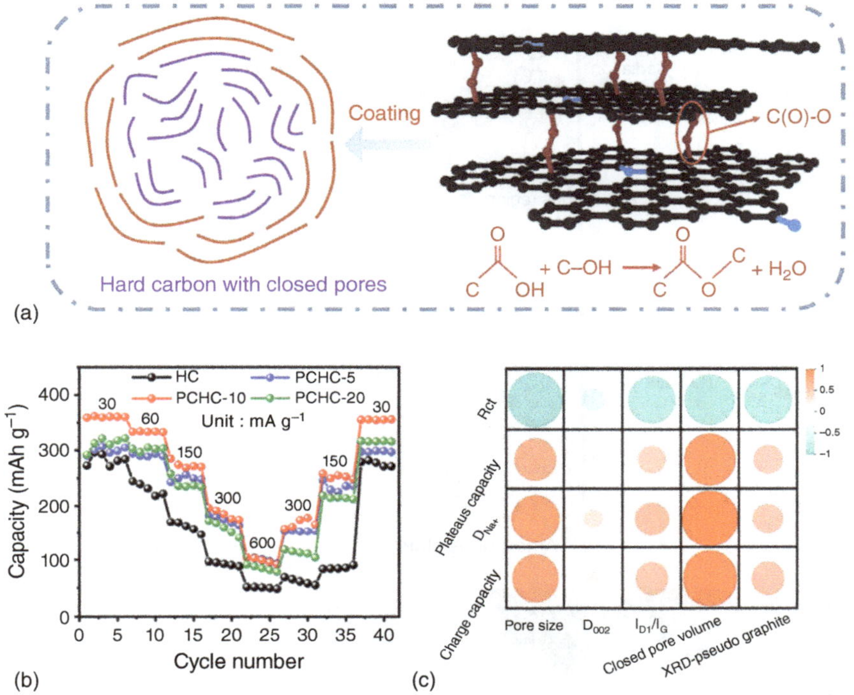

Figure 2.37 (a) Schematic diagram illustrates the synthesis route for HC and PCHC-x, (b) the rate performance is assessed across current densities from 30 to 600 mA g^{-1}, and (c) a heat map displays the correlation analysis between physical properties (including pore size, D002, ID/IG ratio, closed pore volume, and XRD pseudo-graphitic phase) and electrochemical properties (such as R_{ct}, plateau capacity, D_{Na+}, and charge capacity) for HC and PCHC-x samples. Source: [154]. Reproduced with permission of/John Wiley & Sons.

the connectivity of active materials, thus supporting efficient charge transport. From the study, amorphous Na_xSb phases are formed *in situ* without the need for complex synthesis methods or costly reagents. The redox reactions in SIBs exhibit a distinctive two-stage alloying/dealloying mechanism, where polycrystalline and amorphous phases act as intermediates, as shown in Figure 2.38b. The scanning electron microscope (SEM) image in Figure 2.38c reveals that Sb nanoflakes have a consistent and interconnected morphology. In the NaSb system, two primary phases (NaSb and Na_3Sb) have been experimentally identified, with their voltage profiles slightly higher than the discharge voltage curve. However, the amorphization of the Na_xSb phase increases its energy and reduces the resulting voltage plateau, as depicted in Figure 2.38d, leading to better alignment with the discharge voltage. Tin-based alloy anodes are widely researched due to their low reaction potential and high theoretical capacity (847 mAh g^{-1} for elemental Sn). Bismuth-based materials have also attracted considerable attention due to their unique structural characteristics and suitable working potential. Elemental bismuth (Bi) has a large lattice spacing ($d_{(003)}$ = 3.95 Å) along the *c*-axis, and its unique interlayer crystal

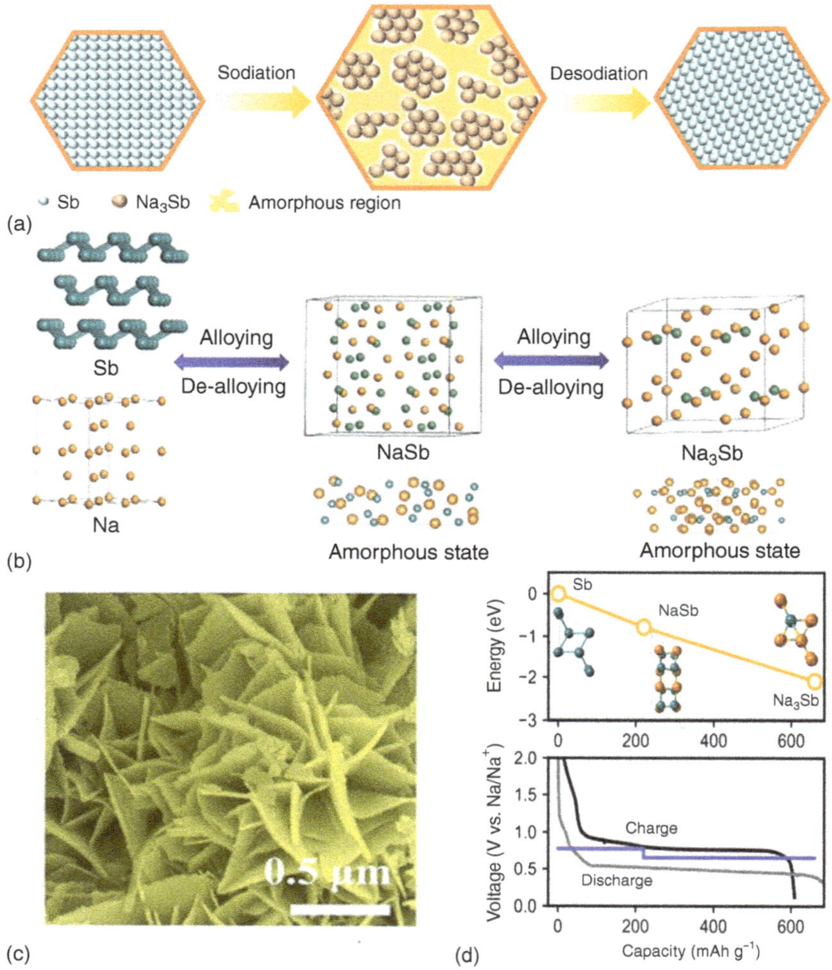

Figure 2.38 (a) Diagrams illustrating the structural changes of Sb nanoflakes during the sodiation and desodiation processes, (b) two distinct reaction mechanisms of the Sb electrode in sodium-ion batteries (SIBs), (c) SEM image showing the morphology, and (d) NaSb convex hull and the voltage profile of the reactions. Source: Reproduced with permission from [155]/John Wiley & Sons.

structure supports sodium-ion storage. Although Bi, as an alloy-type material, offers a high specific capacity (approximately 385 mAh g^{-1}), it also experiences significant volume expansion.

2.6.5 Conversion and Intercalation–Conversion Anode Materials: Balancing Capacity and Cyclability

Transition metal oxides (M_aO_b, where M = Fe, Co, Ni, Cu, Mn, etc.) are extensively explored as conversion-type anode materials due to their high specific capacities and affordability. Despite these advantages, their performance is hampered by significant volume expansion during charge and discharge cycles and their semiconductor nature. This expansion leads to rapid capacity degradation and affects cyclic stability, while the low intrinsic conductivity limits fast charging [156]. Therefore, minimizing volume expansion and enhancing conductivity are key to improving fast-charging performance. Iron oxide materials (FeO, Fe_2O_3, and Fe_3O_4) are particularly researched because of their cost-effectiveness, abundance, and high theoretical capacities (745, 1007, and 928 mAh g^{-1}, respectively) [157]. However, they face challenges such as volume expansion during cycling and low conductivity. Solutions like morphological optimization and composite material design are applied to mitigate these issues. Reducing iron oxides to the nanoscale can significantly decrease ion diffusion paths and relieve expansion stress, thereby boosting capacity and rate performance. For instance, Zhang et al. produced ultrafine Fe_2O_3 nanocrystals (GFe_2O_3) anchored on graphene via a thermal decomposition method. These nanocrystals, approximately 5 nm in size, showed a high discharge capacity of 701 mAh g^{-1} at 0.1 A g^{-1} and impressive rate performance (253 mAh g^{-1} at 2 A g^{-1}) for SIBs. Liu et al. created a Fe_3O_4/Fe/FeS triheterojunction node within a nitrogen-doped carbon nanotube scaffold (FHNCS) through catalytic growth and vulcanization on a modified MIL-88B (Fe) template (see Figure 2.39a). SEM images of this process are depicted in Figures 2.39b–d [158]. Additionally, Li et al. synthesized Fe_3O_4 nanoparticles encapsulated in carbon nanolayers using a one-step pressure-induced vapor synthesis method, which were uniformly dispersed on graphene-based surfaces (Fe_3O_4/C@G), as shown in Figure 2.39e [159].

Professor Xu and his team engineered an innovative composite material for SIBs based on a CoSe/$MoSe_2$ structure enhanced with carbon [160]. This novel electrode material showed promising results in laboratory tests examining the charge–discharge profiles of the CoSe/$MoSe_2$-C composite. Professor Xu and colleagues employed cyclic voltammetry (CV) to investigate sodium-ion storage mechanisms in their novel CoSe/$MoSe_2$-C anode as elaborated in Figure 2.40a. The CV curves, recorded at 0.1 mV s^{-1} between 0.3 and 3.0 V, revealed complex electrochemical processes. During initial discharge, a subtle peak at 1.05 V suggested Na^+ intercalation into CoSe and $MoSe_2$ layers, forming Na_xCoSe and Na_xMoSe_2. Subsequent peaks at 0.96 and 0.81 V indicated the conversion of Na_xCoSe to Co and Na_2Se, coinciding with SEI formation. A peak at 0.59 V marked the conversion of Na_xMoSe_2 to Mo and Na_2Se. The charging process showed distinct peaks at 1.74 and 1.86 V, corresponding to the reformation of $MoSe_2$ and CoSe, respectively. In subsequent cycles, while anodic peaks remained stable, cathodic peaks shifted significantly. A new peak at 1.45 V emerged, associated with $MoSe_2$ intercalation and transformation, while peaks at 1.18 and 1.33 V indicated Na_xCoSe formation. The 0.76 V peak corresponded to Na_2Se and Co production. This comprehensive CV analysis provided crucial insights into the complex sodium storage mechanisms within the composite anode material. In tests at a higher current density, CoSe/$MoSe_2$-C electrode achieved an initial discharge capacity of 453.1 mAh g^{-1}, surpassing both CoSe-C (421.3 mAh g^{-1}) and $MoSe_2$-C (355.4 mAh g^{-1}) electrodes (see Figure 2.40b). While all electrodes exhibited some capacity decay

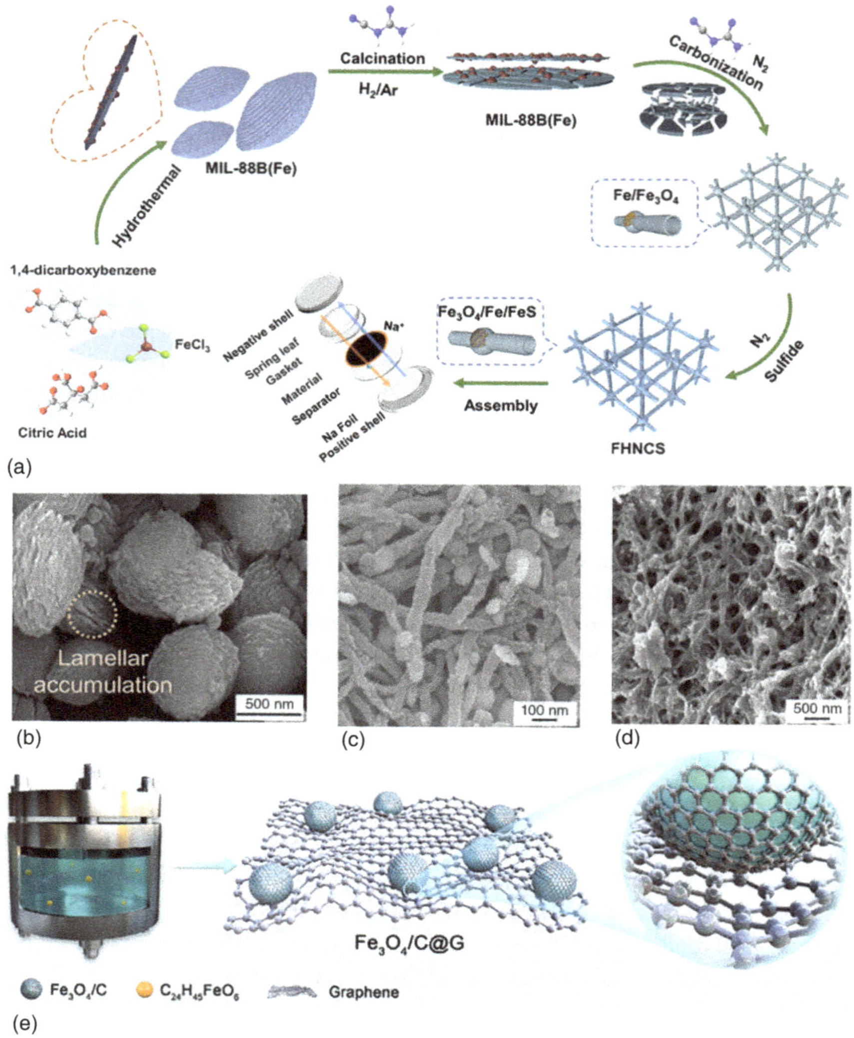

Figure 2.39 (a) Schematic illustrating the preparation process of FHNCS. (b–d) SEM images showcasing the lamellar-stacked MIL-88B(Fe), the Fe-based particle nodes, the resulting N-carbon nanotube scaffold structure, and the final FHNCS. Source: [158]/John Wiley & Sons/CC BY 4.0. (e) Schematic representation of the synthesis process for the samples. Source: [159]/John Wiley & Sons/CC BY 4.0.

in subsequent cycles, likely due to gradual activation processes, side reactions, and SEI development, the composite material demonstrated superior overall performance.

In another achievement, Xu's team engineered an innovative $SnSe_2/NiSe_2$ nanocomposite with a yolk–shell architecture [161]. This heterostructure, featuring selenium vacancies, was embedded in N-doped carbon. Xu employed a straightforward two-phase approach: hydrothermal processing and subsequent selenization. Figure 2.41a demonstrates the CV analysis of the $SnSe_2/NiSe_2$@NC electrode,

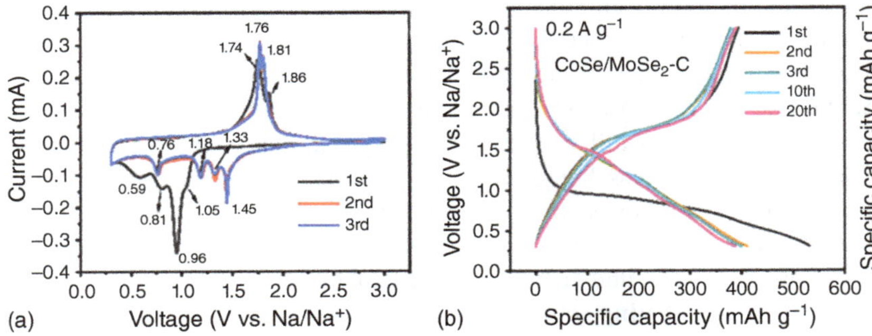

Figure 2.40 (a) CV curves at a scan rate of 0.1 mV s^{-1} and (b) charge–discharge curves. Source: [160]. Reproduce with permission of/John Wiley & Sons.

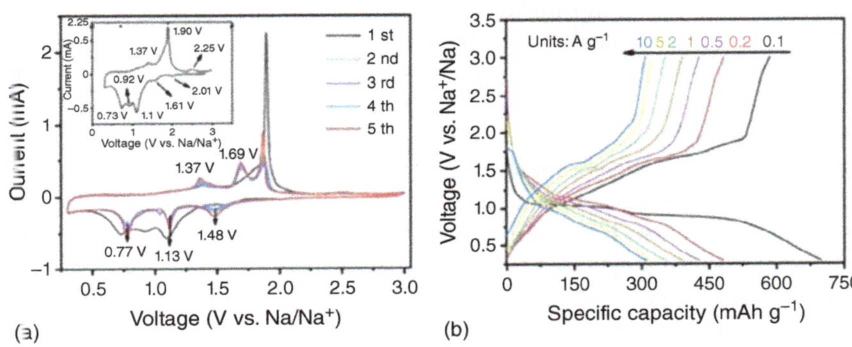

Figure 2.41 (a) Cyclic voltammetry analysis and (b) charge–discharge profiles at numerous current densities. Source: [161]. Reproduced with permission of/John Wiley & Sons.

conducted at 0.2 mV s^{-1} between 0.3 and 3.0 V, revealing complex electrochemical processes. The initial discharge cycle showed weak cathodic peaks at 2.01 and 1.61 V, indicating Na^+ insertion into $NiSe_2$ and $SnSe_2$, respectively. A peak at 1.1 V signified the formation of Sn, Ni, and Na_2Se, while a peak at 0.77 V corresponded to Sn-Na alloying. The broad peak at 0.92 V was attributed to irreversible SEI formation. During charging, peaks at 0.31, 1.37, 1.90, and 2.52 V indicated the dealloying of $Na_{15}Sn_4$ and oxidation of Sn and Ni back to their selenide forms. Subsequent cycles showed stabilized peak positions, with cathodic peaks at 1.48 and 1.13 V for Na^+ ion insertion and selenide conversion, and anodic peaks at 0.31, 1.37, 1.69, and 1.86 V for dealloying and selenide regeneration. The overlapping nature of these peaks in later cycles suggested good reversibility of the $SnSe_2/NiSe_2$@NC electrode. Rate performance tests at current densities from 0.1 to 10 A g^{-1} further confirmed the electrode's stable voltage platform and good reversibility across various charging rates as given in Figure 2.41(b).

2.6.6 2D Materials: Leveraging Unique Structures for Sodium Storage

Transition Metal Sulfide: These materials offer potential advantages for rapid charging applications because of their high theoretical capacity, extended cycle life, and substantial power and energy density. Nevertheless, similar to transition metal oxides, these materials face challenges such as low electrical conductivity and considerable volume changes during charge–discharge processes, which can lead to detrimental side reactions. Many transition metal sulfides are semiconductors with a 2D-layered architecture, exhibiting greater interlayer spacing compared to graphite [162]. During the charge–discharge cycle, the process typically involves an initial intercalation reaction followed by a conversion reaction. Utilizing thin-layered or nanostructured materials with increased interlayer spacing can improve the availability of active sites and reduce volume expansion, improving rate performance. Additionally, creating composite materials with heterogeneous structures is a widely adopted approach to improve electrochemical performance. Transition metals with mixed valence states possess a finely tuned electronic structure, resulting in exceptional reaction activity across various energy applications. For example, ternary copper sulfide compounds, represented by the formula A/Cu/Q (where A = Na, K, Rb, Cs; Q = S, Se, Te), have been widely synthesized and investigated for applications in supercapacitors and ion conductors [163]. Lu et al. developed a 2D metallic anode, KCu_4S_3, adopting a KSCN molten salt method, which facilitates the development of high-performance, fast-charging SIBs as illustrated in Figure 2.42a [164]. *In situ* observations indicate that the sodiation products, $NaCu_2S_2$ and sodium polysulfide, revert to their original phase upon desodiation. These distinctive features enable KCu_4S_3 to achieve exceptional electrochemical performance, including a remarkable capacity of 355 mAh g^{-1} at 20 A g^{-1} and maintaining 100% capacity over 3000 cycles (Figure 2.42b). Additionally, the full cell exhibits a high-energy density of 332 Wh kg^{-1} and maintains 92% of its capacity after 150 cycles at 1 A g^{-1} (Figure 2.42c). The morphology of KCu_4S_3 is demonstrated through SEM images in Figure 2.42d. Xu's team thoroughly studied bimetallic nickel cobalt sulfides using straightforward approach involving precursor construction followed by anion exchange via the solvothermal method. Among these, the coral-like $Ni_{1.01}Co_{1.99}S_4$ (with a Ni/Co ratio of 1 : 2) demonstrated outstanding Na-ion storage capabilities, maintaining over 1000 cycles, highlighting its ultrafast and stable performance [166]. Li et al. reported a significant breakthrough in SIBs with development of a superparamagnetic Fe-doped MoS_2/N,O-codoped carbon structure, designed for wide-temperature-range operation [165]. The Fe doping gives the MoS_2 electrode zero bandgap properties, increased interlayer spacing, and low sodium-ion diffusion energy barriers across a wide temperature range. Remarkably, Fe atoms in the MoS_2 lattice reduce to ≈2 nm superparamagnetic Fe^0 nanocrystals during conversion reactions. *In situ* magnetometry shows these Fe^0 nanocrystals act as electron acceptors, forming space charge zones with Na^+ ion, leading to strong spin-polarized surface capacitance that enables fast sodium-ion storage across a broad temperature range. As a result, the MoS_2 electrode exhibits exceptional fast-charging performance in half and full cells from −40 to 60 °C. Figure 2.42e presents the first charge–discharge profiles of the $FeMoS_2$/C-II electrode at multiple temperatures. At 25 °C, the electrode achieves a charge capacity of 700.2 mAh g^{-1}, while at 60 °C, the capacity increases slightly to 722.5 mAh g^{-1}, primarily as a result of the reduced diffusion barrier for Na ions at elevated temperatures, enabling more

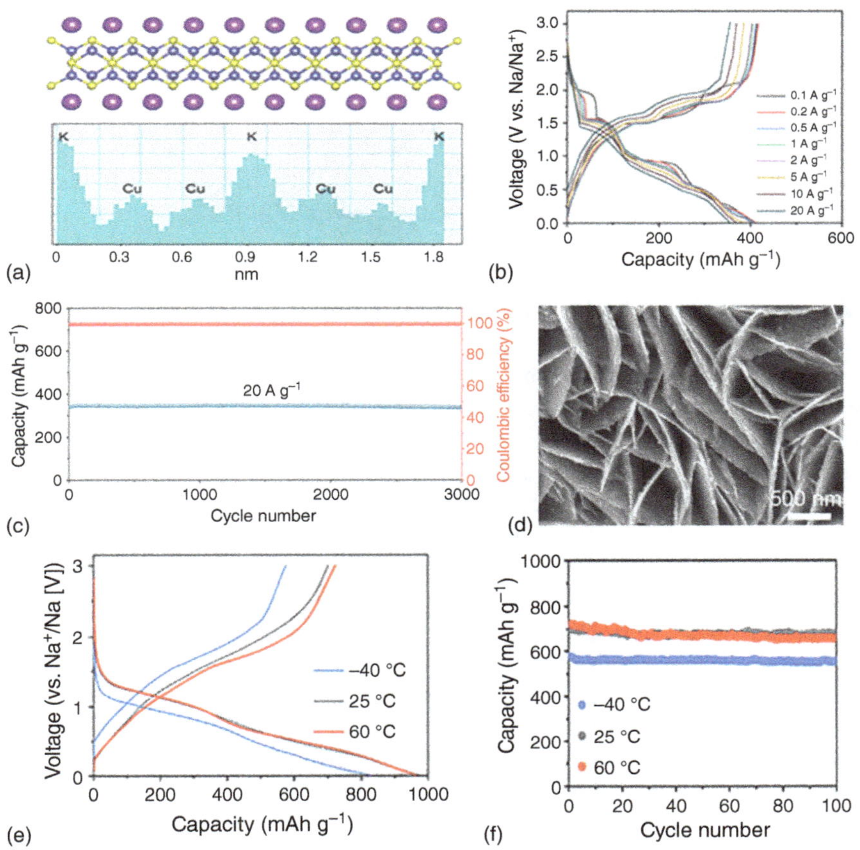

Figure 2.42 (a) Related lattice structure model and atomic intensity image, (b) the associated GCD profiles at different current densities, (c) long-term cycling stability profile at 20 A g^{-1}, and (d) SEM images. Source: Reproduced with permission from [164]/John Wiley & Sons. (e) Voltage profiles for the first cycle at temperatures ranging from −40 to 60 °C and (f) cycling stability at 0.1 C within the same temperature range of −40 to 60 °C. Source: [165]. Reproduced with permission of/John Wiley & Sons.

complete sodiation. Although lower temperatures raise the diffusion barrier, leading to capacity loss, the electrode still delivers a remarkable capacity of 575.7 mAh g^{-1} at −40 °C, that is 82.1% of the capacity at 25 °C. Additionally, as the temperature decreases from 25 to −40 °C, a rise in polarization and a decline in the voltage plateau are noted. It is mainly due to higher electrolyte viscosity and enhanced impedance at the electrode–electrolyte interface, reducing ionic conductivity and increasing polarization. Even after 100 cycles at −40, 25, and 60 °C, the electrode shows minimal capacity loss, demonstrating its robust cycling stability across a wide temperature range (Figure 2.42f).

Furthermore, during cycling, Na^+ ion insertion causes lateral slippage of the S-layers, transitioning MoS_2 from the 2H to the 1T phase. This process introduces crystal defects and breaks the lamellar structure into nanodomains. Although these defects assist in alleviating strain, they can also result in more pronounced structural damage and slower kinetics over time. Engineering phase changes can substantially modify the electronic structure, as well as the physical and chemical properties of materials. Beyond MoS_2, other transition metal sulfides with comparable layered structures, such as WS_2, VS_2, and CoS_2, have also been explored as fast-charging anodes. Mamoor et al. employed first-principles calculations in the DFT framework to investigate WX_2 (X = Se, Te) as anode candidates for SIBs [167].

2.7 Anode-Free SIBs: Design and Working Principle

To compete with the high-energy density of LIBs, substantial advancements in the sodium battery design are necessary. One emerging concept is the anode-free cell design, which eliminates the need for traditional anode materials. Instead of relying on carbon or alloy-based anodes for ion storage during charging, anode-free batteries utilize the direct electrochemical deposition of alkali metals onto the current collector (Figure 2.43a). This approach enhances cell voltage, reduces cost, and boosts energy density by eliminating the anode active material (Figure 2.43b). Despite its potential, several challenges hinder the widespread adoption of anode-free designs in both lithium and sodium batteries. In conventional organic liquid electrolytes, lithium or sodium deposition often results in a porous, mossy structure. Additionally, these electrolytes interact with the deposited metal, resulting in the formation of a SEI, which continues to grow due to ongoing anode morphological changes during cycling, depleting active material [169]. While strategies such as liquid electrolyte modification, [170] current collector adjustments [171], cell stack pressure [172], and cycling protocols [173] have been explored, improvements have been modest, typically resulting in only limited cycle life and low initial Coulombic efficiencies. An effective strategy to enhance anode-free battery design involves utilizing SSEs. The solid-state nature of these electrolytes limits the contact area to a flat interface with the negative electrode current collector. This reduces solid–electrolyte interfacial reactions when compared to the extensive surface area provided by three-dimensional anodes used in liquid electrolyte systems (Figure 2.43a). Moreover, the application of stack pressure is important for creating a compact metal anode, which is vital for avoiding issues such as contact loss, void formation, and potential short circuits [174]. While certain research has demonstrated potential with SSEs in lithium anode-free thin-film batteries [175, 176], areal capacities remain constrained by the thin layers of the cathode. Recent attempts to facilitate anode-free cycling in bulk lithium solid-state cells using oxide-based [177] or sulfide-based [178] electrolytes have shown progress. Strategies such as current collector modifications [178], sacrificial lithium sources [179], or lithiophilic alloying compounds [180] have been explored, but they either fail to provide sufficient reversibility or introduce inactive materials that reduce the benefits of an anode-free design. As of now, there has been just one reported effort to develop a sodium anode-free solid-state battery, and it has yet to demonstrate any electrochemical cycling [181]. For the successful development of an anode-free sodium solid-state battery, it is essential to fulfill four critical requirements (Figure 2.43c): (i) an electrolyte with strong electrochemical stability or high passivation properties to prevent the loss of active sodium caused by SEI formation; (ii) a robust and reliable solid–solid interface connection between the electrolyte and current collector to ensure effective sodium plating and stripping; (iii) an imperfection compact solid-state electrolyte separator to promote metal filament growth and short-circuiting [182]; and (iv) a highly dense current collector, since porous collectors used in liquid electrolyte cells are ineffective in solid-state cells, where trapped sodium lacks conduction pathways during stripping [183].

2.7.1 Electrochemically Stable Electrolyte

Deysher et al. made a significant breakthrough to create effective anode-free SIBs [168]; certain design principles are applied. In one setup, an anode-free half-cell was constructed using the widely used Na_3PS_4 (NPS) solid electrolyte, combined with an aluminum foil as the current collector and a Na_9Sn_4 electrode as the counter electrode. The ICE was observed to be just 4%, largely because of the NPS's reduction at low voltages, which resulted in the formation of Na_3P [184], a well-recognized mixed conductor that results in ongoing growth of the SEI and the irreversible consumption of sodium. To address the issue of electrochemical instability, a solid electrolyte based on sodium borohydride ($Na_4B_{10}H_{10}B_{12}H_{12}$, commonly referred to as NBH) was utilized as the separator. Known for its excellent electrochemical stability when interacting with sodium metal [185], NBH allowed for successful sodium electroplating onto the aluminum foil. After plating, the sodium was efficiently stripped away, leading to a notable enhancement in the ICE, which reached 64% (Figure 2.44a). This outcome emphasizes the importance of employing an electrochemically stable solid electrolyte in anode-free cells. Nonetheless, despite the progress, the efficiency was still lacking, suggesting that other elements of the anode-free design need further refinement. Although metal foils are typically used as current collectors in batteries, only a limited portion of the aluminum foil in this instance was coated with sodium (Figure 2.44a). The restricted plating area indicates that the contact between the solid electrolyte and the aluminum foil current collector is inadequate. Sodium deposits only where Na^+ ions from the solid electrolyte can interact with electrons from the current collector. Furthermore, after the stripping process, some residual sodium was left on the foil, which adversely affected the ICE (Figure 2.44a). The inadequate interface connections between the solid electrolyte, sodium metal, and foil current collector resulted in incomplete sodium stripping, highlighting the inadequacy of conventional aluminum foil for maintaining intimate interface contact and achieving high reversibility. To address this issue and enhance the contact between solid phases, pelletized aluminum was applied to the solid electrolyte separator during cell assembly, using a method analogous to that employed with aluminum foil (Figure 2.44b). The aluminum powder adapts more effectively to the changing surface contours of the solid electrolyte separator layer created during the assembly of the cell (Figure 2.44c).

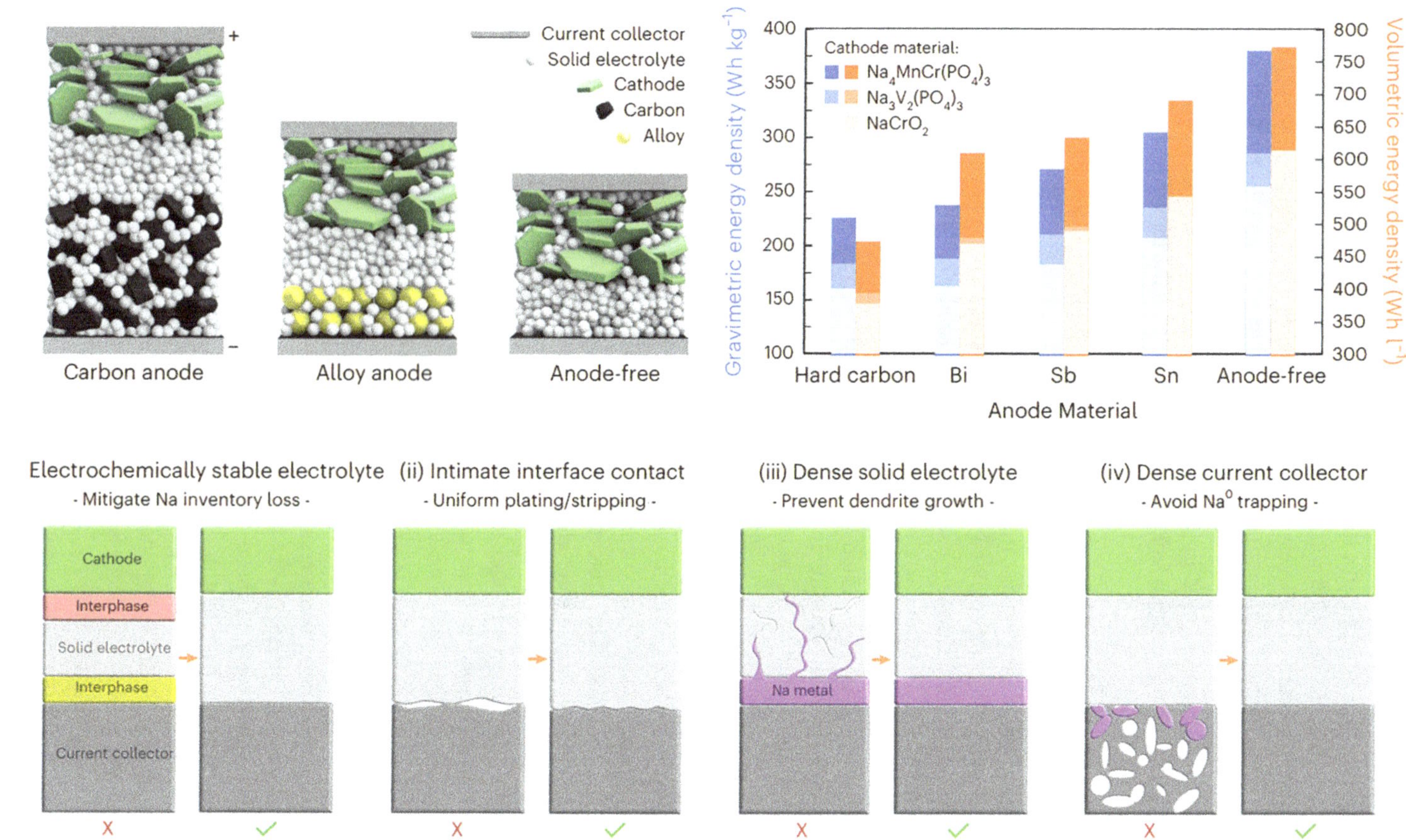

Figure 2.43 Schematic representations and energy density calculations for anode-free systems; (a) cell schematics depicting carbon anodes, alloy anodes, and an anode-free configuration; (b) comparison of theoretical energy densities for various sodium anode materials; and (c) diagram outlining the four essential requirements for an anode-free all-solid-state battery: (i) an electrochemically stable electrolyte, (ii) close interface contact, (iii) a dense solid electrolyte, and (iv) a robust current collector. Source: [168]/Springer Nature.

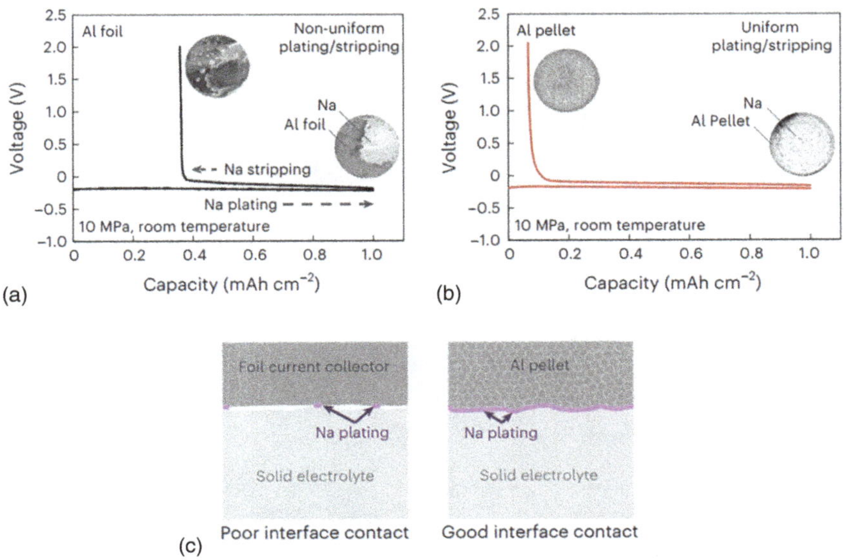

Figure 2.44 (a) Plating/stripping behavior at a current density of 1 mA cm^{-2} for Al foil and (b) aluminum pellet current collectors, and (c) schematic depicting how aluminum powder achieves close contact with the solid electrolyte layer. Source: [168]/Springer Nature.

This approach, referred to as the aluminum pellet current collector, is designed to enhance the interface contact and improve the cell's overall performance.

2.8 All-Climates Sodium-Ion Batteries

The phenomenon of climate change presents a serious peril to the human race. The most prominent catalyst of observed climate change is the greenhouse gases caused by human activities. As a result, it is urgent to restrain the release of greenhouse gases. A crucial method to diminish our carbon footprint is to transition from an economy reliant on fossil fuels to one centered around renewable energy. Certainly, the temperature plays a decisive role in the manifestation of a chemical process, encompassing the requisite activation energy and the velocity at which the chemical reaction transpires, as described by the Arrhenius equation:

$$k = A\exp(-E_a/RT) \tag{2.2}$$

Within this formula, the symbol "k" represents the rate constant, "T" signifies the absolute temperature, "E_a" denotes the activation energy, "A" symbolizes the preexponential factor, and "R" symbolizes the universal gas constant. In the situation of an SIB, the chemical reactions encompass not only the primary redox reaction but also the interphase reactions occurring between the electrodes and electrolyte. This is due to the fact that most electrodes possess redox potentials that lie beyond the electrochemical stability window of the electrolyte solution. As illustrated in Figure 2.45,

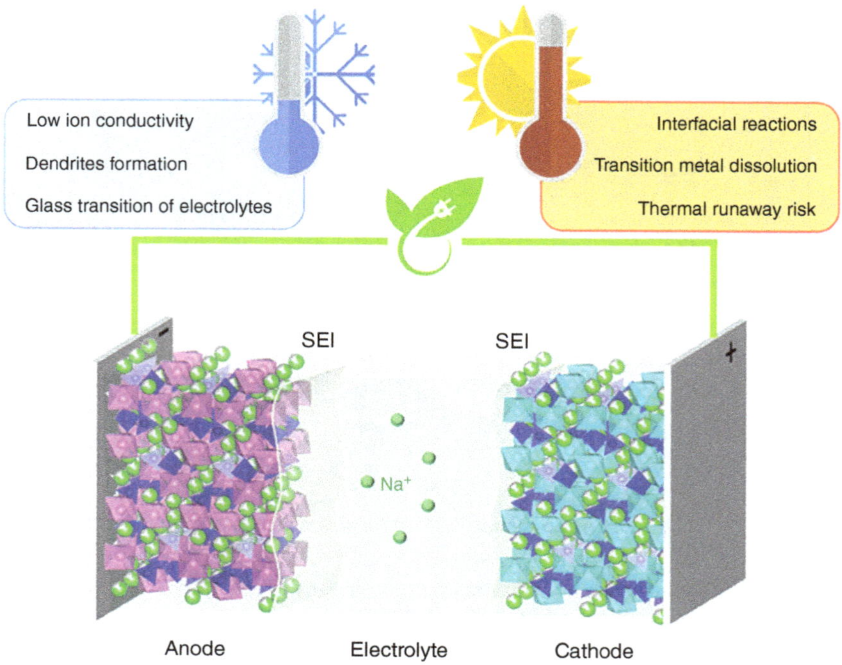

Figure 2.45 Illustrative representation displaying the potential difficulties encountered during extreme temperature conditions. Solid–electrolyte interphase (SEI). Source: [186]/John Wiley & sons/CC BY 4.0.

the temperature reliance of SIBs is manifested through the kinetics, interfacial stability, and safety aspects.

2.8.1 Challenges for SIBs at Extreme Low Temperatures

During the previous years, noteworthy advancements have been made in comprehending the Na storage process and establishing high-performing SIBs with commendable energy density, impressive rate capability, and prolonged cycling longevity. These achievements were accomplished by developing innovative electrode materials and electrolytes. Nonetheless, it is imperative to factor in temperature resilience when assessing the effectiveness of stationary SIBs. Unavoidably, temperature exerts influence on the kinetics and steadfastness of rechargeable batteries, including SIBs. When functioning in colder temperatures, these batteries encounter lethargic electrochemical reactions. Severe polarization of the redox processes hastens the attainment of cutoff voltages [187], leading to reduced usable capacity and the potential emergence of dendrites on the anode. Regarding the rate of reactions in SIBs, the Arrhenius equation illustrates the correlation between temperature and the sodium diffusion coefficient within the electrodes and across the electrolyte:

$$D = D_0 \exp(-E_a/\mathrm{RT}) \tag{2.3}$$

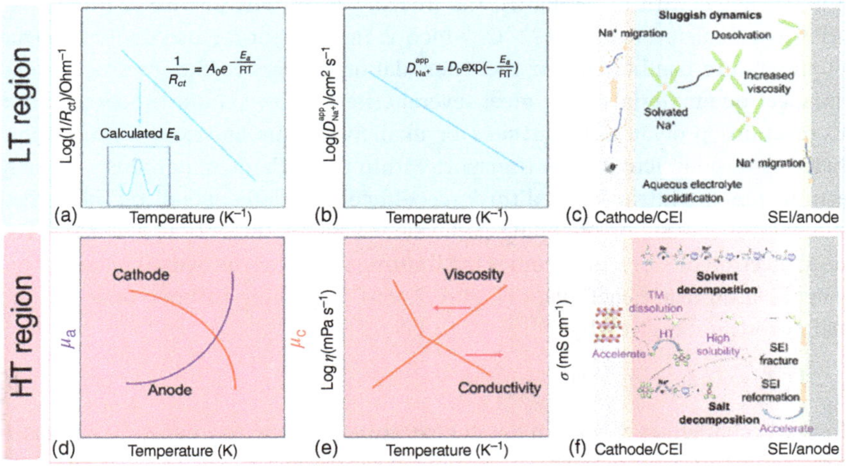

Figure 2.46 (a) Correlation between charge resistance and temperature, (b) intrinsic ion diffusion coefficient as a function of temperature, (c) difficulties encountered by SIBs at low temperatures (LTs), (d) Relationship between electrode potential and temperature, (e) Interplay between viscosity, electrical conductivity, and temperature, and (f) challenges faced by SIBs at high temperatures (HTs). Source: [78]. Reproduced with permission of/John Wiley & Sons.

In this equation, D represents the diffusion coefficient (measured in $m^2\ s^{-1}$), while D_0 symbolizes the maximum diffusion coefficient (at extremely high temperatures; measured in $m^2\ s^{-1}$). Evidently, lower temperatures decelerate the diffusion of Na^+ ions. Factors like crystal structure play a significant role in determining the intrinsic ion conductivity, grain boundaries, and interface properties. Extensive research has explored methods like element doping and material compounding to enhance ion diffusion performance [188]. Another approach involves reducing the diffusion length to decrease diffusion time ($t = D/L^2$); achieving this is possible with the implementation of nanostructured materials. The relationship between diffusion and charge transfer resistance is given as

$$D = \frac{1}{R_{ct}} \tag{2.4}$$

Simultaneously, the increase in charge transfer resistance (R_{ct}), influenced by both the desolvation process and the ion diffusion within the SEI, further impedes ion migration (Figure 2.46a). Efficient Na^+ ion migration, which serves as the charge carrier, is crucial for optimal battery performance and energy storage. The correlation between the freezing point and the solution freezing point is being described by the following formula:

$$T_{water} = T_{solution} + \Delta T_f = K_f \cdot m \tag{2.5}$$

In this context, T_{water} is 0 °C, $T_{solution}$ represents the freezing point of the solution, ΔT_f denotes the freezing point depression, the freezing point depression constant is K_f, and its value for water is set to 1.86 °C kg mol^{-1}, and the molality of the solution is noted with m. Instantly, the freezing point of a 1 M $NaClO_4$ aqueous solution is calculated to be −3.72 °C, which is insufficient for use under extreme low-temperature conditions. For the intercalation process in the electrochemical chain, electrode materials must meet several criteria: they should support feasible redox reactions, provide sufficient sites for alkali metal ions, and exhibit strong ionic and electronic conductivity. Ion transport within the cathode material is primarily determined by the intrinsic ion diffusion coefficient, which depends on the activation barrier of the sodium hopping pathway. When conductivity is low, the resistance to electron movement becomes the limiting factor for the overall macroscopic chemical Na diffusion coefficient (Figure 2.46b). Regarding cathode material, the diffusion constant is given by:

$$D^{\mathrm{app}}_{\mathrm{Na^+}} = D_0 \exp(-E_{\mathrm{a}}/\mathrm{RT}) \tag{2.6}$$

However, challenges remain in low-temperature (LT) environments: (i) increased viscosity and desolvation energy of organic electrolytes; (ii) solidification of aqueous electrolytes and salt solubility issues; and (iii) hindered ion transport kinetics (Figure 2.46c). Notably, these issues become more pronounced below −20 °C.

2.8.2 Challenges for SIBs at High Temperatures

The equilibrium between the energy states of the electrolyte and electrode is determined by thermodynamic balance, and it is typically regarded as the primary factor responsible for the formation of the EEI in batteries, acting as the main driving force for its generation. However, elevated temperatures significantly influence interface reactions, such as SEI dissolution, reconstruction, and gas generation, which accelerates battery degradation [189]. Research indicates that SEI in SIBs exhibits higher solubility compared to LIBs due to its components' increased solubility in electrolytes [190]. For example, NaF and Na_2CO_3 not only dissolve more readily in water than their lithium counterparts but also have greater solubility in organic solvents [191]. Consequently, the SEI in SIBs under high-temperature conditions demonstrates weak stability and is prone to decomposition, which can be modeled using the Arrhenius equation:

$$\frac{d^d_{C_{\mathrm{SEI}}}}{d_t} = A_{\mathrm{SEI}} \cdot c_{\mathrm{SEI}} \cdot \exp\left(-\frac{E_{a,\mathrm{SEI}}}{\mathrm{RT}}\right), (T > T_{\mathrm{onset,SEI}}) \tag{2.7}$$

where $\frac{d^d_{C_{\mathrm{SEI}}}}{d_t}$ is the decomposition rate for SEI, while the normalized concentration of SEI is denoted with c_{SEI}, preexponential factor is A_{SEI}, and onset temperature is represented by $T_{\mathrm{onset,\,SEI}}$ for SEI decomposition. More critically, during repeated cycling, the dissolution of SEI causes its continuous breakdown and reformation, which is linked to the electrolyte's decomposition and side reactions with the cathode and anode materials. These processes significantly deplete Na^+, resulting in irreversible capacity loss, reduced Coulombic efficiency, and shortened cycle life. In high-temperature (HT) environments, electrode materials face two main challenges:

(1) **Increased Risk of Structural Instability**: As temperatures rise, electrode materials with poor HT resistance can decompose and react with electrolyte decomposition products, leading to structural changes and the formation of various by-products.
(2) **Significant Safety Concerns Due to Excessive Heat Generation**: Most of the heat generated during charging and discharging occurs at the electrode (Figure 2.46d). Elevated temperatures trigger chemical reactions at the electrode, producing substantial heat and gas beyond standard operating conditions.

As compared to LT environments, HT enhance ion transport kinetics (Figure 2.46e), which facilitates the reaction processes. However, this also inevitably speeds up electrolyte decomposition. From a thermodynamic standpoint, temperature greatly affects the redox potential of the electrolyte, which can be evaluated using the Nernst equation:

$$V = V^0 - \frac{RT}{nF} \ln \alpha \tag{2.8}$$

Here, V denotes the cell potential, V^0 refers to the standard potential, Faraday constant is represented by F, and the number of electrons transferred per mole is represented by n. Higher temperatures raise the equilibrium potential of the battery, which can cause side reactions within the electrolyte, degradation of electrical/electrochemical junction (EEI), and the development of an unstable interface. In HT environments, several critical issues require urgent attention (see Figure 2.46f).

1) **Insufficient Safety**: HT conditions can easily trigger thermal runaway, posing significant safety risks. Addressing this requires efforts to reduce heat generation and improve heat dissipation within the battery system.
2) **Inadequate Interface Stability**: It is essential to focus on minimizing interfacial side reactions and electrolyte decomposition to ensure the stability and longevity of the battery system.
3) **Transition Metal Ion Dissolution**: The breakdown of transition metal ions in cathode materials is a significant concern that needs to be addressed.

Addressing this issue is vital for preserving the structural integrity and performance of the cathode, making it an area that requires careful consideration and thorough exploration.

2.9 Na-Powered Progress: The Rise of Commercial Sodium-Ion Cells

Sodium (Na) batteries are emerging as strong contenders for large-scale energy storage solutions due to sodium's abundance ranking as the sixth most prevalent element in the Earth's crust and the fourth in the oceans, making it both affordable and widely available. Advancements and innovations in SIBs span over 50 years. The development started with molten sodium batteries, notably the sodium–sulfur (NaS)

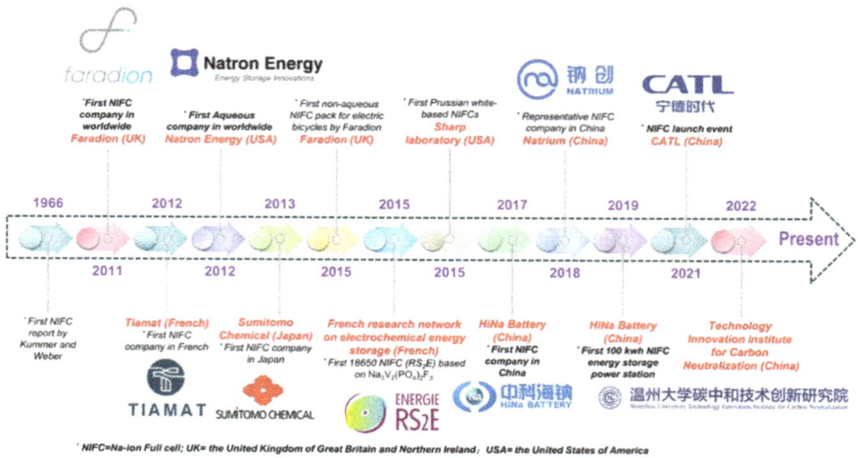

Figure 2.47 Commercial production of SIBs across the world. Source: [70]/John Wiley & Sons/CC BY 4.0.

battery, which emerged in the late 1960s, and aimed at providing a HT power source for vehicle electrification. In the 1970s, the sodium-metal halide (NaMH) battery, often referred to as the ZEBRA battery – short for Zeolite Battery Research Africa Project or Zero Emission Battery Research Activities – emerged as a key development, designed with transportation applications in mind. Concurrently, during the 1980s, SIBs began to be developed alongside LIBs. Despite this progress, SIBs have fallen behind their lithium counterparts, primarily due to constraints related to stable voltage profiles, energy density, charge/discharge rates, and cyclability. Recently, solid-state sodium batteries (SSSBs) have been introduced as promising candidates for industrial use. However, their effectiveness for large-scale, long-duration energy storage applications is still under evaluation. Figure 2.47 illustrates the production of SIBs at the commercial level across the world.

2.9.1 Electrochemical Systems

Sodium is appealing in large-scale energy storage as a result of its widespread availability and worldwide availability. Sodium-based batteries can generally be classified into two primary categories: molten sodium batteries and SIBs. Production and deployment are primarily focused in China and Europe. Key manufacturers such as CATL, AGM Batteries, HiNa Battery Technology Co., Zoolnasm, Faradion, and Tiamat are in the process of planning extensive facilities on a large scale. SIBs are particularly driven by the electrification of small vehicles such as e-bikes, e-scooters, and city cars. Faradion, in the United Kingdom, has installed 10 kW stationary modules in Australia, though such applications are still relatively rare. As technology and manufacturing improve, SIBs are expected to become competitive alternatives to lead–acid and lithium iron phosphate (LFP) batteries for both small-sized vehicle use and stationary storage.

(1) **PBA batteries:** Produced by Natron (Clarios), these are designed for high-power, short-duration needs. They offer lower energy density (~70 Wh kg^{-1}) but are highly valued for their energy capacity and modular expandability. Natron's products include a 48 V, 25 kW unit that can be scaled up to 812 V with a 15-minute charge time, and a smaller 4 kW, 48 V module with a 2-minute duration, with a peak power output of 6 kW and a recharge time of just 8 minutes; both systems are anticipated to endure a minimum of 50 000 cycles and are suitable for applications like peak shaving, data center support, and fast charging of electric vehicles.

(2) **Saltwater batteries:** These use a composite anode made of carbon and titanium, an aqueous electrolyte containing sodium perchlorate, and a cathode of manganese oxide. They were marketed as environmentally friendly with a lifespan of about 15 years and over 5000 cycles. Companies like Aquion Energy and BlueSky Energy attempted to commercialize these but have faced difficulties, and these batteries are not currently active in the market.

(3) **NaMH batteries**: FZSoNick manufactures these Na-$NiCl_2$ molten salt batteries, with over 5.5 million cells installed, totaling over 500 MWh of energy storage capacity. They are used for grid-scale applications, including renewable integration, grid services, backup power, microgrids, and some vehicle applications.

(4) **Solid-state sodium batteries (SSSBs):** Adena Power has successfully demonstrated a 1 kWh SSSB unit and is designed to be safe, recyclable, free from maintenance, and operable over a broad temperature spectrum. Adena aims to reduce price to below \$50 kWh^{-1} for a discharge period ranging from 6 to 18 hours.

2.9.2 The Price of Progress: SIB Baseline Cost Framework Study

Many SIBs are designed and function similar to LIBs, anticipated to secure a significant portion of the industry by 2030. Currently, SIBs have yet to see large-scale commercial applications. Due to the nascent stage of their market introduction, industry-wide forecasts for the specific formulation, cost levels, or performance metrics for 2030 could not be found. This scarcity of data makes it challenging to establish a baseline cost projection without further research and development (R&D) advancements. Nevertheless, its estimates from scholarly research were utilized to offer a preliminary evaluation of SIB costs. SIBs can be categorized based on their cathode types into three groups: TMLO, polyanion compounds, and PBAs each serving distinct roles in SIBs. PBAs are especially well suited for applications requiring high power and extensive cycling capabilities. In contrast, TMLO and polyanion-based SIBs are primarily designed for applications emphasizing energy storage. It is uncertain whether polyanion-based SIBs are currently on the market or still in the development phase. The current estimates (2022) will be used for TMLO-SIB costs and performance. For this analysis, the expense of storage units, lifespan of cycles, overall efficiency, discharge depth, and baseline cost parameters will also apply similar to those used for LFP batteries, such as plant infrastructure, control systems and communication, and project management, assuming similarities in structure and operation between SIBs and LIBs. Based on current data,

SIBs are projected to cost 1.33 times more than LFP batteries for large-scale grid storage, for a 250 kW system, 2-hour battery. This comparison does not account for potential future changes in LFP battery costs, which could vary if LFP batteries become widely used in electric vehicles. To align with the SI 2030 innovations, it is projected that the cost adjustment multiplier (1.33) will stay constant for a 100 MW system, a 10-hour battery. According to 2021 estimates, the price for a 100 MW LFP battery with a 10-hour duration is \$162.32 kWh^{-1}, corresponding to \$215.88 kWh^{-1} for SIBs. This evaluation presumes 1000 cycles with a 20% reduction in capacity throughout the battery's lifespan; aligning with LIB thresholds prior to a replacement of the storage unit becomes necessary.

2.9.3 SIB Cost Reduction: Aiming for \$0.05 kWh^{-1}

A collaborative effort involves seven SIB experts from universities, National Laboratories, and industry to explore innovations in R&D, cost projections, and potential Department of Energy (D-O-E) interventions. The objective was to attain technological improvements and reduce expenses toward the D-O-E's objective to attain a cost of \$0.05 kWh^{-1} for 10 hours, 100 MW energy storage. The research pinpointed 16 possible D-O-E initiatives where support could substantially drive innovation. To avoid plagiarism, you can rephrase the sentence while keeping the technical terms. The various elements of each innovation, including costs, implementation timelines, and efficiency gains, were assessed through a probabilistic simulation model. Further details of this model can be found in the Storage Innovations 2030 Methodology Report. The analysis also examined the influence of the standardized and levelized cost of energy storage (LCOS). This version uses synonyms for key terms and rephrases the sentence structure to maintain originality by combining various innovations into portfolios and evaluating their collective effects. Portfolios were created by considering all possible combinations of two to eight innovations. The simulation results showed that the LCOS for the top 10% of the best-performing portfolios ranged from \$0.23 to \$0.28 kWh^{-1} cycle, highlighting a reduction of 49% to 59%. These values are based on the average outcomes of the simulation, so if the actual impacts of innovations are greater than anticipated, the LCOS reductions could be even more substantial.

2.10 Summary and Prospect

SIBs exhibited satisfactory requirements for large-scale energy systems through their high performance.

Layered oxides have rigid conditions for transportation and storage due to their vulnerable nature toward water and air. Therefore, coating is inevitable, but it requires an expensive, lethal, and atmospherically unsafe organic solution known as *N,N*-dimethylaniline (NMP). A comprehensive investigation is carried out to probe the storage and transportation of layered oxides under normal circumstances and demonstrate the mechanism of utilizing water-soluble binders in succeeding coating procedures. It is also ensured to highlight the relationship among phase transition, structural arrangements, resistance to water and air, and electrochemical performance of crafted transition metal layered oxides for SIBs. Furthermore, it is the necessity of the time to widen the scrutiny of new and novel cathode candidates beyond layered oxides for SIBs. A profound perspective is to model a composite of layered oxide with PBAs and other cathodic materials such as polyanionic materials. Especially, in the case of PBAs, they exhibit outstanding cyclabilities for SIBs with no security issues. If iron and manganese metals are used in such compounds, then cyclability and specific capacity will also be increased. Nevertheless, some efficient and appropriate mechanisms are required to address the aspects; for example, it is essential to develop a high level of understanding to demonstrate a clear relationship for structural relationship in context to attain profound electrochemical performance of PBA. Besides, a unique strategy is still in the infancy stage for electrolyte optimization to achieve meaningful and exceptional performance. It is necessary to assemble pouch/cylindrical full cells with large capacity to estimate PBA as potential practical applications for the commercial market. The way research has boosted to explore PBA in deep is expected to displace $LiFePO_4$/graphite-based LIBs with MnHCF/HC-based SIBs in the near future.

NASICON-type structures composed of high electronegative polyanions are expected to demonstrate high working voltage due to the strong capability of electron-drawn properties. In NASICON, vanadium is an expensive metal that can be replaced and utilized with less expensive elements such as Mn, and Fe. These elements have proved their massive potential by exhibiting high-energy densities. The NASICON structures with splendid element Mn such as $Na_4MnV(PO_4)_3$ and $Na_4MnCr(PO_4)_3$ have unveiled outperformed reversible capacities of more than 150 mAh g^{-1} by leveraging high voltage and have paved their way toward mobile applications. Nonetheless, the large-scale commercial applications have been deterrent due to their poor cyclability. Therefore, a unique and cutting-edge mechanism is essentially needed for Fe and Mn-based NASICON-type cathodes.

Currently, the increasing demands have drawn attention to the focus on exploring organic electrode materials for SIBs. It has become a popular topic to probe and synthesize more sustainable organic electrodes, owing to their superior characteristics such as charge storage flexibility, tunable structural geometry, and fast kinetics; they have evolved their recognition in energy storage applications. A multivalent ion storage system can be well suited for organic electrodes due to their adaptable molecular geometry and weak intermolecular interaction. Additionally, when organic electrodes are hired in the form of solid-state electrodes, then they are considered ideal for LT operating batteries. Organic electrodes generate buffer zone for electron-coupled proton due to their reasonable electrochemical nature. Despite the benefits of organic electrodes, they pose some challenges impeding their way toward commercial applications. The foremost challenge is small organic electrodes quickly dissolve in aqueous and organic electrolytes. So, this is the bottleneck task to refrain organic electrodes from dissolution in alkaline ion batteries. Additionally, insignificant conductivity and reduced tap density of organic electrodes are certainly reducing volumetric energy density. Thus, the new experimental methodology is required to address demonstrated limitations and uphold organic electrode-based batteries to assert their significance in commercial applications.

At the time of designing novel electrode frameworks, it is distinctively crucial to attain a reduced ionic path through low intricacy to accelerate the movement of Na^+ ion and minimize the diffusion in the electrolyte phase. Therefore, it is inevitable to explore novel electrolytes, strategizing significant ionic conductivity, and the envision of enhanced performance of fast charging can be executed by increasing the transfer number of Na^+ ions. Some features must be taken into account while selecting/designing electrolytes including chemical and thermal stability, flame deterrence, and dendritic growth obstruction. These salience features are essential in ionic liquids and high-entropy electrolyte systems. Given the convoluted features of electrolytes, the efforts must be accelerated to explore the existing and obscure more electrolytes. Recently, artificial intelligence (AI) testified striking breakthroughs initiating the probability of its incorporation with the design of electrolytes to expedite the scrutiny process with exclusive mechanisms. It has become viable to predict electrolytes very accurately by exploiting deep machine learning algorithms of AI. Therefore, researchers should develop principles for electrolyte selection/design with high accuracy to adjust them in LT and high-pressure conditions. Besides, it is crucial to classify additives, solvents, and salts on their physiochemical characteristics which is beneficial for the AI model to predict at high accuracy.

For fast charging, the anode materials must possess high areal capacity, high ICE, and exceptional broad temperature progress along with their salient features such as high-rate capability, large reversible capacity, long sustainable cycle life, and meaningful working voltage. These significant properties meet the necessities for practical applications. Therefore, it is inevitable to establish effective strategies for surface modification and coating techniques to attain >92% ICE for commercial applications. Nevertheless, there are some fundamental factors such as electron transfer and ion diffusion, limiting the performance of fast-charging anode candidates. Furthermore, the rate performance directly relies on the reaction mechanism of corresponding anode materials. Thus, the cycling capability is much better in intercalation-type anode materials, but on the other side, these materials exhibit poor kinetics and insignificant energy density. More active sites can be produced to adsorb Na^+ ions through surface modification, revamping microstructure, and doping heteroatoms. In this way, more channels for Na^+ are constructed, shrinking the diffusion paths to promote the rate performance and increase the areal capacity through expansion in layer spacing, whereas higher theoretical capacity is associated with alloy-based materials. HC has emerged as a leading commercial anode material for SIBs, yet its complex sodium storage mechanisms remain a subject of intense study. The interplay between precursor materials, carbonization temperatures, and resulting microstructures significantly influences HC's electrochemical behavior. A precise and definitive relationship is required for sodium storage in HC microstructures, especially, the role of potential buffers against sodium plating in quasi-metallic sodium clusters. There is still further progress needed to improve the rate performance of fast-charging anode materials to meet the standards of commercial applications. Additionally, a number of anode materials still lagging behind the challenges involved in high-energy usage and being economically expensive. It requires unique and optimized novel synthesis methods to boost industrial production. To advance high-performance fast-charging anodes for SIBs, researchers must harness cutting-edge characterization techniques and computational methods, enabling deeper insights into material synthesis, ion transport, and structural evolution during rapid charge–discharge cycles. Innovative synthesis visualization tools like liquid electron microscopy and *in situ* small-angle X-ray scattering allow real-time observation of porous material formation, while multiscale, high-resolution analysis techniques such as *in situ* synchrotron radiation analysis and cryoelectron microscopy track Na^+ ion movement and material transformations. Electrode–electrolyte interface studies using methods like *in situ* XRD and Raman spectroscopy illuminate interface formation and evolution, crucial for optimizing anode–electrolyte interactions. Theoretical modeling approaches, including DFT and advanced multiscale simulations, complement experimental data by offering atomistic and macroscopic insights into ion storage mechanisms, collectively driving the field toward more efficient and durable fast-charging SIBs.

References

1 Balaram, V., Santosh, M., Satyanarayanan, M. et al. (2024). *Geoscience Frontiers* 15: 101868.

2 Rostami, H., Valio, J., Suominen, P. et al. (2024). *Chemical Engineering Journal* 495: 153471.

3 Ahmed, M.D. and Maraz, K.M. (2023). *Materials Engineering Research* 5: 265.

4 Wanison, R., Syahputra, W.N.H., Kammuang-lue, N. et al. (2024). *Journal of Energy Storage* 100: 113497.

5 Duffner, F., Kronemeyer, N., Tübke, J. et al. (2021). *Nature Energy* 6: 123.

6 Ma, J., Feng, X., Wu, Y. et al. (2023). *Journal of Energy Chemistry* 77: 290.

7 Ramasubramanian, B., Ling, J., Jose, R., and Ramakrishna, S. (2024). *Cell Reports Physical Science* 5: 102032.

8 Olivetti, E.A., Ceder, G., Gaustad, G.G., and Fu, X. (2017). *Joule* 1: 229.

9 De Sisternes, F.J., Jenkins, J.D., and Botterud, A. (2016). *Applied Energy* 175: 368.

10 Zhao, L., Zhang, T., Li, W. et al. (2023). *Engineering* 24: 172.

11 Alvira, D., Antorán, D., and Manyà, J.J. (2022). *Journal of Energy Chemistry* 75: 457.

12 Li, Q., Chen, J., Fan, L. et al. (2016). *Green. Energy & Environment* 1: 18.

13 Singh, A.N., Islam, M., Meena, A. et al. (2023). *Advanced Functional Materials* 33: 2304617.

14 Mamoor, M., Li, Y., Wang, L. et al. (2023). *Green Energy and Resources* 1: 100033.

15 Satrughna, J.A.K., Kanwade, A., Srivastava, A. et al. (2023). *Materials Today* S1369702123002067.

16 Tareq, F.K. and Rudra, S. (2024). *Materials Today Communications* 39: 108653.

17 Machín, A., Morant, C., and Márquez, F. (2024). *Batteries* 10: 29.
18 Örüm Aydin, A., Zajonz, F., Günther, T. et al. (2023). *Batteries* 9: 555.
19 Rudola, A., Sayers, R., Wright, C.J., and Barker, J. (2023). *Nature Energy* 8: 215.
20 Gao, Y., Yu, Q., Yang, H. et al. (2024). *Advanced Materials* 2405989.
21 Sandhu, Z.A., Raza, M.A., Awwad, N.S. et al. (2024). *Materials Advances* 5: 30.
22 Xiao, L., Ji, F., Zhang, J. et al. (2023). *Small* 19: 2205732.
23 Zhang, H., Peng, J., Li, L. et al. (2023). *Advanced Functional Materials* 33: 2210725.
24 Liu, S., Wan, J., Ou, M. et al. (2023). *Advanced Energy Materials* 13: 2203521.
25 Han, X., Qiu, T., Li, M. et al. (2023). *Advanced Functional Materials* 33: 2303102.
26 Shadike, Z., Zhao, E., Zhou, Y. et al. (2018). *Advanced Energy Materials* 8: 1702588.
27 Liu, X., Zuo, W., Zheng, B. et al. (2019). *Angewandte Chemie, International Edition* 58: 18086.
28 Li, J., Hu, H., Wang, J., and Xiao, Y. (2022). *Carbon Neutralization* 1: 96.
29 Li, H., Ramakrishnan, S., Freeland, J.W. et al. (2020). *Journal of the American Chemical Society* 142: 8160.
30 Li, S., Xiao, Y., Zhu, Y.-F. et al. (2021). *Chemical Engineering Journal* 412: 128719.
31 Guo, S., Liu, P., Yu, H. et al. (2015). *Angewandte Chemie, International Edition* 54: 5894.
32 Yabuuchi, N., Kubota, K., Dahbi, M., and Komaba, S. (2014). *Chemical Reviews* 114: 11636.
33 Zhao, C., Avdeev, M., Chen, L., and Hu, Y. (2018). *Angewandte Chemie, International Edition* 57: 7056.
34 Wang, J., Liu, W., Luo, G. et al. (2018). *Energy & Environmental Science* 11: 3375.
35 Jin, Y., Zhao, Y., Feng, J. et al. (2022). *ACS Applied Materials & Interfaces* 14: 56715.
36 Hu, Z., Weng, M., Chen, Z. et al. (2021). *Nano Energy* 83: 105834.
37 Schmidt, W., Berthelot, R., Sleight, A.W., and Subramanian, M.A. (2013). *Journal of Solid State Chemistry* 201: 178.
38 Bhange, D.S., Ali, G., Kim, D.-H. et al. (2017). *Journal of Materials Chemistry A* 5: 1300.
39 Yuan, D., Liang, X., Wu, L. et al. (2014). *Advanced Materials* 26: 6301.
40 Mamoor, M., Lian, R., Wang, D. et al. (2019). *Physical Chemistry Chemical Physics* 21: 14612.
41 Li, M., Sun, C., Yuan, X. et al. (2024). *Advanced Functional Materials* 34: 2314019.
42 Li, Z., Sun, C., Li, M. et al. (2024). *Advanced Functional Materials* 34: 2315114.

43 Gu, Z., Zhao, X., Li, K. et al. (2024). *Advanced Materials* 36: 2400690.
44 Mamoor, M., Lian, R., Wang, D. et al. (2021). *Electrochimica Acta* 379: 138157.
45 He, Y., Scivally, E., Shaji, A. et al. (2024). *Advanced Energy Materials* 2403877.
46 Zhu, T., Liu, W., Liao, X. et al. (2024). *Advanced Fiber Materials* 6: 561–569.
47 Liu, J., Liu, J., Tang, M. et al. (2024). *Advanced Functional Materials* 34: 2314167.
48 Jing, Z., Kong, L., Mamoor, M. et al. (2025). *Journal of the American Chemical Society* 147: 4, 3702–3713.
49 Jing, Z., Mamoor, M., Kong, L. et al. (2025). *Angewandte Chemie, International Edition* e202423356.
50 Liu, J., Wang, Y., Jiang, N. et al. (2024). *Angewandte Chemie, International Edition* e202400214.
51 Xing, F., Li, S., Chen, L. et al. (2023). *ACS Nano* 17: 21432.
52 Xu, X., Ren, S., Wu, H. et al. (2024). *Journal of the American Chemical Society* 146: 1619.
53 Wan, Y., Huang, B., Liu, W. et al. (2024). *Advanced Materials* 36: 2404574.
54 Jache, B. and Adelhelm, P. (2014). *Angewandte Chemie, International Edition* 53: 10169.
55 Kim, H., Hong, J., Park, Y. et al. (2015). *Advanced Functional Materials* 25: 534.
56 Tian, Z., Zou, Y., Liu, G. et al. (2022). *Advancement of Science* 9: 2201207.
57 Liu, X., Wang, T., Zhang, T. et al. (2022). *Advanced Energy Materials* 12: 2202388.
58 Kim, Y., An, J., Kim, S. et al. (2022). *Advanced Materials* 34: 2201446.
59 Cao, C.R., Huang, K.Q., Shi, J.A. et al. (1966). *Nature Communications* 2019: 10.
60 Abbas, Z., Labbez, C., Nordholm, S., and Ahlberg, E. (2008). *Journal of Physical Chemistry C* 112: 5715.
61 Tang, Z., Wang, H., Wu, P. et al. (2022). *Angewandte Chemie, International Edition* 61: e202200475.
62 Song, K., Wang, X., Xie, Z. et al. (2023). *Angewandte Chemie, International Edition* 62: e202216450.
63 Liu, M., Wu, F., Gong, Y. et al. (2023). *Advanced Materials* 35: 2300002.
64 Liang, Y., Song, N., Zhang, M. et al. (2023). *Advanced Energy Materials* 13: 2302825.
65 Zuo, W., Liu, R., Ortiz, G.F. et al. (2018). *Journal of Power Sources* 400: 317.
66 Dong, R., Zheng, L., Bai, Y. et al. (2021). *Advanced Materials* 33: 2008810.
67 Shi, J., Ding, L., Wan, Y. et al. (2021). *Journal of Energy Chemistry* 57: 650.
68 Li, H., Zhang, H., Diemant, T. et al. (2021). *Small Structures* 2: 2100035.
69 De Sloovere, D., Vanpoucke, D.E.P., Paulus, A. et al. (2022). *Advanced Energy and Sustainability Research* 3: 2100159.
70 Gao, Y., Zhang, H., Peng, J. et al. (2024). *Carbon Energy* 6: e464.
71 Desai, P., Huang, J., Hijazi, H. et al. (2021). *Advanced Energy Materials* 11: 2101490.

72 Hu, Y., Yu, Q., Tang, W. et al. (2021). *Energy Storage Materials* 41: 738.
73 Hwang, J., Yang, H., Matsumoto, K., and Hagiwara, R. (2021). *Journal of the Electrochemical Society* 168: 030508.
74 Sun, M.-Y., Yu, F.-D., Xia, Y. et al. (2022). *Chemical Engineering Journal* 430: 132750.
75 Liang, H., Gu, Z., Zhao, X. et al. (2021). *Angewandte Chemie, International Edition* 60: 26837.
76 Yang, Z., He, J., Lai, W. et al. (2021). *Angewandte Chemie* 133: 27292.
77 Wang, Z., Xu, Y., Peng, J. et al. (2021). *Small* 17: 2101650.
78 Che, C., Wu, F., Li, Y. et al. (2024). *Advanced Materials* 36: 2402291.
79 He, J., Bhargav, A., Su, L. et al. (2024). *Nature Energy* 9: 446.
80 Huang, Z.-X., Zhang, X.-L., Zhao, X.-X. et al. (2023). *Inorganic Chemistry Frontiers* 10: 37.
81 Qian, Y., Hu, S., Zou, X. et al. (2019). *Energy Storage Materials* 20: 208.
82 Ellis, B.L. and Nazar, L.F. (2012). *Current Opinion in Solid State & Materials Science* 16: 168.
83 Pang, Y., Cao, Y., Chu, Y. et al. (2020). *Advanced Functional Materials* 30: 1906244.
84 Eshetu, G.G., Martinez-Ibañez, M., Sánchez-Diez, E. et al. (2018). *Chemistry - An Asian Journal* 13: 2770.
85 Shipitsyn, V., Antrasian, N., Soni, V. et al. (2023). *Energy Materials* https://doi.org/10.20517/energymater.2023.22.
86 Li, C., Xu, H., Ni, L. et al. (2023). *Advanced Energy Materials* 13: 2301758.
87 Song, B., Xiong, X., Peng, Y. et al. (2024). *Advanced Energy Materials* 14: 2401407.
88 Kim, D.-H., Kang, B., and Lee, H. (2019). *Journal of Power Sources* 423: 137.
89 Dai, P., Shi, C.-G., Huang, Z. et al. (2023). *Energy Storage Materials* 56: 551.
90 Purushotham, U., Takenaka, N., and Nagaoka, M. (2016). *RSC Advances* 6: 65232.
91 Li, B., Xu, M., Li, T. et al. (2012). *Electrochemistry Communications* 17: 92.
92 Wang, Y., Ma, Z., Cao, Z. et al. (2023). *Advanced Functional Materials* 33: 2305974.
93 Nian, Q., Wang, J., Liu, S. et al. (2019). *Angewandte Chemie, International Edition* 58: 16994.
94 Yang, H., Hwang, J., Tonouchi, Y. et al. (2021). *Journal of Materials Chemistry A* 9: 3637.
95 Liu, X., Jiang, X., Zeng, Z. et al. (2018). *ACS Applied Materials & Interfaces* 10: 38141.
96 Song, X., Meng, T., Deng, Y. et al. (2018). *Electrochimica Acta* 281: 370.
97 Jo, J.H., Choi, J.U., Park, Y.J. et al. (2020). *Energy Storage Materials* 32: 281.
98 Li, Y., Li, W., Shimizu, R. et al. (2022). *Advanced Energy Materials* 12: 2103033.
99 Gao, L., Chen, J., Liu, Y. et al. (2018). *Journal of Materials Chemistry A* 6: 12012.
100 Liu, Y., Jiang, R., Xiang, H. et al. (2022). *Electrochimica Acta* 425: 140746.
101 Jiang, R., Hong, L., Liu, Y. et al. (2021). *Energy Storage Materials* 42: 370.

102 Sadan, M.K., Choi, S.-H., Kim, H.H. et al. (2018). *Ionics* 24: 753.

103 Moeez, I., Susanto, D., Chang, W. et al. (2021). *Chemical Engineering Journal* 425: 130547.

104 Chae, M.S., Kim, H.J., Bu, H. et al. (2020). *Advanced Energy Materials* 10: 2000564.

105 Xie, B., Zuo, P., Wang, L. et al. (2019). *Nano Energy* 61: 201.

106 Shao, T., Li, C., Liu, C. et al. (2019). *Journal of Materials Chemistry A* 7: 1749.

107 Che, H., Liu, J., Wang, H. et al. (2017). *Electrochemistry Communications* 83: 20.

108 Manohar, C.V., Forsyth, M., MacFarlane, D.R., and Mitra, S. (2018). *Energy Technology* 6: 2232.

109 Park, M.-S., Choi, J.-Y., Kumar Veerasubramani, G., and Kim, D.-W. (2020). *Electrochemistry Communications* 119: 106829.

110 Xu, X., Zhou, D., Qin, X. et al. (2018). *Nature Communications* 9: 3870.

111 Shi, Q., Zhong, Y., Wu, M. et al. (2018). *Angewandte Chemie, International Edition* 57: 9069.

112 Zhao, S., Wang, C., Du, D. et al. (2021). *Angewandte Chemie, International Edition* 60: 3205.

113 Li, W., Yao, H., Yan, K. et al. (2015). *Nature Communications* 6: 7436.

114 Wang, H., Wang, C., Matios, E., and Li, W. (2018). *Angewandte Chemie, International Edition* 57: 7734.

115 Deng, J., Luo, W., Chou, S. et al. (2018). *Advanced Energy Materials* 8: 1701428.

116 Sun, Y., Guo, S., and Zhou, H. (2019). *Advanced Energy Materials* 9: 1800212.

117 Fang, Y., Luan, D., and David Lou, X.W. (2020). *Advanced Materials* 32: 2002976.

118 Fang, S., Bresser, D., and Passerini, S. (2020). *Advanced Energy Materials* 10: 1902485.

119 Cao, Y., Li, M., Lu, J. et al. (2019). *Nature Nanotechnology* 14: 200.

120 Xie, F., Xu, Z., Guo, Z., and Titirici, M.-M. (2020). *Progress in Energy* 2: 042002.

121 Sada, K., Darga, J., and Manthiram, A. (2023). *Advanced Energy Materials* 13: 2302321.

122 Liu, Y., Zhu, Y., and Cui, Y. (2019). *Nature Energy* 4: 540.

123 Sun, H., Zhu, J., Baumann, D. et al. (2018). *Nature Reviews Materials* 4: 45.

124 Sun, H., Mei, L., Liang, J. et al. (2017). *Three-dimensional holey-graphene/niobia composite architectures for ultrahigh-rate energy storage. Science* 356: 599–604.

125 Chen, Y., Zhao, B., Yang, Y., and Cao, A. (2022). *Advanced Energy Materials* 12: 2201834.

126 Wei, Q., Chang, X., Wang, J. et al. (2022). *Advanced Materials* 34: 2108304.

127 Liu, M., Zhang, P., Qu, Z. et al. (2019). *Nature Communications* 10: 3917.

128 Kitsu Iglesias, L., Antonio, E.N., Martinez, T.D. et al. (2023). *Advanced Energy Materials* 13: 2302171.

129 Kong, Z., Wang, L., Iqbal, S. et al. (2022). *Small* 18: 2107252.

130 Dong, C., Guo, L., Li, H. et al. (2020). *Energy Storage Materials* 25: 679.

131 Li, B., Xi, B., Feng, Z. et al. (2018). *Advanced Materials* 30: 1705788.

132 Lan, K., Liu, L., Zhang, J.-Y. et al. (2021). *Journal of the American Chemical Society* 143: 14097.
133 Liang, H., Liu, L., Wang, N. et al. (2022). *Advanced Materials* 34: 2202873.
134 Ni, J., Fu, S., Wu, C. et al. (2016). *Advanced Energy Materials* 6: 1502568.
135 Xu, L., Liu, Y., Hu, X. et al. (2024). *Accounts of Materials Research* 5: 822.
136 Hu, X., Jia, J., Wang, G. et al. (2018). *Advanced Energy Materials* 8: 1801452.
137 Yang, J., Shen, X., Yang, W., and Kim, J. (2023). *Progress in Materials Science* 133: 101054.
138 Hu, X., Liu, Y., Chen, J. et al. (2019). *Advanced Energy Materials* 9: 1901533.
139 Liu, B., Liu, Y., Hu, X. et al. (2021). *Advanced Functional Materials* 31: 2101066.
140 Wu, L., Li, Y., Fu, Z., and Su, B.-L. (2020). *National Science Review* 7: 1667.
141 Hu, X., Liu, Y., Li, J. et al. (2020). *Advanced Functional Materials* 30: 1907677.
142 Jenik, M., Seifner, A., Lieberzeit, P. et al. (2009). *Pollen-imprinted polyurethanes for QCM allergen sensors. Anal Bioanal Chem* 394: 523–528.
143 Li, J., Hu, X., Zhong, G. et al. (2021). *Nano Micro Letters* 13: 131.
144 Liu, Y., Hu, X., Li, J. et al. (2022). *Nature Communications* 13: 663.
145 Zhao, W., Hu, X., Ci, S. et al. (2019). *Small* 15: 1904054.
146 Zhu, C., Liu, T., Qian, F. et al. (2017). *Nano Today* 15: 107.
147 Cai, J., Fan, Z., Jin, J. et al. (2020). *Nano Energy* 75: 104970.
148 Lewis, J.A. (2006). *Advanced Functional Materials* 16: 2193.
149 Yuan, J., Qiu, M., Chen, J.X. et al. (2022). *Advanced Functional Materials* 32: 2203732.
150 Yu, B., Ji, Y., Hu, X. et al. (2022). *Chemical Engineering Journal* 430: 132993.
151 Ma, J., Zheng, S., Chi, L. et al. (2022). *Advanced Materials* 34: 2205569.
152 He, Z., Zhang, C., Zhu, Z. et al. (2024). *Advanced Functional Materials* 2408285.
153 Zhang, L., Wang, C., Dou, Y. et al. (2019). *Angewandte Chemie, International Edition* 58: 8824.
154 Sun, D., Zhao, L., Sun, P. et al. (2024). *Advanced Functional Materials* 2403642.
155 Shao, R., Sun, Z., Wang, L. et al. (2024). *Angewandte Chemie* 136: e202320183.
156 He, Z., Huang, Y., Liu, H. et al. (2024). *Nano Energy* 129: 109996.
157 Jiang, J., Li, Y., Liu, J. et al. (2012). *Advanced Materials* 24: 5166.
158 Liu, Y., Lin, Q., Chen, X. et al. (2024). *Energy and Environmental Materials* 7: e12684.
159 Li, Z., Han, M., Yu, P., and Yu, J. (2024). *Advancement of Science* 11: 2306992.
160 Li, J., He, Y., Dai, Y. et al. (2024). *Advanced Functional Materials* 2406915.
161 Li, H., He, Y., Wang, Q. et al. (2023). *Advanced Energy Materials* 13: 2302901.
162 Sun, X., Wang, L., Li, C. et al. (2021). *Nano Research* 14: 4696.
163 Chen, H., Rodrigues, J.N.B., Rettie, A.J.E. et al. (2019). *Journal of the American Chemical Society* 141: 635.
164 Lu, C., Liu, L., He, S. et al. (2024). *Advanced Energy Materials* 14: 2401221.
165 Li, Z., Han, M., Wang, J. et al. (2024). *Advanced Functional Materials* 2404263.
166 He, Y., Dong, C., He, S. et al. (2021). *Nano Research* 14: 4014.
167 Mamoor, M., Lian, R., Wu, X. et al. (2022). *Journal of Physics: Condensed Matter* 34: 324001.

168 Deysher, G., Oh, J.A.S., Chen, Y.-T. et al. (2024). *Nature Energy* https://doi.org/10.1038/s41560-024-01569-9.

169 Matios, E., Wang, H., Wang, C., and Li, W. (2019). *Industrial and Engineering Chemistry Research* 58: 9758.

170 Mao, M., Ji, X., Wang, Q. et al. (2023). *Nature Communications* 14: 1082.

171 Lin, C.-C., Chen, Z., Euchner, H. et al. (2023). *Energy Materials* 6: 2140.

172 Louli, A.J., Genovese, M., Weber, R. et al. (2019). *Journal of the Electrochemical Society* 166: A1291.

173 Shin, W. and Manthiram, A. (2022). *Angewandte Chemie, International Edition* 61: e202115909.

174 Spencer Jolly, D., Ning, Z., Darnbrough, J.E. et al. (2020). *ACS Applied Materials & Interfaces* 12: 678.

175 Bates, J. (2000). *Solid State Ionics* 135: 33.

176 Neudecker, B.J., Dudney, N.J., and Bates, J.B. (2000). *Journal of the Electrochemical Society* 147: 517.

177 Wang, M.J., Carmona, E., Gupta, A. et al. (2020). *Nature Communications* 11: 5201.

178 Gu, D., Kim, H., Lee, J.-H., and Park, S. (2022). *Journal of Energy Chemistry* 70: 248.

179 Lewis, J.A., Sandoval, S.E., Liu, Y. et al. (2023). *Advanced Energy Materials* 13: 2204186.

180 Lee, Y.-G., Fujiki, S., Jung, C. et al. (2020). *Nature Energy* 5: 299.

181 Ortmann, T., Fuchs, T., Eckhardt, J.K. et al. (2024). *Advanced Energy Materials* 14: 2302729.

182 Luo, S., Wang, Z., Li, X. et al. (2021). *Nature Communications* 12: 6968.

183 Yun, Q., He, Y., Lv, W. et al. (2016). *Advanced Materials* 28: 6932.

184 Deysher, G., Chen, Y.-T., Sayahpour, B. et al. (2022). *ACS Applied Materials & Interfaces* 14: 47706.

185 Duchêne, L., Kühnel, R.-S., Rentsch, D. et al. (2017). *Chemical Communications* 53: 4195.

186 Zhu, X. and Wang, L. (2020). *EcoMat* 2: e12043.

187 Liu, T., Zhang, Y., Jiang, Z. et al. (2019). *Energy & Environmental Science* 12: 1512.

188 Shao, T., Liu, C., Deng, W. et al. (2019). *Batteries & Supercaps* 2: 403.

189 Ma, L.A., Naylor, A.J., Nyholm, L., and Younesi, R. (2021). *Angewandte Chemie, International Edition* 60: 4855.

190 Eshetu, G.G., Diemant, T., Hekmatfar, M. et al. (2019). *Nano Energy* 55: 327.

191 Mogensen, R., Brandell, D., and Younesi, R. (2016). *ACS Energy Letters* 1: 1173.

3

Research Development on Potassium-Ion Batteries and Potassium–Sulfur Batteries

Yanyan He, Junhui Li, and Yuxin Dai

Key Laboratory of Fine Chemicals in Universities of Shandong, Jinan Engineering Laboratory for Multi-scale Functional Materials, School of Chemistry and Chemical Engineering, Qilu University of Technology (Shandong Academy of Sciences), Jinan, Shandong 250353, P. R. China

3.1 Introduction to Potassium-Ion Batteries

3.1.1 A Brief History of Potassium-Ion Battery Development

Potassium-ion batteries (PIBs) were first conceptualized in 2004 by Ali Eftekhari, who proposed using Prussian blue materials as the cathode [1]. This electrode demonstrated a capacity fade rate of only 12% after 500 cycles. Following this initial work, researchers extensively studied Prussian blue and its analogs. In 2005, potassium hexafluorophosphate (KPF_6) was patented as a promising electrolyte for PIBs. By 2007, the first potassium-powered battery entered the market [2]. However, at that time, specific parameters regarding the safety of PIBs were not provided. Due to this lack of information and concerns about safety, the development of PIBs experienced a hiatus in the subsequent years. Nevertheless, as society has progressed, interest in PIBs has grown [3, 4]. Gradually, researchers have uncovered numerous advantages of PIBs. For example, their low Lewis acidity confers higher ionic conductivity compared to those of Li^+ and Na^+ ions. They exhibit a higher anode redox potential in propylene carbonate (PC) electrolytes compared to lithium-ion batteries (LIBs) and sodium-ion batteries (SIBs). This characteristic enables energy storage reactions with commercial graphite, yielding a specific capacity comparable to that of LIBs (279 mA h g^{-1}). Furthermore, utilizing aluminum foil instead of copper foil as the current collector significantly reduces costs. Research on PIBs, encompassing electrode materials and electrolytes, has also heralded a new era in energy storage technology [5–7].

The Earth is abundant in potassium resources, with a concentration of 1.50% in the Earth's crust. This is comparable to sodium (2.36%), and nearly 900 times greater than lithium resources, which stands at 0.0017%. The plentiful availability of potassium significantly lowers the cost of electrode materials and electrolytes for PIBs. Furthermore, thermodynamic analyses indicate that potassium and aluminum do not react to form potassium–aluminum alloys. As a result, PIBs

Electrode Materials in Energy Storage Technologies: Applications in Lithium-, Sodium-, Potassium-, Sulfur- and Zinc-Based Rechargeable Batteries, First Edition. Edited by Liqiang Xu.

can utilize aluminum foil, which is less expensive than copper foil, as the current collector for the anode, thereby further reducing the overall cost of the batteries [8, 9].

In organic electrolytes, the redox potential of K^+/K is lower than that of Li^+/Li and Na^+/Na. For instance, the standard electrode potentials for lithium, sodium, and potassium are as follows: Na^+/Na (−2.56 V) > Li^+/Li (−2.79 V) > K^+/K (−2.88 V) in a PC solution. The lower potential is advantageous for increasing the operating voltage of the battery and enhancing its energy density. Compared to LIBs with a voltage of 4.5 V and SIBs at 4.2 V, PIBs exhibit the most expansive voltage window at 4.6 V. This also indicates that the potential limit for the anode in PIBs is lower than that in LIBs, thereby enabling a higher operating voltage and increasing energy density [10–12].

Compared to Na^+ and Li^+, K^+ ions have the largest ionic radius; however, they exhibit the smallest Stokes radius in both PC solutions and aqueous solutions. This observation implies that K^+ ions has the fastest diffusion rate in the electrolyte and the highest molar conductivity [13]. The small Stokes radius of K^+ ion results from the weak interaction between K^+ ions and solvent molecules, which can be attributed to the low surface charge density of K^+ ion (i.e. its weak Lewis acidity). Computational results from density functional theory indicate that K^+ ion has a lower desolvation energy in PC solutions. Therefore, considering ionic conductivity and desolvation performance, PIBs have the potential for better rate performance compared to LIBs and SIBs [10, 14].

The anode materials for PIBs can include graphite, which is not a viable option for SIBs. Graphite, already a commercialized anode material for LIBs, enables PIBs to integrate more seamlessly with established LIBs systems, thereby enhancing the feasibility of commercialization. Due to these advantages, PIBs are regarded by the industry as a promising alternative that may replace LIBs to meet the demands of both portable and large-scale energy storage applications [15, 16].

3.1.2 Aqueous Potassium-Ion Batteries

In recent years, aqueous potassium-ion batteries (APIBs) have been progressively developed. Similar to traditional PIBs, APIBs operate on a rocking-chair battery model, which consists of a cathode, an anode, and an aqueous electrolyte (Figure 3.1) [17]. During charging, K^+ ions are released from the cathode, migrate through the electrolyte, and are inserted into the anode, while electrons flow to the anode through an external circuit. This process reverses during discharge [18].

Unlike PIBs, which utilize toxic and flammable organic-based electrolytes, APIBs employ mild aqueous electrolytes, providing advantages such as enhanced safety, reduced costs, and environmental sustainability. Furthermore, the low Lewis acidity of K^+ ions in aqueous electrolytes results in a relatively small hydrated ion radius (3.31 Å), facilitating rapid ion diffusion [19]. These attributes position APIBs as promising candidates for electrochemical energy storage devices. However, APIBs encounter challenges similar to those faced by other aqueous batteries, including limited energy density and short cycle life. The narrow potential window of APIBs, which arises from the hydrogen evolution reaction (HER) and oxygen evolution

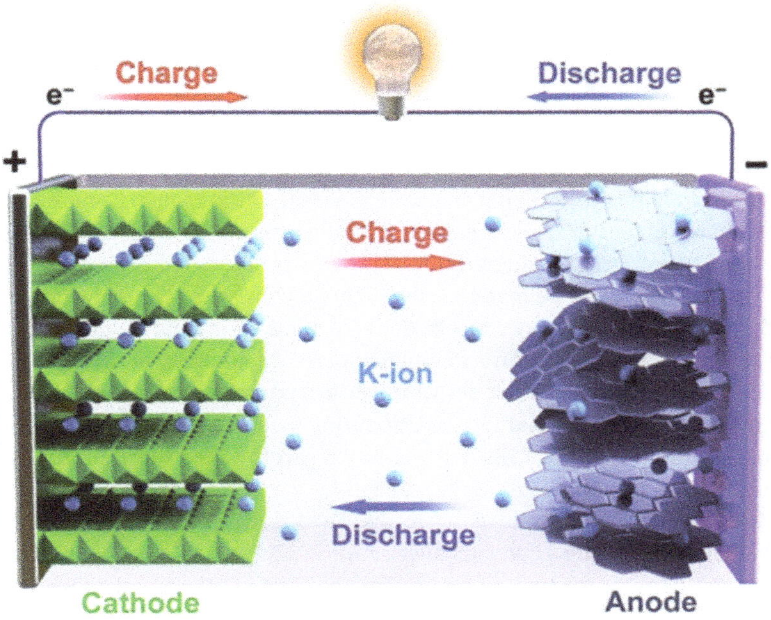

Figure 3.1 Schematic diagram of the working principle of PIBs. Source: [17]. Reproduced with permission/John Wiley & Sons.

reaction (OER) of water, constrains both capacity and energy density. Additionally, side reactions between electrode materials and water or evolved oxygen (O_2) during cycling lead to structural degradation of the electrodes, compromising battery stability. The dissolution of certain active materials in aqueous electrolytes further exacerbates these challenges. Moreover, akin to PIBs, the large radius of K^+ ions can induce material deformation during intercalation, resulting in rapid performance degradation [20, 21].

These challenges significantly limit the application of APIBs. Key strategies to improve their competitiveness among emerging electrochemical energy storage technologies include expanding the electrolyte's electrochemical window, selecting high-performance electrode materials compatible with the electrolyte, and optimizing the overall cell design.

3.2 The Composition and Working Principle of Potassium-Ion Batteries

PIBs operate on a principle similar to that of LIBs, utilizing a "rocking-chair" mechanism that involves the intercalation and deintercalation of K^+ ions within the cathode and anode materials, as illustrated in Figure 3.1 [17]. The construction of PIBs includes a cathode, an anode, an electrolyte, and a separator. The cathode and anode are made from electrode materials and their respective current collectors.

These components collectively facilitate the reversible insertion and extraction of potassium ions. Common cathode materials consist of potassium-containing compounds with higher redox potentials, while anodes typically employ carbon-based materials, such as graphite. The electrolyte is a liquid mixture of potassium salts dissolved in nonaqueous organic solvents, often enhanced with functional additives to improve performance and stability [2, 22].

In potassium-ion half-cells utilizing an anode material, the anode is composed of the prepared active material, while a potassium foil serves as the counter and reference electrode, effectively replacing the cathode. In contrast, for potassium-ion half-cells that employ a cathode material, the cathode consists of the active material, with the counter and reference electrodes still made from a potassium metal block. It is important to note that in a complete potassium-ion cell, both the cathode and anode materials are distinct active materials, and potassium metal foil is not used. This distinction is crucial for understanding the unique characteristics of PIBs in comparison to SIBs and LIBs, particularly regarding their electrochemical performance and potential applications [12, 23].

The cathode typically exists in one of two forms. The first form is a self-supporting electrode that does not require a current collector, such as materials loaded onto carbon cloth, nickel foam, or membrane materials. These materials generally exhibit good flexibility and can reduce costs by eliminating the need for a current collector. The second form consists of an active material, binder, conductive agent, and current collector (either copper foil or aluminum foil). In this configuration, the active material is crucial to the electrochemical reaction, serving as the central element for energy storage. Current research on cathode materials encompasses several major categories, including Prussian blue analogs (PBAs), layered metal oxides, polyanionic compounds, and organic materials [24–26].

The anode, similar to the cathode, is available in two comparable configurations. However, the active material in the anode consists of the prepared anode material, which can be classified into three distinct categories based on their energy storage mechanisms: intercalation, conversion, and alloying types [5, 27, 28].

The separator plays a crucial role in maintaining the integrity of the battery by separating the cathode and anode, thus preventing potential short circuits. The most commonly used separators in polymer electrolyte batteries are glass fiber separators, which are well-known for their effectiveness in enhancing ionic conduction [29].

Binders play a critical role in electrode construction by ensuring active materials' cohesion and enhancing electrical conductivity. The range of binders includes polyvinylidene fluoride (PVDF), polyvinyl fluoride (PVF), polytetrafluoroethylene (PTFE), styrene-butadiene rubber (SBR), sodium carboxymethyl cellulose (CMC), polyvinyl alcohol (PVA), and fluorinated rubbers. In the context of PIBs, PVDF is the preferred binder for the cathode due to its strong adhesion and chemical stability. In contrast, CMC is commonly used for the anode, taking advantage of its compatibility with carbonaceous materials and its ability to create a stable suspension [30, 31].

Conductive agents are essential for enhancing the electrical conductivity of electrode materials in batteries. The range of conductive agents includes acetylene black, conductive carbon black, graphene, carbon nanotubes, and carbon fibers. Among these, acetylene black and conductive carbon black are commonly used in battery formulations due to their outstanding electrical conductivity and ability to create a percolation network within the electrode matrix [13, 32].

Current collectors are essential for establishing effective electrical contact with electrode materials. Copper and aluminum foils are commonly used as current collectors. Aluminum foil is the preferred choice for the cathode due to its lightweight properties and cost-effectiveness. Both aluminum and copper foils are viable options for the anode, although aluminum foil offers distinct advantages. Notably, the lack of side reactions between K^+ ions and aluminum foil at lower voltages enhances the battery's performance and safety profile [33, 34].

Electrolytes play a pivotal role in facilitating charge transfer between the anode and cathode, significantly impacting the performance of the device. Their influence on specific capacity, rate capability, cycle life, and safety is substantial. K^+ ions, with their larger size and lower Lewis's acidity, present unique challenges that distinguish them from other electrolyte components. To enhance our understanding and optimize battery performance, exploring these differences and their implications for overall system design is essential. A comprehensive examination of potassium-based electrolytes will not only contribute to fundamental knowledge but also pave the way for developing advanced, efficient, and safer energy storage solutions [35, 36].

PIBs operate through a series of reversible chemical reactions at their electrodes. Let us focus on the intercalation reaction, which is a crucial process in these batteries. The cathode, composed of a potassium-containing layered transition metal oxide (with the transition metal being manganese, cobalt, nickel, iron, vanadium, etc.), and the graphite anode undergo the following reactions:

Cathode: $KTMO_2 \leftrightarrow K_{1\text{-}x}TMO_2 + xK^+ + xe^-$
Anode: $xK^+ + 8C + xe^- \leftrightarrow K_xC_8$
Overall cell: $KTMO_2 + 8C \leftrightarrow K_{1\text{-}x}TMO_2 + K_xC_8$

During charging, K^+ ions move from the cathode material to the anode, passing through the electrolyte. This movement of ions is accompanied by the flow of electrons from the cathode to the anode through an external circuit. When the battery is discharged, the K^+ ions that have been stored in the graphite anode are released and returned to the cathode. Simultaneously, electrons flow back to the cathode through the external circuit, completing the charging and discharging cycles.

This process is essential for the operation of PIBs, ensuring that they can store and release energy efficiently and reliably. The flow of electrons during charging and discharging powers devices, making these batteries a promising option for energy storage.

3.3 Cathode Materials for Potassium-Ion Batteries

In the PIB systems, the selection of cathode materials critically determines the maximum discharge voltage that can be achieved. Unlike anode materials, cathode materials generally provide lower capacity, which limits the potential improvement of energy density in PIBs. A significant challenge in this endeavor stems from the considerably larger ionic radius of K^+ ions compared to Li^+ and Na^+ ions [13]. This difference necessitates the search for cathode materials that can accommodate larger ions, posing a significant challenge in battery design.

To date, several materials have been reported as potential cathode materials for PIBs, as illustrated in Figure 3.2. These include PBAs [26, 37, 38], layered metal oxides [39, 40], polyanionic compounds [41, 42], and organic compounds [43]. Radar plots depicting the electrochemical properties of the four cathode materials used for PIBs are presented in Figure 3.3 [44]. Among these, layered transition metal oxides are regarded as the most promising cathode materials for LIBs, leading to a growing interest in the application of potassium-containing layered oxides for PIBs. However, materials such as K_xCoO_2 typically demonstrate lower reversible capacities and operating voltages in PIBs compared to their performance in LIBs and SIBs [25]. Insights gained from LIBs and SIBs indicate that optimizing multiphase transition metal systems and doping with inert metals can effectively mitigate these structural challenges. These strategies show promise for the development of transition metal oxide electrode materials with enhanced capacities, increased operating voltages, and extended cycle life [8]. PBAs exhibit improved capacity and operating voltage due to their three-dimensional (3D) open framework, which features numerous channels and voids [45]. These structural characteristics facilitate the efficient diffusion and insertion of K^+ ions, thereby enhancing the overall performance of PBAs in energy storage applications [46]. Polyanionic compounds exhibit a 3D open framework structure. The inductive effect of the polyanionic groups within this crystal structure enables the achievement of higher operating voltages and relatively high energy densities [47, 48]. Simultaneously, the 3D framework facilitates the rapid diffusion of K^+ ions, thereby promoting high power density. Furthermore, this robust framework ensures long-term structural stability. Organic electrode materials can be tailored in both structure and performance through molecular design, allowing for precise control over their redox potential and achieving theoretical specific capacities sufficient to meet the demands for high energy density. Additionally, these materials typically consist of amorphous solids held together by van der Waals forces in a solid-phase stack [43].

3.3.1 Layered Metal Oxides

Layered oxide systems have garnered significant attention as potential cathode materials for PIBs, inspired by the success of widely used cathodes in LIBs, such as $LiCoO_2$ and lithium nickel cobalt manganese oxide. Early research contributions by Hoppe et al. [49] focused on synthesizing and analyzing the structures of layered metal oxides. Their studies particularly highlighted the electrochemical properties of $K_xMnO_2{\cdot}nH_2O$ ($x \sim 0.3$, $0.3 \leq n \leq 0.6$), underscoring its suitability for PIBs.

Layered metal oxides are represented by the chemical formula K_xMO_2, where x denotes the concentration of K^+ ions and M signifies transition metals such as Co, Mn, Ni, or other metals like Mg, and Al. Figure 3.4 illustrates their crystal structure [17]. Within this structure, octahedral MO_6 units connect through shared edges

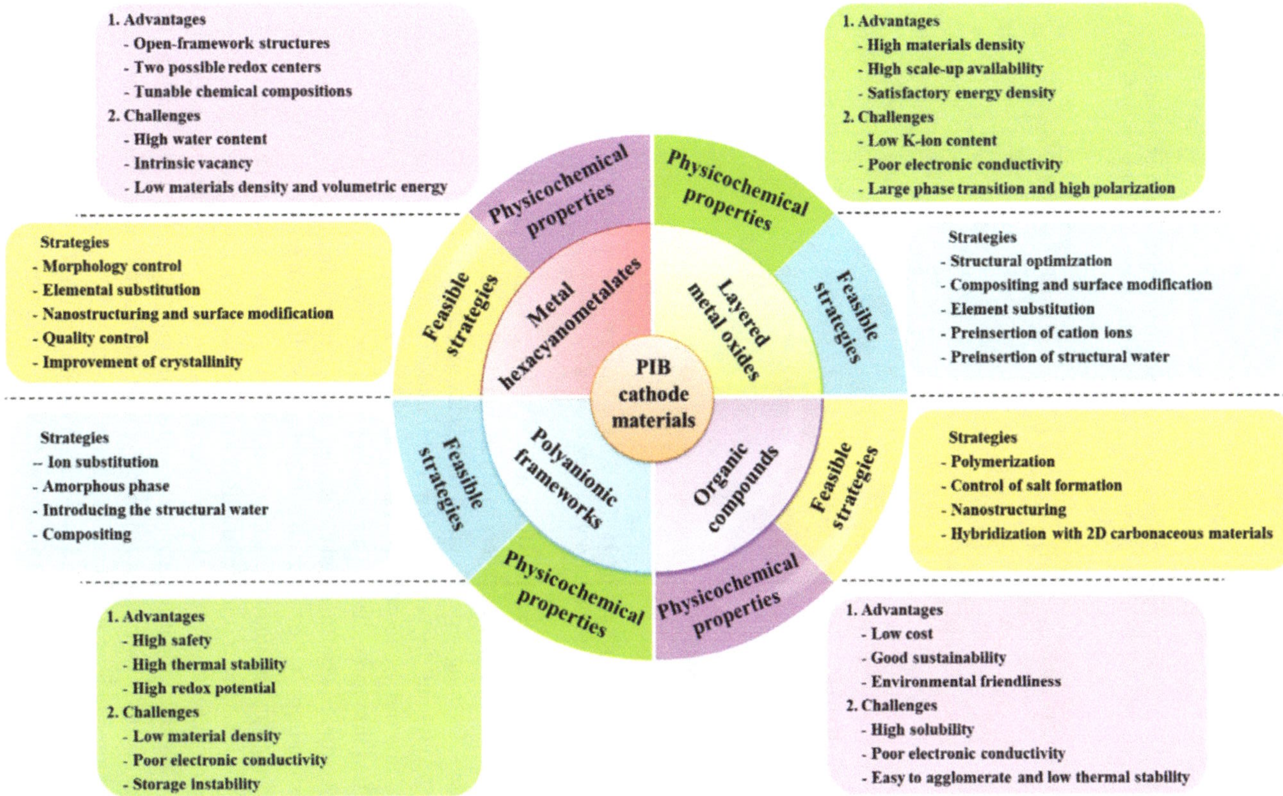

Figure 3.2 Classification of cathode materials for PIBs. Source: [13]. Reproduced with permission/John Wiley & Sons.

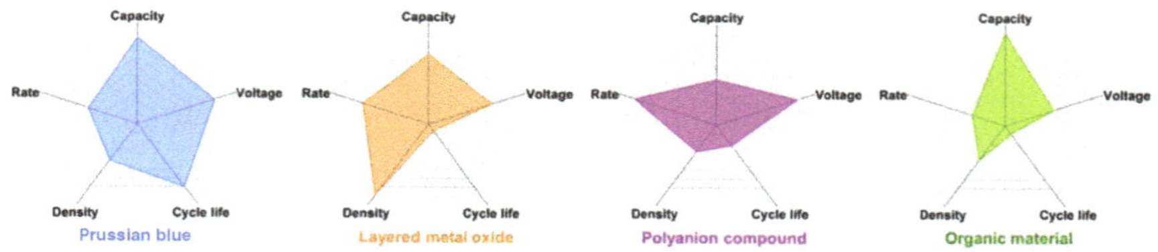

Figure 3.3 Radar charts depicting the electrochemical performance of four cathode materials intended for PIBs. Source: [44]. Reproduced with permission/John Wiley & Sons.

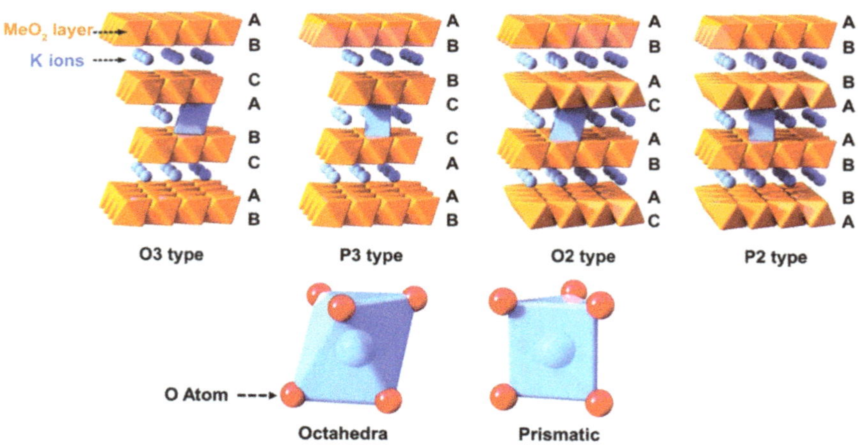

Figure 3.4 Crystal structures of O3-, P3-, O2, and P2-type-layered metal oxides. Source: [17]. Reproduced with permission/John Wiley & Sons.

to form layers composed of transition metals. Interspersed among these layers are K^+ ions, which create separate layers of alkali metals. The K^+ ions interspersed among these layers, forming separate alkali metals layers, and the layers alternate systematically throughout the composition of the material. Based on the arrangement of oxygen atoms and the coordination environment of K^+ ions, layered metal oxides can be classified into three distinct structures [50]. The layered oxide cathode materials that can be directly synthesized experimentally primarily include O3, P3, and P2. In this classification, "O" denotes an octahedral coordination environment for K^+ ion, while "P" signifies a trigonal prismatic coordination environment. The numbers "2" and "3" indicate the minimum number of layers forming the periodic sequence of oxygen atom stacking. When the unit cell is distorted, it is represented as O′3, P′3, and P′2. This categorization provides a systematic framework for understanding the structural diversity within layered metal oxides, facilitating insights into their properties and applications across various fields [51–53].

In O3-type-layered metal oxides, which crystallize in the $R\bar{3}m$ rhombohedral space group, MO_2 planes are stacked along the c-axis in a manner resembling the ABCABC arrangements found in cubic close-packed complexes. Each hexagonal cell in this structure comprises three MO_2 planes. The designation "O3" indicates that the octahedral sites within the structure accommodate K^+ ions, while the "3" signifies the number of MO_2 planes present within each hexagonal cell [54]. The P3-type-layered metal oxides exhibit a laminar structure in which alkali metal ions occupy positions between the MO_2 planes. Oxygen atoms are arranged along the c-axis in an ordered ABBCCA configuration. This structure comprises three MO_2 planes, typically organized within a hexagonal cell that exhibits symmetry described by the *R3m* space group [55]. In contrast, P2-type layered metal oxides have K^+ ions located in prism-like spaces between the MO_2 planes, with oxygen atoms arranged along the c-axis in an ABBAAB sequence. This hexagonal cell structure contains two MO_2 planes and is characterized by the *P63/MMC* space group [56, 57].

3.3.1.1 Challenges and Strategies

Layered metal oxides demonstrate a relatively high specific capacity; however, they face significant challenges. These challenges include susceptibility to phase transitions and structural fragmentation during reactions. A critical issue arises from the large radius of K^+ ions, which generates strong repulsive forces within the K^+ ions layer. Addressing these challenges requires optimization and regulation strategies, including structural design through methods such as doping, compositing, surface modification, and nanostructure adjustment. Additionally, controlling the phase transition of potassium by regulating the voltage range and managing the K content is essential for enhancing the performance of these materials.

3.3.1.2 Chromium-Based Metal Oxide

The O3 phase is the most promising type of layered cathode material structure type in LIBs, characterized by a high alkali metal content (generally $\geq$0.8) in its chemical composition. The initial number of carriers directly determines its theoretical specific capacity. However, these ions tend to occupy more spatially abundant prismatic sites than octahedral sites due to the large radius of K^+ ions. Additionally, the electrostatic repulsion between K^+ ions in the alkali metal layer complicates the incorporation of a significant number of K^+ ions, making it challenging to fabricate stable layered materials with high potassium content in the O3 phase [58, 59].

Kim et al. [40] investigated the thermodynamic stability trends of potassium-filled layered materials KMO_2 (where M = Sc, Ti, V, Cr, Mn, Fe, Co, and Ni) using electronic structure theory. They identified transition metal frameworks capable of accommodating an equal number of K^+ ions as the metal cations. Their results indicated that among the electrochemically active elements, only Cr^{3+} demonstrates ultra-high octahedral coordination field stability. The strong preference of Cr^{3+} for octahedral coordination compensates for the energy loss caused by the strong interactions between closely spaced K^+ ions. Guided by theoretical analysis, the authors successfully synthesized the first O3-type potassium-layered cathode material ($KCrO_2$), which exhibits a discharge capacity of 90 mAh g^{-1} and an average redox potential of 2.73 V; however, it retains only 67% of its capacity after 100 cycles.

O3-K_xCrO_2 undergoes multiple structural transformations during the charge and discharge processes. During electrochemical potassium extraction, the interlayer distance increases, and the crystal structure of O3-$KCrO_2$ sequentially transitions from O3 to O′3 toP′3 to P3 to P′3 within the range of $1 > x > 0.4$ in K_xCrO_2. This process is more complex than the structural changes observed in O3- Na_xCrO_2, which transitions from O3 to O′3 to P′3 within the range of $1 > x > 0.5$. This indicates complex structural changes can be attributed to the strong K^+–K^+ interactions. During electrochemical potassium intercalation in the discharge phase, the potassium-deficient O3-type K_xCrO_2 reversibly transitions sequentially to O3–P3–P′3–P3–P′3–O′3. However, stoichiometric O3-$KCrO_2$ was not obtained even when discharged to 1.5 V relative to K/K^+, likely due to the slow kinetics of the O′3–O3 transition in K_xCrO_2 at $x \approx 1$. The extremely complex potassium storage mechanism led to a series of phase transitions (O3–O′3–P′3–P3–P′3–O3), resulting in a capacity retention rate of only 67% after 100 cycles. Consequently, research on potassium O3 phase cathode materials has stagnated, indicating that the development of high-potassium and high-performance layered oxides continues to face significant challenges.

3.3.1.3 Manganese-Based Metal Oxides

As a cathode material, manganese-based oxides not only facilitate the formation of stable P2 or P3-type structures but also offer advantages such as excellent electrochemical performance and low cost [17, 24]. Passerini et al. were the first to investigate P2-$K_{0.3}MnO_2$ as a cathode material for PIBs [60]. They reported a discharge capacity of 65 mAh g^{-1} within a voltage range of 1.5 to 3.5 V, with a capacity retention rate of 57% after 685 cycles. By extending the cut-off voltage to 4.0 V, the capacity increased to 136 mAh g^{-1}; however, cycling stability was compromised, resulting in only 58% capacity retention after just 50 cycles. This instability is attributed to the close relationship between the reversibility of the deintercalation process and the upper cut-off voltage. During phase transitions, the sliding of adjacent oxygen layers induces irreversible volume changes in the material's lattice structure, leading to rapid capacity decay. Therefore, an effective strategy to mitigate this phase transition is to reduce the upper cut-off voltage. Since then, an increasing amount of research has focused on manganese-based oxides, establishing them as the most extensively studied layered oxide systems [61].

Compared to P2-$K_{0.3}MnO_2$, the P3-$K_{0.5}MnO_2$ material exhibits a reversible capacity of approximately 100 mAh g^{-1} within the voltage range of 1.5 to 3.9 V. After 50 cycles at a current density of 20 mA g^{-1}, $K_{0.5}MnO_2$ retains a reversible specific capacity of about 70 mAh g^{-1}, resulting in a cycle retention rate of 70%. *In situ* X-ray diffraction (XRD) analysis was employed to investigate the potassium storage mechanism of $K_{0.5}MnO_2$ (Figure 3.5a) [62]. During K^+ ions extraction, the (015) peak of K_xMnO_2 disappears, while a new (104) peak emerges at $x \approx 0.41$, indicating a phase transition from P3 to O3. As K^+ ions is further extracted to $x \approx 0.34$, new XRD diffraction peaks appear to the left of the (003) and (006) peaks, suggesting an additional phase transition to an unknown phase, referred to as $K_{0.5}MnO_2$ undergoes a P3-O3-X phase transition process during electrochemical cycling. These multistep phase transitions impede the material's practical application.

Additionally, manganese-based layered oxides face a significant challenge: the unavoidable Jahn-Teller effect, which arises from the odd number of electrons occupying the e_g orbital (Mn^{3+}: $t_{2g}{}^3e_g{}^1$). This effect leads to asymmetric elongation

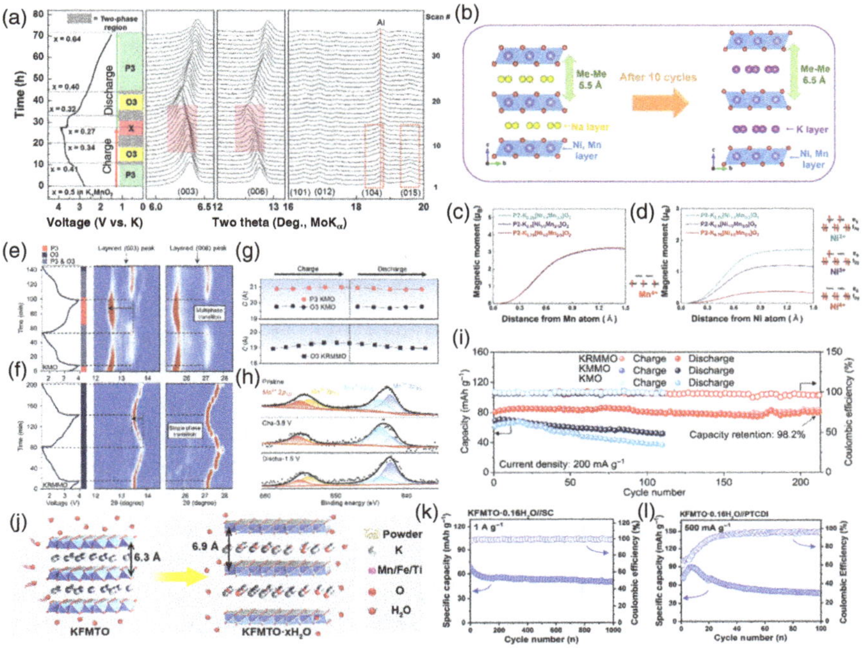

Figure 3.5 (a) *In situ* XRD pattern of P3-type $K_{0.5}MnO_2$ at a current rate of $2\,mA\,g^{-1}$. Source: [62]. Reproduced with permission/John Wiley & Sons. (b) Crystal structure models of P2-$Na_{2/3}[Ni_{1/3}Mn_{2/3}]O_2$ and P2-$K_{0.75}[Ni_{1/3}Mn_{2/3}]O_2$; spin integrations of (c) Mn and (d) Ni ions in P2-$K_{0.75}[Ni_{1/3}Mn_{2/3}]O_2$. Source: [63]. Reproduced with permission/John Wiley & Sons. *In situ* XRD patterns of (e) KMO and (f) KRMMO at $70\,mA\,g^{-1}$ in the range of 1.5–3.9 V; (g) the *c*-lattice parameter of KMO and KRMMO calculated from *in situ* XRD patterns; (h) XPS spectra of Mn 2p at various states; (i) cycling performance of KRMMO at $200\,mA\,g^{-1}$. Source: [64]. Reproduced with permission/John Wiley & Sons. (j) Schematics of KFMTO and KFMTO·xH_2O; (k) Cycling performance of the KFMTO·$0.16H_2O$//soft carbon full cell; (l) cycling stability of KFMTO·$0.16H_2O$//PTCDI full cell. Source: [65]. Reproduced with permission/John Wiley & Sons.

of the Mn—O bond lengths within the MnO_6 octahedra, significantly impacting the cycling stability of these materials [66]. The Jahn-Teller effect refers to the phenomenon in which a nonlinear molecular system in an electronically degenerate state undergoes distortion, thereby removing the degeneracy, reducing symmetry, and ultimately lowering the system's total energy. This effect is particularly pronounced in tetrahedral and octahedral structures, especially those with electronic configurations that include *d*-orbitals [63]. In octahedral geometry, the Jahn-Teller effect results in tetragonal elongation or compression, causing the displacement of ligands along the *z*-axis. This distortion preserves the inversion center while maintaining the equidistant arrangement of the two ligands along the z-axis from the center. The Jahn-Teller effect is especially significant in metal ions coordinated in an octahedral manner, particularly when they possess (high-spin) d^4, (low-spin) d^7, and d^9 electronic configurations, where one of the e_g orbitals contains an unpaired electron. For example, among the 3*d* transition metal ions, the Jahn-Teller effect is most pronounced for high-spin Mn^{3+}, high-spin Fe^{4+}, and Cr^{2+} in octahedral coordination. In contrast, high-spin Mn^{4+}, high-spin Fe^{3+}, and Cr^{3+} in octahedral coordination do not exhibit the Jahn-Teller effect [67].

To address these issues, Chen's research team reported a strategy for adjusting interfacial structures [66]. Specifically, during the charging process, dual interphase layers form *in situ* on the surface of P2-$K_{0.67}MnO_2$ in an ether-based electrolyte (6.0 M KFSI/G2). This layer comprises a solid-electrolyte interphase (SEI) film with high K^+ ion conductivity and a K-deficient spinel intermediate layer.

During charging, K^+ ions are extracted, creating vacancies that are partially occupied by Mn atoms due to the Jahn-Teller distortion and the disproportionation of Mn^{3+}. Mn^{2+} migrates to the 8a sites, coordinating with four oxygen atoms to form MnO_4 tetrahedra. This process is accompanied by the sliding of MnO_2 layers and the rearrangement of K^+ ions and vacancies. Mn^{3+} moves to the 16d sites, coordinating with six oxygen atoms to form MnO_6 octahedra. Concurrently, the sliding of MnO_2 layers results in an ABCABC oxygen sequence, with oxygen atoms located at the 32e sites. This arrangement is influenced by the strong electronegativity of aggregated FSI^- and the adsorption of G2. This unique double interfacial layer not only suppresses Jahn-Teller distortions and mitigates Mn dissolution but also facilitates K^+ diffusion associated with the reversible phase transition. Consequently, when used as a cathode material for PIBs, P2-$K_{0.67}MnO_2$ exhibits an impressive capacity retention rate of up to 90.5% after 300 cycles at 50 mA g^{-1}.

Similarly, doping metal ions into the transition metal layer can effectively inhibit Jahn-Teller deformation and enhance cycling stability. For example, the P3-$K_{0.54}[Co_{0.5}Mn_{0.5}]O_2$ cathode material, which has an oxidation state of +3.5 for Mn, effectively mitigates the Jahn-Teller effect and prevents manganese dissolution. This material exhibits a discharge capacity of 78 mAh g^{-1} at a current density of 500 mA g^{-1} [68]. Hyeon Jo et al. designed and synthesized a $K_{0.75}[Ni_{1/3}Mn_{2/3}]O_2$ cathode material using an ion exchange method (Figure 3.5b) [63]. The potassium storage mechanism of this material involves a single-phase reaction. First-principles calculations indicate that during K^+ intercalation/deintercalation, only the Ni^{4+}/Ni^{2+} redox reaction occurs, while Mn^{4+} remains inactive (Figure 3.5c and d). Remarkably, after 500 cycles at a high current density of 1400 mA g^{-1}, the capacity retention rate reached 83%.

Ti^{4+} and Mn^{4+} share the same valence state and have similar ionic radii (0.605 Å for Ti^{4+} vs. 0.53 Å for Mn^{4+}). However, Ti^{4+} remains electrochemically inactive, which contributes to stability. Introducing Ti^{4+} into the P2-type K_xMnO_2 framework can effectively reduce the sliding of TMO_2 layers during charge and discharge processes. P2-$K_{5/9}Mn_{7/9}Ti_{2/9}O_2$ doped with Ti^{4+} exhibits a highly reversible P2-OP_4 phase transition during K^+ ion intercalation and deintercalation, significantly enhancing its capacity to 68 mAh g^{-1} after 100 cycles. The incorporation of Ti^{4+} at Mn^{3+} sites suppress the Jahn-Teller effect. First-principles calculations based on density functional theory also confirm that the formation energy of the OP_4 phase is lower when $K_xMn_{7/9}Ti_{2/9}O_2$ is in a low potassium content state. This indicates that the OP_4 phase is the thermodynamically stable phase under high-voltage, potassium-deficient conditions [69].

The concentration of introduced metal ions significantly influences the electrochemical performance of manganese-based materials. For instance, the introduction of Fe^{3+} shows that when the Fe content in the molecular formula exceeds 0.2, the electrode demonstrates optimal cycling stability within the voltage range of 1.5–4.0 V. The P3-$K_{0.45}Mn_{0.8}Fe_{0.2}O_2$ cathode material exhibits a reversible capacity of 82 mAh g^{-1} and retains 83% of its capacity after 100 cycles at 20 mA g^{-1} [70].

In addition to binary transition metal oxides, Xiao et al. designed and synthesized a ternary manganese layered oxide, $K_{0.5}Mn_{0.7}Co_{0.2}Fe_{0.1}O_2$, which was controllably doped with Co and Fe to suppress Jahn-Teller distortion [9]. Wang et al. synthesized a novel P3-type ternary layered oxide, $K_{0.67}Ni_{0.17}Co_{0.17}Mn_{0.66}O_2$, using coprecipitation and solid-phase synthesis methods [71]. This material exhibits a discharge capacity of 76.5 mAh g^{-1} at an average operating voltage of 3.1 V and retains 87% of its capacity after 100 cycles.

Lu et al. further compared the potassium storage performance of a ternary metal oxide (P3-type $K_{0.45}Rb_{0.05}Mn_{0.85}Mg_{0.15}O_2$, KRMMO) with that of a single-metal oxide (P3-$K_{0.5}MnO_2$) and explored the underlying reasons for the observed differences [64]. As shown in Figure 3.5e and f, *in situ* XRD analysis was conducted to investigate the structural evolution of KRMMO and P3-$K_{0.5}MnO_2$ during the potassium storage process. During charging, the intensities of the characteristic peaks (003) and (006) weakened, and their angles shifted. By the end of the charging process, P3-$K_{0.5}MnO_2$ transformed into an O3-type structure. Such a phase transformation can significantly reduce the K^+ content in the layered structure and weaken the electrostatic shielding between adjacent transition metal (TM) layers, resulting in the sliding of TM layers. During subsequent discharging, the phase reverted from O3 back to P3. Furthermore, the P3 and O3 structures continuously interconvert during repeated cycling. Unlike multiphase transition materials, KRMMO maintains a single-phase structure throughout cycling. The authors propose that P2-type KRMMO undergoes structural self-evolution, transforming into the O3-type structure before cycling. *In situ* XRD patterns of KRMMO exhibit consistent diffraction peaks during the charge/discharge cycles of the O3-type, confirming that KRMMO functions as a phase-transition-free cathode material for PIBs (Figure 3.5g). Figure 3.5h demonstrates the valence changes of Mn during cycling, as confirmed by ex situ X-ray photoelectron spectroscopy (XPS) analysis. This observation further supports the excellent reversibility of the material, highlighting its stable electrochemical behavior throughout the charge–discharge process. As shown in Figure 3.5i, it is evident that the KRMMO exhibits excellent specific capacity and cycling stability. This stability is attributed to the effects of Rb and Mg, which enhance the structural integrity of KRMMO. The large atomic radius of Rb cations (1.52 Å) expands the interlayer spacing, facilitating rapid K^+ ions diffusion and suppressing phase transitions. Meanwhile, the small Mg cations (0.72 Å) at TM sites reduce the Jahn-Teller distortion of Mn^{3+} ions and minimize disproportionation reactions. Compared to materials with multiple phase transitions, a phase-transition-free structure provides a more straightforward ion transport pathway, significantly improving transport efficiency.

In addition to the aforementioned applications, layered manganese-based oxides have also been utilized in APIBs. Zeng et al. developed a high-entropy-doped layered manganese-based oxide (HE-KMO) for APIBs by incorporating iron (Fe), cobalt (Co), nickel (Ni), zinc (Zn), and magnesium (Mg) into the manganese oxide framework [72]. This doping introduces local disorder due to the uneven distribution of electron density among the various elements, which facilitates potassium-ion intercalation and mitigates stress changes in the material during cycling. As a result, HE-KMO demonstrates exceptional performance in APIBs, retaining nearly 100% of its capacity after 500 cycles, in contrast to only 48.6% for standard KMO materials. These findings underscore the effectiveness of high-entropy doping in significantly enhancing material stability, presenting a promising strategy for the development of high-stability materials for PIBs.

Zhang et al. employed a straightforward hydration strategy to incorporate H_2O molecules into the layers of P3-$K_{0.4}Fe_{0.1}Mn_{0.8}Ti_{0.1}O_2$ (KFMTO), resulting in a structural transformation from hexagonal to monoclinic symmetry (KFMTO·$0.16H_2O$), as shown in Figure 3.5j [65]. The resulting hydrated material offers several advantages: (i) excellent air stability (maintaining its original structure even after 60 days of exposure to air); (ii) a large interlayer spacing of 6.9 Å, which facilitates K^+ ion intercalation; and (iii) the presence of crystalline water, which mitigates electrostatic repulsion between K^+ ions. Due to its high stability in air and water, the hydrated layered material is compatible with both organic and aqueous electrolytes. As shown in Figure 3.5k, in an organic electrolyte (0.8 m KPF_6 in EC/DEC [1 : 1 by volume]), using KFMTO·$0.16H_2O$ as the cathode and soft carbon as the anode with a voltage window of 1.8–4.0 V, the full cell retains 75% of its capacity after 1000 cycles at 1 A g^{-1}. In aqueous K^+ ions full cell (Figure 3.5l), utilizing 0.5 M K_2SO_4 + 0.05 M $KHCO_3$ as the electrolyte and PTCDI as the anode, it demonstrates a capacity of 71 mAh g^{-1} after 100 cycles at 500 mA g^{-1} within a voltage range of 0–1.9 V. These results indicate that KFMTO·$0.16H_2O$ can achieve safe and reliable energy storage with high compatibility in both aqueous and organic systems.

3.3.1.4 Cobalt-Based Metal Oxide

Cobalt-based layered oxide cathode materials have been widely commercialized for LIBs due to their exceptional rate performance and stable discharge voltage. They have also made significant progress in PIBs. The crystal structure of K_xCoO_2 varies with different potassium content. When x is 0.5, the layered K_xCoO_2 adopts the P3 type ($R\bar{3}m$); when x is 0.67, the layered K_xCoO_2 is of the P2 type ($P63/mmc$).

Kim et al. first reported the use of P2-$K_{0.6}CoO_2$ for PIBs. This electrode material exhibited a capacity of 80 mAh g^{-1} at a current density of 2 mA g^{-1} across a wide voltage range of 1.7 to 4.0 V (Figure 3.6a) [73]. However, the cycling performance was unsatisfactory, with a capacity retention rate of only 60% after 120 cycles. Similar to manganese-based layered oxides, $K_{0.6}CoO_2$ displayed numerous redox peaks in the cyclic voltammetry (CV) curves and multiple voltage plateaus in the charge–discharge curves (Figure 3.6b). These features were attributed to the K^+/vacancy ordering structure. To gain insight into the potassium storage mechanism in P2-type K_xCoO_2, *in situ* XRD was conducted during electrochemical cycling.

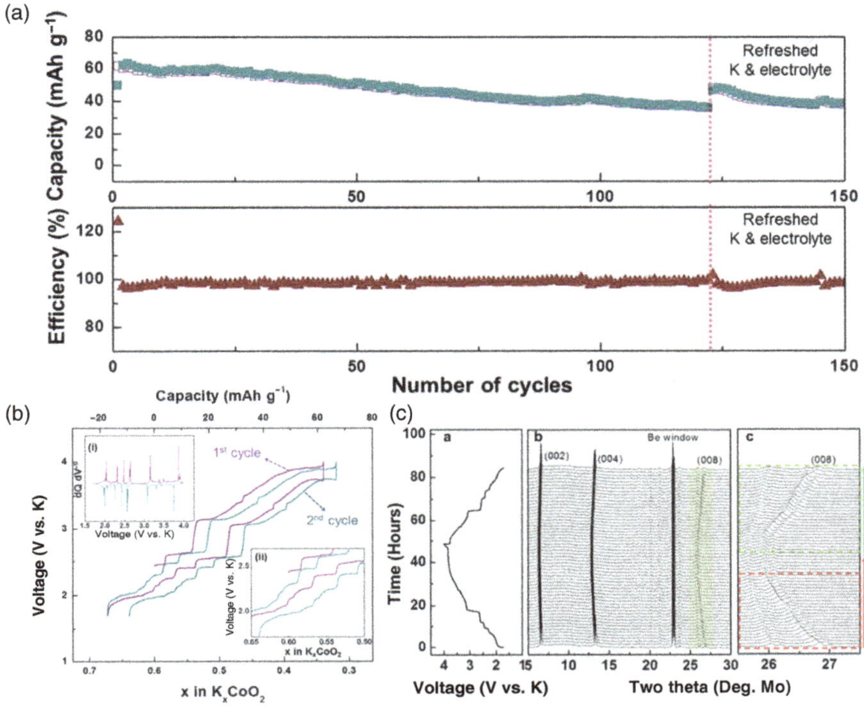

Figure 3.6 (a) Cycle stability and coulombic efficiency of P2-type P2-$K_{0.6}CoO_2$ at a current rate of 100 mA g^{-1} in 1.7–4.0 V; (b) CV curves and charge–discharge curves of P2-$K_{0.6}CoO_2$; (c) *in situ* XRD patterns of P2-$K_{0.6}CoO_2$. Source: [73]. Reproduced with permission/John Wiley & Sons.

During charging, the (008) peak shifts to a lower angle while retaining the P2-type structure, as illustrated in Figure 3.6c. This shift suggests an increase in interlayer spacing due to enhanced repulsion between oxygen atoms when alkali ions are de-intercalated. Additionally, varying degrees of peak shifts and asymmetric peak evolution were observed, further indicating multiple phase transitions in K_xCoO_2 during potassium-ion extraction. These phase transitions are likely influenced by K^+/vacancy ordering.

Hironaka et al. reported on the potassium storage performance of P3-type $K_{2/3}CoO_2$ and P2-type $K_{0.41}CoO_2$. Both materials exhibited similar capacity and voltage curves [74]. They demonstrated a reversible capacity of approximately 60 mAh g^{-1} within a voltage range of 2.0–3.9 V. A further analysis of P2-$K_{0.41}CoO_2$ revealed five distinct voltage plateaus during the passivation process at 3.10 V, 2.59 V, 2.55 V, 2.42 V, and 2.33 V. During the appearance of these plateaus, the (002), (004), and (006) peaks showed varying degrees of shift and asymmetric evolution, characteristic of a two-phase reaction. Six single-phase regions were observed. Despite the voltage drop, the diffraction lines continued to correspond to the P63/mmc space group, indicating the maintenance of the original P2 structure. Interestingly, the P3-type $K_{2/3}CoO_2$ displayed a nearly identical phase transition process as the

P2-type material. This complex phase transition was attributed to the rearrangement of the K^+/vacancy ordered structure rather than the sliding of CoO_2 layers. Professor Wang's research team synthesized P2-$K_{0.6}CoO_2$ microspheres using a two-step self-templating method [57]. This unique structure facilitates the insertion and extraction of K^+ ions while minimizing side reactions, thereby enhancing the energy density of the battery. P2-$K_{0.6}CoO_2$ microspheres exhibit a high reversible capacity of 82 mAh g^{-1} at 10 mA g^{-1}, a high-rate performance of 65 mAh g^{-1} at 100 mA g^{-1}, and a long cycle life with 87% capacity retention after 300 cycles.

Although cobalt-based layered oxide materials exhibit excellent electrochemical performance, the limited availability of cobalt resources leads to high costs. Therefore, it is essential to explore new materials to mitigate the pressure on cobalt resources. In some cases, cobalt can be introduced as a doping element into layered oxides to improve electrochemical performance, as discussed in Section 2.1.2 regarding the application of Co as a doping agent.

Summary and Prospects This section reviews layered metal oxide cathode materials for PIBs, particularly cobalt-based, manganese-based, and chromium-based oxides, along with their electrochemical performance and phase transitions during potassium storage. Layered metal oxides are promising cathode candidates for PIBs due to their high theoretical capacity and low cost. However, they face challenges similar to those in SIBs. Their relatively compact lattice structure cannot accommodate large K^+ ions, making it difficult to restore the layered oxide structure during charge and discharge cycles. This limitation can lead to a low insertion potential, rapid capacity decay, and complex phase transitions. Furthermore, changes in material morphology and structure can significantly affect electrochemical performance. To address these challenges, various strategies have been proposed, including metal ion substitution, structural design, morphology control, reversible oxygen redox processes, and the use of composite carbon materials. Nevertheless, research on the potassium storage mechanisms of layered oxides is still in its early stages. Future studies should focus on gaining a deeper understanding of these mechanisms and developing targeted strategies to enhance their potassium storage performance.

3.3.2 Polyanionic Compounds

Polyanionic compounds exhibit a 3D open framework structure that effectively mitigates the repulsive forces between K^+ ions, thereby enhancing their capacity to accommodate K^+ ions. These compounds are represented by the chemical formula $K_xM_y(XO_4)_n$, where M denotes transition metals such as Fe, Mn, and Co, while X represents nonmetallic elements like sulfur (S), phosphorus (P), silicon (Si), arsenic (As), and others. The strong covalent bonds between the X and O atoms result in relatively isolated valence electrons within the polyanionic structure. This characteristic imparts unique properties to the transition metals, including elevated redox potentials, distinct crystal structures, and stable chemical profiles. Furthermore, polyanionic compounds exhibit notable attributes such as high thermal stability and safety [48, 75]. Polyanionic compounds encompass a variety of substances, among which phosphates, pyrophosphates, sulfates, fluorophosphates, fluorosulfates, and oxyphosphates are the most prominent. Within this group, phosphates and fluorophosphates have garnered the most extensive research attention.

3.3.2.1 Challenges and Strategies

Polyanionic compounds offer high redox potentials; however, they pose a significant challenge: the electrolyte decomposes at elevated voltages, which considerably diminishes battery performance. Additionally, these compounds have several limitations, including low volumetric energy density, restricted specific discharge capacity, and poor electronic conductivity. Their framework can become unstable due to complex multielectron reactions. Moreover, their crystal structure is unfavorable; for instance, the large size of K^+ ions hinders effective incorporation within the octahedral $FePO_4$ structure, resulting in structural phase transitions [76, 77].

To address the aforementioned challenges, several strategies can be employed. First, investigating electrolytes that can withstand high voltages and identifying suitable additives are crucial steps. Second, modifying and enhancing electrode surfaces is essential for optimizing performance. Third, it is necessary to adapt polyanionic compounds originally developed for LIBs and SIBs to suit PIBs. Additionally, employing *in situ* methods to thoroughly study potassium storage mechanisms and phase transitions during the reaction process is vital. Finally, restructuring molecular configurations through ion substitution and introducing water of crystallization can significantly enhance performance and stability. These targeted efforts will propel the advancement of polyanionic compounds, facilitating their broader application in energy storage technologies.

3.3.2.2 Phosphates

$FePO_4$ $FePO_4$ is utilized as a cathode material in rechargeable batteries, primarily existing in two forms: heterosite-$FePO_4$ and amorphous-$FePO_4$. In LIBs, the most prevalent polyanionic compound electrode material is triphylite $LiFePO_4$. $LiFePO_4$ typically exhibits a voltage plateau around 3.45 V, which signifies a two-phase reaction. When lithium is extracted from $LiFePO_4$, it transforms into heterosite-$FePO_4$. In SIBs, $NaFePO_4$ displays two distinct voltage plateaus at 3.0 V and 3.2 V during the insertion and detachment of Na^+ ions. This indicates the presence of an intermediate product, although the final product remains heterosite-$FePO_4$. This behavior suggests that heterosite-$FePO_4$ can reversibly accommodate both Li^+ ions and Na^+ ions [41, 78].

However, heterosite-$FePO_4$ is not suitable for the storage of K^+ ions. As illustrated in Figure 3.7a, $FePO_4$ electrodes exhibit charge/discharge curves with steeper slopes compared to those in LIBs and SIBs, and the average voltage of the K cells is significantly lower at 2.3 V. Additionally, the reversible capacity decreases substantially after multiple cycles, with capacities of 160 mAh g^{-1} in LIBs, 151 mAh g^{-1} in SIBs, and 120 mAh g^{-1} in PIBs. To investigate the underlying reasons for the unsuitability of heterosite-$FePO_4$ in PIBs, Hosaka et al. conducted ex situ XRD analyses comparing $FePO_4$ in SIBs and PIBs [41]. As shown in Figure 3.7b, all peaks of $FePO_4$ shifted to a lower angle after Na^+ insertion and returned to their original positions after Na^+

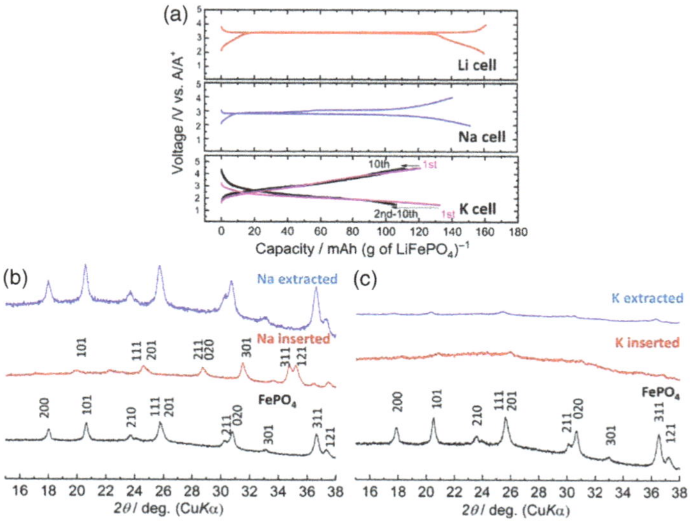

Figure 3.7 (a) Charge–discharge curves of heterosite-$FePO_4$ electrode in coin-type Li, Na, and K-half-cells; (b) comparison of *ex situ* XRD patterns of heterosite-$FePO_4$ electrodes in Na, and (c) K-half-cells. Source: [41]. Reproduced with permission/John Wiley & Sons.

detachment, indicating that heterosite-$FePO_4$ is reversible for Na^+ ion insertion and detachment. In contrast, weaker and broader peaks appeared in the XRD spectra of $FePO_4$ after K^+ ions insertion, and the peak intensities did not recover after K^+ removal. This suggests that the larger size of K^+ ion prevents them from fitting into the interstitial spaces of heterosite-$FePO_4$.

The potassium storage properties of amorphous-$FePO_4$ were investigated by Mathew et al. The team conducted electrochemical tests on amorphous $FePO_4$ in K^+ half-cells [79]. Compared to heterosite-$FePO_4$, amorphous-$FePO_4$ exhibited a higher specific capacity of 160 mAh g^{-1}, but its operating voltage was lower, ranging from 2.1 to 2.4 V. Additionally, *ex situ* XRD analyses of the amorphous $FePO_4$ electrode in K^+ ion half-cell, as presented in Figure 3.7c, revealed the generation of $KFe_2(PO_4)_2$ upon K^+ ion insertion, a reversible process.

The findings indicate that heterosite-$FePO_4$ is unsuitable for accommodating K^+ ions, while amorphous-$FePO_4$ and heterosite-$FePO_4$ with low crystallinity can reversibly accommodate K^+ ions. However, the low operating voltage of $FePO_4$ electrodes limits their potential as cathodes for high-voltage PIBs.

***$KMPO_4$* (Fe and Mn)** In the realm of LIBs, lithium iron phosphate is recognized as the leading polyanionic compound due to its robust covalent structure, strong inductive effects, low diffusion barrier for Li^+ ions, and excellent cycling stability. These characteristics make it highly desirable for use as a cathode material. Similarly, polyanionic compounds are being explored for their potential as cathode materials in PIBs, capitalizing on the advantageous properties observed in LIBs applications [80, 81]. $KFePO_4$, which belongs to the space group $P2_1/n$, and $KMnPO_4$, classified under the space group $P\bar{1}$, are materials derived from $LiFePO_4$. The structure of $KFePO_4$

comprises FeO_4 tetrahedra, FeO_5 polyhedra, and PO_4 tetrahedra, while $KMnPO_4$ consists of MnO_4 tetrahedra and PO_4 tetrahedra. Although the theoretical capacity of both $KFePO_4$ and $KMnPO_4$ can reach up to 141 mAh g^{-1}, their actual capacities are significantly lower [41].

Komaba et al. investigated the electrochemical properties of $KFePO_4$/C and $KMnPO_4$/C and found that the discharge capacities were only 25 mAh g^{-1} and 30 mAh g^{-1}, respectively [41]. Despite the relatively high theoretical capacities of $KFePO_4$ and $KMnPO_4$ (both at 141 mAh g^{-1}), their actual capacities are significantly lower. This discrepancy is likely due to the crystalline state of these materials, which results in poor electrical conductivity and ion diffusion. Consequently, K^+ ion migration through $KFePO_4$ and $KMnPO_4$ is impeded. This indicates that $KFePO_4$ and $KMnPO_4$ are favorable for the storage of K^+ ions.

$K_3M_2(PO_4)_3$ (V and Ti) Compared to $KFePO_4$ and $KMnPO_4$, the $K_3V_2(PO_4)_3$ material exhibits superior electrochemical activity and shows greater potential as an cathode material for PIBs. The $K_3V_2(PO_4)_3$ crystal structure adopts a NASICON-type structure (space group: $R\bar{3}c$) [82], which consists of VO_6 octahedra linked via corner-sharing PO_4 tetrahedra. K^+ ions are transported through the voids within the 3D framework. The substantial size of these voids along the c-axis facilitates rapid K^+ ion transport and enhances the stability of the structure during the K^+ ion insertion and extraction processes. Mai's group was the first to synthesize carbon-coated $K_3V_2(PO_4)_3$ nanowires as cathode materials for ion batteries [17]. It exhibited a specific capacity of 119 mAh g^{-1} within the voltage range of 1.5–4.0 V. However, it did not display a significant voltage plateau, and the average voltage decreased to approximately 2.5 V. Subsequently, Han et al. synthesized porous carbon-coated $K_3V_2(PO_4)_3$ through freeze-drying and tested it in PIBs for the first time. This material demonstrated a specific capacity of 56 mAh g^{-1} and an average voltage of 3.6 V [83].

The $K_3V_2(PO_4)_3$ electrode exhibits high lattice reversibility during the insertion and extraction of K^+ ions. This characteristic is primarily attributed to its NASICON-type structure and the PO_4 tetrahedral network, which provide robust structural support, thereby contributing to the electrode's exceptional cycling stability. In comparison to $Na_3V_2(PO_4)_3$ (NVP), $K_3V_2(PO_4)_3$ (KVP) has a larger unit cell, particularly along the c-axis. This increased size facilitates fast K^+ ions insertion/extraction, enhancing overall structural stability during cycling. Additionally, the 3D PO_4 network in $K_3V_2(PO_4)_3$ reinforces its structural integrity, improving its cycling performance [82].

Modification of $K_3V_2(PO_4)_3$ is an effective strategy for enhancing its electrochemical performance. Chen et al. investigated the impact of Rb doping on this high-voltage cathode material [86]. Using the sol–gel process, they synthesized Rb-doped $K_3V_2(PO_4)_3$, identifying an optimal doping level of 0.05. The resulting $K_{2.95}Rb_{0.05}V_2(PO_4)_3$/C material exhibited a specific capacity of 56 mAh g^{-1} at 2.5–4.6 V and 20 mA g^{-1}. Compared to the original material, the specific capacity after 50 charge/discharge cycles significantly improved from 32.1 to 53 mAh g^{-1}, demonstrating enhanced cycling stability.

Wang et al. incorporated Mn^{2+} into $K_3V_2(PO_4)_3$ and combined it with a carbonaceous material to synthesize $K_{3.2}V_{2.8}Mn_{0.2}(PO_4)_4/C$ [84]. *In situ* XRD analysis revealed that during charge and discharge cycles, the (240), (122), and (260) diffraction peaks showed no significant shifts, indicating high structural stability (Figure 3.8a). No peaks disappeared, nor did any new peaks emerge, suggesting that a biphasic reaction may not be involved. Figure 3.8b presents a schematic of the crystal structure of KMVP-0.2 during charge and discharge. When charged to 4.0 V, the diffraction peak corresponding to the (400) plane slightly shifted to a higher angle. This shift can be attributed to the extraction of K^+ ions from $K_{3.2}V_{2.8}Mn_{0.2}(PO_4)_4/C$, where the strong electronegativity of the polyanion groups in the material causes a tendency for the unit cell to contract inward, resulting in the peak moving to a higher angle, which corresponds to a decrease in the (400) interplanar spacing. Conversely, during discharge, the crystal structure returns to its original state, with the diffraction peak shifting back to its initial position. This minor volume change may contribute to the high stability of the material. $K_{3.2}V_{2.8}Mn_{0.2}(PO_4)_4/C$ demonstrated a capacity retention of 80.8% after 1500 cycles at 2 C, attributed to the enhanced conductivity resulting from Mn^{2+} doping and the structural stability ensured by the incorporation of carbonaceous materials (Figure 3.8c). The potassium storage properties of titanium-based materials also merit in-depth exploration. In 2016, researchers developed a novel nanocubic structure of sucrose-carbon-coated $KTi_2(PO_4)_3$ using the hydrothermal method, employing it for the first time as an anode material for PIBs. The final charging product was $K_3Ti_2(PO_4)_3$. The electrochemical storage mechanism involved embedded potassium storage, enabling up to 100 cycles with no significant degradation in specific capacity.

In 2020, Kim et al. designed and presented a rhombohedral $KTi_2(PO_4)_3$ cathode material modified by a conductive carbon network [85]. This composite demonstrated a high specific capacity of 126 mAh g^{-1} (approximately 98.5% of the theoretical capacity) and exhibited a capacity retention of 89% after 500 cycles. As illustrated in Figure 3.8d, the insertion and extraction process of K^+ ions in this electrode involves a two-phase reaction between $KTi^{4+}{}_2(PO_4)_3$ and $K_3Ti^{3+}{}_2(PO_4)_3$ within the voltage range of 1.0–4.0 V, as analyzed by *in situ* XRD results. The valence transition of titanium ions is the source of charge transfer. Additionally, as shown in Figure 3.8e, during K^+ ion intercalation, $K_{1+x}Ti_2(PO_4)_3$ exhibited an increase in all lattice parameters (e.g. a, b, and c), resulting in a total volume change of 11.68%.

3.3.2.3 Fluorophosphates

The introduction of strongly electronegative fluorine can enhance the operating voltage of phosphate cathode materials, resulting in higher energy density. Consequently, fluorophosphates have attracted significant attention as cathode materials for SIBs, and a similar analogy can be drawn for PIBs. Incorporating highly electronegative fluorine anions into polyanionic materials to form fluorine-containing derivatives has proven effective in increasing redox potential. The presence of

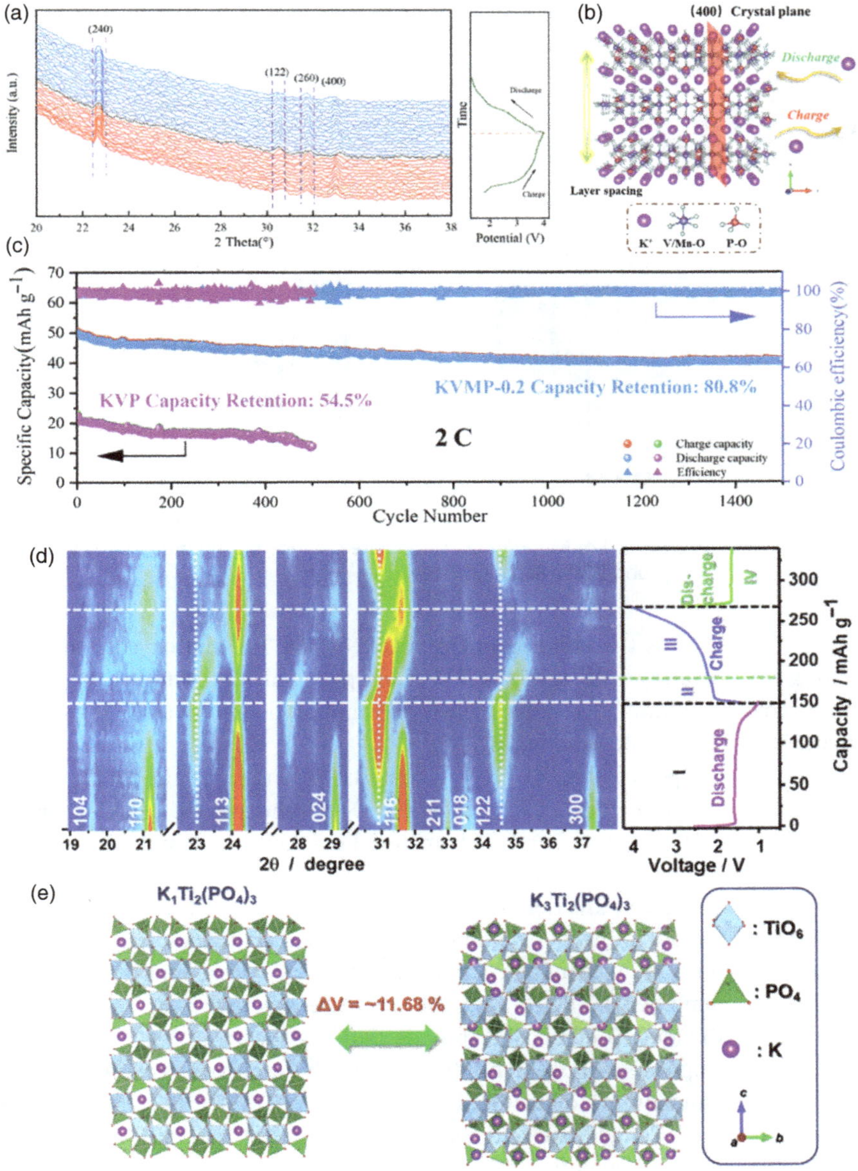

Figure 3.8 (a) *In situ* XRD patterns collected during the initial cycle for $K_{3.2}V_{2.8}Mn_{0.2}(PO_4)_4/C$ at 0.2 C from 1.3 to 4.0 V; (b) schematic diagram of the crystal structure of KVMP-0.2 material (400) plane; (c) long-term cycling at 2 C of the $K_{3.2}V_{2.8}Mn_{0.2}(PO_4)_4/C$ and $K_3V_2(PO_4)_3$ cathode. Source: [84]. Reproduced with permission/John Wiley & Sons. (d) *In situ* XRD of $KTi_2(PO_4)_3$ for PIBs; (e) comparison of predicted crystal structures of $K_1Ti_2(PO_4)_3$ and $K_3Ti_2(PO_4)_3$. Source: [85]. Reproduced with permission/John Wiley & Sons.

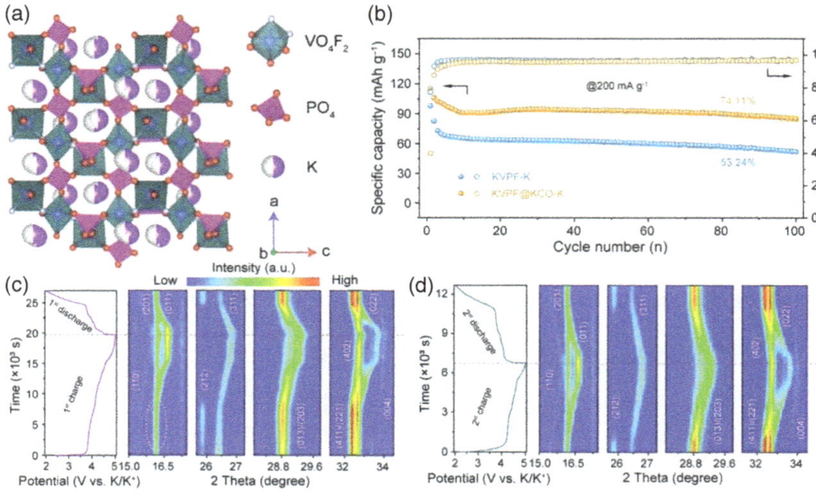

Figure 3.9 (a) Crystal structure of $KVPO_4F$ and (b) cycling performance at 200 mA g^{-1} of KVPF@KCO; (c, d) *in situ* XRD of KVPF@KCO composite cathode during initial two cycles at 50 mA g^{-1}. Source: [90]/John Wiley & sons/CC BY 3.0.

fluorine ions weakens the bonding energy between the TM and oxygen, thereby elevating the operating voltage of the TM redox pair. This characteristic suggests that fluorine-containing derivatives have substantial potential as high-voltage cathodes for rechargeable batteries [87–89].

Komaba et al. reported that $KVPO_4F$, a material belonging to the $Pna2_1$ space group, features VO_4F_2 octahedra connected to PO_4 tetrahedral and can be utilized as a cathode for PIBs (Figure 3.9a). This material exhibits a reversible capacity of 92 mAh g^{-1} and an average discharge voltage of 4.13 V, which exceeds the working potential of most layered transition oxides. Its energy density reaches 380.0 Wh kg^{-1}. However, the $KVPO_4F$ cathode experiences poor rate capability and rapid capacity decay due to low electronic conductivity and significant side reactions [91].

Heng et al. combined $KVPO_4F$ with the electrode additive $K_2C_4O_4$ (KCO) to synthesize a KVPF@KCO composite cathode [90]. As shown in Figure 3.9b, the KVPF@KCO composite cathode exhibits significantly improved cycling stability, retaining approximately 74% of its initial capacity after 100 cycles. In stark contrast, KVPF without the KCO additive retained only about 53% of its capacity. To better elucidate the structural evolution of KVPF@KCO during the passivation process, *in situ* XRD analysis was performed during the first two cycles, as illustrated in Figure 3.9c and d. Specifically, at the beginning of the first charging process, only the diffraction peak of the (212) crystal plane gradually weakens and disappears, indicating a two-step structural evolution. As charging progresses and more K^+ ions are extracted from the KVPF lattice, the corresponding diffraction peaks shift to higher angles. Upon reaching a fully charged state, new diffraction peaks emerge, signifying a two-phase transition reaction. During discharge, these peaks nearly revert to their original state. Similar structural evolution processes occur during subsequent charge–discharge cycles, further demonstrating the excellent structural reversibility of this material.

Lin et al. synthesized $K_3V_2(PO_4)_2F_3$ by electrochemically removing sodium and incorporating potassium in an ionic electrolyte, using $Na_3V_2(PO_4)_2F_3$ as the precursor material [92]. Unlike $Na_3V_2(PO_4)_2F_3$, this material cannot be directly synthesized using a high-temperature solid-phase method, likely due to its instability at elevated temperatures. Its framework structure comprises PO_4 tetrahedra and VO_4F_2 octahedra, exhibiting an average operating voltage of approximately 3.7 V and a specific capacity of up to 100 mAh g^{-1}. $K_3V_2(PO_4)_2F_3$ demonstrates an energy density of 400 Wh kg^{-1}, making it competitive among cathode materials. Additionally, the material experiences a volume change of 6.2% during charging and discharging, while retaining 97% of its capacity after 100 cycles. The authors concluded that $K_3V_2(PO_4)_2F_3$ undergoes two phase transitions during the de-embedding of K^+ ions. The first transition occurs from monoclinic $K_3V_2(PO_4)_2F_3$ (space group *Cmcm*) to tetragonal $K_2V_2(PO_4)_2F_3$ (space group *I*4/*mmm*) when charged to 4.2 V. Further charging to 4.6 V results in a transition from the tetragonal phase $K_2V_2(PO_4)_2F_3$ to the monoclinic phase $K_1V_2(PO_4)_2F_3$ (space group $Cmc2_1$). The physical phase changes of $K_3V_2(PO_4)_2F_3$ in PIBs are consistent with those of $Na_3V_2(PO_4)_2F_3$ in SIBs, indicating the potential application of $K_3V_2(PO_4)_2F_3$ in PIBs. However, the challenge of directly synthesizing $K_3V_2(PO_4)_2F_3$ remains to be addressed.

3.3.2.4 Pyrophosphates

Pyrophosphates are also common polyanionic compounds that have been extensively studied as cathode materials for LIBs and SIBs. To investigate the potential application of pyrophosphates in PIBs, Park et al. screened six possible PIB pyrophosphate cathode materials using a six-step method [93]. This process involved selecting from the elements potassium (K), titanium (Ti), vanadium (V), chromium (Cr), manganese (Mn), iron (Fe), cobalt (Co), nickel (Ni), molybdenum (Mo), phosphorus (P), and oxygen (O) (3–5 elements), as well as determining the crystal position and valence state of the TM (M), theoretical capacity (> 80 mAh g^{-1}), K^+ ion diffusion channel size (>1.8 Å) and type (1D channel). They subsequently assessed the redox potential of each material using CV at 0.1 mV s^{-1} (Figure 3.10a) and discussed the results based on the color of the oxidized materials. The findings indicated that $KTiP_2O_7$, KVP_2O_7, and $KMoP_2O_7$ are effective for K^+ ions storage, while $KFeP_2O_7$, $KCrP_2O_7$, and $K_2MnP_2O_7$ are not. These six pyrophosphates belong to the $P2_1/c$ space group consisting of MO_6 octahedra and PO_4 tetrahedra connected at their vertices, as illustrated in Figure 3.10b, using the crystal structure of $KFeP_2O_7$ as an example.

For KFe(III)P_2O_7, the density functional theory (DFT) calculations indicate that the Fe^{3+}/Fe^{4+} redox potential in KFe(III)P_2O_7 is exceptionally high (>5.0 V), possibly due to the Fe^{3+}/Fe^{4+} redox center and the inductive effect of FeO_6. In contrast, K_2Fe(II)P_2O_7 exhibits a lower redox potential (< 3.0 V) and demonstrates good reversible capacity (58 mAh g^{-1}). K_2Fe(II)P_2O_7 belongs to the $P\bar{4}2_1/m$ space group, featuring FeO_4 tetrahedra and PO_4 tetrahedra connected at their vertices (Figure 3.10c). The FeO_4 tetrahedral environment has a weaker inductive effect

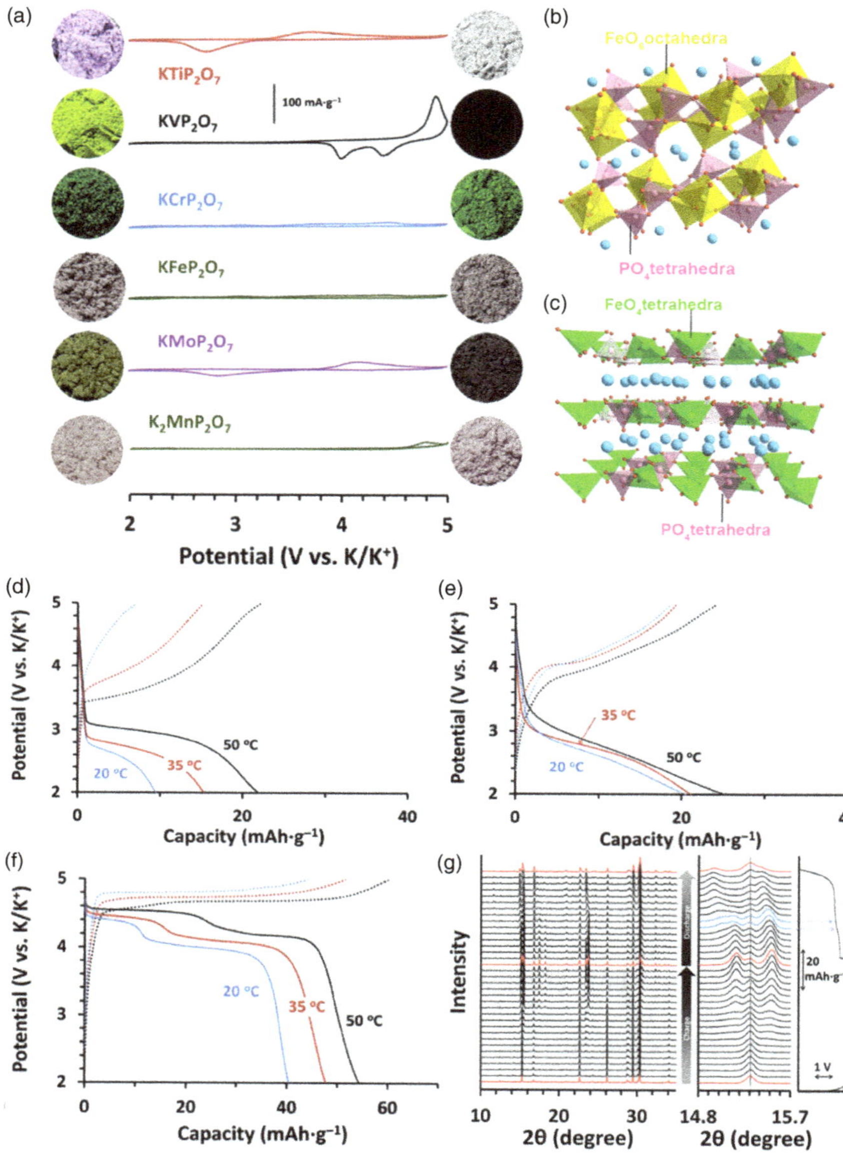

Figure 3.10 (a) Cyclic voltammograms of 6 selected pyrophosphates tested at 20 °C and the scan rate 0.1 mV s^{-1}. Source: Reproduced with permission from [93]/John Wiley & Sons. Photos compare the color of the (left) as-prepared and (right) chemically oxidized samples; schematic crystal structure of (b) $KFe(III)P_2O_7$ and (c) $K_2Fe(II)P_2O_7$. Charge/discharge profiles of (d) $KTiP_2O_7$, (e) $KMoP_2O_7$, and (f) KVP_2O_7 at 20, 35, and 50 °C; (g) *in situ* XRD of KVP_2O_7 during a full charge/discharge cycle (0.25 C, 20 °C). Source: [93]. Reproduced with permission/John Wiley & Sons.

compared to the FeO_6 octahedral environment, and the Fe^{2+}/Fe^{3+} redox center leads to a lower redox potential.

For $KTiP_2O_7$, KVP_2O_7, and $KMoP_2O_7$, which exhibit potassium storage activity, their charging and discharging curves at different temperatures (20 °C, 35 °C, and 50 °C) at 1 C current density are shown in Figure 3.10d–f. At room temperature (20 °C), the specific capacities of $KTiP_2O_7$ and $KMoP_2O_7$ are only 9.2 and 20 mAh g^{-1}, respectively, while KVP_2O_7 can reach 40 mAh g^{-1}. Additionally, KVP_2O_7 outperforms the other two materials at elevated temperatures. Therefore, although all three materials demonstrate electrochemical activity, only KVP_2O_7 possesses both high operating voltage (~ 4 V) and a commendable specific capacity. Furthermore, the phase transitions of KVP_2O_7 during potassium storage are reversible. As shown in Figure 3.10g, *in situ* XRD reveals that during charging, the intensity of the monoclinic KVP_2O_7 ($P2_1/c$) (110) characteristic peak ($2\theta = 15.30°$) gradually diminishes, ultimately becoming a faint shoulder peak at full charge. Simultaneously, the (010) and (100) diffraction peaks ($2\theta = 15.15°$ and $15.51°$) of the triclinic $K_{1-x}VP_2O_7$ ($x \approx 0.6$) ($P\bar{1}$) emerge and gradually intensify. During the subsequent discharge, these changes reverse, leading to the gradual reformation of monoclinic KVP_2O_7. This observation demonstrates the reversibility of the reaction and suggests that incomplete reactions during the phase transition may account for the relatively low specific capacity of the material.

Building on this, He et al. designed a carbon matrix potassium vanadium pyrophosphate powder (KVP_2O_7@C) using a spray drying process [94]. The particles exhibit a wrinkled pea-like shape, which significantly improved the electrochemical performance. It has a capacity of 60 mAh g^{-1} and features two discharge plateaus at 4.1 and 4.5 V (vs. K/K^+), similar to KVP_2O_7. However, for the application of KVP_2O_7 in PIBs, its single-electron reaction and the large molecular weight of the P_2O_7 group result in a relatively low theoretical specific capacity. Therefore, further efforts are necessary to improve the energy density of this compound.

3.3.2.5 Oxyphosphates

$VOPO_4$ can intercalate small cations such as H^+, Li^+, Na^+, K^+, and various small molecules, making it widely utilized in PIBs and SIBs. Hyoung et al. reported a layered $VOPO_4·2H_2O$ cathode for PIBs [95]. In $VOPO_4·2H_2O$, there are two types of water molecules: structural water and crystal water, as illustrated in Figure 3.11a. Structural water is directly bonded to the vanadium atoms, positioned opposite to the dangling oxygen (Figure 3.11b). Although the V—O_{water} bond length is longer than the typical V—O bond, it still exhibits significant bonding strength. The crystallization water is located in the interlayer between two PO_4 tetrahedra (Figure 3.11c). To investigate the electrochemical performance of $VOPO_4·2H_2O$, electrochemical tests were conducted using activated carbon as the counter electrode. Due to the presence of structural and crystallization water in the material, 0.1 M of water was added to the electrolyte to maintain structural stability. At a current density of 0.2 C, the initial discharge and charge-specific capacities are 88.3 mAh g^{-1} and 93.8 mAh g^{-1}, respectively, with a capacity retention rate of 86.2% after 100 cycles (Figure 3.11d).

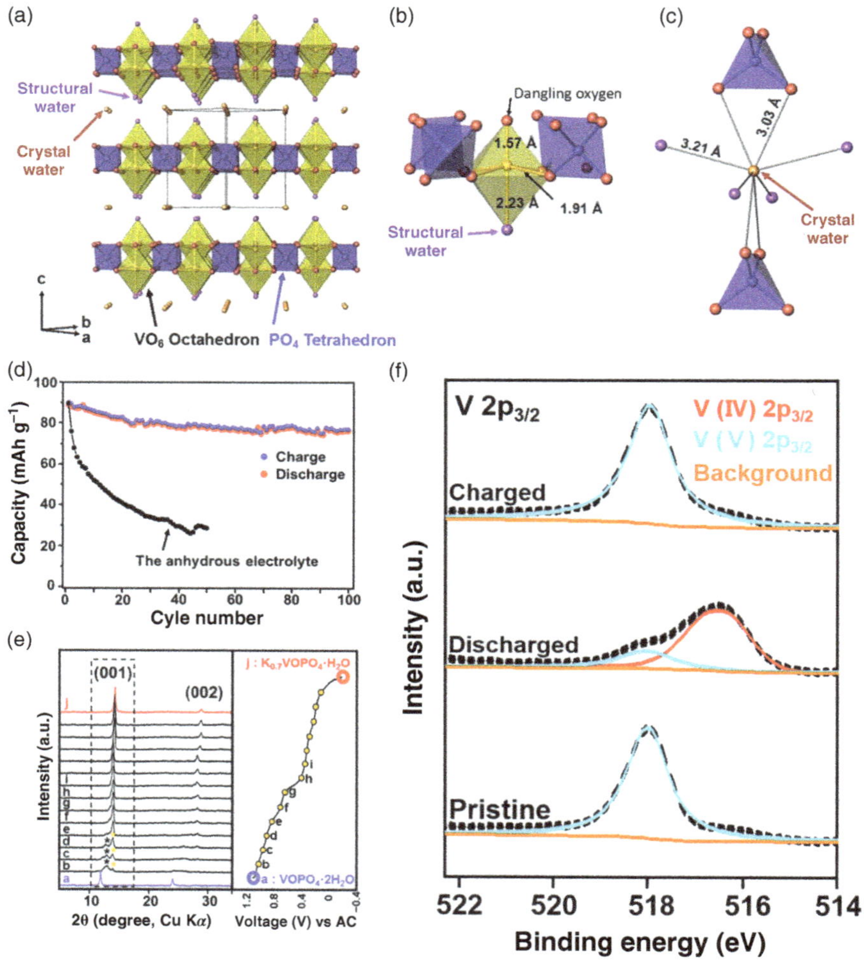

Figure 3.11 (a) Crystal structure of $VOPO_4 \cdot 2H_2O$, (b) local structure of VO_6 octahedron, and (c) local structure surrounding the crystal water in the interlayer in $VOPO_4 \cdot 2H_2O$. Hydrogen atoms are not shown for clarity; (d) cycling performance at 0.2 C rate, (e) operando XRD profiles during the initial discharge of $VOPO_4 \cdot 2H_2O$ at 0.2 C, and (f) XPS spectra of different stages. Source: [95]. Reproduced with permission/John Wiley & Sons.

The structural evolution of the $VOPO_4 \cdot 2H_2O$ cathode material during the first discharge cycle was analyzed using *in situ* XRD (Figure 3.11e). The analysis reveals a two-phase reaction in the b-i region and a single-phase reaction in the i-j region. XPS analysis was conducted to investigate the changes in vanadium oxidation states during the charge–discharge process, as illustrated in Figure 3.11f. The binding energies of V^{4+} $2p_{3/2}$ and V^{5+} $2p_{3/2}$ are measured at 516.5 and 518.0 eV, respectively. During discharge, the average oxidation state of vanadium transitions from pure V^{5+} to a mixture of V^{5+} and V^{4+}, and it reverts to its original state upon charging, confirming the reversibility of the process.

The structure of $KVOPO_4$ is similar to $KVPO_4F$, both belonging to the $Pna2_1$ space group. It features an open framework composed of VO_6 octahedra and PO_4 tetrahedra sharing oxygen atoms. K^+ ions are transported through the gaps between VO_6 and PO_4. The environments of potassium atoms differ slightly, with the K1–O bond length (2.861 Å) generally shorter than the K2–O bond length (2.924 Å). Shorter bond lengths indicate greater stability for the potassium atoms [96]. $KVOPO_4$ was first reported by Chihara et al. [91] Within a voltage range of 5.0–2.0 V, the specific capacity is approximately 80 mAh g^{-1}, with a voltage plateau at 4.0 V. However, during cycling, severe electrolyte decomposition and insufficient coulombic efficiency are observed. Compounds prepared by high-temperature solid-phase sintering contain inherent impurity defects. The high operating voltage of 5.0 V may lead to electrolyte decomposition, resulting in poor cycling stability of the $KVOPO_4$ phase. Consequently, despite the excellent structural stability of $KVOPO_4$, its cycling stability under high-voltage conditions remains inadequate.

Li et al. [97] introduced a carbothermal reduction method to synthesize a pure $KVOPO_4$ phase (KVP) and improved electrode interface stability by adjusting the operating voltage. The layered $VOPO_4$ has a large surface area, allowing K^+ ions to fully intercalate into the $VOPO_4$ structure, ensuring impurity-free products after a complete reaction. This approach solves the limitations of uneven raw materials and incomplete reactions found in the solid-phase sintering method. In electrochemical tests, reducing the cut-off voltage to 4.6 V, while sacrificing some capacity, prevents electrolyte decomposition, thus protecting the electrode interface. In a voltage range of 2.0–4.6 V and a current density of C/10, the first cycle discharge-specific capacity of KVP is about 70 mAh g^{-1}, and about 65 mAh g^{-1} after 100 cycles. Under C/2 conditions, the specific capacity drops from 73 mAh g^{-1} in the first cycle to 63 mAh g^{-1} after 1000 cycles, with a capacity decay rate of about 0.013% per cycle.

The valence state of V in $KVOPO_4$ during K^+ ion extraction was studied. During extraction, the binding energy of V 2*p* increases to 517.3 eV ($2p_{3/2}$) and 524.7 eV ($2p_{1/2}$). Upon further charging to 4.6 V, the V 2*p* peaks continue to shift positively, indicating the oxidation of V^{4+} to V^{5+} [97]. The deintercalation mechanism of K^+ ions in $KVOPO_4$ is quite complex, involving three distinct steps [91]. The first, termed by the authors as a "single-phase transition," occurs without changes in lattice parameters or the formation of new phases. This is followed by a single-phase reaction that results in changes to lattice parameters and is completed through a two-stage transition. The extraction mechanism of K^+ ions from $KVOPO_4$ was studied using first-principles calculations, identifying four distinct stages [98]. In the first stage ($0 \leq x \leq 0.5$), K2 is fully extracted, causing the nearby K1 atom to rearrange to a new position, Kc, located between the K1 and K2 sites. Energy calculations suggest that the energy of $K1_{0.5\text{-}x}K2_{0.5\text{-}x}Kc_xVOPO_4$ is lower than $K1_{0.5\text{-}x}K2_{0.5}VOPO_4$, indicating that K2 extraction is the dominant process. After half of the K^+ ions are extracted, the phase changes into $Kc_{0.5}VOPO_4$. In the second stage ($0.5 \leq x \leq 0.625$), the most stable phase is $Kc_{0.375}VOPO_4$, with all K atoms occupying the Kc site, resulting in a solid-solution mechanism. During the third stage ($0.625 \leq x \leq 0.75$), $K2_{0.25}VOPO_4$ is the most stable phase, with potassium atoms migrating from the Kc site to the K2 site, following a two-phase mechanism. In the final stage ($0.75 \leq x \leq 1.0$), the residual K ions, primarily located in the K2 site, are extracted.

3.3.2.6 Sulfates and Fluorosulfates

Sulfates typically exhibit higher theoretical redox potentials due to the greater electronegativity of SO_4^{2-} compared to other polyanionic groups. For example, the yavapaiite-type $KFe(SO_4)_2$ (Y–$KFe(SO_4)_2$) belongs to the $C2/m$ space group, composed of FeO_6 octahedra and SO_4 tetrahedra that share oxygen atoms. These are aligned along the ab-plane, providing ample space for K ions. Theoretically, each formula unit of the Y–$KFe(SO_4)_2$ structure can accommodate an additional 2 mol of K^+ ions. Moreover, this material has abundant channels that facilitate rapid K^+ ion diffusion. Consequently, $KFe(SO_4)_2$ demonstrates a discharge-specific capacity of 93 mAh g^{-1} at C/20 and retains 80% of its capacity after 300 cycles at 2 C. The structure of Y-$K_xFe(SO_4)_2$ continuously evolves throughout the discharge process, enabling reversible K^+ ion intercalation without significant structural degradation. The total volume change during cycling is minimal, around 1.83%, which accounts for the high capacity retention of the Y-$K_xFe(SO_4)_2$ structure.

$Fe_2(SO_4)_3$ also exhibits excellent performance as a cathode material for PIBs. This material delivers a capacity of approximately 100 mAh g^{-1}, with an average operating voltage of around 3.3 V, and retains about 80% of its capacity at 5 C. Additionally, after 300 cycles at 2 C, approximately 83% of the initial specific capacity is retained [99].

For fluorosulfate compounds, Tarascon et al. investigated both the orthorhombic form (o-$KFeSO_4F$) and the monoclinic form (m-$KFeSO_4F$) of $KFeSO_4F$ [100]. Studies revealed that o-$KFeSO_4F$ has a loose structure, larger ion channels, and a lower K+ diffusion barrier, facilitating faster potassium-ion diffusion than m-$KFeSO_4F$. When used in PIBs (with K metal as the anode), o-$KFeSO_4F$ demonstrated a higher initial capacity, higher operating voltage, and better capacity retention than m-$KFeSO_4F$.

To further improve the electrochemical performance of $KFeSO_4F$, researchers developed a graphene network of $KFeSO_4F$ particles ($KFeSO_4F$@G). The conductive graphene enhances K^+ ion diffusion kinetics, reduces particle size, and creates more reaction sites. Consequently, $KFeSO_4F$@G exhibits a discharge capacity of 111.51 mAh g^{-1} and maintains a high capacity of 71.1 mAh g^{-1} even at a high current density of 10 C [101].

Additionally, carbon nanotubes (CNTs) were combined with $KFeSO_4F$ to produce $KFeSO_4F$@CNT [102]. This material achieves a capacity retention rate of 93.9% after 2000 cycles at 3 C, with an average operating voltage of 3.73 V. The uniform interpenetration of the extensively exposed (100) plane and CNTs within the $KFeSO_4F$ microspheres results in an exceptionally stable structure and high K^+ ion diffusivity.

Summary and Prospects In summary, polyanionic compounds provide a stable framework with large channels for K^+ ion transport, leading to high operating voltages and good stability. However, their high average operating voltage can cause electrolyte decomposition at elevated levels, and currently, no suitable electrolytes are available to address this, which presents a challenge for practical applications. Moreover, these

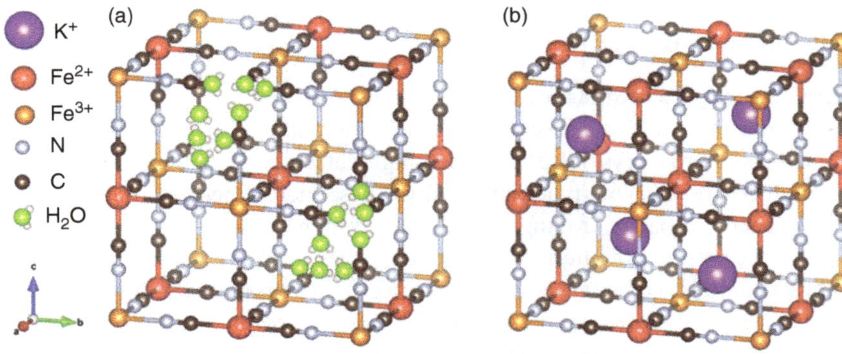

Figure 3.12 Schematic crystal structures of PBAs. Source: [104]. Reproduced with permission/John Wiley & Sons.

materials tend to have poor conductivity and slow reaction kinetics. Future research should prioritize the discovery of new potassium-based polyanionic compounds with high specific capacity or focus on developing effective methods to enhance existing ones. Strategies such as particle size reduction, carbon coating, and metal element doping could substantially improve their electrochemical performance.

3.3.3 Prussian Blue Analog

Prussian blue (PB) and its analogs (PBAs) can be uniformly represented by the chemical formula $A_xM_2[M_1(CN)_6]_y{\cdot}zH_2O$, where A represents alkali metals such as Li, Na, K), and M_1 and M_2 are transition metals like Fe, Ni, Co, Mn, and Ti, with $0 < x < 2$. Both M_1 and M_2 can act as redox-active sites, providing the system with a theoretical capacity of up to 155 mAh g^{-1} [103]. Prussian blue compounds generally exhibit a face-centered cubic structure, belonging to the space group $Fm\bar{3}m$. The high-spin M_1 ions are located in the M_1N_6 octahedra, while the low-spin M_2 ions reside in the M_2C_6 octahedra. As illustrated in Figure 3.12, each metal ion occupies the corners of a cubic structure, connected through -C≡N- groups. The number of interstitial cations, used for charge compensation, depends on the charge of the metal ions coordinated with nitrogen and carbon. These interstitial ions may include the same metal ions, coordinated with nitrogen and carbon, or any alkali metal ions. Additionally, the open 3D rigid framework of PBAs facilitates the rapid migration of charge carriers along the (100) crystal direction. This structure also alleviates lattice stress caused by volume expansion, ensuring high structural stability [104].

Prussian white (PW), Prussian blue (PB), and Berlin green (BG) are classified according to their potassium content. PW has the highest potassium content, such as in $K_2Fe^{II}[Fe^{II}(CN)_6]{\cdot}nH_2O$. PB contains some potassium, as seen in $KFe^{III}[Fe^{II}(CN)_6]{\cdot}nH_2O$, while BG has negligible or no potassium content, for example, in $Fe^{III}[Fe^{III}(CN)_6]{\cdot}nH_2O$.

PBAs were the first materials developed for cathodes of PIBs, gaining significant attention due to their excellent rate performance and cycling stability. Current research focuses primarily on KFeFe-PBA and KMnFe-PBA, as these materials consist of abundant, and nontoxic elements. Their dual redox reactions efficiently leverage the host structure for K^+ ions storage.

3.3.3.1 Challenges and Strategies

PBAs cathode materials face several challenges when used as cathode materials for PIBs, such as internal crystallization water, high defect levels caused by vacancies, poor conductivity, and low material density. These factors negatively impact the battery's specific capacity, cycling stability, power density, and energy density. In particular, the presence of internal crystallization water can lead to undesirable side reactions and, in severe cases, pose safety concerns. To promote the further development of PBAs, several measures should be considered: improving the crystallinity of the material, designing appropriate stoichiometry, optimizing nanostructures and particle sizes, compositing with other materials, surface modification, and introducing K^+ ions to reduce the crystallization water content.

3.3.3.2 KFeFe-Prussian Blue Analogs

Iron-based Prussian blue (KFeFe-PBA) was the first to gain attention as a fundamental PBA. Eftekhari pioneered the application of $KFeFe(CN)_6$ in PIBs [1]. Despite its relatively low capacity of 78 mAh g^{-1}, this electrode material demonstrates good cycling stability, retaining 88% of its capacity after 500 cycles. This research laid the foundation for using Prussian blue materials as cathodes in PIBs.

In a study by Paige et al., Prussian blue (PB) was synthesized by varying the concentrations of ferric nitrate and potassium ferricyanide ($K_3Fe(CN)_6$) [105]. They confirmed the phase transitions among PB, BG, and PW during charge and discharge cycles:

$$K_2Fe^{II}Fe^{II}(CN)_6 \leftrightarrow K^+ + KFe^{III}Fe^{II}(CN)_6 + e^- \leftrightarrow 2\,K^+ + Fe^{III}Fe^{III}(CN)_6 + 2e^-$$

PBAs are typically synthesized using traditional coprecipitation methods, which result in numerous $[Fe(CN)_6]$ vacancies within the lattice. These vacancies lead to the adhesion of water molecules to the lattice structure, occupying sites intended for alkali metal ions. The presence of bound water and vacancies significantly impairs the material's electrochemical performance. First, the release of water molecules from the material into the organic electrolyte during charge–discharge cycles poses significant safety risks. Second, these vacancies hinder electron conduction within the PBA framework, increasing internal resistance. Third, the random distribution of vacancies destabilizes the PBA structure, causing it to collapse during alkali metal ion insertion and extraction. Fourth, water molecules occupying alkali metal ion sites reduce the ion content in the material, leading to capacity loss. Lastly, the presence of vacancies allows more water molecules to enter the lattice structure. To address these issues, researchers have explored various strategies [105, 106].

Incorporating chelating agents during the synthesis process can significantly enhance the crystallinity of PBAs and reduce the presence of crystallization water. For instance, using EDTA-2 K·4H_2O as a chelating agent has been shown to produce highly crystalline $K_2Fe[Fe(CN)_6]$ (KFeHCF-E) [106]. As depicted in Figure 3.13a, under weak acidic conditions, $K_4[Fe(CN)_6]\cdot 3H_2O$ decomposes to produce $[Fe(CN)_6]^{4-}$. EDTA easily complexes with Fe^{2+} to form $[Fe(EDTA)]^{2-}$. This complex

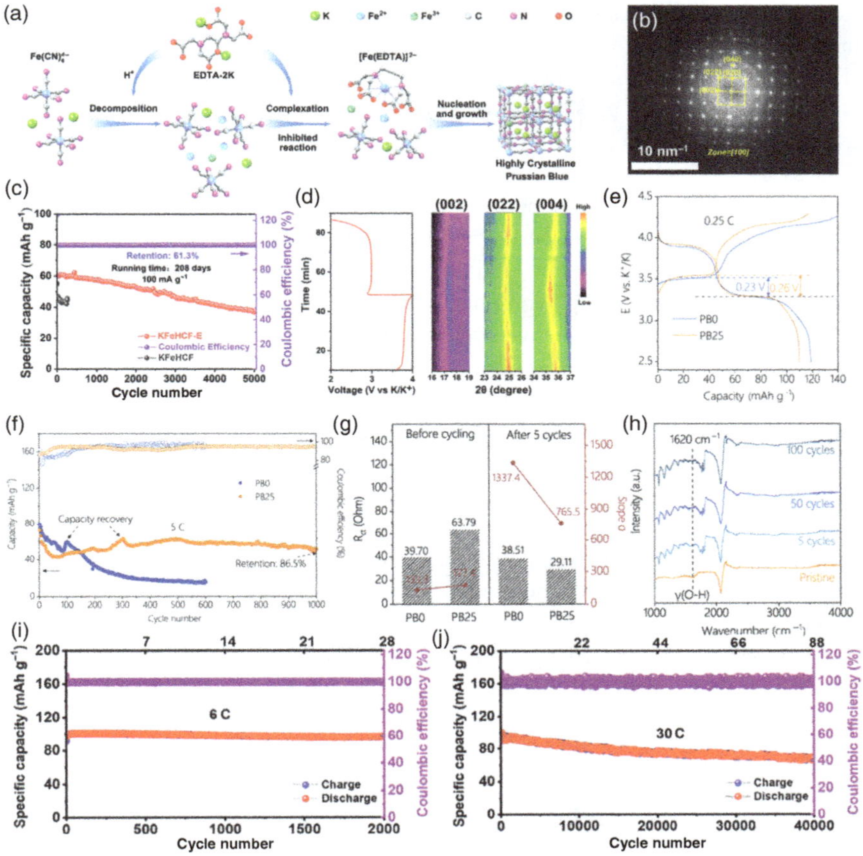

Figure 3.13 (a) Schematic illustration of coprecipitation synthesis of KFeHCF-E with EDTA chelating agent; (b) FFT pattern of KFeHCF-E; (c) cycling comparison performance at 100 mA g^{-1} of KFeHCF-E; (d) *in situ* XRD patterns of KFeHCF-E. Source: Reproduced with permission from [106]/John Wiley & Sons. (e) The first cycle charge/discharge voltage profiles of PB0 and PB25 at a current density of 0.25 C; (f) long-cycle performances of PB0 and PB25; (g) the charge-transfer resistances (R_{ct}) and the ionic diffusion resistance (σ) of PB0 and PB25 before and after five cycles; (h) FTIR spectrum of PB25 electrode at different cycles. Source: [107]/John Wiley & Sons/CC BY 3.0. Cycling performance at (i) 6 C and (j) 30 C of HEPBA in the half-cells within 0–1.25 V (vs. Ag/AgCl). Source: [108]. Reproduced with permission of/John Wiley & Sons.

slowly releases Fe^{2+}, which decelerates the reaction between Fe^{2+} and $[Fe(CN)6]^{4-}$, favoring the formation of highly crystalline PB. Figure 3.13b illustrates the resulting well-defined lattice structure of the material. It exhibits exceptional cycling stability, maintaining 60% of its capacity even after 5000 cycles (Figure 3.13c). *In situ* XRD tests (Figure 3.13d) further confirm that the material remains stable and retains high crystallinity after one charge–discharge cycle. The use of the EDTA-2 K chelating agent not only promotes the growth and morphology of PBAs but also increases the initial potassium content, thereby enhancing the material's overall capacity.

Regulating the synthesis temperature can also control the number of defects. Zhai et al. demonstrated this by synthesizing PBAs with varying levels of vacancies and crystallization water content through temperature adjustments [107]. PB0 was synthesized at 0 °C as $K_{1.36}Fe[Fe(CN)_6]_{0.74}·0.48H_2O$, while PB25 was synthesized at 25 °C as $K_{1.43}Fe[Fe(CN)_6]_{0.94}·0.42H_2O$. Figure 3.13e and f illustrate that PB0 possesses a higher initial capacity; however, PB25, with fewer defects and reduced polarization, exhibits superior cycling stability. Electrochemical impedance spectroscopy (EIS) tests were performed before and after cycling for both materials (Figure. 3.13g). The results indicate that before cycling, PB0 had smaller charge-transfer resistance (R_{ct}) and ionic diffusion resistance (σ). However, the R_{ct} and σ of PB25 decreased after five cycles, indicating enhanced K^+ ion diffusion. FTIR analysis of PB25 before and after cycling reveals that the water-related signal peak (~1620 cm^{-1}) gradually diminishes during cycling, indicating the removal of lattice water (Figure 3.13h). Once the water is extracted, PB25 forms a relatively ideal lattice with only 0.06 vacancies per unit cell. In contrast, PB0 retains 0.27 vacancies per unit cell after cycling. This reduction in vacancies elucidates the superior performance of PB25.

Adjusting the pH value can also influence the crystallinity of PBAs. For instance, the concentration of HCl in the aqueous solution significantly impacts crystallinity. Different concentrations of HCl solution (0, 0.2, 0.5, 1.0, 1.5, 2.0 mol l^{-1}) result in varying defect densities in KFeFe-PBA. Among these, the product obtained at 1.0 mol l^{-1} HCl concentration possesses the highest K content and the fewest defects ($K_{1.93}Fe[Fe(CN)_6]_{0.97}·1.82H_2O$), yielding the best performance (88.2 mAh g^{-1} after 300 cycles at 1500 mA g^{-1}) [109].

Utilizing novel synthesis methods can also reduce defects caused by traditional coprecipitation techniques. For instance, reacting iron powder with citric acid produces ferrous citrate, which then reacts with ferrocyanide to form $K_{1.62}Fe[Fe(CN)_6]_{0.92}·0.33H_2O$. The hydrogen gas generated during the reaction between iron powder and citric acid effectively prevents oxidation of the final product [110].

Doping with TM ions can effectively enhance the electrochemical performance and structural stability of materials by reducing defects. Chong et al. synthesized PBAs with varying nickel content ($K_2Ni_xFe_{1-x}[Fe(CN)_6]$ ($0 \leq x \leq 1$)). When x is 1/2, the resulting compound, $K_{1.90}Ni_{0.5}Fe_{0.5}[Fe(CN)_6]_{0.89}·0.42H_2O$ (KNFHCF-1/2), exhibited the superior performance [111]. At a current density of 500 mA g^{-1}, it demonstrated excellent rate performance (achieving 53.4 mAh g^{-1}) and retained 82.3% of its capacity after 1000 cycles at 100 mAh g^{-1}. During the charge–discharge cycles, the Fe 2*p* spectrum shifts to higher binding energies during charging, indicating that Fe^{2+} is oxidized to Fe^{3+} when charged to 4.5 V. After full charging, the peak returns to its original position, demonstrating a highly reversible redox reaction. In contrast, the oxidation state of Ni^{2+} remains unchanged during K^+ ion insertion/extraction, indicating its electrochemical inertness and its role in enhancing structural stability. Calculations indicate that KNFHCF-1/2 has a lower K^+ ionic diffusion barrier compared to undoped KNFHCF. Similarly, as a PIBs cathode material, $K_{0.3}Ti_{0.75}Fe_{0.25}[Fe(CN)_6]_{0.95}·2.8H_2O$ exhibits excellent rate performance with a high reversible capacity of 58.9 mAh g^{-1} at a current density of 500 mA g^{-1} [112].

Constructing composite materials is a promising strategy to enhance the potassium storage performance of PBAs. Integrating PBAs with reduced graphene oxide (rGO) significantly boosts their electrochemical performance. However, during synthesis, the oxygen-containing functional groups on rGO may be lost, which weakens the interaction between PBAs and rGO, thus reducing the electrochemical performance of the PB/rGO composite. Wei et al. demonstrated that polyvinylpyrrolidone (PVP) can effectively restore this weakened interaction [113]. PB is anchored to the rGO surface through stable covalent bonds with the amide groups in PVP. The PB/PVP-rGO composite exhibited excellent rate capability, achieving 72 mAh g^{-1} at 10 C and demonstrated improved capacity retention of 96.5% after 800 cycles at 5 C. The enhanced interaction between PB and rGO resulted in a higher capacity retention of 91.8% after 300 cycles at 5 C, compared to 69.7% for the PB/rGO composite without PVP.

Enhancing the particle morphology of PBAs can significantly improve their electrochemical performance. Nazar et al. manipulated the particle size of PBAs by using potassium citrate as a chelating agent with Fe^{2+} [114]. This slowed the reaction rate between Fe^{2+} and ferrocyanide, resulting in more uniform particles. Smaller particles exhibit better electrochemical performance in PIBs due to shorter K^+ diffusion paths. Additionally, modulating the acidity of the hydrothermal reaction can influence particle size – the higher the acidity, the smaller the particles. However, excessive acidity can deplete the potassium-ion content in the material, negatively impacting electrochemical performance [109]. Mai et al. fabricated flower-like Prussian blue particles composed of nanosheets using a low-temperature hydrothermal method. With increased aging time, the morphology transformed from nanospheres to nanosheets and nanoflowers. This unique structure effectively accommodates volume changes during charge–discharge cycles, thereby improving cycling stability [115].

PBAs are also extensively utilized in APIBs. Ma et al. synthesized a high-entropy PBA (HEPBA) as a cathode material for APIBs [108]. Theoretical calculations indicate that, compared to traditional PBAs, HEPBA possesses lower formation energy, a narrower band gap, reduced ion diffusion barriers, and enhanced covalent bonding characteristics. These properties substantially improve its electrochemical performance. Electrochemical tests demonstrate that HEPBA exhibits impressive reversible specific capacities of 102.4 and 92.5 mAh g^{-1} at discharge rates of 1 C and 30 C, respectively. Even under stricter conditions, at 6C and 30 C, HEPBA maintains remarkable capacity retention of 96.9% and 73.8% after 2000 and 40 000 cycles, respectively (Figure 3.13i and j). During the charge–discharge process, HEPBA exhibits strong resistance to dissolution and undergoes minimal volume change, with only a 0.872% variation.

3.3.3.3 KMnFe-Prussian Blue Analogs

Manganese-based PBAs (KMnFe-PBAs) offer enhanced cycling stability and higher operating potentials compared to conventional KFeFe-PBAs due to the substitution of iron with manganese. Goodenough et al. reported on the $K_{1.89}Mn[Fe(CN)_6]_{0.92}\cdot 0.75H_2O$ cathode material, which can accommodate two K^+ ions per structural unit, featuring a theoretical specific capacity of 156 mAh g^{-1} and an actual specific capacity of 142 mAh g^{-1} [116].

The crystal structure of KMnFe-PBA closely resembles that of other PBAs. In this structure, manganese and iron are coordinated with nitrogen and carbon, respectively, while vacancies in the $[Fe(CN)_6]$ framework create a 3D lattice that facilitates the unimpeded movement of alkali metal cations. The cyano groups link two octahedral units centered on TM elements to form a face-centered cubic structure. However, in KMnFe-PBA, one of these octahedral units is manganese-centered, distinguishing it from classical PBAs where both octahedral units are iron-centered.

The unique Jahn-Teller effect associated with manganese results in a notably distinct structure in KMnFe-PBA materials compared to classical KFeFe-PBA. In KMnFe-PBA, Mn^{3+} preferentially coordinates with nitrogen, while Fe^{2+} tends to coordinate with carbon atoms. The weaker ligand field of the nitrogen atom leads to low d-orbital splitting energy for the central atom. Consequently, with only three electrons filling the t_{2g} orbital at half capacity, the remaining electron is likely to occupy the e_g orbital due to the lower splitting energy relative to the electron pairing energy, placing Mn^{3+} in a high-spin state. To further reduce the total energy, the e_g and t_{2g} orbitals split into b_{1g}, a_{1g}, c_g, and b_{2g} orbitals. This process reduces orbital and electron degeneracy, alters the shape of the electron cloud, and ultimately leads to geometric distortions.

Researchers employed XRD analysis to investigate lattice structure changes in the lattice structure of KMnFe-PBA ($K_{1.75}Mn[Fe(CN)_6]_{0.93}\cdot 0.16H_2O$) during charging and discharging [117]. During the first oxidation reaction upon charging, the lattice transitions from a monoclinic structure ($P2_1/n$ space group) to a cubic structure ($Fm\bar{3}m$ space group). As charging continues and a second oxidation reaction occurs, the structure shifts from cubic to tetragonal ($I\bar{4}m_2$ space group). Upon discharging, the structure reverts from tetragonal back to monoclinic. This transformation is driven by the Jahn-Teller effect induced by Mn^{3+}, which distorts the MnN_6 octahedron and elongation along the z-axis, facilitating the transition from cubic to tetragonal. This conversion reduces the overall energy of the material and provides additional space for potassium-ion transport. In contrast, if Fe coordinates with N instead of Mn, the Jahn-Teller effect does not manifest, allowing the cubic structure to remain stable without transitioning to tetragonal upon further charging.

The KMnFe-PBA cathode demonstrates superior kinetic properties and high structural stability. Similar to KFeFe-PBA, its energy density and lifespan can be further enhanced through chemical composition optimization, microstructure construction, and careful modulation of crystallization water content.

3.3.3.4 Other Prussian Blue Analogs

In the structure of PBAs, the M_2 atom can be substituted with various TM elements such as Ni, Co, and Cu. Liu et al. synthesized $K_{1.72}Cu[Fe(CN)_6]_{0.83}\cdot 0.887H_2O$ (KCuFe-PBA) and $K_{1.81}Ni[Fe(CN)_6]_{0.97}\cdot 0.086H_2O$ (KNiFe-PBA) using a hydrothermal synthesis method [118]. Ni^{2+} and Cu^{2+} are electrochemically inert ions, and their incorporation enhances the overall structural stability of Prussian blue analogs. At a current density of 10 mA g^{-1}, the initial reversible discharge capacity of KNiFe-PBA is 57.0 mAh g^{-1}, while KCuFe-PBA exhibits a capacity of 51.2 mAh g^{-1}. KNiFe-PBA demonstrates exceptional cycling stability, with a mere 0.0052% capacity loss per cycle over 8000 cycles at a current density of 200 mA g^{-1}. Moreover, partial doping of KFeFe-PBA with Ni not only preserves the high discharge capacity of KFeFe-PBA but also significantly enhances its cycling performance [111].

Theoretically, Co-based Prussian blue (KCoFe-PBAs) should exhibit dual redox reactions during charge and discharge, similar to KMnFe-PBA and KFeFe-PBAs. However, this material only undergoes a single redox reaction, resulting in a discharge capacity of just 60 mAh g^{-1}. This is due to the strong binding energy between the K^+ ion and -C≡N-, which causes a shift in the electron cloud, elevating the redox potential of Co beyond the upper cut-off voltage for PIBs. Consequently, this prevents the release of the Co^{2+}/Co^{3+} capacity [119]. Doping with Ni can modify the electron cloud structure of Fe^{2+}/Fe^{3+} coordinated with carbon, enhancing the electrochemical performance of KCoFe-PBAs [120].

Summary and Prospects In summary, PBAs present several advantages as cathode materials for PIBs, including facile synthesis, abundant availability, high redox potential, good thermal stability, and wide K^+ ions diffusion channels. However, certain drawbacks continue to limit their electrochemical performance. Most PBAs are synthesized through liquid-phase methods, which introduce lattice water and vacancies in the $Fe(CN)_6$ framework. This leads to diminished rate performance and specific capacity and raises safety concerns. Therefore, minimizing both the excess water content and vacancy concentration in $Fe(CN)_6$ is crucial to enhance the electrochemical performance of PBAs cathode materials.

3.3.4 Organic Cathode Materials

Organic materials, composed of abundant and environmentally friendly light elements (specifically C, H, and occasionally including heteroatoms such as O, N, S, and P), offer several notable advantages. These materials are inherently green and eco-friendly (easy to recycle), featuring low production costs, and can be synthesized under simple and mild conditions. Moreover, organic electrode materials typically exhibit an amorphous structure, stacking in the solid phase through van der Waals forces, which creates ample void space to accommodate the (de)intercalation of large radius K^+ ions. Consequently, they have garnered significant attention as promising cathode materials for PIBs. In the context of LIBs and SIBs, compounds such as quinones, carboxylates, anhydrides, radical compounds, and conductive polymers have been extensively studied, and similar research progress has been observed for certain materials in PIBs.

3.3.4.1 Challenges and Strategies

The structure and performance of organic electrode materials can be adjusted and enhanced through molecular structure design, allowing for controllable redox potentials and theoretical specific capacities sufficient to meet high-energy-density requirements. However, these organic electrode materials face significant challenges due to their high solubility in electrolytes, which leads to rapid deactivation and capacity decline. This issue can be mitigated through various strategies, including structural design (including doping, compositing with 2D carbon materials, and nanostructure adjustments), electrolyte engineering (involving the development of new electrolytes, electrolyte additives, and SEI film interface engineering), polymerization, salification (the introduction of polar groups and formation of alkali salts), *in situ* characterization (for better understanding potassium storage mechanisms and controlling irreversible phase transitions), and the exploration of new metal-organic compounds.

3.3.4.2 Organic Small Molecules and Polymers

Hu and colleagues were the first to propose 3,4,9,10-perylene tetracarboxylic dianhydride (PTCDA) for K^+ ions storage [125]. When discharged to 0.01 V, each unit of PTCDA can accommodate approximately 11 K^+ ions. This cathode material demonstrates a high discharge capacity of 131 mAh g^{-1} within the voltage range of 1.5–3.5 V. In a subsequent study by Fan et al., the electrochemical properties of PTCDA were improved through annealing at 450 °C, a temperature significantly lower than its decomposition point [126]. This annealing process improved the material's electronic conductivity and increased the interlayer spacing. As a result, the modified PTCDA electrode exhibited a reversible capacity of 123 mAh g^{-1} after 140 cycles at 100 mA g^{-1}, indicating good capacity retention. Additionally, it maintained a highly reversible capacity of 98 mAh g^{-1} even after 1000 cycles at a current density of 1000 mA g^{-1}.

Despite the advantages of organic cathode materials, they still encounter several challenges, including poor conductivity, high solubility in electrolytes, and agglomeration issues [127]. Polymerization has proven to be an effective strategy to mitigate the dissolution of organic materials in electrolytes. Tong et al. synthesized PTCDA-based polymers by conducting polymerization reactions between PTCDA and diamines of varying chain lengths. Their findings revealed that incorporating shorter alkyl chains between the monomers significantly enhanced the electrical conductivity of the PTCDA-based polymers (Figure 3.14a). This enhancement facilitated the conversion between radical anions and divalent ions, thereby improving ion transport efficiency and the charge–discharge rate performance of the battery (Figure 3.14b) [121].

To further enhance the potassium storage performance of PTCDA, Hu et al. reacted 1 mol of PTCDA with 2 mol of anthraquinone to synthesize [N, N′-bis(2-anthraquinone)]-perylene-3,4,9,10-tetracarboxydiimide (PTCDI-DAQ) [122]. Due to the increased molecular weight, PTCDI-DAQ exhibits low solubility in most organic liquid electrolytes. Remarkably, even after 900 cycles at a current density of 3 A g^{-1}, it retains a capacity of 147 mAh g^{-1}. To investigate the potassium storage mechanism of PTCDI-DAQ, *ex situ* FTIR testing was conducted, as shown in Figure 3.14c and d. During the initial reduction process, the intensity of the peak associated with the C=O bond vibrations in the imide group (≈1700 cm^{-1}) gradually weakened until it disappeared; however, this peak (≈1700 cm^{-1}) reappeared at its

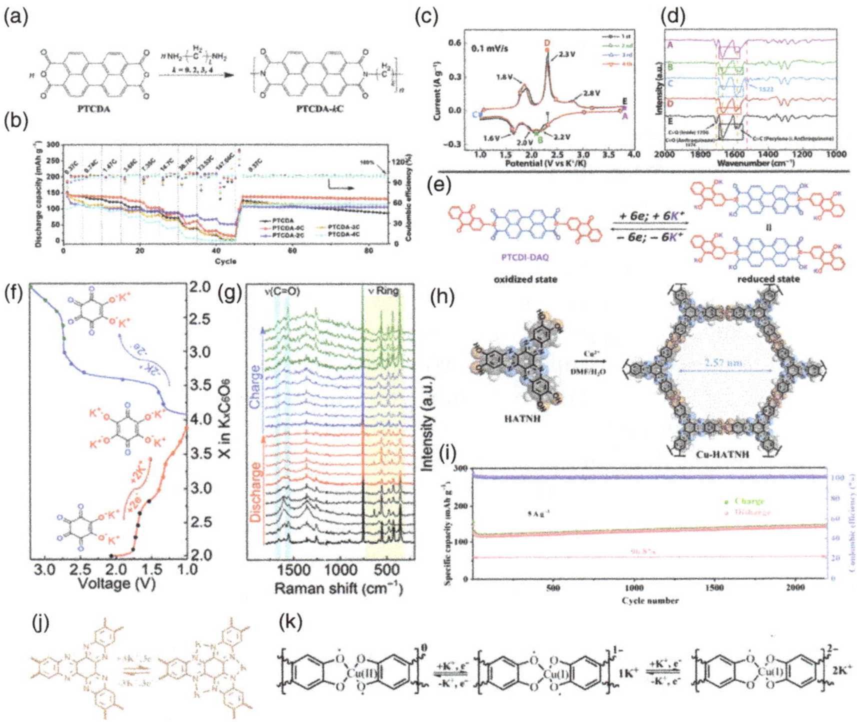

Figure 3.14 (a) Synthesis route and (b) rate performance of PTCDA-based polymers for potassium storage. Source: [121]. Reproduced with permission/John Wiley & Sons. (c) The multiple CV curves of PTCDI-DAQ in half-cells (scan rate was 0.1 mV s^{-1}); d) the *ex-situ* FTIR spectroscopy of PTCDI-DAQ in half-cells; (e) the proposed redox mechanism for PTCDI-DAQ in PIBs. Source: [122]. Reproduced with permission/John Wiley & Sons. (f) Discharge/charge curves with marked points for *in situ* Raman tests and (g) corresponding *in situ* Raman tests spectra of $K_2C_6O_6$ electrode. Source: [123]. Reproduced with permission/John Wiley & Sons. (h) The synthetic route of Cu-HATNH; (i) cycle performance of Cu-HATNH@CNT at 5 A g^{-1}; (j) schematic of the K^+ ions storage process in the HATNH units; (k) schematic of the K^+ ions storage process in the [CuO_4] units. Source: [124]. Reproduced with permission/John Wiley & Sons.

initial state after full charging. Simultaneously, a new peak (≈1522 cm^{-1}) emerged after complete discharge but vanished during other charge–discharge states. Furthermore, the relative intensities of two characteristic peaks (≈1674 cm^{-1} and ≈1586 cm^{-1}) varied with different charge–discharge states, further indicating the highly reversible nature of the reaction. The potassium storage mechanism, as illustrated in Figure 3.14e, indicates that 1 mol of PTCDI-DAQ can store 6 mol of K^+ ions.

Tian et al. synthesized three types of organic polymers – polyimide (PI), polyquinone imide (PQI), and two-dimensional conjugated microporous polymer (PI-CMP) – by condensing NTCDA with three different organic compounds [128]. They utilized these polymers as cathode materials for PIBs. DFT calculations revealed that the carbonyl group serves as the binding site for K^+ ions. Due to the high carbonyl content, PQI exhibited a higher initial capacity than the other two electrode materials. However, the densely packed carbonyl groups resulted in a rapid capacity decline, leading to poor cycling stability. In contrast, PI-CMP demonstrated superior cycling stability and charge–discharge rate performance, attributed to its extended π-conjugated structure and higher porosity, retaining a high specific capacity of 80.8 mAh g^{-1} after 1000 cycles at 1 A g^{-1}. The authors further analyzed the potassium storage mechanism of PI-CMP using FTIR and XPS under different states. The characteristic peak of –C=O in PI-CMP disappeared after discharging to 1.5 V and reappeared after fully charging, indicating the reversible formation of –C–OK. Additionally, the shift of the C=C characteristic peak during the charge–discharge process confirmed the reversible changes in the aromatic environment of PI-CMP. The appearance of the K_{2p} peak further validated the potassiation process after the first discharge. Chen et al. employed poly(pentacenetetrone thioether) (PPTS) as a cathode material for PIBs, capable of undergoing a four-electron transfer process during K^+ ions diffusion [129]. This material demonstrated an impressive specific capacity of 257.9 mAh g^{-1} at a current density of 5 A g^{-1}. Even after 3000 cycles, it maintained a high capacity of 190 mAh g^{-1}.

To address the solubility challenges of organic materials, introducing inactive alkali metal ion-containing groups can enhance polarity, thus reducing solubility while improving potassium storage performance. For instance, in anthraquinone-1,5-disulfonic acid sodium salt (AQDS), the presence of -SO_3Na polar groups significantly decreases its solubility in electrolytes, leading to improved electrochemical performance [130]. During discharge process, the –C=O groups gain electrons and bind with K^+ ions to form –C–OK. Additionally, incorporating non-electrochemically active groups can adjust the cut-off voltage and further mitigate the solubility of small molecules. For example, enolate salts, with their lower molecular weight, have higher theoretical capacity. Zhao et al. synthesized $K_2C_6O_6$ as a cathode material for PIBs [123], where the presence of two –OK groups increased the cut-off voltage and reduced solubility. $K_2C_6O_6$ demonstrated a high capacity of 164 mAh g^{-1} at 10 C within a voltage window of 1.0–3.2 V. The potassium storage mechanism of $K_2C_6O_6$, illustrated in Figure 3.14f, shows that each mole of $K_2C_6O_6$ can store 2 mol of K^+ ions. *In situ* Raman spectroscopy confirmed this mechanism (Figure 3.14g) by demonstrating that the characteristic peak of –C=O gradually disappeared during discharge and reappeared during charge, indicating a reversible potassium storage process. The continuous vibration of the ring structure throughout the charge–discharge cycle indicates its stability, confirming the material retains structural integrity.

Metal-organic frameworks (MOFs) can also serve as cathode materials for PIBs. Wang et al. synthesized a two-dimensional π-d conjugated metal-organic framework based on hexaazatrinaphthalene (HATN), as shown in Figure 3.14h [124]. This framework features dual active centers (Cu-HATNH) and was integrated with carbon nanotubes (Cu-HATNH@CNT) to create an effective cathode material for PIBs. As illustrated in Figure 3.14i, Cu-HATNH@CNT retains a capacity of 143.3 mAh g^{-1} even after 2200 cycles at a current density of 5 A g^{-1}. The potassium storage mechanism of this material differs from that of other organic materials. First, as shown in Figure 3.14j, the C=N bonds in the organic framework HATNH undergo multi-electron reactions, enabling reversible K^+ ion storage. Additionally, the $[CuO_4]$ units also participate in redox processes, allowing for reversible conversion between C=O (C—O• radical) and C—O bonds. Consequently, the neutral $[CuO_4]$ structural unit transforms into a new negatively charged state (Figure 3.14k), which can effectively act as a receptor for subsequent K^+ ion storage.

Summary and Prospects Compared to inorganic electrode materials, organic materials present several advantages. They can be synthesized through mild methods, such as neutral reactions and polymerization at room or moderate temperatures. This eliminates the need for high-temperature calcination typically required for inorganic electrodes. This approach significantly reduces energy consumption and carbon dioxide emissions. Furthermore, many organic compounds can be derived directly from natural renewable resources or synthesized from their derivatives. Organic electrode materials rely on redox-active centers for their reaction mechanisms, enabling them to accommodate larger metal ions. Numerous organic compounds that demonstrate electrochemical activity in LIBs and SIBs have also shown promise in PIBs.

However, challenges persist. A notable issue is that organic carbonyl compounds often dissolve in organic electrolytes, resulting in rapid capacity fading, which is a significant barrier to the development of organic electrodes. Since the research on PIB electrode materials builds on existing systems from LIBs and SIBs, this issue of dissolution-induced capacity fading is also prevalent in PIBs. Thus, the development of new organic electrode materials that offer high capacity and long-cycle stability is crucial for advancing PIB technology.

3.4 Anode Materials for Potassium-Ion Batteries

The anode materials play a crucial and defining role in the performance of PIBs. Potassium metal, being more reactive than lithium, generates unstable passivation layers in the organic electrolytes, making it unsuitable for use as an anode material in PIBs [131, 132]. Consequently, the development of low-potential, high-capacity, and structurally stable anode materials is the key to advancing PIB technology. Due to the large size and high atomic weight of K^+ ions, anode materials face more stringent structural requirements. Therefore, research into PIB anode materials has experienced rapid growth in recent years. Outpacing that of cathode materials for PIBs.

Figure 3.15 summarizes the specific capacities and voltage windows of various anode materials used in PIBs [133]. Based on different storage mechanisms, anode materials are mainly classified into three main categories: intercalation reaction materials, alloying reaction materials, and conversion reaction materials [132, 134–137]. Intercalation-type anode materials primarily include carbon materials and $K_xTi_yO_z$ materials, and K^+ ions can be reversibly embedded and extracted from the interlayer spaces. Meanwhile, conversion-type anode encompasses materials such as TM oxides, metal chalcogenides, and metal phosphides. Whereas alloy-type anodes consist of elements like phosphorus (P), tin (Sn), germanium

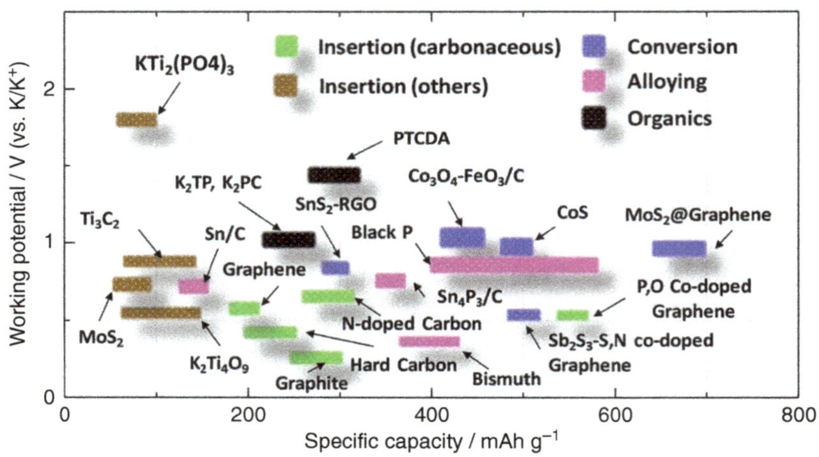

Figure 3.15 Specific capacities and voltage windows of the anode materials for PIBs. Source: [133]. Reproduced with permission/John Wiley & Sons.

(Ge), antimony (Sb), and bismuth (Bi), which can form stable alloys with potassium [138–141]. Additionally, certain organic anode materials undergo redox reactions between potassium ions and specific functional groups of organic compounds. Later in this chapter, we will provide a detailed introduction to the anode materials of PIBs based on the reaction type.

3.4.1 Intercalation-Type Anodes

The primary mechanism in intercalation-type anode materials involves the insertion of potassium ions into the layered structure of the anode, enabling the reversible storage of K^+ ions. For these materials to function effectively, their crystal or microstructure must remain stable during repeated cycles of potassium-ion insertion and extraction. These materials generally possess low theoretical specific capacity, resulting in minimal volume changes during charge and discharge cycles, which is essential to achieve stable potassium storage performance. While intercalation-type anode materials primarily consist of carbon-based materials and other specialized compounds. Carbon-based materials are the most typical example and are classified into graphitic carbon and amorphous carbon. Due to their naturally abundant availability, low cost, and environmental sustainability, carbon-based anodes are considered highly valuable for KIBs. K^+ ions can form stable potassium-graphite intercalation compounds, achieving a theoretical capacity of up to 279 mAh g^{-1} [12, 142]. Although the thermodynamics of potassium–graphite intercalation compounds are favorable, in recent years researchers have developed a comprehensive understanding of the electrochemical behavior of K-insertion into graphite layers.

3.4.1.1 Challenges and Strategies

Interlayer anode materials are renowned for their excellent cycling stability, often characterized by a stable low potential plateau. This feature, when paired with appropriate cathode materials, can substantially enhance the overall energy density of the battery. However, a notable drawback is their relatively low specific capacity, typically constrained to below 300 mAh g^{-1}, which poses a challenge in maximizing energy storage capacity. To overcome this bottleneck problem, the scientific community is actively pursuing exploration and development of high specific capacity materials. Atomic doping and surface coating are two widely adopted and effective strategies aimed at enhancing the performance of existing embedded and extruded materials. Doping introduces impurities like ruthenium (Ru) and silicon (Si) to modify the electronic structure and chemical properties of the material, thereby enhancing the structural stability of electrode materials and promoting increased capacitance behavior, ultimately leading to improved specific capacity to a certain extent. On the other hand, surface coating involves constructing a protective layer to prevent direct contact between electrode materials and electrolytes. This barrier helps to reduce harmful side reactions and further extend cycle life.

3.4.1.2 Graphitic Carbon Materials

Graphite is the most commercially successful anode material in LIBs, and also serves as a promising host for potassium ions, making it a feasible anode candidate in PIBs. Though K-graphite intercalation compounds (K-GICs) were first synthesized as early as 1932, Liu et al. used electrochemical methods to embed potassium ions into graphite interlayer compounds in 2009 [143]. In 2013, Nobuhara et al. conducted the first theoretical calculations on alkali metal-graphite intercalation compounds (Li-GIC, Na-GIC, and K-GIC) using DFT [144]. The findings revealed that Li^+ ions and K^+ ions could embed into graphite, but Na^+ ions struggled hard to be embedded into graphite because the energies of Li-GIC and K-GIC compounds were stable. However, the energy of Na-GIC compounds remained unstable even at low sodium densities (NaC_{16}). In addition, the calculated transition values of lithium ions, sodium ions, and potassium ions in graphite indicate that ions with larger radii diffuse more smoothly in graphite. Ji et al. first reported the electrochemical process of K^+ ions insertion into graphite at room temperature [15] and this investigation was followed by the work of Komaba and Hu et al. in exploring the potassium storage behavior of graphite [28, 145]. Additionally, Xu et al. observed that the discharge and charge platforms during K^+ ion insertion and extraction in graphite occurred at 0.1–0.17 V and 0.27–0.32 V, respectively, which are higher than that of Li^+ ion insertion and extraction [146]. This higher voltage reduces the risk of potassium metal deposition and precipitation during high-rate discharging or low-temperature conditions, further enhancing battery safety.

Qian et al. synthesized a highly [002]-oriented mesoporous graphitic carbon nanocoil (OGCS) through a hydrothermal process, facilitating the formation of both a highly [002]-oriented structure and graphitic carbon nanocoils (Figure 3.16a) [147]. This design also promotes the development of a mesoporous architecture, effectively shortening ion diffusion paths and enhancing electrochemical performance. The coiled structure helps alleviate the volume strain typically experienced during the charge–discharge process in graphite carbon. As demonstrated in Figure 3.16b, OGCS retains a capacity of 99.9 mAh g^{-1} even after 10 000 cycles at a current density of 2000 mA g^{-1}. Additionally, Figure 3.16c presents in situ Raman spectra of OGCS electrodes at 50 mA g−1. During the discharge process, the D and G bands of OGCS weaken and eventually disappear due to the formation of KCx, while these bands reappear during charging. XPS analysis was employed to examine the de-potassiation process of OGCS (Figure 3.16d). Significant changes were observed in the C 1s peak, and the K 2*p* peak became prominent after potassiation, indicating the formation of a K-C compound. After de-potassiation, a weak K 2*p* peak was still observable, likely due to the formation of an SEI layer and residual potassium ions within the OGCS framework, contributing to the initial irreversible capacity. As illustrated in Figure 3.16e and f, compared to conventional graphite anodes, OGCS offers a greater number of active sites for the storage of K^+ ions, effectively shortening the ion diffusion path. Moreover, the unique nanocoil structure of OGCS electrodes significantly mitigates volume changes during cycling.

3.4.1.3 Non-graphitic Carbon Materials

Non-graphitic carbon materials refer to carbon materials with microcrystalline and amorphous structures of graphite. These materials have been extensively studied as anode materials for PIBs due to their disordered or partially disordered structures, which facilitate ion diffusion and electron transport. Among them, non-graphitic carbon materials can be categorized into hard carbon, soft carbon, and heteroatom-doped non-graphitic carbon. The microstructures of graphite, soft carbon, and hard carbon differ significantly [27]. Soft carbon is a carbon material whose degree of graphitization can be adjusted through high-temperature treatment. Due to its controllable interlayer distance and high ionic conductivity, soft carbon materials exhibit high-rate performance when utilized as anodes in PIBs [148, 149].

Hard carbon is characterized by randomly distributed graphitized microdomains. Since precursors with cross-linked structures maintain their amorphous nature during the formation of hard carbon, the development of graphite structures is inhibited, making it difficult for hard carbon to undergo graphitization [150, 151]. The morphology design of hard carbon, particularly the morphology control of hollow carbon nanostructures, plays a crucial role in preventing structure deformation during the embedding and removal process of potassium ions. Such structural control allows hard carbon materials to exhibit excellent cycling performance when used as anodes in PIBs [152]. Building on this, Ju et al. introduced a hard soft composite carbon composed of 20 wt% soft carbon dispersed within the matrix phase of hard carbon microspheres [153]. When employed as an anode material for PIBs, this composite material exhibits good cycling stability and high-rate performance (with a capacity of 200 mAh g^{-1} at 1 C). In addition, carbon materials such as graphene and carbon nanotubes have also been extensively explored for PIBs anodes. In short, carbon materials remain a focal point of research object for PIBs anode development.

Soft carbon materials exhibit excellent cycling stability and tailorable structural properties, making them as highly promising anode materials for PIBs. Compared to highly graphitized carbon nanocages (Figure 3.17a and b) [154], the honeycomb

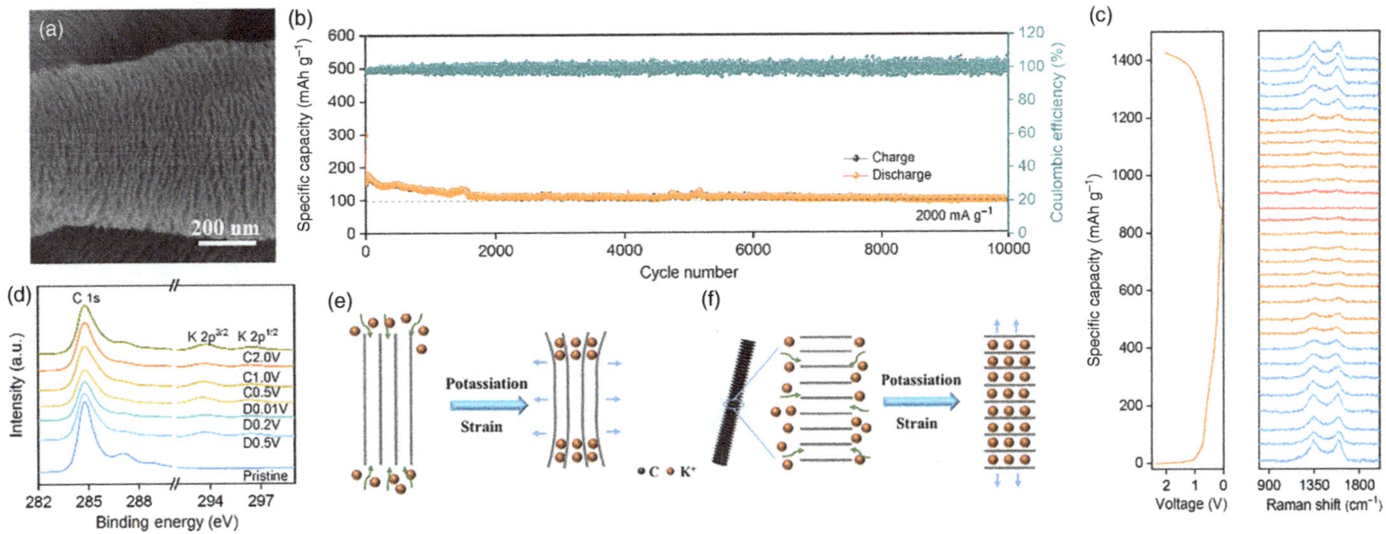

Figure 3.16 (a) SEM images of OGCS. (b) The cycling performance at a high current rate of 2000 mA g^{-1}. (c) *In situ* Raman spectra in the first discharge–charge process of OGCS at 50 mA g^{-1}. (d) *Ex-situ* XPS spectra at different discharge–charge states of OGCS. Schematic illustration of the potassiation process of (e) graphite and (f) OGCS electrodes. Source: Reproduced with permission from [147]/John Wiley & Sons.

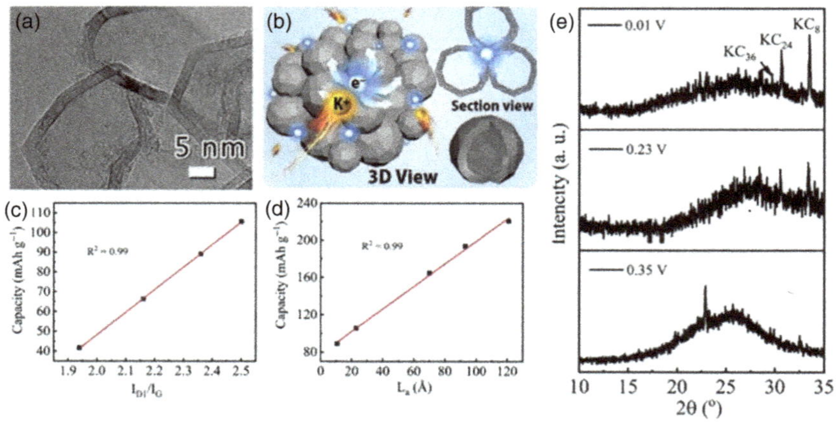

Figure 3.17 (a) HRTEM images and (b) schematic illustration of the structure of CNC. Source: Reproduced with permission from [154]/John Wiley & Sons. (c) The correlation between the capacity between 0.45 and 1.1 V and the amount of edge sites is expressed by the value of ID1/IG. (d) The correlation between the capacity below 0.45 V and the average size of the crystallites L_a. (e) *Ex situ* XRD patterns at various stages of discharge of soft carbon. Source: [155]. Reproduced with permission/John Wiley & Sons.

carbon microsphere material designed and synthesized by Ding et al. demonstrates superior potassium storage performance [156]. This unique structure showcases excellent electrochemical stability, with a capacity of 248 mAh g^{-1} at 500 mA g^{-1}. After 2100 cycles at 100 mA g^{-1}, the reversible capacity of the material was retained at 226 mAh g^{-1}. Liu et al. utilized an ordered MgO template to create N-doped soft carbon clusters with increased interlayer spacing [157]. The harnessed material was used as the anode electrode in PIBs and exhibited capacity after 500 cycles at 1 A g^{-1} was 165 mAh g^{-1}. Li et al. synthesized honeycomb-shaped nitrogen-doped carbon materials from MOF templates, which delivered excellent rate performance (91.8 mAh g^{-1}, 10 A g^{-1}) and long cycle stability (143 mAh g^{-1}, 1 A g^{-1}, 2000 cycles) [158].

Wu et al. proposed an adsorption-intercalation mechanism for potassium storage in soft carbon. In the voltage range of 0.45–1.1 V, K^+ ions are primarily adsorbed at edge defect sites, while capacities below 0.45 V are driven by intercalation of K^+ ions between layers [155]. As illustrated in Figure 3.17c, the variation in ionic diffusion coefficient aligns with changes in the Raman spectrum's I_{D1}/I_G ratio, establishing a correlation between the changes in I_{D1}/I_G and capacity. The results revealed a strong linear relationship ($R^2 = 0.99$) between them, indicating that in the 0.45–1.1 V range, K^+ ion storage in soft carbon is dominated by adsorption at edge defect sites. Furthermore, Figure 3.17d shows a linear increase in capacity ($R^2 = 0.99$) with interlayer spacing when the voltage is below 0.45 V, signifying the onset of intercalation reactions. *Ex situ* XRD analysis further confirms the formation of intercalation compounds KC_{36}, KC_{24}, and KC_8 after complete discharge (Figure 3.17e).

In addition, the introduction of more active sites through regulating the morphology or doping elements can significantly improve the potassium storage performance and cycling stability of the materials. Chen et al. were the first to explore layered porous nitrogen-doped hard carbon microspheres derived from natural polymers (chitin) as raw materials [159]. This unique structure effectively boosts the potassium-ion adsorption capacity and electron/ion conductivity and further improves the electrochemical performance of the material. The energy density analysis further confirmed the ionic features of the bond between K^+ ions and carbon, showing the transferred electron density accumulates around the N-doped sites. Nan et al. synthesized dumbbell-shaped hard carbon (SDHC) doped with sulfur atoms using sucrose as a carbon source. SDHC-700 carbonized at 700 °C exhibited the highest sulfur doping degree (9.3 at.%) and significant reversible capacity [160]. At a high current density of 5.0 A g^{-1}, the specific capacity still reached 174 mAh g^{-1}. Similarly, the nitrogen and sulfur co-doped hard carbon material synthesized by Liu et al. [161]. Featured numerous defect sites, delivering a reversible capacity of 213.7 mAh g^{-1} after 500 cycles at a current density of 0.1 A g^{-1} in a potassium-ion battery. Even after 1200 cycles at 3 A g^{-1}, the reversible capacity of 144.9 mAh g^{-1} was detected. Additionally, Chong et al. successfully constructed N/O-co-doped porous hard carbon nanospheres. Whose core–shell structure mitigated significant volume expansion and promoted electrochemical kinetics [162].

3.4.1.4 Ti-Based Materials

Titanium-based materials are extensively used in numerous energy storage systems due to their advantageous properties such as suitable working voltage, interlayer spacing, low cost, and environmental friendliness. Early theoretical calculations have revealed that MXene, a material with a unique layered structure and excellent physical and chemical properties, holds significant promise as an anode material for PIBs [163, 164].

In a notable study, Shenoy et al. demonstrated through their research that Ti_3C_2 exhibits a theoretical capacity for storing potassium at 191.8 mAh g^{-1} [165]. Furthermore, the diffusion barrier for K^+ ion insertion into Ti_3C_2 is 0.103 eV, which is significantly lower than the diffusion barrier for Li^+ ion insertion into graphite (0.3 eV). Under normal circumstances, lower diffusion barriers typically correlate with higher rate performance, which is more conducive to electrochemistry, highlighting that Ti_3C_2 exhibits superior rate performance. Huang et al. investigated the potassium storage performance of carbon-loaded MXene/$MoSe_2$, reporting that pure MXene achieved a stable capacity of 42 mAh g^{-1} at a current density of 200 mA g^{-1} [166]. Additionally, Alshareef et al. prepared porous K-V_2C nanosheets through acid/alkali treatment [167]. These porous K-V_2C nanosheets maintained a specific capacity of 195 mAh g^{-1} after electrochemical testing at 50 mA g^{-1}. This finding underscores the potential of rational design and modification of MXene materials in significantly enhancing the electrochemical performance of PIBs.

In addition to the previously discussed strategies, experimental studies have demonstrated that under specific conditions, synthesizing various high-performance

MXene-based derivatives from MXene precursors can effectively improve the electrochemical performance of PIBs. Zhou et al. introduced a composite that combined MXene with Super-P particles as a binder and conductive matrix [168]. This synthesis design promotes rapid ion and electron transport, inhibits the solubility of PTCDA, and enhances potassium adsorption, achieving a high capacity of 462 mAh g^{-1} at 50 mA g^{-1}. Guo et al. took a step ahead by designing SQ@MA composite material, anchored on $Ti_3C_2T_x$ MXene aerogel, incorporating antimony monoatom (SA) and quantum dot (QD). $Ti_3C_2T_x$, as a matrix, stabilizes the active materials due to its excellent conductivity, mechanical properties, and widely exposed surface [169]. Besides, Sb quantum dots also enhance potassium storage capacity and charge-transfer kinetics. The introduction of graphene oxide further enhances structural stability by preventing restacking of MXene. SbSQ@MA composite anode exhibits remarkable reversible capacity (521 mAh g^{-1} at 0.1 A g^{-1}) and exceptional cycling stability (94% capacity retention after 1000 cycles).

Cao et al. utilized $Ti_3C_2T_x$ nanosheets as fast electron/potassium-ion conductors and combined them with GNF to construct GNFM electrodes (Figure 3.18a–c) [170]. By leveraging the two-dimensional structure, conductivity, and flexibility, they integrated multiple functions such as binders and conductive agents, significantly improving potassium storage capacity, cycling performance, and rate performance. XRD analysis revealed a new phase transition mechanism in the graphite anode, while theoretical calculations confirmed the enhanced potassium adsorption/migration at the GNF/MXene interface (Figure 3.18e). This work further demonstrated that the constructed non-flammable potassium-ion battery exhibited excellent cycling performance and high energy/power density (Figure 3.18d), highlighting the enormous potential of MXene in the field of electrochemical potassium storage (Figure 3.18f). Finally, Qin et al. developed a method to introduce ultra-high P-doped graphene through MXene [171]. The harnessed N/P co-doping introduced defects and improved K^+ ion storage performance. The prepared ultra-high phosphorus/nitrogen co-doped flexible reduced graphene oxide/MXE (NPrGM) electrode displayed high reversible capacity, with a decay of 0.0005% per cycle after 10 000 cycles at 2 A g^{-1}.

Summary and Prospects Graphite anodes are not comparable to soft and hard carbon, particularly regarding cycle life and high-rate performance. Hard carbon exhibits a good cycle life, owing to its excellent mechanical stability, while soft carbon has excellent capacity retention at high rates. In addition, recent advancements in carbon-based anode for PIBs, including hollow carbon, mesoporous carbon, carbon nanotubes, carbon microspheres, and carbon derived from MOFs, have driven good electrochemical performance. Heteroatoms have also been utilized as dopants to modify carbonaceous electrodes and deliver an improved electrochemical performance. However, these strategies still need to overcome fundamental and technological challenges in designing carbon-based anodes with a reasonable degree of graphitization, fine-tuned pore structure, and selected surface functionality to develop anode materials for practical PIBs.

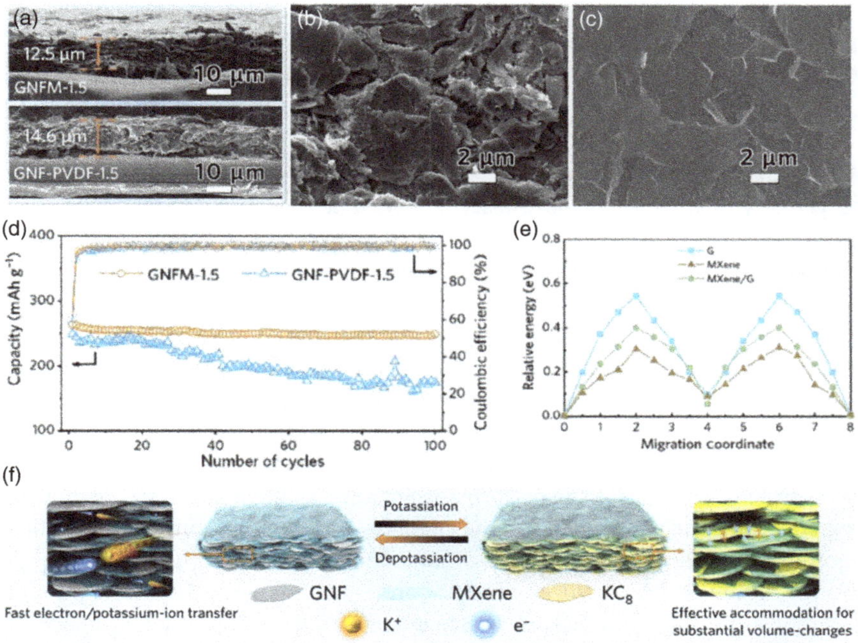

Figure 3.18 (a) SEM images of GNFM-1.5 and GNF-PVDF-1.5 cross sections and top views of (b) GNF-PVDF-1.5 and (c) GNFM-1.5. (d) Cycle performances measured at 50 mA g^{-1}. (e) Potassium diffusion paths on the corresponding diffusion barrier energy. (f) Schematic illustration for GNFM electrodes during potassium storage. Source: Reproduced with permission from [170]/John Wiley & Sons.

3.4.2 Alloying Reaction-Type Anodes

Compared to carbon-based anodes, potassium (K) alloys containing elements such as silicon (Si), tin (Sn), antimony (Sb), lead (Pb), and germanium (Ge) offer the potential for significantly higher specific capacity [172]. The underlying oxidation–reduction reaction can be expressed as follows:

$$M + nK^{+} + ne^{-} \rightleftharpoons K_{n}M$$

Among them, M represents an alloying element. Unlike intercalation-type anodes, the ability of certain elements to alloy with K metal generally involves a higher electron transfer number, resulting in a substantial increase in capacity.

Alloy materials are gaining attention due to their high conductivity and high theoretical capacity. They mainly rely on alloying reactions with potassium to achieve energy storage, though they tend to undergo considerable volume expansion during the cycling process. Recently, Kim et al. calculated and compared the capacities of various anode materials with K and Na alloys using first-principles-based DFT computations, drawing two key findings: typically, the capacity of K alloy materials is smaller than that of Na alloy materials; the average voltage of K is higher than that of Na, indicating the loss of energy density in the K system [173]. Due to the relatively large radius of K^+ ions, the volume expansion of the K systems is more pronounced than in Na systems, leading to capacity decay. To mitigate the aforementioned issues, researchers have optimized the structure of alloy materials through strategies such as designing unique structures and surface coating protective layers [141, 174, 175].

3.4.2.1 Challenges and Strategies

The primary advantage of alloy-based anode materials is their relatively large theoretical specific capacity, making them promising potential for future battery technology. However, these materials face several significant challenges, with key drawbacks, including severe volume expansion, non-wave interface reactions, voltage hysteresis, multiple phase transitions accompanying complex reactions, intense electrode polarization phenomena, and unstable SEI films, all of which severely impact the electrochemical performance of batteries and demand in-depth exploration and innovative solutions from the scientific research community. To overcome the above challenges, researchers have concentrated their efforts on several aspects to achieve a leap in the performance of alloy anode materials: (i) structural design optimization to alleviate volume changes. For example, constructing hollow or porous structures and designing reasonable nanostructures. (ii) Interface engineering to enhance the electrochemical performance and safety of batteries through interface control strategies such as surface modification and formation of stable SEI layers. (iii) Phase change control, by adjusting the reaction conditions, selecting appropriate alloying elements or bulk materials, etc., achieves optimized control of the phase change process and reduces its adverse effects on battery performance. (iv). The improvement of cycle stability involves but is not limited to optimizing the synthesis process of materials, controlling the large particle adsorption distribution, incorporating appropriate amounts of electrical additives, and other measures to enhance the structural stability, leading to better electrochemical activity.

3.4.2.2 P-Based Anodes

Researchers have demonstrated that red phosphorus/black phosphorus can undergo alloying reactions with alkali metals [176–178]. Given the high theoretical capacity of P in LIBs and SIBs ($2596\,mAh\,g^{-1}$), exploring the electrochemical performance of P anode materials in PIBs has become a topic of great interest [176]. Xiong et al. reported a red phosphorus@carbon composite material for the first time. The unique structure of this composite material enhanced the electron transfer rate, shortened K^+ ion transfer distance, and increased active surface area. As a anode material for PIBs, the P@CN composite electrode exhibited a high discharge capacity of $655\,mAh\,g^{-1}$ with solid rate performance. It proposed that the final reaction product is PK, corresponding to a theoretical capacity of $843\,mAh\,g^{-1}$, making P a promising PIB anode material. However, its cycling performance is still not particularly ideal and researchers need to continue their efforts [178, 179].

Phosphorus exists in three primary forms: white phosphorus, red phosphorus, and black phosphorus. Each of the three has its advantages and disadvantages, with unstable white phosphorus and poor conductivity of red phosphorus, and black phosphorus is expensive. Researchers have identified KP and K_4P_3 as the most stable compounds for phosphorus alloying, delivering high theoretical specific capacities of 865 mA h g^{-1} and 1154 mA h g^{-1}, respectively [173, 176]. Moreover, K_4P_3 exhibits a lower voltage plateau than that of KP, indicating that K_4P_3 possesses a higher energy density. Despite the advantages, phosphorus-type alloy materials still face numerous shortcomings, including low conductivity of phosphorus, large volume expansion, and unstable SEI film, which inevitably lead to poor electrochemical properties. The current methods to improve phosphorus with various carbon materials, design excellent nanostructures, or convert elemental phosphorus into metal phosphides (Sn_4P_3, GeP_5, CoP), which to some extent optimize the electrochemical performance. However, further progress requires focusing on the following aspects for improvement: designing simple and low-cost preparation methods, conducting in-depth research on reaction mechanisms, especially metal phosphide energy storage processes, and modifying P elements to some extent (black P, small molecule P).

Red phosphorus and black phosphorus are increasingly utilized in electrode materials, particularly for PIBs. Phosphorus possesses low redox potential and can form potassium phosphide-rich crystal forms with a high theoretical capacity (KP) of 865 mAh g^{-1} [173, 180]. Kang et al. synthesized MoP@NP-HCNTs composite using the solvothermal method as the anode of PIBs, delivering a relatively high reversible capacity, but the decay rate is too fast to meet expectations [181]. Subsequently, Sultana et al. prepared nanocomposites featuring black phosphorus as the main active component through ball milling, showing a reversible capacity of 617 mAh g^{-1} in the first cycle when used as the anode of PIBs [182]. Unfortunately, the rapid decline in capacity is still problematic and needs to be overcome. Therefore, Xu et al. designed an innovative approach to synthesize a novel nanocomposite anode material by anchoring red P NPs onto a 3D CNS framework (Figure 3.19a) [178]. This nanocomposite material showcased exceptional electrochemical performance, achieving a high reversible capacity of 655 mAh g^{-1} at 100 mA g^{-1} and a high specific capacity of 323.7 mAh g^{-1} at 2000 mA g^{-1} (Figure 3.19b and c). Through theoretical calculations and other testing analyses, it was found that the reaction mechanism could be expressed as $P + K + e^- \rightarrow KP$, corresponding to a potassium storage capacity of 843 mAh g^{-1}. Additionally, Liu et al. first proposed the concept of restricted phosphorus within carbon nanotube-backboned mesoporous carbon (TBMC) materials. In PIBs, P@TBMC exhibited a stable capacity of 244 mAh g^{-1} after 200 cycles at 0.5 A g^{-1}. The findings added the results that the ability to quickly transfer electrons, and the unique structure of the composite material provides ample space to minimize the volume changes during the potassium oxidation process [183].

3.4.2.3 Sn-Based Anodes

Alloy-based materials have garnered a significant focus of research in the realm of alkali metal ion batteries due to their favorable conductivity and high theoretical capacity. Among alloy materials, tin (Sn) stands out for much attention due to its environmentally friendly nature, abundant reserves, and cost-effectiveness. The potassium storage mechanism of Sn primarily involves an alloying reaction of potassium, but there are problems such as volume expansion during the charging and discharging processes, which can cause rapid capacity decay. To mitigate these

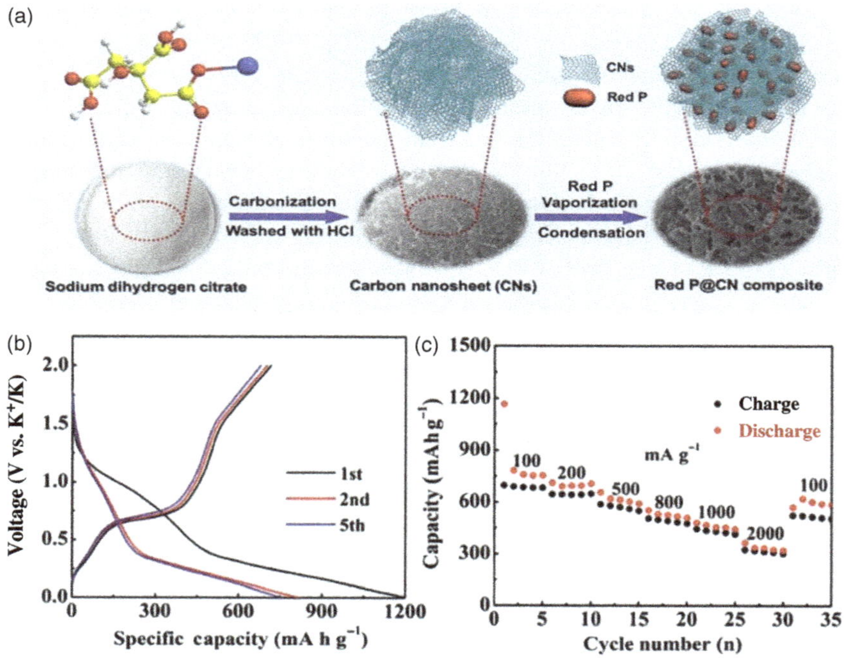

Figure 3.19 (a) Schematic illustration of P@CN. (b) Discharge–charge profiles and (c) rate performance of the P@CN electrode. Source: Reproduced with permission from [178]/John Wiley & Sons.

issues, researchers have adopted a method of reducing the size of tin-based materials and combining them with carbon materials to shorten the electron/ion transport path, thereby alleviating material volume expansion and improving electrochemical performance.

Sultana et al. were pioneers in investigating the electrochemical behavior of Sn-based materials in PIBs. However, the significant volume change that Sn undergoes during the cycling process results in poor electrochemical performance [182]. Research indicates that tin participates in two-step alloying reactions in PIBs, forming K_4Sn_9 and KSn, respectively, with a theoretical capacity of 197 mAh g^{-1}. The alloying reaction of the system was studied by *in situ* transmission electron microscopy (TEM), revealing the electrochemical behavior of Sn nanoparticles in PIBs [184]. Moreover, Sn nanoparticles suffer severe pulverization during the cycling process, resulting in poor cycling stability. To address this issue, Huang et al. developed 3D-layered porous carbon/Sn composites using *in situ*-generated NaCl crystals as templates [185]. A high reversible capacity of 276.4 mAh g^{-1} was achieved after 100 cycles at 50 mA g^{-1}. Compared with previous work, adjusting Sn nanoparticles to an appropriate size of 22–94 nm and embedding them into carbon thin films with 3D network structures can effectively alleviate the problem of structural collapse caused by volume expansion, thereby improving electrochemical performance. Additionally, researchers have prepared Sn nanoparticles coated with

N-doped carbon materials (Sn/NPC), and the excellent carbon coating has improved the specific capacity and rate performance [186]. Xing et al. produced a Sn@rGO composite, exhibiting a high specific capacity of 200 mAh g^{-1} after 50 cycles at a current density of 100 mA g^{-1} [187]. Nonetheless, after 300 cycles, the capacity retention rate is only 25%, likely due to the pulverization of submicron-sized tin particles during repeated potassium/potassium removal processes.

Phosphorus exhibits high theoretical capacity and excellent performance as a PIBs anode material. Guo et al. were the first to utilize Sn_4P_3/C as an anode material for PIBs [181], achieving the reversible capacity of Sn_4P_3/C was 384.8 mAh g^{-1} at 50 mA g^{-1}. Subsequently, they also proposed the reaction mechanism of $Sn4P_3$/C composite material in PIBs. Next, they combined Sn_4P_3 with N-doped carbon nanotubes, which can effectively alleviate material volume expansion [180]. The results indicate that the cycling stability and specific capacity have been significantly improved.

In another study, Ci et al. synthesized a layered porous carbon-loaded system Sn_4P_3@C that maintained a high reversible capacity of 181.5 mAh g^{-1} after 800 cycles at 0.5 A g^{-1} [188]. Zhao et al. synthesized $SnP_{0.94}$/graphene oxide nanocomposites using a simple thermal injection method. At current densities of 25, 50, and 100 mA g^{-1}, the specific capacities of the composites were 245, 190, and 162 mAh g^{-1}, respectively. The excellent electrochemical performance is attributed to the ability of graphene oxide to alleviate volume expansion, accelerate electron and charge transfer, and reduce the contact between $SnP_{0.94}$ and the electrolyte [189].

In summary, nanosizing Sn-based materials and integrating them with various carbon materials such as carbon, graphite, and graphene can significantly improve the electrochemical performance of PIBs.

3.4.2.4 Sb-Based Anodes

Antimony is recognized for its high theoretical capacity (660 mAh g^{-1}, corresponding to K_3Sb) and low reaction potential used in PIBs, which also possesses a high specific capacity, exhibiting significant volume changes (400% volume expansion rate). To mitigate these volume changes, common strategies include composite with conductive carbon materials and preparation of metal antimonides. For example, Professor Mai Liqiang's research group prepared Sb nanoparticles *in situ* enclosed in a 3D carbon framework (3D-SbNPs@C) [190]. Additionally, researchers also conducted experiments to fabricate hollow Sb_2Se_3@C microtubes, leveraging both conversion energy storage and alloying energy storage to enhance the specific capacity of batteries [191].Although Sn-based/Sb-based alloying materials possess high specific capacity, the critical problem remains volume expansion. While simple carbon coating provides some relief, researchers need further improvements in the aspects of structural design, electrolyte, and energy storage mechanisms.

Zhang et al. focused on Sb as the research object and employed an experimental method combining electrospinning and ion exchange to prepare carbon nanofiber composite materials with core–shell structure (Sb@CNFs) [192]. The nanosized active materials help to reduce the volume changes during charge and discharge cycles, while the hollow structures accommodate volume expansion. The Sb@CNFs electrode demonstrates exceptional performance after 2000 cycles at $1\,A\,g^{-1}$, the electrode still exhibits a high reversible specific capacity of $227\,mAh\,g^{-1}$, demonstrating excellent potassium storage electrochemical performance. Tian et al. utilized an electrodeposition method to deposit a multilevel structure of Sb single element assembled from nanoparticles on the surface of MXene through electrochemical methods, resulting in a self-supporting material MXene@Sb composite materials. This material achieved a reversible specific capacity of $271\,mAh\,g^{-1}$ at $500\,mA\,g^{-1}$ [193]. The electrode's specific capacity is also well restored, showcasing good reversibility and stability.

The specific capacity of Sb-based materials is $660\,mAh\,g^{-1}$, but their significant volume expansion is large, resulting in a rapid capacity decay of the material [194]. Additionally, the continuous reconstruction of the SEI film during each electrochemical cycle further exacerbates the structural deterioration of the anode material.

To address the aforementioned issues, researchers have optimized the structure of alloy materials through strategies such as designing unique structures and surface coating protective layers [141, 175]. Cheng et al. designed and synthesized Sb nanoparticles embedded in porous carbon anode materials using Sb MOFs as templates. After 100 cycles at a current density of $100\,mA\,g^{-1}$, this material achieved a specific capacity of $497\,mAh\,g^{-1}$ and an average coulombic efficiency of 99.1% [195].

Sb/3D porous carbon composites exhibit high capacity (discharge capacity$\sim 478\,mAh\,g^{-1}$ at $200\,mA\,g^{-1}$) and rate performance (discharge capacity$\sim 200\,mAh\,g^{-1}$ at $1\,A\,g^{-1}$) in PIBs [190]. To mitigate the volume change of Sb during charge and discharge processes, An et al. synthesized microscale nanoporous Sb (NP Sb) through vacuum distillation [196]. The NP Sb anode material for PIBs exhibited discharge capacities of 560 and $265\,mAh\,g^{-1}$ at current densities of 50 and $500\,mA\,g^{-1}$, respectively; however, its cycling performance was not satisfactory. Recently, Zheng et al. synthesized Sb@SCN composite materials [197], displaying a discharge capacity of $504\,mAh\,g^{-1}$ after 200 cycles in PIBs. This excellent performance can be attributed to the uniform embedding of Sb nanoparticles in carbon nanospheres and the optimization of the electrolyte. These results underscore the potential of Sb as a promising anode material for PIBs.

Lu et al. embedded Sb clusters into graphite layers using ultrasound and ball milling to create composite materials of Sb and graphite. Based on its novel structure, Sb graphite composite material ($Sb_7@G_3$) [198] effectively mitigates the volume expansion during cycling and has an excellent cycling performance of $238\,mAh\,g^{-1}$ at $2\,A\,g^{-1}$. Furthermore, Han et al. conducted *in situ* XRD analysis on the $Sb_7@G_3$, elucidating the charging and discharging reaction mechanism of Sb $\leftrightarrow$ KSb $\leftrightarrow$ K3Sb. Han et al. prepared Sb nanocrystals embedded in ultra-thin CNSs utilizing a one-step solvothermal decomposition reaction [199]. After 600 cycles at $0.2\,A\,g^{-1}$, the nanocrystals displayed a high stable capacity of $247\,mAh\,g^{-1}$ and demonstrated excellent rate performance. In addition, Cheng et al. utilized Sb MOF as a precursor and embedded Sb NPs into a porous carbon network through carbonization. After cycling for 100 cycles at $100\,mA\,g^{-1}$, the material maintained a high specific capacity of $497\,mAh\,g^{-1}$ (with a capacity decay of only 0.054% per cycle) [195]. Additionally, Zhang et al. reported a novel anode material that embeds Sb NPs after 1500 cycles at 1 A g^{-1} in nitrogen and phosphorus co-doped mesoporous carbon nanofibers, which sustained a capacity of 130 mAh g^{-1} [200]. Designing and optimizing the structure of Sb-based materials is crucial for solving the problems, such as (i) reducing the size of anode materials to the nanometer level; (ii) composite with carbon-based or modified carbon-based materials to form structures such as nanoparticles, hollow spheres, nanotubes, and nanosheets; (iii) increase the active sites on the surface of the material. These strategies enable composite materials to accommodate the volume changes of Sb during cycling, promote rapid electron/ion conduction, and thus endow PIBs with excellent electrochemical performance.

3.4.2.5 Bi-Based Anodes

Bi belongs to the same main group (VA) as Sb, shares similar chemical properties, and is recognized for its nontoxic nature, known by the title of "green metal" in materials science [201]. Bismuth-based materials are particularly attractive to considerable attention for PIBs due to several features such as the large interlayer spacing along the c-axis (d(003) = 0.395 nm), which facilitates ion insertion, along with low thermal conductivity, a low melting point, and a high theoretical capacity (384 mAh g^{-1}) [202].

Gabaudan et al. investigated the K^+ ion storage mechanism of Bi, showing that during the potassium transformation process, Bi undergoes phase evolution forming KBi_2, K_3Bi_2, and K_3Bi. Conversely, these phases reverse during the potassium removal process [194]. The complete conversion to K_3Bi induces a dramatic volume expansion of up to 411%, resulting in pulverization and electrode failure. This issue can be mitigated through electrode engineering. Li and Chen et al., in early 2018, described a Bi anode with high energy density [202]. They discovered that the interaction between Bi atoms on the surface and dimethoxymethane (DME) molecules in the electrolyte led to the formation of a porous network in commercial Bi during battery operation. Benefiting from this porous structure, it provides fast K^+ ion transfer kinetics and alleviates the volume change of Bi. At 800 mA g^{-1}, the initial coulombic efficiency is 87.2%, the initial capacity is 371.4 mAh g^{-1}, and the subsequent coulombic efficiency is as high as 99.6%. After 300 cycles, the capacity retention rate is 86.9%. These results provide insights into the application of alloy anodes in PIBs.

After this research, subsequent studies focused on modifying the morphology and Bi-based nanostructures to alleviate the increase in volume during the potassium embedding process, enhancing electrochemical performance. An efficient approach is to couple Bi with 2D carbonaceous materials, permitting ultrafine Bi nanoparticles (~3 nm) to be loaded onto 2D graphene [203]. A few bismuth layers less than 2 nm thickness can be formed through a direct method involved in constructing Bi nanostructures with 2D morphology [204, 205], which is attractively conducive to the diffusion of potassium ions, the penetration of electrolyte, and the buffering of the c-axis volume changes. These nanostructures achieved 423 mAh g^{-1} at 2500 mA g^{-1} and retained a remarkable capacity of 201 mAh g^{-1} after 2500 cycles at 20000 mA g^{-1}.

Dispersing Bi nanoparticles within 3D carbonaceous matrices is another effective strategy for profound electrochemical performance. Yu et al. reported Bi nanospheres (10–150 nm) embedded in large pores of a 3D porous graphene framework [206], which maintained structure integrity of ultra-small Bi nanospheres and attained a reversible capacity of approximately 163 mAh g^{-1} after 400 cycles at 1000 mA g^{-1}. Similar to Sb-based anode materials, carbonaceous substrates can be formed into CuBi bimetallic materials by conductive metal substrates (such as copper) to replace and overcome sluggish potassium-ion diffusion and electrode expansion [207].

Su et al. proposed an innovative 3D architecture by utilizing Bi-based metal-organic framework precursors. In this structure, Bi nanoparticles are uniformly dispersed along carbon nanorods and coated with ultra-thin carbon on the surface, improving structural stability, electrode conductivity, and mitigating volume expansion [208]. After 600 cycles at 100 mA g^{-1}, the capacity decay of the material is only 0.038%, indicating high Coulombic efficiency. The few-atom-thick layered nanosheets facilitate the mitigation of volume changes, accelerate ion diffusion, and offer abundant active sites due to their expanded surface area. In the following study, Yang et al. established a superior multi-core shell system Bi@N-doped 3D carbon nanospheres [209]. This uniform carbon coating successfully prevents the formation of SEI directly on Bi nanoparticles, thereby reducing the significant volume change caused by particle fracture benefiting from retaining a capacity of 152 mAh g^{-1} at 10 A g^{-1}.

Subsequently, the construction of additional 3D carbonaceous matrices was also reported, including carbon-coated bilayer Bi, Bi@N-doped carbon nanocages, and Bi@3D porous carbon frameworks [210, 211]. The Bi@3D porous carbon framework was prepared through a straightforward effective electrochemical milling process, delivering 285 mAh g^{-1} at 20 000 mA g^{-1}, maintaining 92.8% of its initial capacity. These findings highlight its ability to shrink charge diffusion paths and enable high-speed dynamic processes [210].

Zhu et al. reported flower-shaped Bi_2WO_6 with a stable capacity of 216 mAh g^{-1} after 300 cycles at 1000 mA g^{-1} [212]. Zhu et al. reported $Bi_{12}O_{17}Cl_2$ bilayer nanosheets, which exhibit improved electrochemical performance in terms of morphology, with an initial capacity of 420 mAh g^{-1} at 10 000 mA g^{-1} [213].

In addition, Wang et al. developed an innovative flower-shaped graded bismuth subcarbonate (BCO) as the anode material for PIBs (Figure 3.20a and b) [214]. Its two-dimensional nanosheet structure endows it with high theoretical capacity (631 mAh g^{-1}) and excellent K^+ ion diffusion performance. The specially designed 5 M KFSI-G2 electrolyte forms a robust and inorganic-rich SEI through the coupling of long-chain G2 and KFSI, significantly enhancing battery stability. BCO exhibits excellent long-cycle stability during the alloying/dealloying process, which maintains a capacity of 300 mAh g^{-1} over 1500 cycles at 0.2 A g^{-1} with an attenuation rate of only 0.007% (Figure 3.20c), showcasing an extremely rare among alloy anode materials. The distinct voltage platform of the BCO anode (0.3 V/0.55 V)

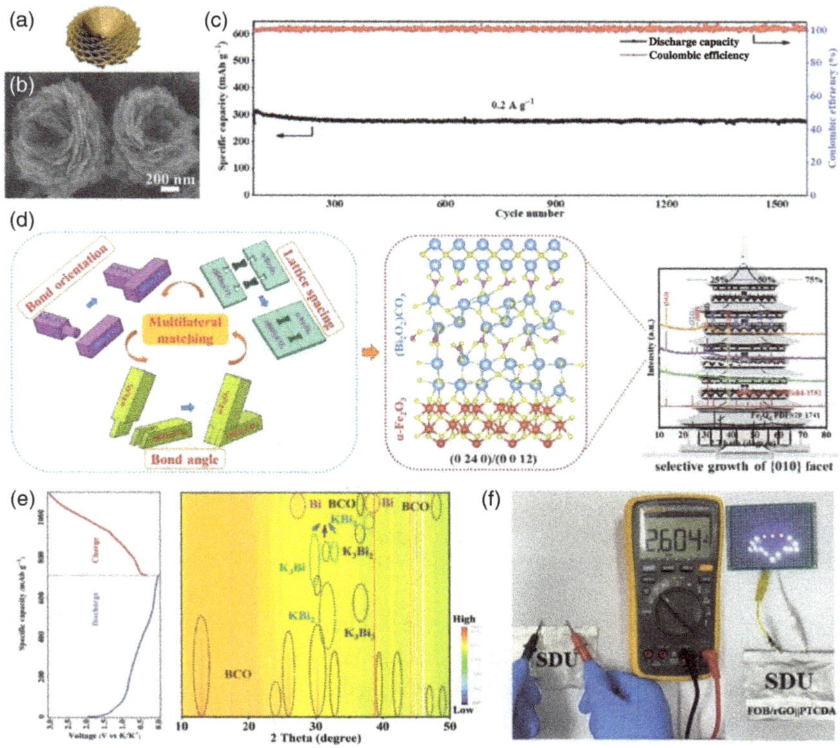

Figure 3.20 Structure and composition of the obtained BCO: (a) demonstration. (b) HAADF-STEM image. (c) Long cycle performance at 0.2 A g^{-1}. Source: Reproduced with permission from [214]/John Wiley & Sons. (d) Schematic illustration of multilateral matching of α-Fe_2O_3 to trigger preferred growth of facets in $(BiO)_2CO_3$. (e) Constant current charge–discharge curve and corresponding contour map in *in situ* XRD measurement. (f) Its pouch cell is employed in lighting LED models. Source: Reproduced with permission from [215]/John Wiley & Sons.

allows it to be combined with the PTCDA cathode to achieve high-performance full PIBs (energy efficiency of about 82%). The integrated full battery performs well at different current densities, especially with stable cycling of over a hundred times at 0.03 A g^{-1}. Furthermore, Wang et al. developed a preferred facet growth strategy for $(BiO)_2CO_3$ in a two-phase system, optimizing the FO/BCO composite material (FOB-50) and its rGO composite structure and controlling its rGO composite structure and nanoparticle size between 5 and 10 nm (Figure 3.20d) [215]. The analytical findings were revealed through DFT computations and experimental characterization, indicate that the high lattice matching between α-Fe_2O_3 and $(BiO)_2CO_3$, similar to the traditional Chinese mortise and tenon structure, promotes the preferential growth of the $(BiO)_2CO_3$ plane. The FOB-50 electrode exhibits excellent cycling stability, retaining 70% of its capacity after 32 months. With the

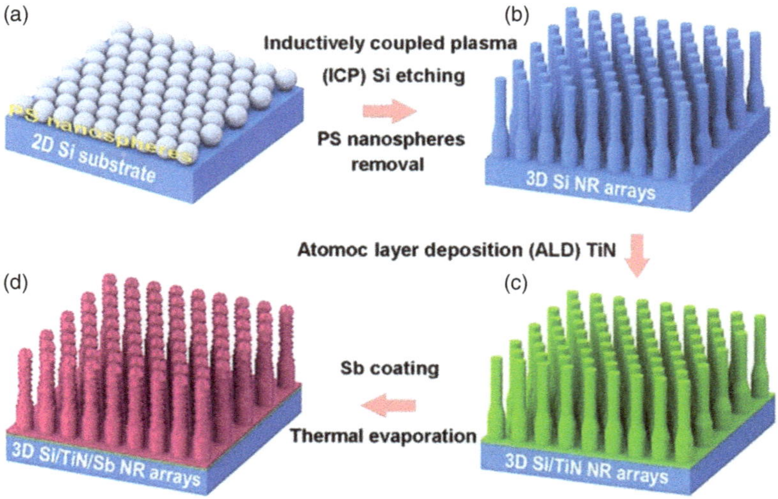

Figure 3.21 (a–d) Illustration of the fabrication processes of 3D Si/TiN/Sb nanoelectrodes. Source: [218]. Reproduced with permission/John Wiley & Sons.

addition of rGO, the FOB-50/rGO electrode exhibited fast K^+ ion diffusion kinetics and high specific capacity (590 mAh g^{-1}), maintaining stability over long cycles (1500 cycles, 400 mAh g^{-1}). *In situ* XRD revealed reversible transformation and alloying mechanisms (Figure 3.20e). The full battery (paired with PTCDA cathode) performed excellent performance, with an energy efficiency of up to 75%. As shown in Figure 3.20f, a practical pouch battery assembly was used to power an LED model, further verifying its commercial applications. This study not only advances the preferred surface growth technology of nanomaterials but also deepens valuable insights into the correlation between crystal structure and potassium-ion energy storage performance, paving the way for future high-performance battery design.

3.4.2.6 Si-Based Anodes

Silicon is an environmentally friendly and abundant material that has garnered attention as an alloy-type anode material with extremely high specific capacity (3579 mAh g^{-1}) in LIBs. However, in the case of PIBs, the potassium oxidation of c-Si is thermodynamically disadvantageous because the formation energy of KSi is positive. An investigation by Sultana et al. on electrochemical potassium mineralization of silicon found no clear evidence of silicon potassium mineralization [216]. XRD results further confirm the absence of K_xSi_y compounds after testing the electrochemical potassium fertilization of silicon in PIBs. These results suggest that c-Si is unfavorable for electrochemical alloying with potassium. On the other hand, theoretical predictions indicate the possibility of dealloying between a-Si and K alloys. It is suggested that each Si atom in a-Si can store 1.1 K ions ($K_{1.1}Si$), offering a capacity of 1049 mAh g^{-1}. Despite the large size of the K^+ ion, it diffuses rapidly within amorphous silicon at a rate of 5.3×10^{-8} cm^2 s^{-1} [217]. Figure 3.21 illustrates a schematic diagram of the preparation process of 3D Si/TiN/Sb nanoelectrodes. The high ion diffusion rate of K^+ ions in amorphous KSi is primarily attributed to weak electrostatic K-Si attraction, relatively high K^+ ion concentration, and the formation of isolated small Si clusters during the potassium oxidation process. In conclusion, due to its low capacity, silicon fails to exhibit similar electrochemical properties in SIBs/PIBs as it does in LIBs [218]. Currently, there are very limited literature reports on pure silicon alloy anodes for PIBs, and this opens an opportunity for further experimental research to understand the alloying mechanism of pure silicon with K in PIBs.

To enhance the efficiency of silicon anodes, researchers have explored and developed various strategies, including nanostructures, doping of elements, surface modification, composite materials, and alloying of silicon with metal materials, metal oxides, and carbon-based materials. Sun et al. synthesized crystalline silicene from Zintl phase $CaSi_2$ and demonstrated that potassium ions could be effectively inserted/extracted, exhibiting enhanced cycling stability (the capacity remained stable after 3000 cycles, and the coulombic efficiency sustained above 99.4%) [219]. Contrary to the established view of "inert silicon," the reversible kinetic controlled K-Si phase transition occurring in silicene was elucidated by *in situ* synchrotron radiation XRD, and KSi became the main discharge product, achieving a reliable reversible potassium storage capacity of 180.1 mA h g^{-1}.

Yang et al. successfully constructed a hierarchical porous C/SiO_x composite material using brown coal as the precursor through simple carbonization and activation steps, making it a promising anode material for PIBs [220]. SiO_x nanoparticles are effectively embedded into the carbon matrix, significantly enhancing the electrical conductivity of SiO_x and effectively alleviating the stress caused by volume changes during charge–discharge cycles, thus preventing particle aggregation. The graded porous properties exhibited by C/SiO_x composite materials provide ideal channels for the smooth penetration of electrolytes and the rapid intercalation and deintercalation of K^+ ions, thereby optimizing the electrochemical performance of the battery. The graded porous C/SiO_x anode prepared at 600 °C maintains a reversible capacity of up to 370 mAh g^{-1} after 100 cycles at a current density of 0.1 A g^{-1}, achieving an astonishing capacity retention rate of 90%. Even after 1200 cycles of testing at a higher current density (1.0 A g^{-1}), the anode still exhibited a stable capacity of 208 mAh g^{-1}, demonstrating its excellent long-term cycling stability.

3.4.2.7 Ge-Based Anodes

Germanium (Ge) is an attractive semiconductor material that is favored by researchers for its high electron/hole mobility, nontoxic and harmless properties, and abundant reserves. While the crystalline phase of Ge is inactive toward potassium ions, its amorphous phase along with alloy product KGe offers a theoretical specific capacity of 369 mA h g^{-1}, surpassing that of KPb [221]. Like other alloy materials, the key challenges for Ge electrodes include the deactivation of active materials and significant volume changes. To promote the further development of Ge electrodes in the PIBs field, researchers require further improvements in several aspects: designing reasonable nanostructures to control volume changes, exploring suitable electrolytes/electrolyte additives to maintain SEI film stability, designing germanides, and appropriate conductive material composites. Kohandehghan et al. demonstrated amorphous Ge could lower the nucleation barrier of the Na_xGe phase, improving solid-state ion diffusion performance, and enhancing sodium-ion storage performance [222]. However, the production of amorphous Ge remains a problem that involves complex methods, for instance, reducing $GeCl_4$ into Ge through utilizing strong reducing alkali metals such as Li, Na, K, or organic alkali reagents, and needs to be resolved through straightforward methods. Yi et al. synthesized amorphous Ge using a simpler acid etching method at room temperature, and formed amorphous Na-Ge and K-Ge alloys with sodium and potassium during discharge, exhibiting better electrochemical performance than crystalline Ge [223].

In addition, designing nanostructures incorporating with carbon materials proves an effective method to improve the electrochemical performance of germanium-based materials in PIBs. Liu et al. synthesized a dual carbon structure with germanium encapsulated internally [224]. The prepared dual carbon matrix consists of mesoporous carbon and amorphous carbon layers. Using this structure, dual carbon effectively alleviates the expansion of germanium. The use of active or inactive elements to form germanium-based binary or composite materials is another effective approach for improving electrochemical performance. Inactive metals like cobalt (Co) and copper (Cu), when alloyed with Ge, can improve conductivity [225, 226]. The active materials such as silicon (Si), tin (Sn), antimony (Sb), tellurium (Te), and selenium (Se) used to incorporate with Ge-based compounds, demonstrating high theoretical capacities [227, 228]. As mentioned above, phosphorus exhibits the highest theoretical capacity in PIBs and can augment the overall capacity by forming GeP_x. Zhang et al. prepared GeP_5, which exhibited a stable capacity of 213.7 $mAh\,g^{-1}$ in PIBs at a current density of 500 $mA\,g^{-1}$ for 2000 cycles [229]. Active selenium metal can form layered metal selenides with germanium, characterized by an interlayer distance of 5.41 Å. The GeSe/CNT composite material, synthesized by a simple ball milling method, maintained a stable cycling performance of 311 $mAh\,g^{-1}$ after 400 cycles. In addition, the electrode offers a capacity of approximately 200 $mAh\,g^{-1}$ at a high current density of 5 $A\,g^{-1}$ [230]. Due to the large size of potassium ions, the PIBs volume of Ge-based anode materials changes significantly compared to their LIBs and SIBs counterparts. Integrating carbon materials such as carbon spheres, graphene, or amorphous carbon to construct 3D porous and yolk-shell structures can effectively mitigate volume change and enhance electrochemical performance.

Summary and Prospects The primary drawback of alloy materials is their structural instability during cycling, which hinders their inability to maintain high capacity. Additionally, alloy materials often experience various side reactions, leading to poor Coulombic efficiency. Furthermore, many compounds also demonstrate a lower number of reversible potassium ions compared to sodium or lithium ions. Despite these challenges, the high theoretical capacities of alloy materials present significant opportunities for further in-depth research and development, leveraging the advantages to create functional PIBs using the high theoretical capacity of alloy materials.

3.4.3 Conversion Reaction-Type Anodes

In conversion reaction-type anode materials, the active material interacts with alkali metal ions from the electrolyte, forming at least one metal and alkali ion-based intermediate compound at the anode through an oxidation reaction. Because the total reduction amount of TMs relates to the metal state during the electrochemical reaction process. This results in a much higher theoretical capacity of conversion reaction-type anode materials compared to intercalation-type anode materials. The general conversion reaction of anode materials in PIBs is represented by the following molecular formula:

$$\text{MxAy} + (\text{yn})\text{K}^+ + (\text{yn})\text{e}^- \leftrightarrow \text{xM} + \text{yKnA}$$

In this formula, M is a TM, such as iron (Fe), cobalt (Co), molybdenum (Mo), titanium (Ti), and other metals; A is an anion, such as oxygen (O), sulfur (S), selenium (Se), tellurium (Te), and other elements; N is the oxidation state of A.

3.4.3.1 Challenges and Strategies

The conversion-type anode materials for PIBs mainly consist of metal oxides, metal phosphides, and TM chalcogenides. Similar to alloy-type anodes, conversion-type anodes usually experience severe volume expansion. In addition, conversion-type reaction materials tend to suffer from poor rate performance and low conductivity, resulting in unstable long cycling stability. Therefore, researchers focus on material modifications, including designing unique morphological structures, compositing with high conductivity/high conductivity ionic materials, and multiple composite modifications. Lai et al. prepared ultrafine CoM_x (M = S, O, Se, Te) nanocrystals embedded in multi-layered porous carbon nanofibers through a conventional method [231]. This cobalt-based anode material exhibits excellent reversible capacity and rate performance in PIBs. Liang et al. designed and synthesized a porous CoV_2O_6 nanosphere@graphene oxide potassium anode material. The composite material exhibits an extremely high initial discharge-specific capacity ($264\,mA\,h\,g^{-1}$, $1\,A\,g^{-1}$), and a reversible capacity of up to $97\,mA\,h\,g^{-1}$ at $10\,A\,g^{-1}$. The DFT computations have demonstrated the potassium storage mechanism of the material involves both intercalation and transformation reactions [232].

3.4.3.2 Metal Oxides

TM oxides have attracted widespread attention as high theoretical capacity containing anode materials for PIBs. In 2017, Sultana et al. prepared a composite material of Co_3O_4-Fe_2O_3 nanoparticles combined with a Super P carbon matrix by salt melting method and evaluated it as the anode of PIBs [233]. After continuous cycling for 50 cycles at low current density, the capacity of the composite material remained at $220\,mAh\,g^{-1}$, with progressive cycling performance. The Co_3O_4-Fe_2O_3 material underwent substantial volume changes and caused powdering of the electrode, disrupting the good contact between the material and the collector electrode, and thereby degrading battery capacity. Thus, further modifications are necessary for transition metal oxides to serve as high-performance PIB anode materials.

Liu et al. prepared hollow spherical MoO_2/reduced graphene oxide (MoO_2/rGO) sub-microsphere composites using a simple hydrothermal method and heat treatment process [234]. When treated as an anode for PIBs, the synthesized MoO_2/rGO composite displayed an initial charge-specific capacity of 367.2 $mAh\,g^{-1}$ at 50 $mA\,g^{-1}$, and its reversible capacity was 218.9 $mAh\,g^{-1}$ after 200 cycles. Even after cycling at 500 $mA\,g^{-1}$, a high specific capacity of 104.2 $mAh\,g^{-1}$ was achieved after 500 cycles. Its excellent cycling and rate performance can be attributed to the synergistic effect of reduced graphene oxide and hollow MoO_2 spheres, enhancing the conductivity of the composite material and alleviating strain generated during discharge/charge processes. Liu et al. proposed a simple chemical blow molding method to construct a composite of Fe_3O_4 nanoparticles and N-doped porous graphene skeleton [235]. The loose and porous geometry of the internal structure helped buffer volume changes during the insertion and extraction of embedded potassium ions, minimizing steric hindrance and facilitating the diffusion of electrons/ions with profound cycling performance. After 500 cycles at a high current density of 1 $A\,g^{-1}$, the capacity was about 154.6 $mAh\,g^{-1}$. Wang et al. conducted the first study on the electrochemical properties of Sb_2MoO_6 containing bimetallic oxides and its composite with RGO as an anode material for PIBs [236]. The composite exhibited a capacity of 402 $mAh\,g^{-1}$ at 100 $mA\,g^{-1}$ and 381 $mAh\,g^{-1}$ at 200 $mA\,g^{-1}$, with minimal capacity variation and good rate performance at a current density of 500 $mA\,g^{-1}$, whereas a capacity of 247 $mAh\,g^{-1}$ was retained after 100 cycles, indicating good cycling stability.

Researchers traditionally believed that Sn oxides do not undergo reversible conversion reactions; however, Shimizu et al. discovered only SnO exhibits a reversible conversion reaction when using SnO and SnO_2 as anode materials for PIBs [237]. The reversible capacity of the SnO electrode was 183 $mAh\,g^{-1}$, with a capacity retention rate of 80% at the 30th cycle, while the capacity of the Sn electrode rapidly decreased due to significant volume changes during the K-Sn alloying/dealloying process. The reason for the improved cycling stability of SnO electrodes may be that K_2O inhibited the aggregation of Sn and acted as a buffer against Sn volume changes. Li et al. demonstrated that the introduction of surface amorphization can further improve the cycling stability of materials [238]. By etching the surface of rod-shaped VO_2 with a strong reducing agent to form an amorphous state, the etched material can stably cycle for more than 500 cycles. In contrast, the capacity of the original rod-shaped VO_2 decays to zero after about 200 cycles. Theoretical calculations reveal that the heterojunction of crystallization/amorphization can reduce the diffusion barrier of potassium ions, improve efficiency, and keep the volume unchanged after charging and discharging are completed.

Metal oxides, as anode materials for PIBs, generate nonconductive K_2O, which results in poor reversibility and generally low capacity of the electrode. Therefore, it is necessary to construct an excellent electron/ion diffusion channel to ensure stable cycling.

3.4.3.3 Metal Phosphides

Elemental phosphorous (P) exists in three main allotropes: white phosphorus, red phosphorus, and black phosphorus. Among them, white phosphorus has a lower ignition point and can self-ignite in the air, with high toxicity, and is not suitable as an electrode material. Red phosphorus is an amorphous structure with relatively poor conductivity (10^{-14} S cm^{-1}). Black phosphorus has a layered structure with superior conductivity (0.2–3.3 S cm^{-1}); however, the production is difficult and expensive [239]. Therefore, reacting P with other compounds to prepare transition metal phosphides (TMPs) is a common strategy to improve potassium storage performance.

The potassium storage mechanism in TMPs can particularly be classified into two types or a combination of both: intercalation type: K_yMP_x; conversion type: K_xP, where the volume expansion of TMPs for the conversion reaction mechanism is much smaller than that of elemental P (300%) [240]. So far, research on TMP as an anode material for PIBs is still in its early stages and faces several problems. Urgent solution: due to the conversion reaction mechanism during the cycling process, problems such as volume expansion are inevitable, leading to electrode material crushing, battery short circuits, etc. [241, 242] Improvements can be made in the following aspects: in terms of mechanism, by studying and calculating the reaction mechanism between TMPs and potassium, adjusting the structure and composition of the material, and determining the properties of a certain material. In terms of volume expansion, carbon sources and TMPs composite materials can be introduced to reduce the volume expansion of the material, provide a fast channel for ion transport, and ensure the cyclic integrity of the material during the process.

TMPs have attracted attention due to their higher conductivity, excellent thermal stability, and higher theoretical specific capacity compared to metal oxides and sulfides. However, in addition to the common issues of PIBs anodes mentioned above, there are also problems such as cumbersome and complex preparation. Researchers have employed many strategies to prepare TMP nanocomposites such as ball milling, solid-phase method, and hydrothermal method, which have greatly improved the electrochemical K^+ ion storage performance to a certain extent. However, research progress on TMPs as anodes for PIBs has been somewhat slow, and most of these preparation methods are complex and cumbersome. Most studies focus on regulating the morphology and structure to improve the electrochemical K^+ ion storage performance. However, the problem of poor rate performance due to the sluggish reaction kinetics of K^+ ions in the solid-state interface and the underlying K^+ storage reaction mechanism remains unresolved. Therefore, in addition to optimizing the structure of TMPs, sample preparation methods, exploring its reaction mechanism, and improving the reaction kinetics of electrodes all require further efforts from researchers to explore.

Nowadays, the three types of TMPs, Fe, Co, and Ni, have received more attention due to their high theoretical specific capacity and diverse compounds and crystal structures.

Fe-based phosphides: Fe-based phosphides are among the most promising TMPs materials due to their high theoretical specific capacity and relatively straightforward synthesis methods. Zhang et al. used MOF materials wrapped in 3D graphene as precursors, and the confinement effect of 3DG can promote the micro-phase separation of MOF nanoparticles at the nanoscale [200]. Hollow FeP spheres were prepared under KirKendall-induced heat treatment. Analysis revealed

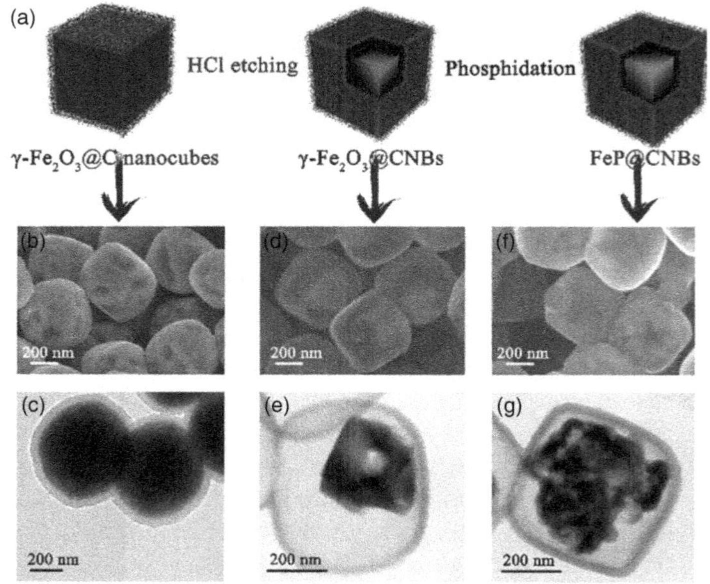

Figure 3.22 (a) Schematic illustration of the synthesis process for the FeP@CNBs; (b, d, and f) SEM and (c, e, and g) TEM images of Fe_2O_3@C nanocubes (b and c), Fe_2O_3@CNBs (d and e), and FeP@CNBs (f and g). Source: Reproduced with permission from [241]/John Wiley & Sons.

that the material stores potassium through intercalation and conversion reactions. The combination of hollow FeP nanostructures, 3DG skeleton encapsulation, and the synergistic effect of the overall structure delivers extremely excellent potassium storage performance of the material. At a high current density of $5\,A\,g^{-1}$, it still exhibits a cycling-specific capacity of $101\,mAh\,g^{-1}$, and the cycling performance is highly stable. After 2000 cycles at a current density of $2\,A\,g^{-1}$, the capacity retention rate is 97.6%. To enhance the structural stability of TMP anode materials and address their inherently poor conductivity, it is a common design to introduce a solid morphology structure and efficient ion transport by introducing a main body with certain spatial gaps and high conductivity.

Yang et al. reported a yolk-shell structure FeP@C as an anode material for PIBs. The nanobox features FeP nanoparticles encapsulated within a thin carbon shell (Figure 3.22a–g) [241]. There is a certain gap between the carbon shell and the internal particles as a buffer zone for volume change. This unique yolk-shell configuration effectively mitigates the problem of FeP nanoparticles expanding and powdering, while ensuring the carbon shell remains intact, thus preserving the electrode structure. In addition, the presence of highly conductive carbon shells enhances the conductivity of the entire electrode material and increases the ion transfer rate. The material has a specific capacity of $264\,mAh\,g^{-1}$ at a current density of $100\,mA\,g^{-1}$.

Co-based metal phosphides: Cobalt and phosphorus can generate various compounds such as CoP, Co_2P, CoP_2, CoP_3, and CoP_4, which encounter challenges such as volume expansion and poor conductivity during electrochemical reactions.

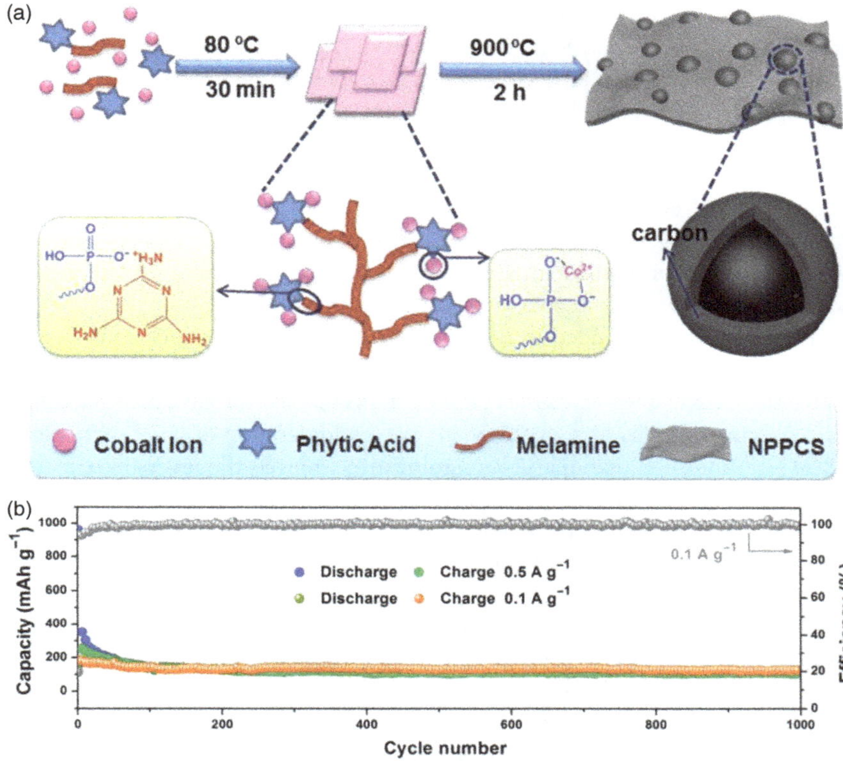

Figure 3.23 (a) Schematic diagram of the preparation process of CoP/NPPCS composite material. (b) Cycling performance at current densities of 0.1 and 0.5 A g^{-1} and the CE at a current density of 0.1 A g^{-1}. Source: [242]. Reproduced with permission/John Wiley & Sons.

Therefore, it is crucial to design the structure of the material reasonably to increase its cycling stability.

Jing Bai et al. designed a core-shell CoP nanoparticle composite of N and P co-doped porous carbon sheets (CoP/NPPCS) using a one-step method to enhance potassium storage (Figure 3.23a) [242]. A single precursor was self-assembled using the special coordination ability of melamine and cobalt ions, as well as the synergistic effect of melamine and phytic acid. This layered porous carbon structure can effectively improve ion transport efficiency and mitigate volume change generated by cobalt phosphide during electrode reactions. The doping of N and P introduced vacancies and dangling bonds around it, leading to local electronic defects and the generation of local active sites, making the material more affinity for potassium ions. After 1000 cycles at 0.1 and 0.5 A g^{-1}, the electrodes demonstrated reversible capacities of 127 and 114 mAh g^{-1}, respectively (Figure 3.23b).

Dan Zhou et al. encapsulated ultrafine CoP nanoparticles in N-doped porous carbon to enhance the storage capacity of K^+ ions [243]. This material exhibits strong conductivity, few effective transmission pathways, and abundant active sites, thus achieving efficient potassium storage. At a current density of 500 mA g^{-1}, after

200 cycles, the material retained a specific capacity of 162.7 mAh g^{-1}. MOFs are compounds with specific structures composed of organic ligands and metal ions. They have a large specific surface area, good structural tunability, and uniform doping of heteroatoms, making them ideal for use as precursors or sacrificial templates for anode materials in PIBs. Jian et al. employed CoP derived from ZIF-67 supported by rGO as electrode material to achieve high-performance potassium storage. After 2000 cycles, a specific capacity of 177 mAh g^{-1} was still achieved at a current density of 1 A g^{-1} [244].

Ni-based metal phosphides: Nickel-based phosphides exhibit diverse crystal structures and possess certain potassium storage activity. Ye et al. prepared ultrafine Ni_5P_4 nanoparticles embedded in 3D porous carbon foam as anode material for PIBs [245]. This specially designed porous carbon foam structure not only provides enough buffer space for the possible volume expansion of materials during charging and discharging but also effectively avoids the agglomeration of Ni_5P_4 nanoparticles. The ultrafine Ni_5P_4 nanoparticles significantly improve the reversible capacity due to their short ion transport pathway and abundant active sites. Therefore, it exhibits excellent cycling performance in PIBs. Specifically, its specific capacity remains stable at 145.6 mAh g^{-1} at 200 mA g^{-1}, and even at a high current density of 5000 mA g^{-1}, it can achieve a high-rate performance of 115.5 mAh g^{-1}. In addition, the authors also used *ex situ* XRD technology to deeply analyze the insertion and extraction mechanisms of K^+ ions, further confirming the feasibility and potential of metal phosphides anodes for PIBs.

Additionally, Zhang Yang et al. embedded dispersed Ni_2P nanodots into porous carbon co-doped with N and P as the anode material for PIBs [204]. Using environmentally friendly gelatin as the raw material, the chelation of Ni^{2+} and gelatin promoted the generation of smaller-sized Ni_2P, making Ni_2P more evenly distributed in the porous carbon. The material maintained a specific capacity of 212 mAh g^{-1} after 5000 cycles at a current density of 1 A g^{-1}.

Ge-based metal phosphides: Germanium phosphide (GeP_5) is the layered material with the highest P content in the binary metal phosphide family, its unique electronic structure endows it with extraordinary metallic conductivity, surpassing traditional black P and is even comparable to graphite [246]. Due to its ultra-high theoretical capacity, excellent charge-transfer rate, and outstanding ion diffusion properties, GeP_5 has attracted widespread research interest as an anode material for PIBs. Guo's research team successfully prepared GeP_5 particles through ball milling technology and comprehensively evaluated the effects of electrolyte potassium salt and additives on the potassium storage performance of GeP_5 anode [247]. Research demonstrated that the use of KFSI as a potassium salt electrolyte, supplemented with carbonate solvent (EC/DEC mixture, volume ratio 1 : 1), can significantly promote the formation of a stable and uniform SEI, effectively suppress the occurrence of side reactions, and greatly improve CE and cycling stability. The capacity at 50 mA g^{-1} was 495.1 mAh g^{-1}, demonstrating remarkable cycling stability. Excitingly, the GeP_5 anode also exhibited extraordinary rate performance, maintaining a capacity of 213.7 mAh g^{-1} even after 2000 cycles at a current density of up to 500 mA g^{-1}. This outstanding performance contributes to the formation of germanium nanocrystals during the discharge process, which serves as highly conductive electronic channels, accelerates the charge-transfer process, and effectively alleviates the force caused by the volume shrinkage of the K_4P_3 phase as an elastic matrix. In addition, the K_4P_3 phase and KGe alloy, as potassium products, can jointly cope with the huge volume changes during the potassium addition/removal process through a mutual buffering effect, further enhancing the structural stability and cycle life of the GeP_5 anode. Therefore, if the complete potassium conversion of GeP_5 can be achieved, that is, it can be converted into KGe and K_4P_3 compounds, the theoretical capacity will reach an unprecedented level. This undoubtedly opens up new possibilities for the development of anode materials for PIBs, making GeP_5 a highly promising candidate material.

Sn-based metal phosphides: Similar to the GeP_5 anode design in PIBs, the combination of metal Sn with red P element to form Sn_4P_3 compound, the theoretical capacity is significantly improved, the electron transfer efficiency is enhanced, and significant volume fluctuations caused by P element during potassium/potassium removal cycles are effectively addressed. Meanwhile, the potassium products of P and K-P alloys help to prevent the aggregation, dispersion, and alleviation of structural stress of Sn particles. However, despite these advantages, the challenge of volume expansion caused by alloying and transformation reaction mechanisms remains unresolved [248].

To address this, researchers have focused on the design of nanostructures and the introduction of conductive buffer matrices. Zhang et al. attempted to embed micrometer-sized Sn_4P_3 particles into an amorphous carbon matrix, but due to particle aggregation and structural damage during cycling, the anode material exhibited rapid capacity degradation after 30 cycles [249]. Subsequently, they made significant progress through electrospinning technology, which allowed for the uniform embedding of nanoscale Sn_4P_3 particles (with a particle size less than 350 nm) into N-doped carbon fibers [250], exhibiting a current of 50 mA g^{-1} in a 1 M KFSI + EC/DEC electrolyte Sn_4P_3@carbon. The fiber electrode exhibited very stable cycling performance, maintaining a capacity of 403.1 mAh g^{-1} after 200 cycles and demonstrating long-term cycling stability of over 1000 cycles.

In addition, Li et al. successfully prepared Sn_4P_3@C composite using a solvothermal method, cleverly embedding Sn_4P_3 particles into a hierarchical porous carbon structure [251]. This design not only enhances the overall electronic conductivity performance through porous carbon but also effectively accommodates the volume change of Sn_4P_3 in the pore space, thus achieving a high capacity of 431 mAh g^{-1} upon initial charging at a current density of 100 mA g^{-1} in a 0.8 m KPF_6 + EC/DEC electrolyte, and maintaining 73% capacity after 100 cycles. Its excellent performance can be attributed to first that the graded porous carbon matrix enhances the conductivity of the composite material; Second, the rich pore structure provides sufficient buffering space for the volume expansion of Sn_4P_3, ensuring long-term cycling stability.

3.4.3.4 Metal Chalcogenides

Compared to metal oxides, metal sulfides (MS_x) express better conductivity and higher theoretical capacity. Furthermore, the weaker M–S bond facilitates faster diffusion kinetics of alkali metal ions as compared to the M–O bond [252, 253]. Metal sulfides involve multiple reaction mechanisms during ion storage, mainly including diffusion control processes dominated by embedding, transformation, and alloying reactions, as well as pseudocapacitance control processes dominated by intercalation pseudocapacitance and redox pseudocapacitance. These diverse reaction mechanisms contribute to the high specific capacity of metal sulfides [254]. Although metal sulfides have many advantages, they still have many shortcomings in practical applications. For example, low conductivity and ion diffusion coefficient, slow interface/electrode surface reaction kinetics, and significant volume expansion during continuous potassium storage all limit the further application of metal sulfides in PIBs.

In response to the research bottleneck problems of metal sulfides mentioned above, researchers have proposed numerous improvement or modification strategies (Figure 3.24) [255]. First, by constructing unique morphological structures to shorten ion diffusion channels and promote charge transfer, volume expansion can

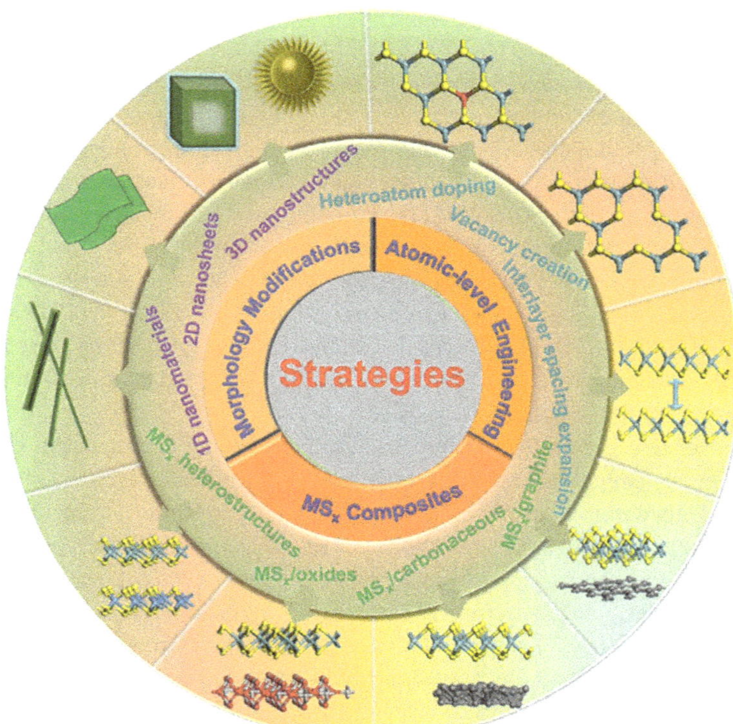

Figure 3.24 Design strategies of MS_x for the enhanced battery performance. Source: [255]. Reproduced with permission/John Wiley & Sons.

be alleviated. Meanwhile, this structure generally has a larger specific surface area, which can effectively promote contact between the electrode and the electrolyte, providing additional active sites for interface reactions. Second, atomic-level engineering can reduce ion diffusion energy barriers by increasing interlayer spacing and enhance ion embedding and diffusion kinetics by providing additional active sites and channels. Third, metal sulfides are usually combined with other materials, such as other metal sulfides, metal oxides, carbon-based materials, and graphite materials. Metal sulfide heterostructures and metal sulfide/metal oxide heterostructures improve the electrochemical performance of materials by constructing built-in electric fields to enhance interfacial reaction kinetics [256]. By introducing carbon-based materials or graphite materials, the conductivity of the material can be improved. At the same time, it can also alleviate the agglomeration and volume expansion of metal sulfide nanoparticles, thereby enhancing the cycling stability of the material structure.

Metal Sulfides Due to their unique layered structure, weak van der Waals forces between adjacent layers, and strong covalent bonds within each layer, various metal sulfides are used in energy storage systems [257]. Metal sulfides generate potassium polysulfides that are soluble in the electrolyte during the charging and discharging process, which deteriorates the cycling performance.

Among numerous sulfides, SnS_2 is particularly widely studied. Fang et al. obtained a very thin composite of SnS_2 and RGO through simple oil bath reflux, which exhibited high reversible capacity as an anode material for PIBs [258]. The SnS_2 nanosheets underwent sequential transformation reactions (SnS_2 to Sn) and multiple alloying reactions (Sn to K_4Sn_{23}, KSn) during potassium insertion/removal, resulting in high specific capacity; at the same time, ultra-thin nanosheets enable the surface capacitance charge storage mechanism to dominate and effectively alleviate volume changes, enabling rapid potassium storage kinetics and maintaining structural integrity due to the rapid transfer of electrons/ions. Cao et al. synthesized sheet-like SnS_2/S and N-C by one-step calcination of tetraphenyl tin and sulfur powder in an inert atmosphere [259]. The material exhibited impressive potassium storage capacity and good rate performance, with a reversible capacity of 614.8 mAh g^{-1} at a current density of 0.1 A g^{-1} and a capacity of 104.5 mAh g^{-1} at a high current density of 2 A g^{-1}. Carbon materials doped with S and N can not only serve as conductive and volume-reducing substrates but also inhibit the dissolution of potassium polysulfide in electrolytes.

Xie et al. first utilized rGO-encapsulated MOFs to derive FeS_2 hollow nanocages as anode materials for PIBs [260]. Due to the synergistic effect of FeS_2 nanocages and rGO shell, the material exhibits excellent electrochemical performance, showcasing a high capacity of 264 mAh g^{-1} after 50 cycles at 50 mA g^{-1}. Even at a high current density of 500 mA g^{-1}, it reaches 123 mAh g^{-1} after 420 cycles. Lqbal et al. innovatively prepared composite materials with $Co_{1-x}S$/ZnS hollow nanocube heterostructures modified on reduced graphene oxide (rGO) substrate using a simple hydrothermal combined with subsequent thermal sulfurization technology ($Co_{1-x}S$/ZnS@rGO) (Figure 3.25a–c) [261]. This unique composite

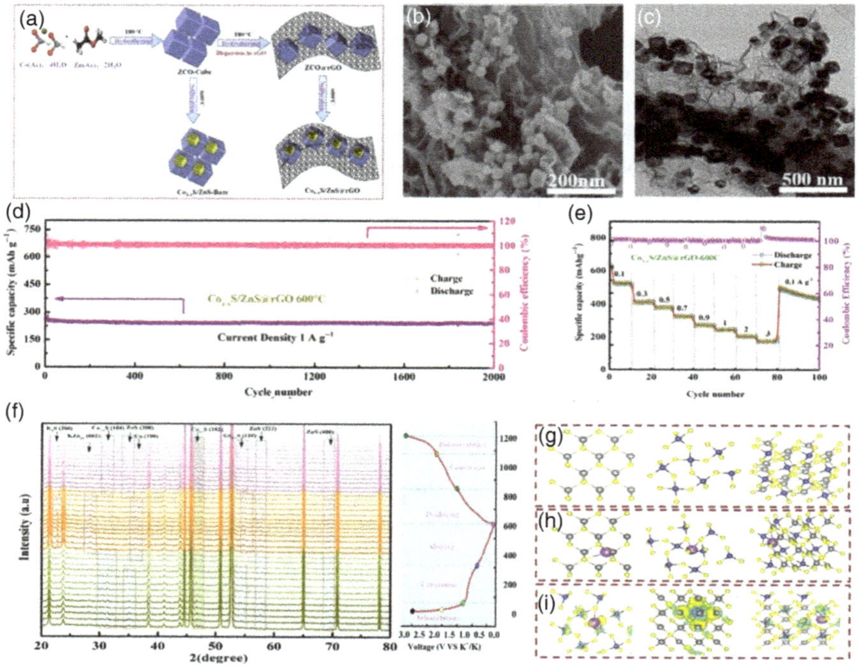

Figure 3.25 (a) Schematic illustration process for the preparation of hollow nanocubes $Co_{1-x}S/ZnS@rGO$ heterostructures obtained at 600 °C. (b) FESEM and (c) TEM images of $Co_{1-x}S/ZnS@rGO$ 600 °C nanocomposites. (d) Long cycling performance at 1 A g^{-1} for 2000 cycles. (e) Rate performance at various current densities. (f) *In situ* XRD patterns of $Co_{1-x}S/ZnS@rGO$ 600 °C electrode. The computational optimized structures: (g) before potassium-ion insertion, (h) after potassium-ion insertion, and (i) charge density structures of ZnS, $Co_{1-x}S$, and $Co_{1-x}S/ZnS$ heterostructure. Source: Reproduced with permission from [261]/John Wiley & Sons.

material design not only constructs a more stable and efficient conductive network but also effectively shortens the diffusion path of ions inside the material. The experimental results demonstrated that the composite material exhibited an initial charging capacity of 638.5 mA h g^{-1} in SIBs, while it reached 606 mA h g^{-1} in PIBs. Of particular note is that in the testing of PIBs, the material exhibited excellent long-term cycling stability, with an initial capacity of 274 mA h g^{-1}. Even after 2000 cycles, the charging capacity remained stable at 245 mA h g^{-1} at a current density of up to 1 A g^{-1} (Figure 3.25d and e). To further explore the mechanism behind this excellent performance, the research team also conducted a series of *in situ* and *ex situ* tests, combined with first-principle calculations, to further confirm the strong adsorption capacity and efficient diffusion kinetics of $Co_{1-x}S/ZnS$ anode material for potassium ions (Figure 3.25f and g).

Chong et al. synthesized nano rose-shaped MoS_2 confined to reduced graphene oxide [262]. As an anode material for PIBs, the initial capacity was 438.5 mAh g^{-1}, and it exhibited excellent cycling stability at a current density of 100 mA g^{-1}. After 200 cycles, the capacity retention rate was 95.0%, delivering an excellent rate performance, with a capacity of 196.8 mAh g^{-1} at a high current density of 2 A g^{-1}; At 500 mA g^{-1}, no significant attenuation was observed within the ultra-long lifespan of 600 cycles. Good stability benefits from a larger specific surface area to achieve excellent electronic conductivity and strong potassium-ion diffusion ability, and a stable 3D conductive network is provided through the strengthened chemical bonds between MoS_2 and rGO (Mo-C and Mo-O-C), which can reduce mechanical stress and ensure the stability of the structure during cycling.

Jia et al. proposed a bamboo-shaped MoS_2/N-doped carbon hollow tube with extremely high capacity at low current densities [263]. After 50 cycles at a current density of 50 mA g^{-1}, the capacity was 330 mAh g^{-1}, possessing high capacity retention at high current density, with a capacity of 151 mAh g^{-1} after 500 cycles at a current density of 500 mA g^{-1}. Wang et al. synthesized Sb_2S_3 nanoflowers via solvothermal reaction and self-assembled them onto MXene (Ti_3C_2) [264]. High conductivity two-dimensional Ti_3C_2 not only promotes charge-transfer kinetics but also effectively buffers the volume expansion of Sb_2S_3. The Ti_3C_2-Sb_2S_3 anode exhibits high reversible capacity, with a reversible capacity of 461 mAh g^{-1} at a current density of 100 mA g^{-1}. It has a long cycle life, with a capacity retention rate of up to 79% after 500 cycles, and excellent rate performance, with a capacity of 102 mAh g^{-1} at a high current density of 2 A g^{-1}.

Compared to metal oxides, metal sulfides have higher reversibility and much higher capacity. Especially sulfides of antimony (Sb), bismuth (Bi), and tin (Sn) demonstrated higher capacity due to the combination of alloying reactions in addition to the capacity contribution from the conversion reaction, and the synergistic effect of two reactions improving the reversibility of the reaction.

Lei et al. prepared WS_2 material and used it as an anode for PIBs, and the specific capacity was 103 mAh g^{-1} after 100 cycles at 100 mA g^{-1} [265]. When the current density increases to 800 mA g^{-1}, the capacity retention rate is 57%, and stable cycling ability is demonstrated at 500 mA g^{-1}. Dynamics studies have demonstrated that the main reason for promoting the migration ability of potassium ions is that the material is controlled by intercalation reactions.

Metal Selenides Metal selenides exhibit higher tap densities, contributing to improved volumetric capacity. Compared to sulfides, selenides experience smaller volume expansion, weaker M-Se, and larger interlayer spacing. The above advantages make selenides a highly promising anode material for PIBs. Although a large number of studies have demonstrated the potential of metal selenides as anode materials for PIBs, it is still an emerging research field, facing challenges that need to be overcome. The inherent low electronic and ionic conductivity of metal selenides leads to sluggish reaction kinetics, high polarization, and low ion storage capacity. During the conversion reaction of metal selenides, significant volume expansion occurs due to the generation of multiple intermediate phases, leading to rapid capacity decay and severe structural pulverization. To address the above challenges and promote the commercial application of high-performance metal selenides in PIBs, a large number of strategies have been adopted, such as designing unique nanostructures, element doping, and constructing heterostructures.

Wang et al. synthesized a novel happy-fruit shell-like structure $MoSe_2$@C core-shell nanostructure (PMC) [266]. Molybdenum selenide is an agglomerate with a layer spacing of approximately 0.85 nm, which facilitates the insertion/extraction of potassium ions. The thin amorphous carbon shell restricts the volume expansion of active molybdenum selenide nanosheets during cycling, maintaining high structural stability. The most important advantage of PMC is that it reduces excess hollow internal space, increases filling density, buffers volume expansion during K^+ ion intercalation, and further improves stability. Therefore, in 100 cycles, PMC exhibited a very high capacity of 322 mAh g^{-1} at 0.2 A g^{-1} and was still able to maintain 226 mAh g^{-1} at 1.0 A g^{-1} after 1000 cycles. Lin et al. synthesized a cubic CuSe with a columnar-like morphology assembled from nanosheets and used it as an anode material for PIBs [267]. CuSe can quickly store potassium ions and has good cycling stability, providing a specific capacity of 280 mAh g^{-1} at a current density of 5 A g^{-1}. It can maintain nearly 100% high Coulombic efficiency in 1200 and 340 cycles, showing a high initial Coulombic efficiency of 92.4%, indicating that the cyclic reversibility of cubic CuSe is extremely high. Chen et al. innovatively selenized WO_3@ppy precursor obtained WSe_2@N-doped C core-shell nanotubes as anode materials for PIBs (Figure 3.26a) [268]. Its core-shell structure prevents WSe_2 aggregation and provides volume change buffering; large interlayer spacing promotes K^+ migration; nitrogen-doped carbon shell enhances conductivity and optimizes electronic structure. The experiment demonstrated that a capacity of 122.3 mAh g^{-1} was maintained after 1300 cycles at 500 mA g^{-1}. Theoretical calculations reveal that K^+ ion adsorption causes the Fermi level of WSe_2 to shift toward the conduction band, demonstrating that K^+ ion insertion can effectively enhance the electronic conductivity of WSe_2 (Figure 3.26b and c). Zhang et al. synthesized layered carbon-coated $MoSe_2$/MXene composite materials ($MoSe_2$) using a simple hydrothermal method/ MXene@C [166]. According to electrochemical testing, $MoSe_2$/ MXene@C, the electrode possesses a specific capacity of approximately 355 mAh g^{-1} after 100 cycles at 200 mA g^{-1}, exhibiting outstanding rate performance (183 mAh g^{-1} at 10 A g^{-1}) and excellent long-cycle life. Research elucidates that $MoSe_2$/MXene@C exhibits excellent potassium storage performance of composite materials, attributed to the following: (i) having high electronic conductivity MXene substrate can further alleviate the volume expansion of $MoSe_2$ nanosheets and improve electronic conductivity; (ii) the carbon layer can further enhance the overall conductivity of the composite structure; (iii) $MoSe_2$ nanosheets and MXene flakes have strong chemical interactions at the interface, which helps to promote reaction kinetics and improve the structural stability of composite materials.

Guo's research group successfully made strides in developing ultra-thin graphene-like materials (V_5Se_8@C) for the first time by combining solid-state selenization, a simple ion intercalation and exfoliation strategy, and solvent thermal treatment [272]. Research has exhibited that the material has a NiAs-type superstructure with multidimensional electron/ion transport pathways and pseudocapacitive controlled behavior. In addition, carbon coating can effectively prevent the aggregation of active substances, adapt to the huge volume changes during charge and discharge processes, and provide more active sites to promote the

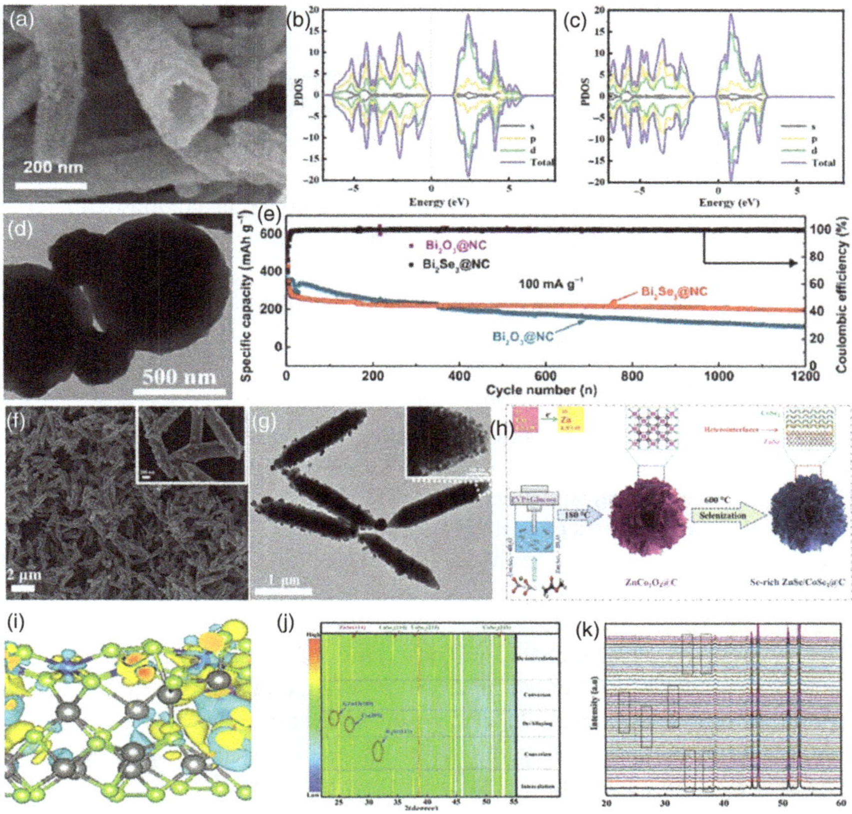

Figure 3.26 (a) Enlarged SEM images of WSe_2@N-doped C nanorods. Calculated spin-polarized projected density of states (DOS) for WSe_2 (b) before and (c) after adsorbing K. Source: Reproduced with permission from [268]/John Wiley & Sons. (d) TEM images of Bi_2Se_3@NC and (e) long cycle stability of Bi_2Se_3@NC and Bi_2O_3@NC at a current density of 100 mA g^{-1}. Source: Reproduced with permission from [269]/Elsevier. (f) SEM image and (g) TEM image of $NF_{11}S$/C. Source: Reproduced with permission from [270]/John Wiley & Sons. (h) Schematic representation process of 2D Se-rich ZnSe/$CoSe_2$@C heterostructure obtained at 600 °C during selenization. (i) The computational charge density models of ZnSe/$CoSe_2$. (j, k) *In situ* XRD patterns of ZnSe/$CoSe_2$@C-2. Source: [271]. Reproduced with permission/John Wiley & Sons.

adsorption of potassium ions. Given the advantages of the synergistic effects mentioned above, the V_5Se_8@C electrode exhibits satisfactory charging specific capacity (at 0.1 A g^{-1}, after 50 cycles, the specific capacity remains at 327 mAh g^{-1}), outstanding rate performance (at 4 A g^{-1}, the specific capacity still reaches 162 mAh g^{-1}), and long cycle life (at 4 A g^{-1}, after 800 cycles, the capacity retention rate is 82.9%). Additionally, the material was assembled with the pretreated $K_{0.5}MnO_2$ cathode to form a potassium-ion full cell, demonstrating excellent performance. Experimental results unleashed the findings that V_5Se_8@C nanoflakes have enormous potential for application in large-scale power grid energy storage in the future.

Sun et al. utilized a fully chemical vapor deposition method, employing 3D MXene as a carrier to directly synthesize nitrogen-doped graphene (NG)/rhenium diselenide ($ReSe_2$) hybrid materials (NG/$ReSe_2$/MXene) [273]. In this innovative design, 1T′-$ReSe_2$ nanoparticles are sandwiched between NG coating and $ReSe_2$ through strong interfacial interactions MXene frameworks, which promotes rapid diffusion of potassium ions, improves the overall conductivity of the material and enhances the junction of electrode structural stability. The composite material exhibits outstanding rate performance (with a specific capacity of up to 138 mAh g^{-1} at 10 A g^{-1}) and excellent cycle life.

Yi et al. encapsulated Sb_2Se_3 nanoparticles (<50 nm) within self-wrinkled reduced graphene oxide sheets through hydrothermal reaction and subsequent selenization process [274]. The inherent high elasticity of self-folding graphene allows it to accommodate the volume expansion of Sb_2Se_3 nanoparticles during electrochemical reactions with potassium ions, leading to improved potassium storage performance. The reversible capacity of the material at 100 mA g^{-1} exhibites 391.4 mAh g^{-1}, and the capacity after 460 cycles at a current density of 0.5 A g^{-1} is 203.6 mAh g^{-1}. Sun et al. successfully prepared a unique lollipop-like Bi_2Se_3@NC compound through simple hydrothermal synthesis and subsequent selenization treatment technology (Figure 3.26d) [269]. The clever combination between Bi_2Se_3 and nitrogen-doped carbon shell produces significant synergistic effects, greatly enhancing the electrochemical performance of the material in PIBs. When Bi_2Se_3@NC was treated as an anode material for PIBs, it maintained a specific capacity of up to 201.2 mA h g^{-1} after 1200 charge–discharge cycles at a current density of 100 mA g^{-1}. From the 10th cycle to the 1200th cycle, the average capacity decay rate per cycle is extremely low, only 0.024%, demonstrating excellent cycling stability (Figure 3.26e). Liu et al. reported a $FeSe_2$/NC structure consisting of nitrogen-doped carbon-coated $FeSe_2$ nanoparticles anchored on a thin carbon nanosheet substrate as an anode material for PIBs [275]. Thanks to the nanoparticles that reduce ion diffusion length, the protective carbon coating that reduces volume fluctuations, and the carbon nanosheet matrix that promotes electron transfer, the $FeSe_2$/NC electrode provides a high capacity of 434 mAh g^{-1} after 70 cycles at 0.1 A g^{-1}, and 301 mAh g^{-1} after 250 cycles at 1.0 A g^{-1}. Kong et al. innovatively constructed a unique porous composite material by selenization treatment of MOFs precursors [270]. The core of the composite material is a nickel-doped $FeSe_2$/Fe_3Se_4 heterojunction (Figure 3.26f and g). This composite design not only ensures high-density active sites inside the material, greatly promoting the penetration of electrolyte ions and rapid electron transfer but also enhances the number of lattice defects and active sites through strategic doping of nickel elements, thereby significantly improving the electrochemical performance of the material. The experimental findings illustrate that the composite material maintains an astonishing 177.3 mAh g^{-1} capacity even after 3050 cycles at a current density of up to 2 A g^{-1} in PIBs.

Yu et al. reported a composite featuring metal octahedral $CoSe_2$ and N-doped carbon nanotubes as a flexible framework, positioning it as an advanced anode material for PIBs [276]. The metallic properties of $CoSe_2$, together with highly conductive

N-doped carbon nanotubes, greatly accelerate electron transfer and improve rate performance. Carbon nanotubes are the backbone that inhibits aggregation, anchors active materials, and stabilizes the overall structure. Each octahedral $CoSe_2$ particle is arranged in an orderly manner along carbon nanotubes, and the serrated void space can accommodate the volume expansion during cycling, thereby improving cycling stability. This unique structure provides high capacity, with a capacity of 253 $mAh\,g^{-1}$ after 100 cycles at a current density of 0.2 $A\,g^{-1}$, exhibiting good cycling performance, with a capacity of up to 173 $mAh\,g^{-1}$ after 600 cycles at a high current of 2.0 $A\,g^{-1}$. Dong et al. reported an innovative synthesis strategy that successfully prepared hydrogen-doped carbon-coated ZnSe composite materials with a unique leaf-like structure using a simple solvothermal reaction combined with subsequent selenization technology (ZnSe@NC) [277]. This greatly increases the exposed area of active sites and effectively promotes the rapid transfer of electrons and ions inside the material, thereby significantly improving electrochemical performance in PIBs. After the initial 550 cycles, its capacity remained at 106.5 $mAh\,g^{-1}$, and even after a longer 1500 cycle test (with a current density of 200 $mA\,g^{-1}$), the capacity remained at 71.4 $mAh\,g^{-1}$. He et al. presented a carefully designed ZnSe nanoparticle for decorating ammonia-doped hollow carbon composite materials (ZnSeNP@NHC) [278]. ZnSeNP@NHC composite materials achieved uniform distribution of ZnSe nanoparticles on nitrogen-doped hollow carbon frameworks through unique structural design, which not only enhanced the conductivity of the material but also created abundant active sites, facilitating electrochemical reactions. To delve deeper into ZnSe NP@NHC, the reaction mechanism of composite materials in electrochemical processes was studied by the research team employing *ex situ* XRD patterns combined with CV testing, successfully revealing the structural evolution and phase transition process of the materials during the first charge–discharge process. Furthermore, through quantitative kinetic analysis, the importance of capacitive contribution in the electrochemical process of composite materials was discovered, providing a strong explanation for their high-rate performance.

Iqbal et al. successfully prepared a two-dimensional selenium-rich ZnSe/$CoSe_2$@C heterogeneous composite through a straightforward and efficient hydrothermal synthesis method (Figure 3.26h) [271]. The excellent capacity performance of this composite material is mainly attributed to the synergistic effect of the conversion reaction initiated by K^+ ions inside the material and the alloying/dealloying process. Specifically, selenium-rich ZnSe and $CoSe_2$@C structures demonstrate highly insertable channels for K^+ ions, greatly promoting rapid ion transport and reaction efficiency. In PIBs testing, an excellent reversible charging capacity of 214 $mA\,h\,g^{-1}$ was maintained after 4000 charge and discharge cycles at a current density of 1 $A\,g^{-1}$. To further explore the electrochemical conversion mechanism behind this excellent performance, various advanced characterization techniques such as DFT theoretical calculations and *in situ* XRD were employed (Figure 3.26i–k). These studies not only reveal the detailed interaction mechanism between K^+ ions and electrode materials during charge and discharge processes but also visually demonstrate the dynamic changes in material structure during cycling.

Metal Tellurides Researchers have conducted extensive exploration in the field of anode materials, many of which have demonstrated competitive mass-to-capacity ratios. However, for large-scale energy storage applications, volumetric capacity and power density are equally critical factors in determining battery performance. Metal sulfides have shown promise as anode materials with practical application prospects due to their notable mass/volume-specific capacity. However, among the chalcogenides, tellurium ($6.24\,g\,cm^{-3}$) has a higher density compared to sulfur ($2\,g\,cm^{-3}$) and selenium ($4.81\,g\,cm^{-3}$), making tellurium ($2625\,mAh\,cm^{-1}$) and tellurides easier to produce small-sized energy storage devices in terms of volumetric capacity. In addition, tellurium (with a conductivity of $2 \times 10^2\,S\,m^{-1}$) has the closest characteristics to metals and is several orders of magnitude higher in conductivity than sulfur ($5 \times 10^{-16}\,S\,m^{-1}$) and selenium ($1 \times 10^{-4}\,S\,m^{-1}$) [279, 280], therefore exhibiting higher conductivity makes it easier to achieve excellent rate performance. Metal tellurides also have several other advantages, making them a strong candidate anode materials for PIBs. First, its favorable diffusion kinetics and K^+ ion insertion properties enable it to exhibit excellent electrochemical performance in PIBs. Second, its crystal structure typically exhibits high stability, which is crucial for maintaining the structural integrity of the electrode during repeated charging and discharging cycles. Finally, the unique electronic properties of metal tellurides contribute to their high-capacity K^+ ion storage. The ability of metal tellurides to reversibly embed and de-intercalate K^+ ions is crucial for achieving high energy density in PIBs. At present, metal tellurides that have been studied in K^+ ion storage anode electrodes mainly include $CoTe_2$, $FeTe_2$, ZnTe, Bi_2Te_3, etc. [281–283]

$CoTe_2$: $CoTe_2$ shows great potential for potassium storage capacity, large interlayer spacing, and narrow bandgap, which accelerates the ion/electron diffusion rate of the electrode in redox reactions and lowers the energy barrier. However, volume expansion from potassium-ion deintercalation during the cyclic charging and discharging process is still an unavoidable challenge, making it difficult to achieve high reversible capacity and long cycle life. Yang et al. synthesized cobalt tellurium-C ($CoTe_2$-C) composite microspheres via spray pyrolysis and utilized them as anode materials for PIBs. The cyclic performance of $CoTe_2$-C composite material at a current density of $0.2\,A\,g^{-1}$ demonstrated a slight decrease in capacity observed in the first 10 cycles due to partial structural damage that occurred during the cycling process. Afterward, $CoTe_2$-C exhibited stable cycling performance, producing a high discharge capacity of $295.8\,mA\,h\,g^{-1}$ after 100 cycles, with a capacity retention rate of 84% measured at the second cycle [281].

Element doping has further been investigated to enhance the kinetics of potassium-ion transport. For example, Zhou et al. constructed ultrafine $CoTe_2$ quantum rods coated with graphene and nitrogen-doped carbon using hydrothermal and carbonization methods ($CoTe_2$@rGO@NC) [282]. The dual physical and chemical constraints and quantum size effect enhance the electrochemical kinetics, boosting K^+ ion adsorption and diffusion capabilities. Used as an anode electrode material for PIBs, $CoTe_2$@rGO@NC delivered a high initial capacity of $237.6\,mAh\,g^{-1}$ at $0.2\,A\,g^{-1}$ and a long lifespan after 500 cycles, with capacity decay as low as 0.10% per cycle. Xu et al. [283] utilized Te-rich vacancy defects to prepare

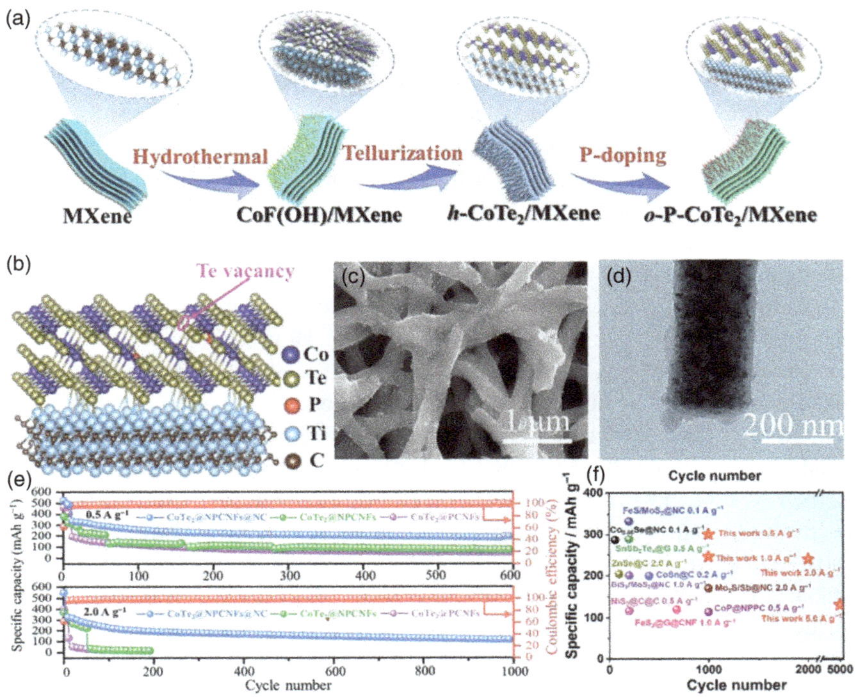

Figure 3.27 (a) Schematic illustration of the preparation process of *o*-P-$CoTe_2$/MXene. (b) Side view of the crystal structure of *o*-P-$CoTe_2$/MXene interface. Source: [283]. Reproduced with permission/John Wiley & Sons. (c) SEM and (d) TEM of $CoTe_2$@NPCNFs@NC. (e) Cycling performance at 0.5 and 2 A g^{-1}. Source: Reproduced with permission from [284]/John Wiley & Sons. (f) Comparison of selected materials for high-rate PIBs. Source: [285]. Reproduced with permission/John Wiley & Sons.

P-doped orthogonal $CoTe_2$ nanowires (o-P-$CoTe_2$/MXene) loaded on MXene, as illustrated in Figure 3.27a and b. After 200 cycles at a current density of 0.2 A g^{-1}, the specific capacity achieved about 373.7 mAh g^{-1}, and at a high current density of 2 A g^{-1}, with a minimal capacity lost by 0.011% after each cycle, exhibiting excellent long cycle performance. Li et al. synthesized a dual-type nitrogen-doped carbon-constrained $CoTe_2$ composite material ($CoTe_2$@NPCNFs@NC). Figure 3.27c and d illuminates SEM and TEM images of the composite material [284]. Thanks to the synergistic effect of dual-type N-doped carbon confinement, hierarchical nanostructure, conductive network, and uniform distribution of $CoTe_2$ nanoparticles in the carbon skeleton, the significant volume changes of $CoTe_2$ and soluble polymetallic intermediates were effectively suppressed. Therefore, when used as an anode electrode material for PIBs, the electrode delivered an excellent specific capacity of 526 mAh g^{-1} at 0.05 A g^{-1}, with the ability to cycle up to 1000 times at a high current density of 2.0 A g^{-1}, demonstrating excellent cycling stability (Figure 3.27e).

In recent years, metal-organic framework materials have gained prominent importance in PIBs due to their large specific surface area and suitable pore structure. ZIF-67 is a uniformly porous material composed of cobalt ions conjugated with organic compounds and contains suitable nitrogen doping, thus it is often used as a modification scheme for $CoTe_2$. Huang et al. successfully constructed a porous grid-like nitrogen-doped carbon-constrained $CoTe_2$ ($CoTe_2$@T-NC) as an anode electrode material for PIBs [286]. The porous grid-like carbon matrix plays a role in inhibiting the aggregation of $CoTe_2$ during cycling, reducing volume expansion, and increasing ion transport channels, which greatly enhances reaction kinetics. Optimized $CoTe_2$@T-NC electrode has a specific capacity of 428.8 mA h g^{-1} at 0.05 A g^{-1} and can endure 1500 cycles at 2.0 A g^{-1}, showcasing excellent cycling performance and stability. In further advancement, Debasish Sarkar et al. synthesized cobalt telluride in nitrogen-rich carbon dodecahedron based on ZIF-67 MOFs through further pyrolysis (CoTe@NCD) [287]. When used as an anode electrode material for PIBs, CoTe@NCD electrode exhibited specific capacities of 340, 208, 174, 136, and 58 mAh g^{-1} at current densities of 0.1, 0.3, 0.5, 1, and 2 A g^{-1}, respectively. At higher currents, maintaining the steady-state CE above 99%. Additionally, Li et al. conducted various *in situ/ex situ* techniques and DFT calculations to elucidate the chemical reaction evolution process in PIBs [288]. Moreover, Wang et al. designed a dual carbon-constrained $CoTe_{2-x}$ composite material in a 3D hollow grid S, N-co-doped dual carbon structure composite material by combining structural design. This design incorporated heteroatom doping, Te vacancy modulation, and physicochemical dual anchoring strategy ($CoTe_{2-x}$@3DPSNDC) [289]. A well-designed composite material provides fast ion/electron transport and stability and verifies the excellent physical confinement and chemical adsorption ability of the S, N-co-doped dual carbon skeleton on K_5Te_3 and K_2Te_3, achieving ultra-stable K^+ ion storage with an astonishing capacity of 508.1 mAh cm^{-3} at 5.0 A g^{-1}.

Heterogeneous interfaces are formed by introducing new metal elements into single metal selenides. These interfaces can not only introduce a large number of lattice distortions and defects but also help improve the conductivity of ions or electrons in the local area of metal selenides, boosting reaction kinetics [290]. Zeng et al. synthesized bimetallic molybdenum cobalt telluride composite materials through a straightforward hydrothermal process followed by the tellurization method, yielding MCTNC@rGO composite material [291]. Through DFT calculations, the synergistic influence was revealed between Mo-Co heterostructures, promoting conductivity and boosting K^+ ion adsorption. As an anode material for PIBs, it exhibited excellent cycling stability and sustained 2000 cycles at a current density of 1.0 A g^{-1}. Zhang et al. designed and synthesized ZnTe/$CoTe_2$@NC bimetallic tellurides using a template-assisted strategy [285]. In heterostructures, the electron-rich Te sites and the built-in electric field between ZnTe and $CoTe_2$ provide abundant adsorption sites for K^+ ions, thus improving the interface electron transfer rate. Consequently, the ZnTe/$CoTe_2$@NC electrode exhibited excellent electrochemical performance, with an initial charge-specific capacity of 486.1 mAh g^{-1} at 0.1 A g^{-1} and 5000 cycles at a high current density of 5 A g^{-1}, with a coulombic efficiency of nearly 100%. This composite compound outperformed TM chalcogenides reported for PIBs anode electrode materials, performing better than all selected anode electrodes at high current density and long cycling (Figure 3.27f).

FeTe/FeTe$_2$: As a material with high theoretical capacity, environmental sustainability, and extensive resources, $FeTe_2$ has aroused great interest among battery researchers. Gi Dae Park et al. prepared iron telluride nanocrystals ($FeTe_2$-C) with yolk-shell structure by infiltration and cellularization techniques (Figure 3.28a–c) [292]. The hollow carbon nanosphere structure provided ample space to accommodate volume changes during the reaction process, enhancing characteristics. The carbon content within the composite material was analyzed by Raman and thermogravimetric measurements, thus exhibiting good structural stability. After one cycle, the $FeTe_2$-C electrode forms a heterojunction of $FeTe_{1.1}$ and quasi-metallic Te, producing a synergistic effect and exhibiting excellent electrochemical performance. At $0.5\,A\,g^{-1}$, the specific capacity after 500 cycles is $171\,mAh\,g^{-1}$. Wang et al. synthesized FeTe/reduced graphene oxide (FeTe/RGO) nanocomposites using a straightforward hydrothermal and annealing process, which allowed FeTe to deposit on the surface of graphene oxide, alleviating the volume change during the FeTe reaction process, which enhanced conductivity [294]. Therefore, it exhibited favorable cycle reversibility (with a reversible capacity of $320\,mAh\,g^{-1}$ after 100 cycles at $0.1\,A\,g^{-1}$).

ZnTe: Among various semiconductor materials, zinc telluride stands out for its lowest electron affinity and ultrafast charge separation ability, which helps to improve charge-transfer kinetics. In addition, ZnTe possesses a relatively high density, which contributes to an increased volumetric capacity. Hu et al. utilized MOF as the starting material, by dispersing ZnTe within a carbon matrix, and then anchoring it on $Ti_3C_2T_x$ MXene nanosheets to form ZnTe@C/$Ti_3C_2T_x$ (ZCT) nanohybrid material, and the porous structure of ZCT material was demonstrated by N_2 adsorption desorption technique (Figure 3.28d) [293]. $Ti_3C_2T_x$ MXene is known for its metallic conductivity, excellent hydrophilicity, and flexibility. The dual confinement of 0D carbon coating and 2D $Ti_3C_2T_x$ mitigated the volume expansion of ZnTe during chemical reactions. In addition, through a series of theoretical calculations and characterizations, it has been demonstrated for the first time that ZnTe is an advanced anode material for PIBs with a conversion reaction mechanism ($yZnTe + xK + xe^- \rightarrow yZn + K_xTe_y$) (Figure 3.28d and e). The ZCT electrode exhibited highly stable cycling ability (with a specific capacity of $230.2\,mA\,h\,g^{-1}$ after 3500 cycles at $1.0\,A\,g^{-1}$) [293].

Pan et al. employed heterostructure engineering to anchor the dual TM complex $CoTe_2$/ZnTe onto $Ti_3C_2T_x$ MXene nanosheets, significantly improving the performance of PIBs [295]. This design enhances K^+ ion diffusion and adsorption by adjusting the interface electronic structure, and leveraging the built-in electric field induced by the different work functions between $CoTe_2$/ZnTe to promote charge transfer. The conductivity and elasticity of $Ti_3C_2T_x$ enhance the electrode conductivity and reduce volume changes. Consequently, the $CoTe_2$/ZnTe/$Ti_3C_2T_x$ composite material exhibited excellent rate and cycling performance, with a specific capacity of $175.3\,mAh\,g^{-1}$ after 4000 cycles at a high current density of $3.0\,A\,g^{-1}$, and a capacity retention rate of up to 89.4%. Based on the testing of CV curves at

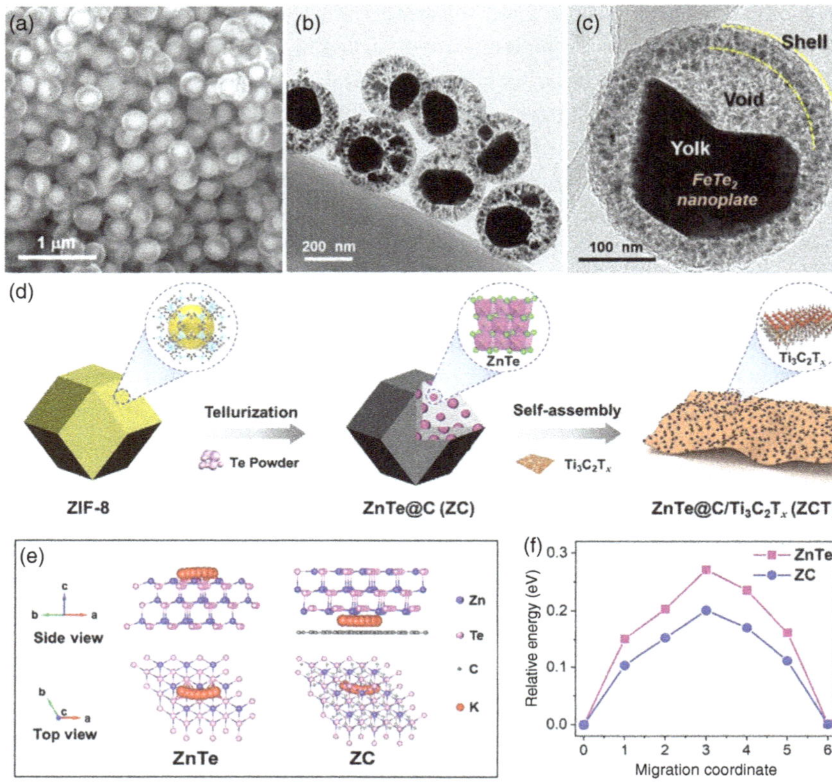

Figure 3.28 Morphologies images of $FeTe_2$-C nanospheres: (a) SEM image and (b,c) TEM images. Source: Reproduced with permission from [292]/John Wiley & Sons. (d) Illustration of preparation process of ZCT nanohybrids. (e) Configuration and (f) relative energy of K^+ ion diffusion in ZnTe and ZC. Source: [293]. Reproduced with permission/John Wiley & Sons.

different scan rates, the influence of heterostructures on K^+ ion storage kinetics was also examined.

Others: In addition to the previously discussed TM tellurides, several relatively less-explored anode candidates such as Cu_2Te, $MoTe_2$, and WTe_2 show promise for PIBs. Chang et al. prepared copper telluride nanotubes by thermal injection method and combined them with copper to improve tellurium utilization efficiency, achieving high capacity, high rate capability, and stable cycling of K^+ ion batteries [296]. Cu_2Te nanotubes have a capacity of 300 mAh g^{-1} at 50 mA g^{-1}. The synergistic effect of high conductivity and copper generation reduces the shuttle effect and electrode resistance. In addition, optimizing design metal ratio to boost performance, offering key insights for PIBs anode materials.

The theoretical mass density of $MoTe_2$ (7.78 g cm^{-3}) is approximately 3.5 times that of graphite (2.09–2.23 g cm^{-3}), and it has a high theoretical specific capacity of 305 mAh g^{-1}, showing potential as an anode material of PIBs. Wu et al. successfully synthesized submicron level single crystal $MoTe_2$ and systematically studied the potassium storage performance of $MoTe_2$ [297]. After initial discharge/charge, a gradual transition between K_2Te_3-KTe-K_5Te_3 was achieved, and the $MoTe_2$ electrode had a specific capacity of 210 mAh g^{-1} at a current density of 0.1 A g^{-1}. The contribution of pseudocapacitance at different scan rates was demonstrated, reflecting the dominance of surface control in high scan rates.

1T′-$ReSe_2$ features weak van der Waals interactions (18 MeV) between adjacent layers and a larger interlayer spacing of 6.37 Å, making it a potential host for reversible and rapid K^+ ion storage [290, 298–300]. Zhou et al. reported a vacancy engineering strategy to prepare a Te-doped 1T′-$ReSe_2$ followed by the anchoring process to form Te-$ReSe_2$/MXene [301]. By utilizing the synergistic effect of a defective Te-$ReSe_2$ array with an extended intermediate layer and elastic MXene nanosheets, resulting in abundant selenium vacancies, high specific surface area, extended interlayer, strong interfacial interactions, and multidimensional ion diffusion channels (Figure 3.28f), the Te-$ReSe_2$/MXene electrode exhibited excellent electrochemical performance (361.1 mA h g^{-1} at 0.1 A g^{-1}).

In addition, Davi Marcelo Soares et al. reported Td-WTe_2 as an anode material for PIBs [302]. Electrochemical studies have demonstrated that Td-WTe_2 maintains a stable semimetallic phase (Td) with high reversibility. Differential capacity analysis and rate capacity testing have demonstrated high reversibility. In addition, it also displays high specific capacity and reasonable cycling stability, achieving a specific charging capacity of WTe_2 electrode at 0.05 A g^{-1} is 238.8 mAh g^{-1}.

Bi_2Te_3: Due to its low oxidation/reduction potential and large interlayer spacing of 10.2 Å, Bi^{3+}/Bi^0 is particularly suitable for use as an anode material in PIBs. Conrad H.R. Gillard et al. synthesized Bi_2Te_3 ferroelectric materials using the solid-phase method and tested the electrochemical performance of Bi_2Te_3 as an anode electrode material for PIBs, exhibiting a first discharge capacity of 456 mA h g^{-1} [303].

Graphene often forms composite materials with metal tellurides due to its significant conductivity, excellent physicochemical stability, and large specific surface area. Jeong Keun Ko et al. prepared a Bi_2Te_3 composite material modified with nanocarbon by ball milling method (Bi_2Te_3@C) [304]. The conductivity was increased from approximately 10^{-7} S cm^{-1} to 10^{-4} S cm^{-1}, and the material was observed to undergo a transformation reaction through *in-situ* XRD. Although the Te element did not participate in the reaction, it promoted the formation of the intermediate product K_2Te. In addition, nano carbon filled the gaps between Bi_2Te_3 particles and expanded the active area, resulting in highly reversible and excellent electrochemical performance of the electrode (with a reversible capacity of 311 mAh g^{-1} after 200 cycles at 0.1 A g^{-1}).

Chong et al. designed and constructed bismuth telluride layered encapsulated by reduced graphene oxide and nitrogen-doped carbon (Bi_2Te_3@rGO@NC) nanocomposites (Figure 3.29a–c) [305]. Generated Bi_2Te_3@rGO@NC structure has strong Bi-O-C chemical bonds, tight physical and chemical confinement effects, typical conductive properties, and enhanced K^+ ion adsorption ability. Using Bi as the redox site, it undergoes a dual potassium storage mechanism in the conversion-alloying, resulting in excellent electrochemical kinetics and excellent morphology and structural stability. As can be seen from the 2D slice pictures

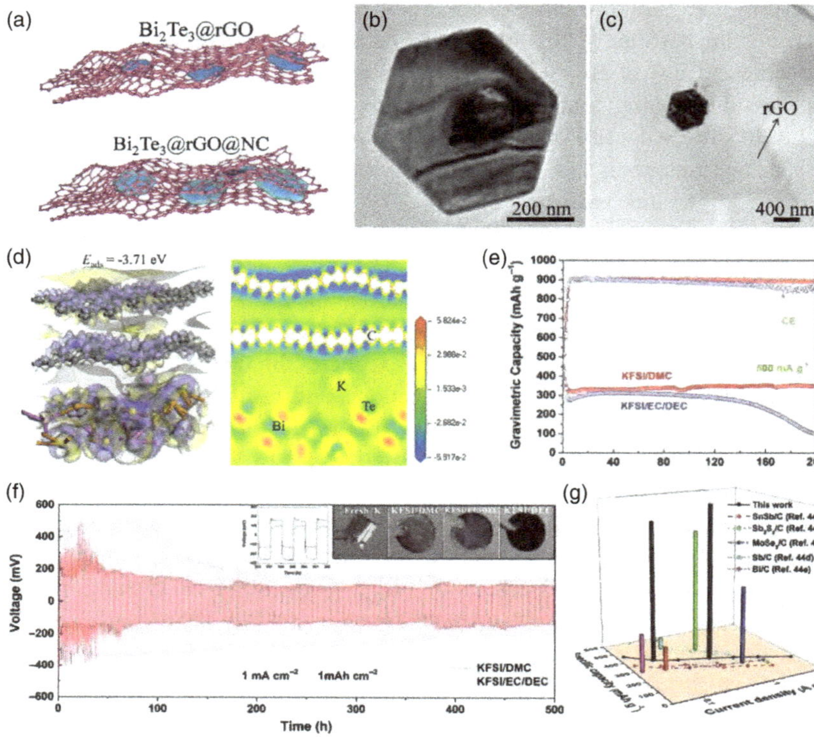

Figure 3.29 (a) Structural and morphological characterizations. (b) TEM images of Bi_2Te_3, (c) Bi_2Te_3@rGO, and (d) the 2D slice pictures of Bi_2Te_3@rGO@NC. Source: Reproduced with permission from [305]/John Wiley & Sons. (e) The cycling performance of the $SnSb_2Te_4$/G electrode for PIBs was measured at 500 mA g^{-1} with different electrolytes. (f) Galvanostatic cycling of symmetric foil cells in different electrolytes at a current density of 1 mA cm^{-2} with a stripping/plating capacity of 1 mAh cm^{-2}. (g) A comparison of cycling and rate performances of the $SnSb_2Te_4$/G composite with those recently published works in PIBs. Source: [306]. Reproduced with permission/John Wiley & Sons.

(Figure 3.29d), Bi_2Te_3@rGO@NC offers lower K^+ ion adsorption energy and, as an anode electrode material for PIBs, exhibits higher cycling performance (specific capacity of 322.70 mAh g^{-1} at 0.05 A g^{-1}).

Element doping is one of the typical material modification methods. Yang et al. synthesized P-doped Bi_2Te_3/MXene type superstructures [307], being anode of PIBs, the material exhibited excellent electrochemical performance due to its abundant Te vacancies, large specific surface area, strong electronic conductivity, unique adaptability, and strong K^+ ion diffusion ability (reversible capacity of 323.1 mAh g^{-1} after 200 cycles at 0.2 A g^{-1}). Wang et al. successfully prepared ultrafine Bi_2Te_{3-x} nanocrystals rich in Te vacancies embedded in nitrogen-doped porous carbon nanofibers using the polyvinylpyrrolidone etching method (Bi_2Te_{3-x}@NPCNFs) [308]. The unique porous structure and abundant Te vacancies not only suppress the deformation of the material structure but also increase the storage capacity of

K, with lower diffusion energy barriers, and improve the diffusion rate of K^+ ions. The Bi_2Te_{3-x}@NPCNFs electrode exhibits excellent rate performance and long-term cycling performance at 10.0 A g^{-1}, even after 1600 cycles, the capacity degradation of Bi_2Te_{3-x}@NPCNF//AC is only 0.02% per cycle. It has been demonstrated through DFT calculations and kinetic analysis that further confirms that Bi_2Te_{3-x}@NPCNFs electrode exhibits excellent pseudocapacitive properties, K^+ ion diffusion ability, and strong K^+ ion adsorption.

SnTe: Tin telluride with high theoretical capacity has received increasing attention. At present, the discussion on tin telluride mainly focuses on SnTe and the construction of heterostructures. Du et al. confined SnTe in an ordered manner within the nanostructure of multilayer graphene sheets (SnTe/MGS), resulting in a smaller size and uniform dispersion of SnTe in the composite material, providing more reaction sites [309]. In addition, MGS layers can effectively alleviate volume collapse and improve the structural stability of materials. As a result, the SnTe/MGS electrode exhibits excellent initial reversible capacity (345.7 mAh g^{-1} at 0.1 A g^{-1}) and ultra-high cycling stability (279.4 mAh g^{-1} at 0.5 A g^{-1} after 1000 cycles). Li et al. synthesized SnTe nanoparticles coated with nitrogen-doped C-layer and graphene oxide using hydrothermal and carbonization methods (SnTe@rGO@NC). Density of states (DOS), molecular dynamics simulation, and the sliced 2D contour maps for SnTe@rGO@NC, indicated that the physical and chemical confinement effect of dual carbon can improve conductivity, reduce K^+ ion transport barriers, and suppress large lattice stresses caused by volume changes. In addition, various *ex situ* characterizations demonstrate that SnTe undergoes a dual potassium storage mechanism of conversion-alloying with Sn as the redox site, exhibiting excellent structural stability [310]. As anode electrode material for PIBs, it exhibits an exceptional charging specific capacity of 243.9 mAh g^{-1} at a current density of 0.05 A g^{-1}. In addition to structural construction, element doping and constructing heterostructure are commonly used methods for material modification, offering broad prospects for developing high-performance rechargeable battery electrode materials. Wu et al. synthesized $SnSb_2Te_4$ nanomaterials ($SnSb_2Te_4$/G) anchored on several layers of graphene using the ball mode method [306]. A P-N heterojunction was formed between single-atom-doped graphene and thin $SnSb_2Te_4$ nanodots, thereby accelerating ion transport efficiency. Importantly, the capacity retention rate reached 91.2% in 200 cycles at 500 mA g^{-1} (Figure 3.29f). Symmetric foil batteries were cycled at a constant current density of 1 mA cm^{-2} in different electrolytes, with a stripping/plating capacity of 1 mAh cm^{-2} (Figure 3.29g). The excellent rate and cycling performance of $SnSb_2Te_4$/G composite material used as PIBs anode are comparable to other recently reported works.

Sb_2Te_3: Sb_2Te_3 is a layered compound that can contribute the highest theoretical specific capacity compared to Bi_2Te_3 and SnTe. Conrad Gillard et al. synthesized antimony tetroxide-type Sb_2Te_3 by solid-phase method and verified the chemical reaction mechanism through *ex situ* powder XRD (PXRD) combined with electrochemical measurements. When used as an anode electrode material for PIBs, the Sb_2Te_3 electrode exhibited an initial capacity of 906 mAh g^{-1}. However, in all cases, the electrode suffered from rapid capacity decay, with a decrease of 98 mAh g^{-1} in the second discharge capacity. This declined performance is usually due to the large volume changes and irreversibility involved in chemical reaction processes [311].

Chong et al. prepared Sb_2Te_3 hexagonal nanosheets constrained by graphene and N-doped C using solvothermal and pyrolysis strategies (Sb_2Te_3@rGO@NC) [312]. By using a dual carbon matrix as a buffer zone to alleviate the volume changes generated during chemical reactions, the strong Sb-C chemical bond can effectively enhance electron transfer, making the Sb_2Te_3@rGO@NC electrode exhibit excellent electrochemical performance (with an initial charge-specific capacity of 384.9 mAh g^{-1} at 0.05 A g^{-1}). In addition, DFT calculations and kinetic studies have confirmed that physical and chemical confinement effects can greatly enhance electron transport, K^+ ion adsorption, and transfer capabilities. It has been confirmed through *ex situ* XPS and TEM techniques that the insertion/extraction of K^+ ions occurs through a dual mechanism of conversion-alloying, allowing for the reversible transfer of 12 electrons based on Sb ion redox sites. Therefore, excellent K^+ ion storage performance can be achieved by Sb_2Te_3@rGO@NC electrode, including a high first specific capacity of 384.9 mAh g^{-1} at 50 mA g^{-1}, excellent rate performance, and outstanding cycling stability.

Summary and Prospects The theoretical specific capacity of converted anode materials is remarkable, but in practical applications, the complexity of electrochemical energy storage mechanisms brings many challenges. These challenges include sluggish reaction kinetics and insufficient cycling stability caused by significant volume changes, anode impact of shuttle effect formed by polysulfides/selenides/polysulfides at low voltage on performance, instability of SEI film caused by volume expansion, and reduction of battery power density under the limitation of suction pressure platform. In addition, the voltage hysteresis during the charging and discharging process also leads to irreversible capacity loss and lower initial Coulombic efficiency. To overcome these challenges, researchers are committed to optimizing from multiple aspects. Although carbon coating technology has been widely studied, it often comes with the sacrifice of volumetric energy density. Therefore, developing high-capacity carbon coating materials has become crucial. At the same time, innovation in material structure design, optimization of electrolytes, precise control of charge and discharge voltage ranges, and improvement of the reversibility of conversion reactions are all key research directions at present. It is particularly important to have a deep understanding of the energy storage mechanism of conversion materials, reveal their microscopic changes during charge and discharge processes, improve the reversibility of electrochemical reactions, and achieve efficient energy storage. The research in this field is not only full of challenges but also harbors enormous opportunities, which are expected to open up new paths for the development of battery technology.

3.4.4 Organic Anode Materials

Organic anode materials for PIBs have been a focus of research in recent years. Because organic materials have controllable oxidation–reduction potential, it is important to design and synthesize organic anodes with suitable oxidation–reduction potentials. These materials can effectively prevent the formation of active metal dendrites and improve battery safety. Based on their molecular structure, organic anode materials for PIBs can be divided into two categories: carboxylate and framework compound organic anode materials.

3.4.4.1 Challenges and Strategies

Although organic electrode materials have demonstrated unique advantages as anode materials for PIBs, there are still some issues that need to be addressed in their practical applications: (i) currently, the types of high-performance organic anode materials that can stably store potassium are very limited. (ii) Due to the easy solubility of organic materials in organic electrolytes, rapid capacity decay occurs during the cycling process. Among them, organic small molecule materials are often more soluble, which seriously affects their cycling stability. (iii) Organic materials generally have poor electronic conductivity, which affects their rate performance. The addition of a large amount of inactive conductive carbon (20–50%) can lead to an increase in electrode porosity and electrolyte usage, as well as a significant decrease in the proportion of organic electrode active materials, resulting in the inability to fully utilize the energy density of the battery.

To address the problems of high solubility and poor conductivity of organic electrode materials, researchers have implemented various improvement measures, including molecular structure design, composite with carbon materials, addition of additives, and optimization of electrolytes.

3.4.4.2 Carboxylate Organic Anode Materials

Carboxylate organic materials can form ionic bonds with metal ions, which leads to high molecular polarity. As a result, they become difficult to dissolve in commonly used organic electrolytes and can potentially achieve good cycling stability. Carboxylate organic anode materials typically utilize the carbonyl group of carboxylates as the redox center. During the reduction process, the carbonyl group gains electrons to form a single bond, and the negatively charged O atom combines with potassium ions from the electrolyte, forming an ionic bond due to charge neutralization. Simultaneously, the benzene ring will open to form a C=C double bond. The oxidation process is the reverse of the reduction process.

Chen et al. were the first to report a highly efficient organic small molecule anode material called potassium terephthalate (K_2TP) [313]. Its theoretical specific capacity is 220 mAh g^{-1}, and its oxidation–reduction potential is between 0.5 and 0.8 V (vs. K^+/K). The higher potassium insertion (reduction) potential is advantageous in preventing the production of metallic potassium, thus enhancing the safety of PIBs. The optimized K_2TP anode material demonstrates excellent performance, maintaining an actual specific capacity of 200 mAh g^{-1} (500 cycles) under high current charge and discharge test conditions of 1 A g^{-1}. This original research paper has opened up new directions for the development of PIBs. Subsequently, several research groups have made significant improvements around K_2TP [314, 315]. Liao et al. reported on the application of composite materials generated *in situ* from K_2TP and carbon nanotubes (CNT) in potassium-ion full-cell batteries [316]. Mai et al. reported that

K_2TP coated with graphene oxide exhibits better potassium storage capacity and rate performance [317]. Additionally, a series of novel carboxylate organic anode materials, including K_2BPDC [318], Co_2TP [319], K_2NDC [320], and ADAPTS [321] with azo groups as active centers, have also been reported.

3.4.4.3 Organic Anode Materials Based on Framework Compounds

The use of organic anode materials based on framework compounds usually has the characteristics of a large specific surface area, suitable ion channels, and designable electrochemical performance [322–324]. However, so far, research on framework compounds as anode materials for PIBs is still in the early stages of research [325–327].

In 2017, Feng et al. achieved promising performance of PIBs using the metal-organic framework MIL-125 based on titanium (Ti) as the anode material [328]. This material demonstrated a capacity of $208\,mAh\,g^{-1}$ at a current density of $10\,mA\,g^{-1}$, with a capacity retention rate of up to 90.2% after 2000 cycles at a high current density of $200\,mA\,g^{-1}$, along with a high Coulombic efficiency. Feng et al. investigated the K^+ ion storage mechanism through *ex situ* XRD and FTIR techniques, confirming that K^+ ions can reversibly insert into organic unit parts without directly contacting Ti ions. Additionally, Hu et al. developed a stable multilayer metal-organic framework (L-$Co_2[OH]_2BDC$, where BDC is 1,4-p-phenyl dicarboxylate) based on cobalt (II) terephthalate as an anode for PIBs [329]. After various current density tests, L-$Co_2(OH)_2BDC$ exhibited stable performance. In addition, nanostructured $[C_7H_3KNO_4]_n$, microporous Fe-based MOF-235 [330], UCF@CNs@BiN composite material [208], and a series of framework compounds such as covalent triazine framework CTF-0 have also been developed for potential use as anode materials in PIBs [331].

Summary and Prospects Organic anode materials have abundant redox sites and adjustable chemical structures, which enable them to exhibit good ion conductivity and structural stability in PIBs. In terms of performance, organic anode materials have made certain progress. By optimizing molecular structure, introducing conductive additives, or constructing composite materials, researchers have successfully improved the electronic conductivity, ion mobility, and structural stability of organic anode materials. These improvements enable organic anode materials to exhibit higher energy density, power density, and longer cycle life in batteries. Looking ahead, researchers will continue to explore new organic anode materials to further improve their performance. By innovating molecular design and synthesis methods, organic anode materials with higher energy density, better stability, and lower cost can be developed. In addition, the performance optimization of organic anode materials will become a research focus. By optimizing the structure, composition, and preparation process of materials, the electronic conductivity, ion mobility, and structural stability of organic anode materials can be further improved. These improvements will enable organic anode materials to exhibit superior performance in batteries. Finally, with the increasing awareness of sustainable development and environmental protection, the commercial application of organic anode materials will gradually accelerate. By reducing production costs, improving production efficiency, and expanding application areas, the widespread application and industrialization process of organic anode materials can be promoted.

In summary, the application of organic anode materials in PIB anode materials has broad development prospects and enormous application potential. In the future, with the continuous efforts and explorations of researchers, it is believed that organic anode materials will play a more important role in the field of batteries.

3.5 Electrolyte for Potassium-Ion Batteries

The electrolyte is an indispensable part of battery system operation, serving as an ion-conducting medium in the battery structure, allowing charges to move freely between the cathode and anode. Electrolytes are typically composed of salts and organic solvents in a certain ratio, sometimes requiring the addition of appropriate additives. In potassium-ion batteries, commonly used electrolytes include aqueous electrolytes, organic electrolytes, ionic electrolytes, and solid electrolytes.

3.5.1 Challenges and Strategies

In PIBs, the electrolyte not only plays a critical role in the formation of SEI, K^+ ion diffusion, and compatibility with electrode materials but also influences the electrochemical performance of the anode electrode. The anode electrode undergoes significant volume changes during potassium insertion/extraction, especially for alloy-type anode electrodes. In non-electrolyte engineering, significant volume changes can severely damage anode electrode materials, exposing new electrode surfaces to the electrolyte and inducing further electrolyte decomposition and SEI formation, which often leads to electrode failure. However, using electrolyte engineering can maintain the integrity of the electrode. Therefore, when selecting electrolytes, it is necessary to consider the following aspects: compatibility with electrode materials, electrochemical/chemical/thermal stability, wide operating voltage window, ion conductivity, dendrites, and safety. The properties of electrolytes are influenced by solutes, potassium salts, and electrolyte additives. Existing PIBs electrolytes can particularly be classified into four categories: aqueous electrolytes, organic electrolytes, ionic electrolytes, and solid electrolytes, as summarized in Figure 3.30 [35].

3.5.2 Water-Based Electrolyte

Water-based electrolytes are attractive for large-scale energy storage systems (power grids) due to their inherent nontoxic nature, safety (non-flammable), stability, cost-effectiveness, high conductivity, and fast ion diffusion advantages. However, water-based electrolytes demonstrate relatively high requirements for electrode materials, requiring their operating voltage to be lower than 1.23 V, because water will decompose into oxygen and hydrogen gas at voltages higher than 1.23 V.

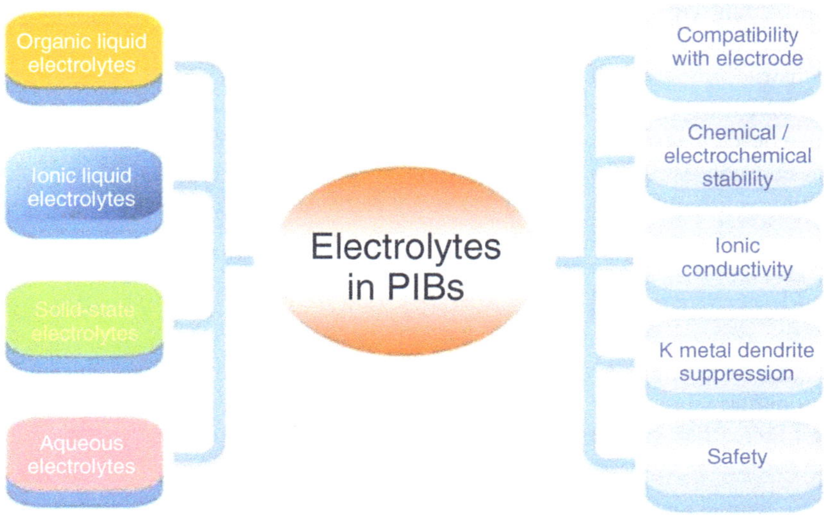

Figure 3.30 Classification and requirements of PIBs electrolyte. Source: [35]. Reproduced with permission/John Wiley & Sons.

The most extensively studied aqueous electrolytes currently include the following types: potassium hydroxide (KOH), potassium nitrate (KNO_3), potassium sulfate (K_2SO_4), and potassium chloride (KCl). In 2011, Professor Cui Yi's research group found that nickel-iron cyanide materials could achieve 5000 cycles under high current density in a water system of 1 mol l^{-1} KNO_3, with a specific capacity retention rate of up to 66% [332]. Moreover, at low current density, the Coulombic efficiency was as high as 99%. Although electrode materials exhibit good electrochemical performance in aqueous electrolytes, their low-voltage plateau results in low energy density and electrode polarization occurs during cycling, constraining their further development. Later, researchers further explored and found that adding high concentrations of salts to aqueous solutions can introduce suitable ions, form a dense interface phase on the anode surface, and prevent water from being reduced [333]. Based on this, researchers prepared a water-based potassium-ion full cell using 22 mol KCF_3SO_3 salt as the electrolyte and found that this electrolyte can effectively avoid the dissolution of electrode materials, improve the working voltage of the battery, and achieve an energy density of 80 Wh kg^{-1} for the full cell [334].

The development of water-based electrolytes has shown great potential in large-scale energy storage devices but still faces ongoing serious issues like the cost of high-concentration salt electrolytes, low operating voltage, low energy density, electrode material dissolution, and unstable SEI films all require further improvement by researchers. Generally, these issues can be considered from the following aspects: appropriate electrolyte additives, SEI film interface engineering, appropriate electrolyte salts, and reasonable passivation electrode surfaces, to construct water-based PIBs with high energy density and excellent cycling stability.

3.5.3 Organic Electrolyte

Organic electrolytes are increasingly studied electrolytes by researchers due to their wide operating voltage window, stable electrochemical performance, good compatibility with electrode materials, high ion conductivity, and good temperature performance. Organic electrolytes are generally composed of potassium salts and solvents, and sometimes additives are added to build a more stable SEI film.

The commonly used potassium salts nowadays include potassium trifluoromethanesulfonate (KCF_3SO_3), potassium bis (trifluoromethanesulfonyl) imide (KFSI), potassium bis (trifluoromethanesulfonyl) imide (KTFSI), potassium perchlorate ($KClO_4$), and potassium hexafluorophosphate (KPF_6). However, $KClO_4$ and KPF_6 cannot completely dissolve in carbonate electrolytes, and KFSI-based electrolytes exhibit higher conductivity and cycling stability than KPF_6-based electrolytes. The most extensively studied solvents are PC, diethyl carbonate (DEC), dimethyl carbonate (DMC), ethylene carbonate (EC), methyl ethyl carbonate (EMC), dioxane (DOL), and ethylene glycol dimethyl ether (DME), and the battery is more stable when circulating in ether-based (DOL, DME) electrolytes than in carbonate-based electrolytes. The commonly used additive is fluorinated vinyl carbonate (FEC) and the most preferred solvent in carbonate electrolytes is EC, as it can increase the solubility of potassium salts, thereby achieving the goal of improving ion conductivity and reversible specific capacity. EC/DEC is the most common solvent for anode electrodes in PIBs. However, research demonstrates that EC/DEC is prone to reduction and decomposition, resulting in thicker SEI films than those formed in ether solvents and greater charge-transfer resistance. Meanwhile, DME has also been proven to be a superior solvent than PC [335]. KPF_6 offers good solubility and electrochemical stability in ester electrolytes, so KPF_6 and EC are common partners in electrolytes. However, different solvents combined with EC influence different aspects of the electrochemical performance of batteries.

In 2016, Professor Wang Chunsheng's research group conducted an in-depth study on the cycling stability of graphite anodes in three electrolytes: KPF_6 (EC: PC), KPF_6 (EC: DMC), and KPF_6 (EC: DEC). Their findings revealed that batteries containing PC exhibited superior Coulombic efficiency and cycling stability, while DMC displayed the worst [146]. This discrepancy is attributed to the continuous decomposition of DMC and DEC under low voltage, which consumes more electrolytes and is not conducive to the stability of SEI film. This situation will be significantly improved in ether electrolytes, as they demonstrate higher solubility and can strongly bind with salt molecules to form a stable SEI film, improving the cycling stability of the battery. In 2018, researchers used KPF_6 electrolyte with $1\ mol\ l^{-1}$ DME to achieve a reversible specific capacity of $392\ mA\ h\ g^{-1}$ for the metal bismuth anode electrode after 100 cycles, with a specific capacity retention rate of up to 97%, and the battery exhibited an ultra-high first Coulombic efficiency of 83% [336]. In addition, compared with ester electrolytes, the introduction of ether solvents offers lower resistance, faster charge-transfer kinetics, and high wettability for potassium storage. PC is usually mixed with EC as a solvent for electrolytes.

Due to its higher solubility and better cycling stability in ethers and esters compared to KPF_6, KFSI-based electrolytes have received widespread attention in recent years. Lu's research team developed a 1.25 mol l^{-1} KFSI (EMC) electrolyte that exhibited ultra-high cycling stability compared to the conventional 0.8 mol l^{-1} KPF_6 (EC: EMC) electrolyte, achieving almost zero decay rate after 17 months of charge–discharge cycles [337]. This excellence in the stability of the electrode is attributed to the formation of an inorganic KFSI-based SEI film, which possesses stronger mechanical strength. Subsequent studies from the Lu Bing'an research group continued to focus on the impact of electrolytes with different salt concentrations on the electrochemical performance of batteries [338]. Additionally, to improve the cycling performance of batteries through ether electrolytes and KFSI salt electrolytes, reasonable electrolyte additives are another effective approach. FEC is the most commonly used electrolyte additive due to its strong antioxidant properties, high operating voltage, and ability to suppress electrode side reactions. In addition, FEC can achieve stable deposition by increasing the active sites on the electrode surface, preventing the growth of potassium dendrites. FEC is most commonly used in electrolytes containing EC, DEC, and PC to increase the operating voltage of Prussian blue and similar cathode materials. Researchers have found that adding 2 vol% FEC to electrolyte KPF_6 (EC/DEC) can achieve high initial Coulombic efficiency, high specific capacity, and high operating voltage plateau of the battery [339]. This is because the decomposition of FEC facilitates the formation of a stable cathode electrolyte interface (CEI), thereby effectively suppressing the occurrence of detrimental side reactions.

Organic electrolytes, while widely favored among researchers for their volatile and flammable nature, seriously restrict their commercialization process. In addition, organic electrolytes also suffer from the following shortcomings: first, carbonate electrolytes are prone to forming unstable SEI films, which leads to rapid capacity decay; Second, ether electrolytes demonstrate weak antioxidant properties, which is not conducive to constructing cathode electrode materials with high operating voltages; Furthermore, the inorganic SEI film formed by excessive inorganic salts in KFSI (DME) type electrolytes is prone to breakage during cycling. The mechanism of SEI film formation in electrolytes containing different potassium salts and electrolytes is still unclear. To promote the development of organic electrolytes, in addition to improving the ion conductivity and solubility of electrolytes, further developments are required to be made in the following aspects: exploring new electrolytes, constructing organic–inorganic hybrid SEI films, developing new electrolyte additives, using *in situ* methods to deeply analyze the mechanism of electrolyte film formation on electrode surfaces, and modifying electrode surfaces.

3.5.4 Ionic Electrolyte

Ionic electrolytes consist of organic/inorganic anions paired with organic cations, and their biggest highlights are non-flammability (high safety factor) and stable operation under high voltage (high energy/power density), which have been favored by researchers. Moreover, ionic electrolytes possess advantages such as good high-temperature thermal stability, good ionic conductivity, low volatility, and electrochemical stability. However, the ion electrolytes for PIBs are still in their infancy, with the most extensively studied being potassium bis (fluorosulfonyl) imide (KFSI)/potassium bis (trifluoromethanesulfonyl) amide (KTFSA)-based electrolytes and KFSI/KTFSI-based ion electrolytes, mainly used in the research of cathode materials. In 2019, researchers reported PIBs using KTFSA-based ionic electrolyte as the electrolyte and layered $K_2Ni_{2-x}Co_xTeO_6$ as the cathode material, achieving an ultra-high electrochemical working window (6 V) along with good electrochemical stability [340]. Ionic electrolytes are not only used to increase the working voltage of cathode materials but are also applied in the research of anode electrode materials. Researchers investigated the electrochemical performance of KTFSI-based electrolytes for hard carbon anode electrode materials in 2019 and found that the physicochemical properties of ionic electrolytes in PIBs can be compared to those in LIBs/SIBs. They also found that non-proton-based KTFSI-based electrolytes can effectively facilitate the reversible intercalation of potassium ions, while proton-based KTFSI-based electrolytes cannot store potassium reversibly due to cation decomposition [341].

The above-detailed discussion testifies that ionic electrolytes are ideal electrolytes for PIBs, but there remain challenges such as high cost, poor battery cycle life, and unstable SEI film. To promote the further development of ionic electrolytes and to improve the ion conductivity of batteries, researchers are also required to research the following aspects. First, a comprehensive investigation is necessary to explore the solvation effect, electrochemical window, and electrolyte dissociation process of electrolytes by employing calculations and *in situ* methods. Second, adding appropriate electrolyte additives or electrode material surface composites to construct a strong SEI film. Finally, conduct in-depth research on the influence of various ions in the electrolyte on electrochemical stability, especially cations in ionic electrolytes.

3.5.5 Solid Electrolyte

Solid-state electrolytes (SSEs) have attracted significant attention in recent years due to their excellent safety, thermal stability, ease of synthesis, and environmental sustainability and are expected to achieve high safety and high-energy-density PIBs. Polymer solid electrolytes and inorganic solid electrolytes are the two primary components of solid electrolytes. Inorganic solid electrolytes have better ion conductivity at room temperature, and the most commonly used inorganic solid electrolyte is β''-Al_2O_3. A key aspect of β'-Al_2O_3 solid electrolytes is to have good wettability between β'-Al_2O_3 and metallic potassium at room temperature to avoid rupture of β'-Al_2O_3. In 2015, researchers proposed the K^+ ion-conducting β''-Al_2O_3 solid electrolyte (K-β''-Al_2O_3). Thanks to the introduction of potassium ions, the wettability of the solid electrolyte was improved. The prepared potassium–sulfur battery demonstrated no significant decrease in specific capacity after up to 1000 cycles. This electrolyte is more stable than the electrolyte of lithium–sulfur batteries and operates at a lower and safer temperature than the electrolyte of sodium–sulfur batteries [342]. In addition to β''-Al_2O_3, other inorganic solid electrolytes have also emerged in recent years, such as ion crystal solid electrolytes with open frameworks ($K_3Cr_3(AsO_4)_4$, $KAlO_2$, $K_3Bi_2(AsO_4)_3$, $K_3Sb_4O_{10}(BO_3)$, $K_2Fe_4O_7$), and solid electrolytes of new hydrated alkali metal ion conductor types [343–345]. The development of inorganic solid electrolytes is progressing at a slow rate, necessitating increased exploration of solid electrolytes with high wettability and high ion conductivity to enhance the cycling stability of batteries. Polymer electrolytes combine the advantages of liquid phase and inorganic solid electrolytes, with flexible mechanical properties, toughness, easy synthesis, dendrite suppression, high thermal stability, high safety, wide voltage operating window, and high ionic conductivity, gaining favor among researchers. Polymer electrolytes are mainly composed of polymers and salt solutions. Polymers are generally polyethylene oxide (PEO), polyester, and polyether. In recent years, research on polymer electrolytes has mainly focused on electrolyte interface engineering and ion transfer mechanisms. In 2019, researchers found that the electrochemical performance of Ni_3S_2 in polymer solid electrolytes was superior to that in ionic electrolytes. This is because polymer solid electrolytes solve the problem of solubility of polysulfides and avoid the deactivation of active materials. Although solid electrolytes have made good progress in recent years, their ion conductivity, high electrode/solid electrolyte interface resistance, and cycling stability still need to be improved. Therefore, researchers need to make improvements in these aspects: using polymers or metal salts with high conductivity, improving the wettability of electrolytes and electrodes (such as coating other metals on the surface of solid electrolytes), and polymer hybridization (transforming single polymers into multiple polymer composites). Electrolyte is the link between positive and negative electrodes and present a significant influence on various parameters of the battery. It is a focal point and challenge of PIB research, so it is necessary to pay much attention to electrolyte and electrolyte/electrode interface engineering to construct electrolytes with excellent electrochemical performance.

Summary and Prospects The electrolyte serves as a key bridge link between the cathode and the anode and plays a crucial role in battery performance, influencing its profound characteristics on various parameters of the battery. Thus, in the process of promoting the development of PIB technology, the engineering design of electrolytes and their interfaces with electrodes has become a top priority in research. To build electrolyte systems with excellent electrochemical performance, researchers need to focus on two core directions: optimization and innovation of electrolytes, and fine control of electrolyte–electrode interface engineering, thus, improving the cycling stability and safety of the battery.

3.6 Potassium–Sulfur Batteries

3.6.1 Composition, Working Mechanism, and Challenges of Potassium–Sulfur Batteries

Potassium–sulfur (K–S) batteries are an emerging member of the metal-sulfur battery family, currently in the development phase. They have advanced quickly by building on the extensive research already conducted on lithium–sulfur (Li–S) and

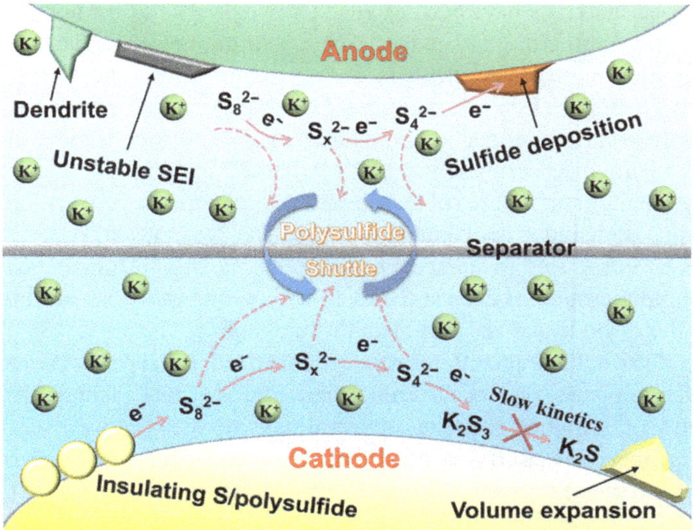

Figure 3.31 Challenges of K-S batteries. Source: [346]/John Wiley & Sons/CC BY 4.0.

sodium–sulfur (Na–S) batteries. K–S batteries have a structure similar to that of Li–S batteries, comprising a cathode, anode, separator, electrolyte, and casing. The key distinction is that K–S batteries use potassium metal as an anode, while the separator is generally a multi-layered fiberglass membrane. Besides ether-based solvents, K–S batteries can also utilize ester-based solvents such as EC and PC. Potassium salts, mainly KPF_6 and $KClO_4$, serve as the solute in the electrolyte. K–S batteries exhibit a comparatively lower energy density (1023 Wh kg^{-1}) and slightly reduced safety compared to sodium-based systems. Consequently, research interest in K–S batteries, as an alternative to other alkali metal-sulfur batteries, has been less prominent compared to the focus on Li–S and Na–S batteries [346–349].

The full multi-electron electrochemical process in potassium–sulfur (K–S) batteries can be described by the following equations:

At the anode: $K \leftrightarrow K^+ + e^-$

At the cathode: $S_n + 2nK^+ + 2ne^- \leftrightarrow nK_2S\ (1 \leq n \leq 8)$

The sulfur reactions at the cathode occur in several stages, illustrating a complex electrochemical pathway, typically represented as $S_8 \leftrightarrow K_2S_8 \leftrightarrow K_2S_6 \leftrightarrow K_2S_4 \leftrightarrow K_2S_2 \leftrightarrow K_2S$.

K–S batteries face challenges similar to lithium–sulfur (Li–S) batteries, including (i) the insulating nature of sulfur and its discharge products; (ii) volume expansion due to the larger ionic radius of potassium; (iii) the "shuttle effect" of soluble potassium polysulfides; and (iv) sluggish reaction kinetics (Figure 3.31).

3.6.2 Sulfur Cathodes

The sulfur cathode is a critical component in the performance of K-S batteries, as it directly influences capacity, energy density, and cycle stability. Like Li-S and Na-S batteries, carbon-based materials have been extensively explored as sulfur hosts in K-S systems due to their numerous advantages. The carbon framework effectively traps active sulfur within its structure, while its porous nature helps to absorb dissolved potassium polysulfides (KPS), minimizing the "shuttle effect," where KPS intermediates migrate toward the anode. Carbon materials also mitigate sulfur's volumetric expansion during reactions, preserving the structural integrity of the electrode. Additionally, the high surface area of carbon improves electrolyte penetration, enhancing K^+ ion diffusion and overall battery performance [348, 350, 351].

In 2014, Chen et al. introduced the concept of K-S batteries, utilizing a composite material of ordered mesoporous carbon (CMK-3) with sulfur and a polyaniline (PANI) coating as the cathode, paired with metallic potassium as the anode [352]. Their investigation of various sulfur contents revealed that the optimal performance of the CMK-3/sulfur composite occurs at a sulfur content of 40.8 wt%, achieving a capacity of 202.3 mAh g^{-1} after 50 cycles at a current density of 50 mA g^{-1}. Additionally, the authors probed the electrochemical reaction mechanisms through electrochemical testing and phase characterization before and after cycling. The results indicated that the K-S battery exhibited two reduction peaks at approximately 2.1 V and 1.8 V, along with an oxidation peak at around 2.2 V. Notably, K_2S_3 was identified as the main discharge product, which can be reversibly charged to split into sulfur and K^+ ions. However, the highly reactive nature of metallic potassium leads to the reduction of dissolved potassium polysulfides (KPS) on its surface, resulting in rapid capacity decay. This electrochemical performance offers insufficient evidence for the practical applications of K-S batteries, highlighting the need for improvements to enhance their viability.

To effectively tackle the issue of KPS dissolution, Xu et al. developed a K-S battery utilizing a composite cathode made of small molecular sulfur confined within microporous carbon [353]. The strong confinement provided by the microporous carbon matrix, along with the unique structure of small molecular sulfur, successfully prevents the formation of soluble polysulfides. This approach achieved a reversible capacity of 1198.3 mAh g^{-1}, with 72.5% capacity retention after 150 cycles. Theoretical calculations suggest that K_2S is the final potassium product, represented by the reaction $2\,K + S \leftrightarrow K_2S$, offering insights into the potassium storage mechanism in sulfur cathode materials.

The aforementioned method primarily relies on physical confinement to reduce KPS dissolution. However, the nonpolar surface of the materials results in weak interactions with K_2S_x intermediates. Thus, a more effective approach is urgently needed – not only to limit KPS dissolution but also to catalyze and accelerate the conversion between K_2S_x and K_2S, thereby fundamentally suppressing the shuttle effect. To achieve this, the development of sulfur hosts with both high conductivity and excellent catalytic activity is essential. These materials can enhance the rapid transfer of K^+ ions and electrons, thus, improving redox kinetics and boosting overall battery performance.

To tackle the slow oxidation kinetics of K_2S, Qiao et al. introduced a catalytic approach using a sulfur host material (S/SA-NC) anchored with single cobalt atoms on nitrogen-doped carbon [354]. This K-S battery demonstrated impressive capacities of 773 mAh g^{-1} and 535 mAh g^{-1} at current densities of 1 C and 2 C,

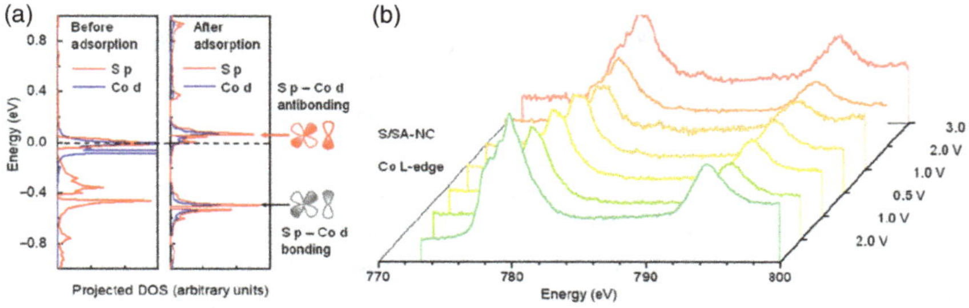

Figure 3.32 (a) DOS of S p and Co d before and following K_2S adsorption on SA-NC. (b) *Ex situ* NEXAFS of the Co L-edge of S/SA-NC in varying discharge and charge states. Source: [354]. Reproduced with permission/American Chemical Society.

respectively. DOS analysis revealed that the hybrid energy levels split into S *p*– Co *d*-antibonding states and S *p* – Co *d*-bonding orbitals, crossing the Fermi level (EF). This hybridization indicates a strong interaction between sulfur and cobalt, which facilitates the oxidation of K_2S (Figure 3.32a). The altered electronic structure highlights the catalytic role of cobalt atoms in the S/SA-NC cathode during the oxidation process. Additionally, high-resolution N 1*s* XPS spectra of the fully discharged S/SA-NC cathode demonstrated that the binding energies of pyridinic and pyrrolic nitrogen shifted to higher energy levels. This shift highlights charge transfer between nitrogen in SA-NC and potassium in potassium polysulfides, leading to the formation of N–K interactions. *Ex-situ* NEXAFS characterization revealed that the oxidation state of Co gradually decreased during discharge, reaching its lowest point at 0.5 V (Figure 3.32b). During the subsequent charging process, the oxidation state of Co increased accordingly, returning to its initial level when the voltage reached 3.0 V. This finding indicates effective adsorption of KPS and the highly reversible nature of the atomic-level Co-S interaction. These results suggest that the synergistic effect of dynamic interactions between Co-S and N-K facilitates the catalysis of K_2S oxidation.

Hu et al. introduced theoretical screening criteria for designing efficient sulfur hosts, leading to the development of a composite host featuring tungsten single atoms and tungsten carbide [355]. This dual-site configuration was engineered to synergistically control the migration and conversion of potassium polysulfides. During carbonization, WO_4^{2-} species transformed into W single atoms and W_2C nanocrystals. In K-S batteries, W_2C catalyzed the conversion of potassium polysulfides, while the W single atoms facilitated polysulfide migration, reducing the accumulation of insulating sulfides and preventing catalytic poisoning at the cathode. K-S batteries with the WSA-W_2C-modified cathode achieved high sulfur utilization of 89.8% (1504 mAh g^{-1}), strong rate performance (1059 mAh g^{-1} at 1675 mA g^{-1}), and long-term stability, maintaining performance for over 200 cycles at 837.5 mA g^{-1}.

3.6.3 Anodes

The stable plating/stripping process of potassium metal is vital for its performance as an anode material in K-S batteries. The stability of the potassium metal anode directly influences the overall capacity of the K-S batteries. The potassium metal anode, due to its highly reactive nature, is prone to severe side reactions with the electrolyte and KPS, which can undergo dendrite growth, leading to a reduced cycle life. Dendrite formation poses a significant challenge in the commercialization of L-S batteries and is expected to be a major issue for K-S batteries as well. During cycling, geometric fluctuations occur between the constantly changing metal and the SEI layer. Dendrite growth of potassium metal accelerates in regions of the SEI layer that exhibit low elasticity, such as pores and cracks. Compared to lithium and sodium, unstable SEI layers are more likely to cause serious dendrite growth issues in K-S batteries. Therefore, it is crucial to form an SEI layer with high conductivity and chemical stability to protect the potassium metal anode [350].

Potassium metal can be effectively encapsulated within a conductive matrix to create a highly conductive network that suppresses dendrite growth. This alkali metal/conductive matrix network is usually formed through techniques like molten injection or electrodeposition. Recently, researchers developed K-ACM by infiltrating molten potassium into aligned carbon nanotube films (Figure 3.33a). The resulting K-ACM anode exhibits stable potassium plating and stripping behavior over 230 hours, with lower polarization compared to pure potassium electrodes (Figure 3.33b). The potassium metal is evenly distributed within the 3D ACM matrix, ensuring a substantial electrode/electrolyte contact area for efficient charge transfer. The stable ACM host provides numerous nucleation sites for potassium, facilitating uniform metal deposition. This K-ACM configuration establishes an effective conductive network with tightly integrated pathways for both electrons and ions, enhancing electrochemical redox reactions [356].

Artificial SEIs have proven their significant potential in suppressing dendrite growth in lithium and sodium metals. The Kang team designed a carbon nanotube (CNT) membrane to serve as an artificial SEI on the surface of potassium metal anodes, successfully maintaining long-term stability and dendrite-free performance (Figure 3.33c). After just 0.5 hours of contact with potassium foil, the tubular structure of the CNTs was well-preserved, with a visible amorphous SEI layer covering their surface. Remarkably, this amorphous SEI layer retained a uniform and consistent thickness of approximately 5.4 nm, maintaining its integrity even after around 2000 hours of cycling (Figure 3.33d). Thanks to these advantages, the CNT-coated potassium metal anode achieved unprecedented cycling stability at a high current density of 5 mA cm^{-2}, along with a competitive areal capacity of 4 mA h cm^{-2} [357].

3.6.4 Electrolytes

K-S batteries currently employ both liquid electrolytes, such as those based on ethers and esters, as well as SSEs, which include polymer and inorganic materials. As a key component of the battery, electrolytes are essential for facilitating ion transport and stabilizing the interface between the electrode and electrolyte, significantly impacting the battery's overall performance and stability [358]. The nucleophilic nature of polysulfides prevents the use of sulfur cathodes in carbonate-based electrolytes. Therefore, K-S systems primarily employ ether-based electrolytes, which form stable solid electrolyte interphases (SEIs) with potassium metal anodes,

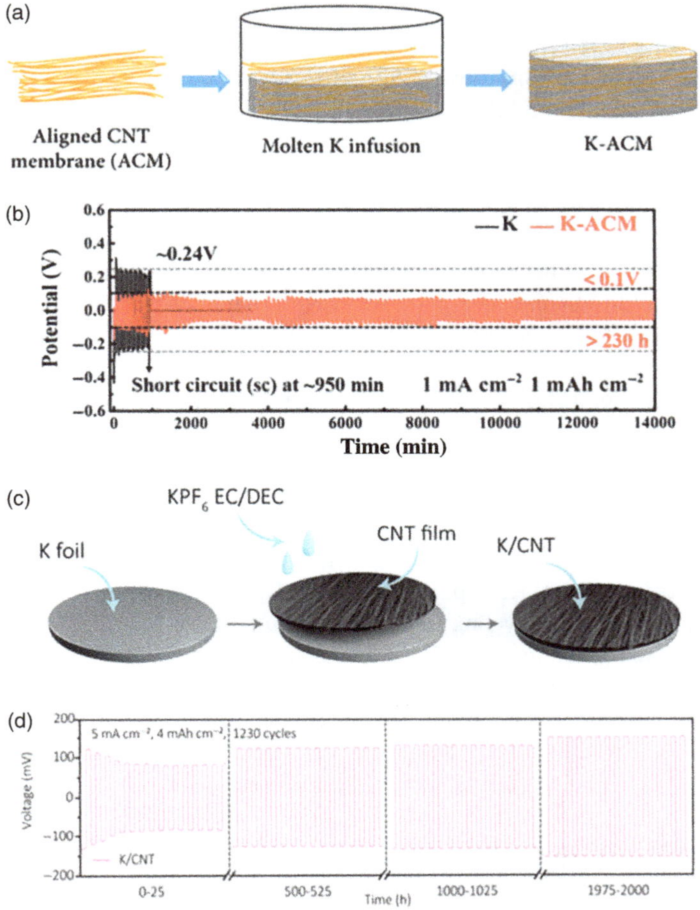

Figure 3.33 (a) Schematic illustration of the fabrication process for K-ACM. (b) Potential–time profiles of K||K symmetric cells with K-ACM and bare K electrodes. Source: [356]. Reproduced with permission/John Wiley & Sons. (c) Schematic illustration of the synthesis process for a K/CNT electrode. (d) Voltage–time profiles of the K||K symmetric cells with a K/CNT electrode. Source: [357]. Reproduced with permission/John Wiley & Sons.

which help mitigate dendrite formation. However, using TEGDME and DEGDME does not completely address the shuttling issue of potassium polysulfides (KPS), such as K_2S_6 and K_2S_4 [359].

In K-S systems, potassium salts such as potassium perchlorate ($KClO_4$), potassium hexafluorophosphate (KPF_6), potassium trifluoromethanesulfonate (KCF_3SO_3), and potassium bis(fluorosulfonyl)imide (KTFSI) are typically dissolved in ether-based electrolytes. KTFSI salts are commonly dissolved in DOL/DME. For instance, Wu et al. examined the solubility and chemical compatibility of K_2S_x ($1 \leq x \leq 3$) in DME electrolyte. Although K_2S_3 has low solubility in DME, it still exhibits a yellow solubility composed of polysulfide species like S_5^{2-}, S_4^{2-}, and S_3^{2-}. As S_5^{2-} further decomposes into S_6^{2-} and S_4^{2-}, it indicates that K_2S_3 breaks down into K_2S_6, K_2S_4, and K_2S_5, with possible formation of K_2S and K_2S_2. However, in the presence of KFSI salt, the solubility of K_2S_3 is suppressed due to the common ion effect resulting from increased K^+ ion concentration. K_2S_2 and K_2S species exhibit almost no absorption, indicating very low solubility. When the solvent contains a high concentration of potassium salts, the solubility of other potassium salts is inevitably suppressed. Additionally, high electrolyte concentrations increase viscosity, which can inhibit the dendrite growth of potassium metal, leading to uniform deposition on the K anode surface [360].

Research on SSEs for K-S batteries remains relatively limited. However, insights from solid-state potassium-ion and Na-S batteries suggest that SSEs offer a promising solution to improve the safety and performance of K-S systems. SSEs help mitigate safety risks, prevent the dissolution of polysulfides, and inhibit the formation of potassium metal dendrites, ultimately enhancing the cycle life and stability of K-S batteries. The Goodenough team has made groundbreaking advances in potassium batteries by developing a novel cross-linked poly(methyl methacrylate) (PMMA) polymer gel electrolyte to replace traditional organic liquid electrolytes. This innovative material not only stabilizes the interface between the electrode and electrolyte but also effectively suppresses the dendrite growth of potassium metal. Compared to batteries using 0.8 M KPF_6 in EC/DEC/FEC liquid electrolytes, the new electrolyte significantly enhances cycling stability and Coulombic efficiency [359].

Beyond SSEs, research on inorganic solid-state electrolytes (ISEs) is also in its early stages. The Wu team has conducted in-depth studies on sulfide-based ISEs (K_3SbS_4), which are potassium superionic conductors with low activation energy. Their research first demonstrated the potential of using non-commercial inorganic electrolytes to suppress the polysulfide shuttling phenomenon [361].

Summary and Prospects K-S batteries have emerged as a next-generation energy storage system due to the abundance of potassium, environmental benefits, and high theoretical energy density. By utilizing a potassium metal anode and a sulfur cathode, K-S batteries demonstrate promising performance through polysulfide (KPS) reactions during electrochemical cycling. However, their development is still hindered by several challenges, including the insulating nature of the sulfur cathode, the polysulfide shuttle effect, dendrite formation on the potassium metal anode, and the instability of the electrolyte–electrode interface.

The sulfur cathode plays a critical role in determining the capacity and cycle life of K-S batteries. Carbon-based materials, extensively studied as sulfur hosts, can effectively mitigate sulfur expansion, adsorb potassium polysulfides (KPS), and reduce the shuttle effect. The stability of the potassium metal anode is crucial to the performance of K-S batteries. Due to its high reactivity, potassium metal readily undergoes side reactions with the electrolyte and KPS, leading to dendrite growth and a shortened cycle life. Electrolytes play a vital role in K-S batteries, affecting ion transport and the stability of the electrode–electrolyte interface. Current liquid electrolytes, primarily based on ether solvents, help reduce potassium dendrite formation to some extent but fail to fully address the KPS shuttle effect. Further improvements in sulfur host materials to enhance polysulfide adsorption and develop hosts with higher catalytic activity are necessary to promote polysulfide conversion, suppress the shuttle effect, and extend battery cycle life. Dendrite formation in potassium metal remains a critical challenge. Future efforts could focus on developing more effective artificial SEI layers and incorporating novel conductive matrix encapsulation techniques to enhance the stability and safety of the anode. Research into SSEs and high-concentration liquid electrolytes offers promising avenues for improving the cycling performance of K-S batteries.

Future studies should focus on optimizing the conductivity and stability of SSEs to prevent dendrite growth and the polysulfide shuttle effect. With continued innovation in materials and structural design, K-S batteries are poised to overcome current technical barriers and become a high-energy-density, long-life, and safe energy storage solution, unlocking new opportunities for large-scale energy storage and electric vehicles.

References

1 Eftekhari, A. (2004). Potassium secondary cell based on Prussian blue cathode. *Journal of Power Sources* 126: 221–228.

2 Rajagopalan, R., Tang, Y., Ji, X. et al. (2020). Advancements and challenges in potassium ion batteries: a comprehensive review. *Advanced Functional Materials* 30: 1909486.

3 Jo, C.-H. and Myung, S.-T. (2024). Role of ether-based electrolytes in enhancing potential of potassium-ion batteries. *Advanced Energy Materials* 14: 2400217.

4 Xin, B.-J. and Wu, X.-L. (2024). Research progresses on metal-organic frameworks for sodium/potassium-ion batteries. *Battery Energy* 3: 20230074.

5 Pramudita, J.C., Sehrawat, D., Goonetilleke, D., and Sharma, N. (2017). An initial review of the status of electrode materials for potassium-ion batteries. *Advanced Energy Materials* 7: 1602911.

6 Ma, R., Wang, Z., Fu, Q. et al. (2023). Dual-salt assisted synergistic synthesis of Prussian white cathode towards high-capacity and long cycle potassium ion battery. *Journal of Energy Chemistry* 83: 16–23.

7 Yang, Z., Zhang, J., Kintner-Meyer, M.C.W. et al. (2011). Electrochemical energy storage for green grid. *Chemical Reviews* 111: 3577–3613.

8 Cong, J., Luo, S.-H., Li, K. et al. (2022). Research progress of manganese-based layered oxides as cathode materials for potassium-ion batteries. *Journal of Electroanalytical Chemistry* 927: 116971.

9 Xiao, Z., Xia, F., Xu, L. et al. (2022). Suppressing the Jahn–Teller effect in Mn-based layered oxide cathode toward long-life potassium-ion batteries. *Advanced Functional Materials* 32: 2108244.

10 Min, X., Xiao, J., Fang, M. et al. (2021). Potassium-ion batteries: outlook on present and future technologies. *Energy & Environmental Science* 14: 2186–2243.

11 Thackeray, M.M., Wolverton, C., and Isaacs, E.D. (2012). Electrical energy storage for transportation—approaching the limits of, and going beyond, lithium-ion batteries. *Energy & Environmental Science* 5: 7854–7863.

12 Hosaka, T., Kubota, K., Hameed, A.S., and Komaba, S. (2020). Research development on K-ion batteries. *Chemical Reviews* 120: 6358–6466.

13 Liu, S., Kang, L., and Jun, S.C. (2021). Challenges and strategies toward cathode materials for rechargeable potassium-ion batteries. *Advanced Materials* 33: 2004689.

14 Zhang, X., Wei, Z., Dinh, K.N. et al. (2020). Layered oxide cathode for potassium-ion battery: recent progress and prospective. *Small* 16: 2002700.

15 Jian, Z., Luo, W., and Ji, X. (2015). Carbon electrodes for K-ion batteries. *Journal of the American Chemical Society* 137: 11566–11569.

16 Zhang, J., Liu, T., Cheng, X. et al. (2019). Development status and future prospect of non-aqueous potassium ion batteries for large scale energy storage. *Nano Energy* 60: 340–361.

17 Wang, X., Xiao, Z., Han, K. et al. (2023). Advances in fine structure optimizations of layered transition-metal oxide cathodes for potassium-ion batteries. *Advanced Energy Materials* 13: 2202861.

18 Yao, Q. and Zhu, C. (2020). Advanced post-potassium-ion batteries as emerging potassium-based alternatives for energy storage. *Advanced Functional Materials* 30: 2005209.

19 Zeng, G., Liu, B., Ali, U. et al. (2024). The local disorder induced by high-entropy doping results in highly stable cathode materials for aqueous potassium-ion batteries. *Applied Catalysis B: Environment and Energy* 351: 123996.

20 Zhang, X., Xiong, T., He, B. et al. (2022). Recent advances and perspectives in aqueous potassium-ion batteries. *Energy & Environmental Science* 15: 3750.

21 Ali, U., Liu, B., Jia, H. et al. (2024). In situ Fe-substituted hexacyanoferrate for high-performance aqueous potassium ion batteries. *Small* 20: 2305866.

22 Li, W., Bi, Z., Zhang, W. et al. (2021). Advanced cathodes for potassium-ion batteries with layered transition metal oxides: a review. *Journal of Materials Chemistry A* 9: 8221–8247.

23 Kubota, K., Dahbi, M., Hosaka, T. et al. (2018). Towards K-ion and Na-ion batteries as "beyond Li-ion". *The Chemical Record* 18: 459–479.

24 Zheng, Y., Xie, H., Li, J. et al. (2024). Insights into the Jahn-teller effect in layered oxide cathode materials for potassium-ion batteries. *Advanced Energy Materials* 14: 2400461.

25 Huang, Z.-X., Gu, Z.-Y., Heng, Y.-L. et al. (2023). Advanced layered oxide cathodes for sodium/potassium-ion batteries: development, challenges and prospects. *Chemical Engineering Journal* 452: 139438.

26 Shu, W., Han, C., and Wang, X. (2024). Prussian blue analogues cathodes for nonaqueous potassium-ion batteries: past, present, and future. *Advanced Functional Materials* 34: 2309636.

27 Xu, Y.-S., Duan, S.-Y., Sun, Y.-G. et al. (2019). Recent developments in electrode materials for potassium-ion batteries. *Journal of Materials Chemistry A* 7: 4334–4352.

28 Komaba, S., Hasegawa, T., Dahbi, M., and Kubota, K. (2015). Potassium intercalation into graphite to realize high-voltage/high-power potassium-ion batteries and potassium-ion capacitors. *Electrochemistry Communications* 60: 172–175.

29 Wang, T., Liu, Q., Zhou, J. et al. (2022). Natural supra-molecular structure engineering for high-performance potassium batteries separator. *Advanced Energy Materials* 12: 2202357.

30 Wang, C., Su, L., Wang, N. et al. (2022). Unravelling binder chemistry in sodium/potassium ion batteries for superior electrochemical performances. *Journal of Materials Chemistry A* 10: 4060–4067.

31 Wu, J., Zhang, Q., Liu, S. et al. (2020). Synergy of binders and electrolytes in enabling microsized alloy anodes for high performance potassium-ion batteries. *Nano Energy* 77: 105118.

32 Ma, J., Liu, M., Wan, Z. et al. (2024). Matching design on conversion anode and conductive agent for potassium-ion storage. *Journal of Energy Storage* 92: 111997.

33 Chen, X., Li, M., Wang, S.-P. et al. (2022). In situ fabrication of cuprous selenide electrode via selenization of copper current collector for high-efficiency potassium-ion and sodium-ion storage. *Advanced Science* 9: 2104630.

34 Zhao, Y., Liu, B., Yi, Y. et al. (2022). An anode-free potassium-metal battery enabled by a directly grown graphene-modulated aluminum current collector. *Advanced Materials* 34: 2202902.

35 Zhou, M., Bai, P., Ji, X. et al. (2021). Electrolytes and interphases in potassium ion batteries. *Advanced Materials* 33: 2003741.

36 Liu, Y., Gao, C., Dai, L. et al. (2020). The features and progress of electrolyte for potassium ion batteries. *Small* 16: 2004096.

37 Song, Q., Li, Z., Gan, S. et al. (2023). Progress and challenges of Prussian blue analogs for potassium-ion batteries: a perspective on redox-active transition metals. *Journal of Materials Chemistry A* 11: 1532–1550.

38 Zhang, C., Xu, Y., Zhou, M. et al. (2017). Potassium Prussian blue nanoparticles: a low-cost cathode material for potassium-ion batteries. *Advanced Functional Materials* 27: 1604307.

39 Nathan, M.G.T., Yu, H., Kim, G.-T. et al. (2022). Recent advances in layered metal-oxide cathodes for application in potassium-ion batteries. *Advanced Science* 9: 2105882.

40 Kim, H., Seo, D.-H., Urban, A. et al. (2018). Stoichiometric layered potassium transition metal oxide for rechargeable potassium batteries. *Chemistry of Materials* 30: 6532–6539.

41 Hosaka, T., Shimamura, T., Kubota, K., and Komaba, S. (2019). Polyanionic compounds for potassium-ion batteries. *The Chemical Record* 19: 735–745.

42 Liao, J., Han, Y., Zhang, Z. et al. (2021). Recent progress and prospects of layered cathode materials for potassium-ion batteries. *Energy & Environmental Materials* 4: 178.

43 Liang, Y., Luo, C., Wang, F. et al. (2019). An organic anode for high temperature potassium-ion batteries. *Advanced Energy Materials* 9: 1802986.

44 Shu, W., Han, C., and Wang, X. (2024). Prussian blue analogues cathodes for nonaqueous potassium-ion batteries: past, present, and future. *Advanced Functional Materials* 34: 2309636.

45 Wang, J., Wang, B., Liu, X. et al. (2020). Prussian blue analogs (PBA) derived porous bimetal (Mn, Fe) selenide with carbon nanotubes as anode materials for sodium and potassium ion batteries. *Chemical Engineering Journal* 382: 123050.

46 Ma, C., Lin, C., Li, N. et al. (2024). A high-entropy prussian blue analog for aqueous potassium-ion batteries. *Small* 20: 2310184.

47 Zhang, K.-Y., Gu, Z.-Y., Ang, E.H. et al. (2022). Advanced polyanionic electrode materials for potassium-ion batteries: progresses, challenges and application prospects. *Materials Today* 54: 189–201.

48 Huang, H., Wu, X., Gao, Y. et al. (2024). Polyanionic cathode materials: a comparison between Na-ion and K-ion batteries. *Advanced Energy Materials* 14: 2304251.

49 Hoppe, R. and Sabrowsky, H. (1965). Oxoscandate der alkalimetalle: $KScO_2$ und $RbScO_2$. *Zeitschrift für anorganische und allgemeine Chemie* 339: 144–154.

50 Kim, J.-M., Zhang, X., Zhang, J.-G. et al. (2021). A review on the stability and surface modification of layered transition-metal oxide cathodes. *Materials Today* 46: 155–182.

51 Gao, R.-M., Zheng, Z.-J., Wang, P.-F. et al. (2020). Recent advances and prospects of layered transition metal oxide cathodes for sodium-ion batteries. *Energy Storage Materials* 30: 9–26.

52 Li, Y.-Q., Shi, H., Wang, S.-B. et al. (2019). Dual-phase nanostructuring of layered metal oxides for high-performance aqueous rechargeable potassium ion microbatteries. *Nature Communications* 10: 4292.

53 Zhang, Q., Didier, C., Pang, W.K. et al. (2019). Structural insight into layer gliding and lattice distortion in layered manganese oxide electrodes for potassium-ion batteries. *Advanced Energy Materials* 9: 1900568.

54 Lin, Y., Luo, S., Cong, J. et al. (2024). Strategies for developing layered oxide cathodes, carbon-based anodes, and electrolytes for potassium ion batteries. *Materials Horizons* 11: 2053.

55 Xiao, Z., Xia, F., Xu, L. et al. (2022). Suppressing the Jahn–Teller effect in Mn-based layered oxide cathode toward long-life potassium-ion batteries. *Advanced Functional Materials* 32: 2108244.

56 Zheng, Y., Li, J., Ji, S. et al. (2023). Zinc-doping strategy on P2-type Mn-based layered oxide cathode for high-performance potassium-ion batteries. *Small* 19: 2302160.

57 Deng, T., Fan, X., Luo, C. et al. (2018). Self-templated formation of P2-type K0. $6CoO_2$ microspheres for high reversible potassium-ion batteries. *Nano Letters* 18: 1522.

58 Zhao, S., Liu, Z., Xie, G. et al. (2022). High-efficiency cathode potassium compensation and interfacial stability improvement enabled by dipotassium squarate for potassium-ion batteries. *Energy & Environmental Science* 15: 3015–3023.

59 Wang, X., Xiao, Z., Han, K. et al. (2023). Advances in fine structure optimizations of layered transition-metal oxide cathodes for potassium-ion batteries. *Advanced Energy Materials* 13: 2202861.

60 Vaalma, C., Giffin, G.A., Buchholz, D., and Passerini, S. (2016). Non-aqueous K-ion battery based on layered $K_{0.3}MnO_2$ and hard carbon/carbon black. *Journal of The Electrochemical Society* 163: A1295.

61 Sun, K., Luo, S.H., Hao, G. et al. (2024). Review on layered manganese-based metal oxides cathode materials for potassium-ion batteries: from preparation to modification. *The Chemical Record* 24: e202300327.

62 Kim, H., Seo, D.H., Kim, J.C. et al. (2017). Investigation of potassium storage in layered P3-type $K_{0.5}MnO_2$ cathode. *Advanced Materials* 29: 1702480.

63 Jo, J.H., Choi, J.U., Park, Y.J. et al. (2020). P2-$K_{0.75}[Ni_{1/3}Mn_{2/3}]O_2$ cathode material for high power and long life potassium-ion batteries. *Advanced Energy Materials* 10: 1903605.

64 Caixiang, Z., Hao, J., Zhou, J. et al. (2023). Interlayer-engineering and surface-substituting manganese-based self-evolution for high-performance potassium cathode. *Advanced Energy Materials* 13: 2203126.

65 Zhang, X., Yang, X., Sun, G. et al. (2022). Hydration enables air-stable and high-performance layered cathode materials for both organic and aqueous potassium-ion batteries. *Advanced Functional Materials* 32: 2204318.

66 Lei, K., Zhu, Z., Yin, Z. et al. (2019). Dual interphase layers in situ formed on a manganese-based oxide cathode enable stable potassium storage. *Chem* 5: 3220–3231.

67 Asl, H.Y. and Manthiram, A. (2020). Reining in dissolved transition-metal ions. *Science* 369: 140–141.

68 Choi, J.U., Kim, J., Hwang, J.-Y. et al. (2019). $K_{0.54}[Co_{0.5}Mn_{0.5}]O_2$: New cathode with high power capability for potassium-ion batteries. *Nano Energy* 61: 284–294.

69 Xu, Y.-S., Zhang, Q.-H., Wang, D. et al. (2020). Enabling reversible phase transition on $K_{5/9}Mn_{7/9}Ti_{2/9}O_2$ for high-performance potassium-ion batteries cathodes. *Energy Storage Materials* 31: 20–26.

70 Liu, C.-L., Luo, S.-H., Huang, H.-B. et al. (2019). Fe-doped layered P3-type $K_{0.45}Mn_{1-x}Fe_xO_2(x \leq 0.5)$ as cathode materials for low-cost potassium-ion batteries. *Chemical Engineering Journal* 378: 122167.

71 Liu, C., Luo, S., Huang, H. et al. (2017). $K_{0.67}Ni_{0.17}Co_{0.17}Mn_{0.66}O_2$: A cathode material for potassium-ion battery. *Electrochemistry Communications* 82: 150–154.

72 Zeng, G., Liu, B., Ali, U. et al. (2024). The local disorder induced by high-entropy doping results in highly stable cathode materials for aqueous potassium-ion batteries. *Applied Catalysis B: Environment and Energy* 351: 123996.

73 Kim, H., Kim, J.C., Bo, S.H. et al. (2017). K-ion batteries based on a P2-type $K_{0.6}CoO_2$ cathode. *Advanced Energy Materials* 7: 1700098.

74 Hironaka, Y., Kubota, K., and Komaba, S. (2017). P2-and P3-K_xCoO_2 as an electrochemical potassium intercalation host. *Chemical Communications* 53: 3693–3696.

75 Lan, Y., Yao, W., He, X. et al. (2020). Mixed polyanionic compounds as positive electrodes for low-cost electrochemical energy storage. *Angewandte Chemie International Edition* 59: 9255–9262.

76 Zhang, Q., Wang, Z., Zhang, S. et al. (2018). Cathode materials for potassium-ion batteries: current status and perspective. *Electrochemical Energy Reviews* 1: 625–658.

77 Xu, J., Dou, S., Cui, X. et al. (2021). Potassium-based electrochemical energy storage devices: development status and future prospect. *Energy Storage Materials* 34: 85–106.

78 Zhang, Z., Du, Y., Wang, Q.C. et al. (2020). A yolk–shell-structured $FePO_4$ cathode for high-rate and long-cycling sodium-ion batteries. *Angewandte Chemie International Edition* 59: 17504–17510.

79 Mathew, V., Kim, S., Kang, J. et al. (2014). Amorphous iron phosphate: potential host for various charge carrier ions. *NPG Asia Materials* 6: e138.

80 Lu, J., Nishimura, S.-I., and Yamada, A. (2017). Polyanionic solid-solution cathodes for rechargeable batteries. *Chemistry of Materials* 29: 3597–3602.

81 Xu, C., Fu, Q., Hua, W. et al. (2024). Overcoming kinetic limitations of polyanionic cathode toward high-performance Na-ion batteries. *ACS Nano* 18: 18758–18768.

82 Zhang, L., Zhang, B., Wang, C. et al. (2019). Constructing the best symmetric full K-ion battery with the NASICON-type $K_3V_2(PO_4)_3$. *Nano Energy* 60: 432–439.

83 Han, J., Li, G.N., Liu, F. et al. Investigation of $K_3V_2(PO_4)_3/C$ nanocomposites as high-potential cathode materials for potassium-ion batteries. *Chemical Communications* 2017: 53, 1805–1808.

84 Wang, S., Xiong, S., Li, Z. et al. (2024). Constructing multi-electron reactions by doping Mn^{2+} to increase capacity and stability in $K_{3.2}V_{2.8}Mn_{0.2}(PO_4)_4/C$ of phosphate cathodes for potassium-ion batteries. *Small* 2308628.

85 Voronina, N., Jo, J.H., Konarov, A. et al. (2020). $KTi_2(PO_4)_3$ electrode with a long cycling stability for potassium-ion batteries. *Small* 16: 2001090.

86 Zheng, S., Cheng, S., Xiao, S. et al. (2020). Partial replacement of K by Rb to improve electrochemical performance of $K_3V_2(PO_4)_3$ cathode material for potassium-ion batteries. *Journal of Alloys and Compounds* 815: 152379.

87 Xu, S., Yang, Y., Tang, F. et al. (2023). Vanadium fluorophosphates: advanced cathode materials for next-generation secondary batteries. *Materials Horizons* 10: 1901–1923.

88 Zhu, Y., Ou, B., Gao, C. et al. (2024). Ligand engineering enables fast kinetics of $KVPO_4F$ cathode for potassium-ion batteries. *ACS Energy Letters* 9: 3212–3218.

89 Palomares, V., Blas, M., Serras, P. et al. (2018). Waste biomass as in situ carbon source for sodium vanadium fluorophosphate/C cathodes for Na-ion batteries. *ACS Sustainable Chemistry & Engineering* 6: 16386–16398.

90 Heng, Y., Gu, Z., Guo, J. et al. (2024). Low-strain and high-energy $KVPO_4F$ cathode with multifunctional stabilizer for advanced potassium-ion batteries. *Energy & Environmental Materials* e12721.

91 Chihara, K., Katogi, A., Kubota, K., and Komaba, S. (2017). $KVPO_4F$ and $KVOPO_4$ toward 4 volt-class potassium-ion batteries. *Chemical Communications* 53: 5208–5211.

92 Lin, X., Huang, J., Tan, H. et al. (2019). $K_3V_2(PO_4)_2F_3$ as a robust cathode for potassium-ion batteries. *Energy Storage Materials* 16: 97–101.

93 Park, W.B., Han, S.C., Park, C. et al. (2018). KVP_2O_7 as a robust high-energy cathode for potassium-ion batteries: pinpointed by a full screening of the inorganic registry under specific search conditions. *Advanced Energy Materials* 8: 1703099.

94 He, X.-D., Liao, J.-Y., Chen, T. et al. (2021). Spray drying derived wrinkled pea-shaped carbon-matrixed KVP_2O_7 as a cathode material for potassium-ion batteries. *Journal of Alloys and Compounds* 884: 161126.

95 Hyoung, J., Heo, J.W., Chae, M.S., and Hong, S.T. (2019). Electrochemical exchange reaction mechanism and the role of additive water to stabilize the structure of $VOPO_4{\cdot}2H_2O$ as a cathode material for potassium-ion batteries. *ChemSusChem* 12: 1069–1075.

96 Fedotov, S.S., Samarin, A.S., and Antipov, E.V. (2020). $KTiOPO_4$-structured electrode materials for metal-ion batteries: a review. *Journal of Power Sources* 480: 228840.

97 Li, J., Zheng, Y., San Hui, K. et al. (2023). Enhanced K-storage performance in ultralong cycle-life potassium-ion batteries achieved via carbothermal-reduction-synthesized $KVOPO_4$ cathode. *Energy Storage Materials* 61: 102852.

98 Lian, R., Wang, D., Ming, X. et al. (2018). Phase transformation, ionic diffusion, and charge transfer mechanisms of $KVOPO_4$ in potassium ion batteries: first-principles calculations. *Journal of Materials Chemistry A* 6: 16228–16234.

99 Kang, J., Ko, W., Park, H. et al. (2020). High-power rhombohedral-$Fe_2(SO_4)_3$ with outstanding cycle-performance as Fe-based cathode for K-ion batteries. *Energy Storage Materials* 33: 276–282.

100 Lander, L., Rousse, G., Abakumov, A.M. et al. (2015). Structural, electrochemical and magnetic properties of a novel $KFeSO_4F$ polymorph. *Journal of Materials Chemistry A* 3: 19754–19764.

101 Dong, J., Liao, J., He, X. et al. (2020). Graphene encircled $KFeSO_4F$ cathode composite for high energy density potassium-ion batteries. *Chemical Communications* 56: 10050–10053.

102 Liao, J., Hu, Q., Du, Y. et al. (2022). Robust carbon nanotube-interwoven $KFeSO_4F$ microspheres as reliable potassium cathodes. *Science Bulletin* 67: 2208–2215.

103 Zhao, S., Guo, Z., Yan, K. et al. (2021). The rise of Prussian blue analogs: challenges and opportunities for high-performance cathode materials in potassium-ion batteries. *Small Structures* 2: 2000054.

104 Qin, Z., Li, Y., and Gu, N. (2018). Progress in applications of Prussian blue nanoparticles in biomedicine. *Advanced Healthcare Materials* 7: 1800347.

105 Padigi, P., Thiebes, J., Swan, M. et al. (2015). Prussian green: a high rate capacity cathode for potassium ion batteries. *Electrochimica Acta* 166: 32–39.

106 Shu, W., Huang, M., Geng, L. et al. (2023). Highly crystalline Prussian blue for kinetics enhanced potassium storage. *Small* 19: e2207080.

107 Wei, Y., Zhang, S., Zhai, D., and Kang, F. (2023). Escape of lattice water in potassium iron hexacyanoferrate for cyclic optimization in potassium-ion batteries. *Battery Energy* 2: 20220027.

108 Ma, C., Lin, C., Li, N. et al. (2024). A high-entropy Prussian blue analog for aqueous potassium-ion batteries. *Small* 20: 2310184.

109 Li, C., Wang, X., Deng, W. et al. (2018). Size engineering and crystallinity control enable high-capacity aqueous potassium-ion storage of Prussian white analogues. *ChemElectroChem* 5: 3887–3892.

110 Dong, J., Lei, Y., Han, D. et al. (2019). Utilizing an autogenously protective atmosphere to synthesize a Prussian white cathode with ultrahigh capacity-retention for potassium-ion batteries. *Chemical Communications* 55: 12555–12558.

111 Chong, S., Yang, J., Sun, L. et al. (2020). Potassium nickel iron hexacyanoferrate as ultra-long-life cathode material for potassium-ion batteries with high energy density. *ACS Nano* 14: 9807–9818.

112 Luo, Y., Shen, B., Guo, B. et al. (2018). Potassium titanium hexacyanoferrate as a cathode material for potassium-ion batteries. *Journal of Physics and Chemistry of Solids* 122: 31–35.

113 Wei, Y., Wang, H., Wang, J. et al. (2021). Polyvinylpyrrolidone-bridged Prussian blue/rGO composite as a high-performance cathode for K-ion batteries. *ACS Applied Materials & Interfaces* 13: 54079–54087.

114 He, G. and Nazar, L.F. (2017). Crystallite size control of Prussian white analogues for nonaqueous potassium-ion batteries. *ACS Energy Letters* 2: 1122–1127.

115 Qin, M., Ren, W., Meng, J. et al. (2019). Realizing superior Prussian blue positive electrode for potassium storage via ultrathin nanosheet assembly. *ACS Sustainable Chemistry & Engineering* 7: 11564–11570.

116 Xue, L., Li, Y., Gao, H. et al. (2017). Low-cost high-energy potassium cathode. *Journal of the American Chemical Society* 139: 2164–2167.

117 Bie, X., Kubota, K., Hosaka, T. et al. (2017). A novel K-ion battery: hexacyanoferrate (II)/graphite cell. *Journal of Materials Chemistry A* 5: 4325–4330.

118 Chong, S., Wu, Y., Guo, S. et al. (2019). Potassium nickel hexacyanoferrate as cathode for high voltage and ultralong life potassium-ion batteries. *Energy Storage Materials* 22: 120–127.

119 Wu, X., Jian, Z., Li, Z., and Ji, X. (2017). Prussian white analogues as promising cathode for non-aqueous potassium-ion batteries. *Electrochemistry Communications* 77: 54–57.

120 Huang, B., Shao, Y., Liu, Y. et al. (2019). Improving potassium-ion batteries by optimizing the composition of Prussian blue cathode. *ACS Applied Energy Materials* 2: 6528–6535.

121 Tong, Z., Tian, S., Wang, H. et al. (2019). Tailored redox kinetics, electronic structures and electrode/electrolyte interfaces for fast and high energy-density potassium-organic battery. *Advanced Functional Materials* 30: 1907656.

122 Hu, Y., Tang, W., Yu, Q. et al. (2020). Novel insoluble organic cathodes for advanced organic K-ion batteries. *Advanced Functional Materials* 30: 2000675.

123 Zhao, Q., Wang, J., Lu, Y. et al. (2016). Oxocarbon salts for fast rechargeable batteries. *Angewandte Chemie International Edition* 55: 12528–12532.

124 Wang, J., Jia, H., Liu, Z. et al. (2024). Anchoring π-d conjugated metal–organic frameworks with dual-active centers on carbon nanotubes for advanced potassium-ion batteries. *Advanced Materials* 36: 2305605.

125 Chen, Y., Luo, W., Carter, M. et al. (2015). Organic electrode for non-aqueous potassium-ion batteries. *Nano Energy* 18: 205–211.

126 Fan, L., Ma, R., Wang, J. et al. (2018). An ultrafast and highly stable potassium–organic battery. *Advanced Materials* 30: 1805486.

127 Shchurik, E.V., Mumyatov, A.V., Zhidkov, I.S. et al. (2025). Potassium batteries for low temperature applications using high energy density organic cathodes. *Journal of Materials Chemistry A* 13: 721–729.

128 Tian, B., Zheng, J., Zhao, C. et al. (2019). Carbonyl-based polyimide and polyquinoneimide for potassium-ion batteries. *Journal of Materials Chemistry A* 7: 9997–10003.

129 Tang, M., Wu, Y., Chen, Y. et al. (2019). An organic cathode with high capacities for fast-charge potassium-ion batteries. *Journal of Materials Chemistry A* 7: 486–492.

130 Zhao, J., Yang, J., Sun, P., and Xu, Y. (2018). Sodium sulfonate groups substituted anthraquinone as an organic cathode for potassium batteries. *Electrochemistry Communications* 86: 34–37.

131 Yang, S., Zhang, F., Ding, H. et al. (2018). Lithium metal extraction from seawater. *Joule* 2: 1648–1651.

132 Xu, Y., Zhang, C., Zhou, M. et al. (2018). Highly nitrogen doped carbon nanofibers with superior rate capability and cyclability for potassium ion batteries. *Nature Communications* 9: 1720.

133 Hwang, J.Y., Myung, S.T., and Sun, Y.K. (2018). Recent progress in rechargeable potassium batteries. *Advanced Functional Materials* 28: 1802938.

134 Kishore, B., Venkatesh, G., and Munichandraiah, N. (2016). $K_2Ti_4O_9$: a promising anode material for potassium ion batteries. *Journal of the Electrochemical Society* 163: A2551.

135 Han, J., Niu, Y., Bao, S.-J. et al. (2016). Nanocubic $KTi_2(PO_4)_3$ electrodes for potassium-ion batteries. *Chemical Communications* 52: 11661–11664.

136 Ren, X., Zhao, Q., McCulloch, W.D., and Wu, Y. (2017). MoS_2 as a long-life host material for potassium ion intercalation. *Nano Research* 10: 1313–1321.

137 Hwang, J.-Y., Myung, S.-T., and Sun, Y.-K. (2018). Recent progress in rechargeable potassium batteries. *Advanced Functional Materials* 28: 1802938.

138 Yuan, F., Li, Z., Zhang, D. et al. (2022). Fundamental understanding and research progress on the interfacial behaviors for potassium-ion battery anode. *Advanced Science* 9: 2200683.

139 Zhou, P., Zhang, M., Wang, L. et al. (2022). MOFs-derived flower-like hierarchically porous Zn-Mn-Se/C Composite for extraordinary rate performance and durable anode of sodium-ion and potassium-ion batteries. *Small* 18: 2203964.

140 Zheng, J., Wu, Y., Sun, Y. et al. (2021). Advanced anode materials of potassium ion batteries: from zero dimension to three dimensions. *Nano-Micro Letters* 13: 1–37.

141 Song, K., Liu, C., Mi, L. et al. (2021). Recent progress on the alloy-based anode for sodium-ion batteries and potassium-ion batteries. *Small* 17: 1903194.

142 Chen, J., Cheng, Y., Zhang, Q. et al. (2021). Designing and understanding the superior potassium storage performance of nitrogen/phosphorus co-doped hollow porous bowl-like carbon anodes. *Advanced Functional Materials* 31: 2007158.

143 Liu, D., Yang, Z., Li, W. et al. (2010). Electrochemical intercalation of potassium into graphite in KF melt. *Electrochimica Acta* 55: 1013–1018.

144 Nobuhara, K., Nakayama, H., Nose, M. et al. (2013). First-principles study of alkali metal-graphite intercalation compounds. *Journal of Power Sources* 243: 585–587.

145 Luo, W., Wan, J., Ozdemir, B. et al. (2015). Potassium ion batteries with graphitic materials. *Nano Letters* 15: 7671–7677.

146 Zhao, J., Zou, X., Zhu, Y. et al. (2016). Electrochemical intercalation of potassium into graphite. *Advanced Functional Materials* 26: 8103–8110.

147 Qian, Y., Jiang, S., Li, Y. et al. (2019). Water-induced growth of a highly oriented mesoporous graphitic carbon nanospring for fast potassium-ion adsorption/intercalation storage. *Angewandte Chemie International Edition* 58: 18108–18115.

148 Luo, W., Jian, Z., Xing, Z. et al. (2015). Electrochemically expandable soft carbon as anodes for Na-ion batteries. *ACS Central Science* 1: 516–522.

149 Tan, W., Wang, L., Liu, K. et al. (2022). Bitumen-derived onion-like soft carbon as high-performance potassium-ion battery anode. *Small* 18: 2203494.

150 Yuan, F., Zhang, D., Li, Z. et al. (2022). Unraveling the intercorrelation between micro/mesopores and K migration behavior in hard carbon. *Small* 18: 2107113.

151 Zhang, K., He, Q., Xiong, F. et al. (2020). Active sites enriched hard carbon porous nanobelts for stable and high-capacity potassium-ion storage. *Nano Energy* 77: 105018.

152 Zhao, L.-F., Hu, Z., Lai, W.-H. et al. (2021). Hard carbon anodes: fundamental understanding and commercial perspectives for Na-ion batteries beyond Li-ion and K-ion counterparts. *Advanced Energy Materials* 11: 2002704.

153 Jian, Z., Hwang, S., Li, Z. et al. (2017). Hard–soft composite carbon as a long-cycling and high-rate anode for potassium-ion batteries. *Advanced Functional Materials* 27: 1700324.

154 Cao, B., Zhang, Q., Liu, H. et al. (2018). Graphitic carbon nanocage as a stable and high power anode for potassium-ion batteries. *Advanced Energy Materials* 8: 1801149.

155 Wu, S., Song, Y., Lu, C. et al. (2022). An adsorption-insertion mechanism of potassium in soft carbon. *Small* 18: 2105275.

156 Ding, H., Zhou, J., Rao, A.M., and Lu, B. (2021). Cell-like-carbon-micro-spheres for robust potassium anode. *National Science Review* 8: nwaa276.

157 Liu, C., Xiao, N., Li, H. et al. (2020). Nitrogen-doped soft carbon frameworks built of well-interconnected nanocapsules enabling a superior potassium-ion batteries anode. *Chemical Engineering Journal* 382: 121759.

158 Li, J., Li, Y., Ma, X. et al. (2020). A honeycomb-like nitrogen-doped carbon as high-performance anode for potassium-ion batteries. *Chemical Engineering Journal* 384: 123328.

159 Chen, C., Wang, Z., Zhang, B. et al. (2017). Nitrogen-rich hard carbon as a highly durable anode for high-power potassium-ion batteries. *Energy Storage Materials* 8: 161–168.

160 Nan, D., Wu, Z., Lu, Z. et al. (2023). Facile preparation of heteroatom S doped dumbbell-like hard carbon for boosting potassium-ion storage capability. *Electrochimica Acta* 465: 143019.

161 Liu, Y., Dai, H., Wu, L. et al. (2019). A large scalable and low-cost sulfur/nitrogen dual-doped hard carbon as the negative electrode material for high-performance potassium-ion batteries. *Advanced Energy Materials* 9: 1901379.

162 Chong, S., Yuan, L., Li, T. et al. (2022). Nitrogen and oxygen Co-doped porous hard carbon nanospheres with core-shell architecture as anode materials for superior potassium-ion storage. *Small* 18: 2104296.

163 Wu, Y., Sun, Y., Zheng, J. et al. (2021). MXenes: advanced materials in potassium ion batteries. *Chemical Engineering Journal* 404: 126565.

164 Dong, Y., Shi, H., and Wu, Z.-S. (2020). Recent advances and promise of MXene-based nanostructures for high-performance metal ion batteries. *Advanced Functional Materials* 30: 2000706.

165 Er, D., Li, J., Naguib, M. et al. (2014). Ti_3C_2 MXene as a high capacity electrode material for metal (Li, Na, K, Ca) ion batteries. *ACS Applied Materials & Interfaces* 6: 11173–11179.

166 Huang, H., Cui, J., Liu, G. et al. (2019). Carbon-coated $MoSe_2$/MXene hybrid nanosheets for superior potassium storage. *ACS Nano* 13: 3448–3456.

167 Ming, F., Liang, H., Zhang, W. et al. (2019). Porous MXenes enable high performance potassium ion capacitors. *Nano Energy* 62: 853–860.

168 Zhou, S., Zhang, P., Li, Y. et al. (2024). Ultrastable organic anode enabled by electrochemically active MXene binder toward advanced potassium ion storage. *ACS Nano* 18: 16027.

169 Guo, X., Gao, H., Wang, S. et al. (2022). MXene-based aerogel anchored with antimony single atoms and quantum dots for high-performance potassium-ion batteries. *Nano Letters* 22: 1225–1232.

170 Cao, B., Liu, H., Zhang, P. et al. (2021). Flexible mxene framework as a fast electron/potassium-ion dual-function conductor boosting stable potassium storage in graphite electrodes. *Advanced Functional Materials* 31: 2102126.

171 Qin, M., Yao, Y., Mao, J. et al. (2024). Pillar effect induced by ultrahigh phosphorous/nitrogen doping enables graphene/MXene film with excellent cycling stability for alkali metal ion storage. *Journal of Energy Chemistry* 93: 146–156.

172 Wang, J., Liu, Z., Zhou, J. et al. (2021). Insights into metal/metalloid-based alloying anodes for potassium ion batteries. *ACS Materials Letters* 3: 1572–1598.

173 Kim, H., Kim, J.C., Bianchini, M. et al. (2018). Recent progress and perspective in electrode materials for K-ion batteries. *Advanced Energy Materials* 8: 1702384.

174 Wang, Z., Dong, K., Wang, D. et al. (2020). Constructing N-doped porous carbon confined FeSb alloy nanocomposite with Fe-NC coordination as a universal anode for advanced Na/K-ion batteries. *Chemical Engineering Journal* 384: 123327.

175 Xiong, P., Wu, J., Zhou, M., and Xu, Y. (2020). Bismuth–antimony alloy nanoparticle@porous carbon nanosheet composite anode for high-performance potassium-ion batteries. *ACS Nano* 14: 1018–1026.

176 Wu, Y., Huang, H.-B., Feng, Y. et al. (2019). The promise and challenge of phosphorus-based composites as anode materials for potassium-ion batteries. *Advanced Materials* 31: 1901414.

177 Sun, J., Lee, H.-W., Pasta, M. et al. (2015). A phosphorene–graphene hybrid material as a high-capacity anode for sodium-ion batteries. *Nature Nanotechnology* 10: 980–985.

178 Xiong, P., Bai, P., Tu, S. et al. (2018). Red phosphorus nanoparticle@3D interconnected carbon nanosheet framework composite for potassium-ion battery anodes. *Small* 14: 1802140.

179 Yu, J., He, Y., Li, J. et al. (2023). In-situ rooting biconical-nanorods-like Co-doped FeP@ carbon architectures toward enhanced lithium storage performance. *Chemical Engineering Journal* 477: 146996.

180 Yan, Z., Huang, Z., Yao, Y. et al. (2021). Monodispersed Ni_2P nanodots embedded in N, P co-doped porous carbon as super stable anode material for potassium-ion batteries. *Journal of Alloys and Compounds* 858: 158203.

181 Wang, Z., Gao, W., Ding, C. et al. (2021). Boosting potassium-ion storage in large-diameter carbon nanotubes/MoP hybrid. *Journal of Colloid and Interface Science* 584: 875–884.

182 Sultana, I., Rahman, M.M., Ramireddy, T. et al. (2017). High capacity potassium-ion battery anodes based on black phosphorus. *Journal of Materials Chemistry A* 5: 23506–23512.

183 Liu, D., Huang, X., Qu, D. et al. (2018). Confined phosphorus in carbon nanotube-backboned mesoporous carbon as superior anode material for sodium/potassium-ion batteries. *Nano Energy* 52: 1–10.

184 Wang, Q., Zhao, X., Ni, C. et al. (2017). Reaction and capacity-fading mechanisms of tin nanoparticles in potassium-ion batteries. *The Journal of Physical Chemistry C* 121: 12652–12657.

185 Huang, K., Xing, Z., Wang, L. et al. (2018). Direct synthesis of 3D hierarchically porous carbon/Sn composites via in situ generated NaCl crystals as templates for potassium-ion batteries anode. *Journal of Materials Chemistry A* 6: 434–442.

186 Yang, Y., Li, D., Zhang, J. et al. (2019). Sn nanoparticles anchored on N doped porous carbon as an anode for potassium ion batteries. *Materials Letters* 256: 126613.

187 Wang, H., Xing, Z., Hu, Z. et al. (2019). Sn-based submicron-particles encapsulated in porous reduced graphene oxide network: advanced anodes for high-rate and long life potassium-ion batteries. *Applied Materials Today* 15: 58–66.

188 Li, D., Zhang, Y., Sun, Q. et al. (2019). Hierarchically porous carbon supported Sn_4P_3 as a superior anode material for potassium-ion batteries. *Energy Storage Materials* 23: 367–374.

189 Zhao, X., Wang, W., Hou, Z. et al. (2019). $SnP_{0.94}$ nanoplates/graphene oxide composite for novel potassium-ion battery anode. *Chemical Engineering Journal* 370: 677–683.

190 Han, C., Han, K., Wang, X. et al. (2018). Three-dimensional carbon network confined antimony nanoparticle anodes for high-capacity K-ion batteries. *Nanoscale* 10: 6820–6826.

191 Yi, Z., Qian, Y., Tian, J. et al. (2019). Self-templating growth of Sb_2Se_3@C microtube: a convention-alloying-type anode material for enhanced K-ion batteries. *Journal of Materials Chemistry A* 7: 12283–12291.

192 Huang, H., Wang, J., Yang, X. et al. (2020). Unveiling the advances of nanostructure design for alloy-type potassium-ion battery anodes via in situ TEM. *Angewandte Chemie International Edition* 59: 14504–14510.

193 Tian, Y., An, Y., Xiong, S. et al. (2019). A general method for constructing robust, flexible and freestanding MXene@ metal anodes for high-performance potassium-ion batteries. *Journal of Materials Chemistry A* 7: 9716–9725.

194 Gabaudan, V., Berthelot, R., Stievano, L., and Monconduit, L. (2018). Inside the alloy mechanism of Sb and Bi electrodes for K-ion batteries. *The Journal of Physical Chemistry C* 122: 18266–18273.

195 Cheng, N., Zhao, J., Fan, L. et al. (2019). Sb-MOFs derived Sb nanoparticles@ porous carbon for high performance potassium-ion batteries anode. *Chemical Communications* 55: 12511–12514.

196 An, Y., Tian, Y., Ci, L. et al. (2018). Micron-sized nanoporous antimony with tunable porosity for high-performance potassium-ion batteries. *ACS Nano* 12: 12932–12940.

197 Zheng, J., Yang, Y., Fan, X. et al. (2019). Extremely stable antimony–carbon composite anodes for potassium-ion batteries. *Energy & Environmental Science* 12: 615–623.

198 Liu, Q., Fan, L., Chen, S. et al. (2019). Antimony–graphite composites for a high-performance potassium-ion battery. *Energy Technology* 7: 1900634.

199 Han, Y., Li, T., Li, Y. et al. (2019). Stabilizing antimony nanocrystals within ultrathin carbon nanosheets for high-performance K-ion storage. *Energy Storage Materials* 20: 46–54.

200 Zhang, W., Miao, W., Liu, X. et al. (2018). High-rate and ultralong-stable potassium-ion batteries based on antimony-nanoparticles encapsulated in nitrogen and phosphorus co-doped mesoporous carbon nanofibers as an anode material. *Journal of Alloys and Compounds* 769: 141–148.

201 Ud Din, M.A., Li, C., Zhang, L. et al. (2021). Recent progress and challenges on the bismuth-based anode for sodium-ion batteries and potassium-ion batteries. *Materials Today Physics* 21: 100486.

202 Lei, K., Wang, C., Liu, L. et al. (2018). A porous network of bismuth used as the anode material for high-energy-density potassium-ion batteries. *Angewandte Chemie International Edition* 130: 4777–4781.

203 Zhao, Y., Ren, X., Xing, Z. et al. (2020). In situ formation of hierarchical bismuth nanodots/graphene nanoarchitectures for ultrahigh-rate and durable potassium-ion storage. *Small* 16: 1905789.

204 Tian, Y., Li, H., Zhang, S. et al. (2021). Polymer "Tape"-assisted ball-milling method fabrication few-atomic-layered bismuth for improving K^+/Na^+ storage. *Energy & Environmental Materials* 4: 421–427.

205 Shen, C., Cheng, T., Liu, C. et al. (2020). Bismuthene from sonoelectrochemistry as a superior anode for potassium-ion batteries. *Journal of Materials Chemistry A* 8: 453–460.

206 Cheng, X., Li, D., Wu, Y. et al. (2019). Bismuth nanospheres embedded in three-dimensional (3D) porous graphene frameworks as high performance anodes for sodium-and potassium-ion batteries. *Journal of Materials Chemistry A* 7: 4913–4921.

207 Wu, X., Zhang, W., Wu, N. et al. (2020). Exploration of nanoporous CuBi binary alloy for potassium storage. *Advanced Functional Materials* 30: 2003838.

208 Su, S., Liu, Q., Wang, J. et al. (2019). Control of SEI formation for stable potassium-ion battery anodes by Bi-MOF-derived nanocomposites. *ACS Applied Materials & Interfaces* 11: 22474–22480.

209 Yang, H., Xu, R., Yao, Y. et al. (2019). Multicore–shell Bi@ N-doped carbon nanospheres for high power density and long cycle life sodium-and potassium-ion anodes. *Advanced Functional Materials* 29: 1809195.

210 Wang, B., Yang, H., Feng, Y. et al. (2021). Boosting low-temperature sodium/potassium storage performance of Bi via novel electrochemical milling process. *Materials Today Energy* 20: 100627.

211 Xie, F., Zhang, L., Chen, B. et al. (2019). Revealing the origin of improved reversible capacity of dual-shell bismuth boxes anode for potassium-ion batteries. *Matter* 1: 1681–1693.

212 Zhu, H., Liu, T., Peng, L. et al. (2021). A compact Bi_2WO_6 microflowers anode for potassium-ion storage: taming a sequential phase evolution toward stable electrochemical cycling. *Nano Energy* 82: 105784.

213 Zhu, J., Fan, J., Cheng, T. et al. (2020). Bilayer nanosheets of unusual stoichiometric bismuth oxychloride for potassium ion storage and CO_2 reduction. *Nano Energy* 75: 104939.

214 Wang, L., Zhang, B., Wang, B. et al. (2021). In-situ nano-crystallization and solvation modulation to promote highly stable anode involving alloy/de-alloy for potassium ion batteries. *Angewandte Chemie International Edition* 60: 15381–15389.

215 Wang, L., Li, Y., Wang, B. et al. (2024). Controllable growth of crystal facets enables superb cycling stability of anode material for potassium ion batteries. *Advanced Functional Materials* 2406988.

216 Sultana, I., Rahman, M.M., Chen, Y., and Glushenkov, A.M. (2018). Potassium-ion battery anode materials operating through the alloying–dealloying reaction mechanism. *Advanced Functional Materials* 28: 1703857.

217 Lee, S., Jung, S.C., and Han, Y.-K. (2019). First-principles molecular dynamics study on ultrafast potassium ion transport in silicon anode. *Journal of Power Sources* 415: 119–125.

218 Yue, C., Wu, M., Cheng, B. et al. (2020). Fabrication of multilayer Si/TiN/Sb NR arrays as anode for 3D Si-based lithium/sodium ion microbatteries. *Advanced Materials Interfaces* 7: 2001043.

219 Sun, Y., Zhao, Y., Liang, Z. et al. (2022). Reversible potassium-ion alloying storage in crystalline silicene. *Chemical Engineering Journal* 435: 134961.

220 Yang, Z., Zhao, S., Jiao, R. et al. (2024). Lignite-based hierarchical porous C/SiOx composites as high-performance anode for potassium-ion batteries. *Energy & Environmental Materials* 7: e12674.

221 Yang, Q., Wang, Z., Xi, W., and He, G. (2019). Tailoring nanoporous structures of Ge anodes for stable potassium-ion batteries. *Electrochemistry Communications* 101: 68–72.

222 Kohandehghan, A., Cui, K., Kupsta, M. et al. (2014). Activation with Li enables facile sodium storage in germanium. *Nano Letters* 14: 5873–5882.

223 Yi, Z., Lin, N., Li, T. et al. (2019). Meso-porous amorphous Ge: Synthesis and mechanism of an anode material for Na and K storage. *Nano Research* 12: 1824–1830.

224 Liu, R., Luo, F., Zeng, L. et al. (2021). Dual carbon decorated germanium-carbon composite as a stable anode for sodium/potassium-ion batteries. *Journal of Colloid and Interface Science* 584: 372–381.

225 Kim, D.-H. and Park, C.-M. (2020). Co–Ge compounds and their electrochemical performance as high-performance Li-ion battery anodes. *Materials Today Energy* 18: 100530.

226 Zhao, Z., Ma, W., Wang, Y. et al. (2021). Boosting the electrochemical performance of nanoporous CuGe anode by regulating the porous structure and solid electrolyte interface layer through Ni-doping. *Applied Surface Science* 558: 149868.

227 Doherty, J., McNulty, D., Biswas, S. et al. (2020). Germanium tin alloy nanowires as anode materials for high performance Li-ion batteries. *Nanotechnology* 31: 165402.

228 Kim, W.S., Vo, T.N., and Kim, I.T. (2020). GeTe-TiC-C composite anodes for li-ion storage. *Materials* 13: 4222.

229 Zhang, Z., Wu, C., Chen, Z. et al. (2020). Spatially confined synthesis of a flexible and hierarchically porous three-dimensional graphene/FeP hollow nanosphere composite anode for highly efficient and ultrastable potassium ion storage. *Journal of Materials Chemistry A* 8: 3369–3378.

230 Lee, G.-W., Choi, Y.J., Kim, Y.H. et al. (2022). Amorphization of germanium selenide driven by chemical interaction with carbon and realization of reversible conversion-alloying reaction for superior K-ion storage. *Chemical Engineering Journal* 430: 132995.

231 Lai, C., Zhang, Z., Xu, Y. et al. (2021). A general strategy for embedding ultrasmall CoM x nanocrystals (M= S, O, Se, and Te) in hierarchical porous carbon nanofibers for high-performance potassium storage. *Journal of Materials Chemistry A* 9: 1487–1494.

232 Liang, H., Zhang, Y., Hao, S. et al. (2021). Fast potassium storage in porous CoV2O6 nanosphere@ graphene oxide towards high-performance potassium-ion capacitors. *Energy Storage Materials* 40: 250–258.

233 Sultana, I., Rahman, M.M., Mateti, S. et al. (2017). K-ion and Na-ion storage performances of Co_3O_4–Fe_2O_3 nanoparticle-decorated super P carbon black prepared by a ball milling process. *Nanoscale* 9: 3646–3654.

234 Liu, C., Luo, S., Huang, H. et al. (2019). Direct growth of MoO_2/reduced graphene oxide hollow sphere composites as advanced anode materials for potassium-ion batteries. *ChemSusChem* 12: 873–880.

235 Liu, Y., He, D., Tan, Q. et al. (2019). A synergetic strategy for an advanced electrode with Fe_3O_4 embedded in a 3D N-doped porous graphene framework and a strong adhesive binder for lithium/potassium ion batteries with an ultralong cycle lifespan. *Journal of Materials Chemistry A* 7: 19430–19441.

236 Wang, J., Wang, B., Liu, Z. et al. (2019). Nature of bimetallic oxide Sb_2MoO_6/rGO anode for high-performance potassium-ion batteries. *Advanced Science* 6: 1900904.

237 Shimizu, M., Yatsuzuka, R., Koya, T. et al. (2018). Tin oxides as a negative electrode material for potassium-ion batteries. *ACS Applied Energy Materials* 1: 6865–6870.

238 Li, Y., Zhang, Q., Yuan, Y. et al. (2020). Surface amorphization of vanadium dioxide (B) for K-ion battery. *Advanced Energy Materials* 10: 2000717.

239 Wu, Y., Huang, H.B., Feng, Y. et al. (2019). Potassium-ion batteries: the promise and challenge of phosphorus-based composites as anode materials for potassium-ion batteries. *Advanced Materials* 31: 1970354.

240 Tan, Q., Han, K., Zhao, W. et al. (2021). Synchronous nesting of hollow FeP nanospheres into a three-dimensional porous carbon scaffold via a salt-template method for performance-enhanced potassium-ion storage. *Sustainable Energy & Fuels* 5: 844–854.

241 Yang, F., Gao, H., Hao, J. et al. (2019). Yolk–shell structured FeP@ C nanoboxes as advanced anode materials for rechargeable lithium-/potassium-ion batteries. *Advanced Functional Materials* 29: 1808291.

242 Bai, J., Xi, B., Mao, H. et al. (2018). One-step construction of N, P-codoped porous carbon sheets/CoP hybrids with enhanced lithium and potassium storage. *Advanced Materials* 30: 1802310.

243 Zhou, D., Yi, J., Zhao, X. et al. (2021). Confining ultrasmall CoP nanoparticles into nitrogen-doped porous carbon via synchronous pyrolysis and phosphorization for enhanced potassium-ion storage. *Chemical Engineering Journal* 413: 127508.

244 Jiang, Y., Liang, J., Yue, L. et al. (2021). Reduced graphene oxide supported ZIF-67 derived CoP enables high-performance potassium ion storage. *Journal of Colloid and Interface Science* 604: 319–326.

245 Ihsan-Ul-Haq, M., Huang, H., Cui, J. et al. (2018). Chemical interactions between red P and functional groups in NiP_3/CNT composite anodes for enhanced sodium storage. *Journal of Materials Chemistry A* 6: 20184–20194.

246 Wei, Y., Chen, J., He, J. et al. (2018). Morphology processing by encapsulating GeP_5 nanoparticles into nanofibers toward enhanced thermo/electrochemical stability. *ACS Applied Materials & Interfaces* 10: 32162–32170.

247 Zhang, W., Wu, Z., Zhang, J. et al. (2018). Unraveling the effect of salt chemistry on long-durability high-phosphorus-concentration anode for potassium ion batteries. *Nano Energy* 53: 967–974.

248 Yang, F.H., Gao, H., Chen, J., and Guo, Z.P. (2017). Phosphorus-based materials as the anode for sodium-ion batteries. *Small Methods* 1: 1700216.

249 Zhang, W., Mao, J., Li, S. et al. (2017). Phosphorus-based alloy materials for advanced potassium-ion battery anode. *Journal of the American Chemical Society* 139: 3316–3319.

250 Zhang, W., Pang, W.K., Sencadas, V., and Guo, Z. (2018). Understanding high-energy-density Sn_4P_3 anodes for potassium-ion batteries. *Joule* 2: 1534–1547.

251 Li, D., Zhang, Y., Sun, Q. et al. (2019). Hierarchically porous carbon supported Sn_4P_3 as a superior anode material for potassium-ion batteries. *Energy Storage Materials* 23: 367–374.

252 He, Y., Dong, C., He, S. et al. (2021). Bimetallic nickel cobalt sulfides with hierarchical coralliform architecture for ultrafast and stable Na-ion storage. *Nano Research* 14: 4014–4024.

253 Li, H., He, Y., Dai, Y. et al. (2022). Bimetallic SnS_2/NiS_2@S-rGO nanocomposite with hierarchical flower-like architecture for superior high rate and ultra-stable half/full sodium-ion batteries. *Chemical Engineering Journal* 427: 131784.

254 Li, H., He, Y., Wang, Q. et al. (2023). $SnSe_2$/$NiSe_2$@N-doped carbon yolk-shell heterostructure construction and selenium vacancies engineering for ultrastable sodium-ion storage. *Advanced Energy Materials* 13: 2302901.

255 Pan, Q., Tong, Z., Su, Y. et al. (2021). Energy storage mechanism, challenge and design strategies of metal sulfides for rechargeable sodium/potassium-ion batteries. *Advanced Functional Materials* 31: 2103912.

256 He, Y.Y., Li, H., Yu, J.X. et al. (2023). Hierarchical mesoporous binary nickel cobalt selenides nanospheres for enhanced sodium-ion storage. *Journal of Alloys and Compounds* 938: 168608.

257 Li, J., He, Y., Dai, Y. et al. (2024). Heterostructure interface construction of cobalt/molybdenum selenides toward ultra-stable sodium-ion half/full batteries. *Advanced Functional Materials* 2406915.

258 Fang, L., Xu, J., Sun, S. et al. (2019). Few-layered tin sulfide nanosheets supported on reduced graphene oxide as a high-performance anode for potassium-ion batteries. *Small* 15: 1804806.

259 Sun, Q., Li, D., Dai, L. et al. (2020). Structural engineering of SnS_2 encapsulated in carbon nanoboxes for high-performance sodium/potassium-ion batteries anodes. *Small* 16: 2005023.

260 Xie, J., Zhu, Y., Zhuang, N. et al. (2018). Rational design of metal organic framework-derived FeS_2 hollow nanocages@ reduced graphene oxide for K-ion storage. *Nanoscale* 10: 17092–17098.

261 Iqbal, S., Wang, L., Kong, Z. et al. (2024). Bimetallic sulfide hollow nanocubes heterostructure promotes dual coupling of conversion and alloying/dealloying reactions to achieve durable potassium-ion battery anode. *Small* 2404194.

262 Chong, S., Sun, L., Shu, C. et al. (2019). Chemical bonding boosts nano-rose-like MoS_2 anchored on reduced graphene oxide for superior potassium-ion storage. *Nano Energy* 63: 103868.

263 Jia, B., Yu, Q., Zhao, Y. et al. (2018). Bamboo-like hollow tubes with MoS_2/N-doped-C interfaces boost potassium-ion storage. *Advanced Functional Materials* 28: 1803409.

264 Wang, T., Shen, D., Liu, H. et al. (2020). A Sb_2S_3 nanoflower/MXene composite as an anode for potassium-ion batteries. *ACS Applied Materials & Interfaces* 12: 57907–57915.

265 Wu, Y., Xu, Y., Li, Y. et al. (2019). Unexpected intercalation-dominated potassium storage in WS_2 as a potassium-ion battery anode. *Nano Research* 12: 2997–3002.

266 Wang, W., Jiang, B., Qian, C. et al. (2018). Pistachio-shuck-like $MoSe_2$/C core/shell nanostructures for high-performance potassium-ion storage. *Advanced Materials* 30: 1801812.

267 Lin, H., Li, M., Yang, X. et al. (2019). Nanosheets-assembled CuSe crystal pillar as a stable and high-power anode for sodium-ion and potassium-ion batteries. *Advanced Energy Materials* 9: 1900323.

268 Chen, X., Muheiyati, H., Sun, X. et al. (2022). Rational design of tungsten selenide@N-doped carbon nanotube for high-stable potassium-ion batteries. *Small* 18: 2104363.

269 Sun, X., Zhang, B., Chen, M. et al. (2022). Space-confined growth of Bi_2Se_3 nanosheets encapsulated in N-doped carbon shell lollipop-like composite for full/half potassium-ion and lithium-ion batteries. *Nano Today* 43: 101408.

270 Kong, Z., Wang, L., Iqbal, S. et al. (2022). Iron selenide-based heterojunction construction and defect engineering for fast potassium/sodium-ion storage. *Small* 18: 2107252.

271 Iqbal, S., Wang, L., Kong, Z. et al. (2024). 2D Se-Rich ZnSe/$CoSe_2$@C heterostructured composite as ultrastable anodes for alkaline-ion batteries. *Small* 2404193.

272 Yang, C., Lv, F., Dong, K. et al. (2021). Carbon-coated ultrathin metallic V_5Se_8 nanosheet for high-energy-density and robust potassium storage. *Energy Storage Materials* 35: 1–11.

273 Xia, Z., Chen, X., Ci, H. et al. (2021). Designing N-doped graphene/$ReSe_2$/Ti_3C_2 MXene heterostructure frameworks as promising anodes for high-rate potassium-ion batteries. *Journal of Energy Chemistry* 53: 155–162.

274 Yi, Z., Qian, Y., Jiang, S. et al. (2020). Self-wrinkled graphene as a mechanical buffer: a rational design to boost the K-ion storage performance of Sb_2Se_3 nanoparticles. *Chemical Engineering Journal* 379: 122352.

275 Liu, Y., Yang, C., Li, Y. et al. (2020). $FeSe_2$/nitrogen-doped carbon as anode material for potassium-ion batteries. *Chemical Engineering Journal* 393: 124590.

276 Yu, Q., Jiang, B., Hu, J. et al. (2018). Metallic octahedral $CoSe_2$ threaded by N-doped carbon nanotubes: a flexible framework for high-performance potassium-ion batteries. *Advanced Science* 5: 1800782.

277 Dong, C., Wu, L., He, Y. et al. (2020). Willow-leaf-like ZnSe@ N-doped carbon nanoarchitecture as a stable and high-performance anode material for sodium-ion and potassium-ion batteries. *Small* 16: 2004580.

278 He, Y., Wang, L., Dong, C. et al. (2019). In-situ rooting ZnSe/N-doped hollow carbon architectures as high-rate and long-life anode materials for half/full sodium-ion and potassium-ion batteries. *Energy Storage Materials* 23: 35–45.

279 Grishanov, D.A., Mikhaylov, A.A., Medvedev, A.G. et al. (2018). Graphene oxide-supported β-tin telluride composite for sodium-and lithium-ion battery anodes. *Energy Technology* 6: 127–133.

280 Han, M.S., Zhou, Z.H., Li, Y. et al. (2021). Highly conductive tellurium and telluride in energy storage. *ChemElectroChem* 8: 4412–4426.

281 Yang, S., Park, G.D., and Kang, Y.C. (2020). Conversion reaction mechanism of cobalt telluride-carbon composite microspheres synthesized by spray pyrolysis process for K-ion storage. *Applied Surface Science* 529: 147140.

282 Zhou, Q., Yuan, L., Li, T. et al. (2023). Boosting cobalt ditelluride quantum-rods anode materials for excellent potassium-ion storage via hierarchical physicochemical encapsulation. *Journal of Colloid and Interface Science* 646: 493–502.

283 Xu, X., Zhang, Y., Sun, H. et al. (2021). Orthorhombic cobalt ditelluride with Te vacancy defects anchoring on elastic MXene enables efficient potassium-ion storage. *Advanced Materials* 33: 2100272.

284 Li, Q., Peng, J., Zhang, W. et al. (2023). Manipulating the polytellurides of metallic telluride for ultra-stable potassium-ion storage: a case study of carbon-confined $CoTe_2$ nanofibers. *Advanced Energy Materials* 13: 2300150.

285 Zhang, C., Li, H., Zeng, X. et al. (2022). Accelerated diffusion kinetics in ZnTe/$CoTe_2$ heterojunctions for high rate potassium storage. *Advanced Energy Materials* 12: 2202577.

286 Huang, Z., Xiong, S., Zhang, W. et al. (2024). Design of grid-like triple-carbon matrix confined ultrafine $CoTe_2$ nanocrystals toward durable and fast potassium storage. *Journal of Materials Chemistry A* 12: 15044.

287 Sarkar, D., Das, D., Nagarajan, S., and Mitlin, D. (2022). Thermally fabricated cobalt telluride in nitrogen-rich carbon dodecahedra as high-rate potassium and sodium ion battery anodes. *Sustainable Energy & Fuels* 6: 3582–3590.

288 Li, Q., Yu, D., Peng, J. et al. (2024). Efficient polytelluride anchoring for ultralong-life potassium storage: combined physical barrier and chemisorption in nanogrid-in-nanofiber. *Nano-Micro Letters* 16: 77.

289 Wang, G., Peng, J., Zhang, W. et al. (2024). Te-vacancy-rich $CoTe_{2-x}$ anodes for efficient potassium-ion storage. *Chemical Engineering Journal* 483: 149436.

290 Fang, Y.J., Luan, D.Y., and Lou, X.W. (2020). Recent advances on mixed metal sulfides for advanced sodium-ion batteries. *Advanced Materials* 32: 2002976.

291 Zeng, Z., Zhang, W., Ma, H. et al. (2023). Strengthening ion uptake and mitigating volume change via bimetallic telluride heterojunction for ultrastable K-ion hybrid capacitors. *Chemical Engineering Journal* 455: 140907.

292 Park, G.D. and Kang, Y.C. (2020). Conversion reaction mechanism for yolk-shell-structured iron telluride-C nanospheres and exploration of their electrochemical performance as an anode material for potassium-ion batteries. *Small Methods* 4: 2000556.

293 Hu, R., Sha, D., Cao, X. et al. (2022). Anchoring metal-organic framework-derived ZnTe@C onto elastic $Ti_3C_2T_x$ MXene with 0D/2D dual confinement for ultrastable potassium-ion storage. *Advanced Energy Materials* 12: 2203118.

294 Wang, X., Chen, X., Hou, R., and Zhang, Q. (2022). Facile fabrication of FeTe/reduced graphene oxide nanocomposites for sodium (potassium) storage. *Journal of Materials Science* 57: 5167–5176.

295 Pan, L., Hu, R., Zhang, Y. et al. (2023). Built-in electric field-driven ultrahigh-rate K-ion storage via heterostructure engineering of dual tellurides integrated with $Ti_3C_2T_x$ MXene. *Nano-Micro Letters* 15: 225.

296 Chang, C.-B., Tsai, H.-Y., and Tuan, H.-Y. (2024). Tubular design and metal ratio refinement of copper telluride electrodes for superior volumetric capacity in potassium-ion batteries. *Journal of Energy Storage* 90: 111929.

297 Wu, B., Luxa, J., Kovalska, E. et al. (2021). Sub-millimetre scale Van der Waals single-crystal $MoTe_2$ for potassium storage: electrochemical properties, and its failure and structure evolution mechanisms. *Energy Storage Materials* 43: 284–292.

298 Liu, Z.J., Ou, X.W., Zhuang, M.H. et al. (2019). Confinement-enhanced rapid interlayer diffusion within graphene-supported anisotropic $ReSe_2$ electrodes. *ACS Applied Materials & Interfaces* 11: 31147–31154.

299 Hafeez, M., Gan, L., Li, H.Q. et al. (2016). Chemical vapor deposition synthesis of ultrathin hexagonal $ReSe_2$ flakes for anisotropic Raman property and optoelectronic application. *Advanced Materials* 28: 8296–8301.

300 Zhang, Q., Tan, S.J., Mendes, R.G. et al. (2016). Extremely weak van der waals coupling in vertical ReS_2 nanowalls for high-current-density lithium-ion batteries. *Advanced Materials* 28: 2616–2623.

301 Zhou, J., Zhang, Y., Liu, Z. et al. (2022). Defective 1T′-$ReSe_2$ nanosheets vertically grown on elastic MXene for fast and stable potassium ion storage. *Science China Materials* 65: 3418–3427.

302 Soares, D.M. and Singh, G. (2020). Superior electrochemical performance of layered WTe_2 as potassium-ion battery electrode. *Nanotechnology* 31: 455406.

303 Gillard, C.H.R., Jana, P.P., Rawal, A., and Sharma, N. (2021). Electrochemical phase evolution of tetradymite-type Bi_2Te_3 in lithium, sodium and potassium ion half cells. *Journal of Alloys and Compounds* 854: 155621.

304 Ko, J.K., Jo, J.H., Kim, H.J. et al. (2021). Bismuth telluride anode boosting highly reversible electrochemical activity for potassium storage. *Energy Storage Materials* 43: 411–421.

305 Chong, S., Yuan, L., Zhou, Q. et al. (2023). Bismuth telluride nanoplates hierarchically confined by graphene and N-doped C as conversion-alloying anode materials for potassium-ion batteries. *Small* 19: 2303985.

306 Wu, Z., Liang, G., Pang, W.K. et al. (2020). Coupling topological insulator $SnSb_2Te_4$ nanodots with highly doped graphene for high-rate energy storage. *Advanced Materials* 32: 1905632.

307 Yang, F., Liu, Z., Wang, D. et al. (2023). Preparation and properties of P-Bi_2Te_3/MXene superstructure-based anode for potassium-ion battery. *Acta Physico-Chimica Sinica* 40: 2303006.

308 Wang, G., Li, Q., Zhang, W. et al. (2024). Unveiling the synergy of architecture and anion vacancy on Bi_2Te_{3-x}@ NPCNFs for fast and stable potassium ion storage. *ACS Applied Materials & Interfaces* 16: 13858–13868.

309 Du, Y., Yi, Z., Zhang, Z. et al. (2022). A highly stable potassium-ion battery anode enabled by multilayer graphene sheets embedded with SnTe nanoparticles. *Chemical Engineering Journal* 435: 135100.

310 Li, T., Wang, Y., Zhou, Q. et al. (2023). SnTe nanoparticles physicochemically encapsulated by double carbon as conversion-alloying anode materials for superior potassium-ion batteries. *Journal of Materials Science & Technology* 158: 86–95.

311 Gillard, C., Jana, P.P., Avdeev, M., and Sharma, N. (2021). The structural evolution of tetradymite-type Sb_2Te_3 in alkali ion batteries. *Journal of Alloys and Compounds* 871: 159378.

312 Chong, S., Qiao, S., Yuan, L. et al. (2023). Sb_2Te_3 hexagonal nanoplates as conversion-alloying anode materials for superior potassium-ion storage via physicochemical confinement effect of dual carbon matrix. *Chemical Engineering Journal* 461: 141957.

313 Lei, K., Li, F., Mu, C. et al. (2017). High K-storage performance based on the synergy of dipotassium terephthalate and ether-based electrolytes. *Energy & Environmental Science* 10: 552–557.

314 Deng, Q., Tian, C., and Luo, Z. (2020). Atomic layer deposition of Al_2O_3 on organic potassium terephthalate with enhanced K-storage behavior for K-ion batteries. *Ionics* 26: 1805–1812.

315 Deng, Q., Pei, J., Fan, C. et al. (2017). Potassium salts of para-aromatic dicarboxylates as the highly efficient organic anodes for low-cost K-ion batteries. *Nano Energy* 33: 350–355.

316 Liao, J., Hu, Q., Yu, Y. et al. (2017). A potassium-rich iron hexacyanoferrate/dipotassium terephthalate@ carbon nanotube composite used for K-ion full-cells with an optimized electrolyte. *Journal of Materials Chemistry A* 5: 19017–19024.

317 Wang, X., Han, K., Wang, C. et al. (2018). Graphene oxide-wrapped dipotassium terephthalate hollow microrods for enhanced potassium storage. *Chemical Communications* 54: 11029–11032.

318 Li, C., Deng, Q., Tan, H. et al. (2017). Para-conjugated dicarboxylates with extended aromatic skeletons as the highly advanced organic anodes for K-ion battery. *ACS Applied Materials & Interfaces* 9: 27414–27420.

319 Fan, C., Zhao, M., Li, C. et al. (2017). Investigating the electrochemical behavior of cobalt (II) terephthalate ($CoC_8H_4O_4$) as the organic anode in K-ion battery. *Electrochimica Acta* 253: 333–338.

320 Li, C., Xue, J., Ma, J., and Li, J. (2018). Conjugated dicarboxylate with extended naphthyl skeleton as an advanced organic anode for potassium-ion battery. *Journal of The Electrochemical Society* 166: A5221.

321 Liang, Y., Luo, C., Wang, F. et al. (2019). An organic anode for high temperature potassium-ion batteries. *Advanced Energy Materials* 9: 1802986.

322 Li, Y., Yang, C., Zheng, F. et al. (2018). High pyridine N-doped porous carbon derived from metal–organic frameworks for boosting potassium-ion storage. *Journal of Materials Chemistry A* 6: 17959–17966.

323 Xiong, P., Zhao, X., and Xu, Y. (2018). Nitrogen-doped carbon nanotubes derived from metal–organic frameworks for potassium-ion battery anodes. *ChemSusChem* 11: 202–208.

324 Zhang, Q., Wei, H., Wang, L. et al. (2019). Accessible COF-based functional materials for potassium-ion batteries and aluminum batteries. *ACS Applied Materials & Interfaces* 11: 44352–44359.

325 Chen, X., Zhang, H., Ci, C. et al. (2019). Few-layered boronic ester based covalent organic frameworks/carbon nanotube composites for high-performance K-organic batteries. *ACS Nano* 13: 3600–3607.

326 Li, C., Wang, K., Li, J., and Zhang, Q. (2020). Nanostructured potassium–organic framework as an effective anode for potassium-ion batteries with a long cycle life. *Nanoscale* 12: 7870–7874.

327 Li, S.-Y., Li, W.-H., Wu, X.-L. et al. (2019). Pore-size dominated electrochemical properties of covalent triazine frameworks as anode materials for K-ion batteries. *Chemical Science* 10: 7695–7701.

328 An, Y., Fei, H., Zhang, Z. et al. (2017). A titanium-based metal–organic framework as an ultralong cycle-life anode for PIBs. *Chemical Communications* 53: 8360–8363.

329 Li, C., Hu, X., and Hu, B. (2017). Cobalt (II) dicarboxylate-based metal-organic framework for long-cycling and high-rate potassium-ion battery anode. *Electrochimica Acta* 253: 439–444.

330 Deng, Q., Luo, Z., Liu, H. et al. (2020). Facile synthesis of Fe-based metal-organic framework and graphene composite as an anode material for K-ion batteries. *Ionics* 26: 5565–5573.

331 Zhang, C., Qiao, Y., Xiong, P. et al. (2019). Conjugated microporous polymers with tunable electronic structure for high-performance potassium-ion batteries. *ACS Nano* 13: 745–754.

332 Wessells, C.D., Peddada, S.V., Huggins, R.A., and Cui, Y. (2011). Nickel hexacyanoferrate nanoparticle electrodes for aqueous sodium and potassium ion batteries. *Nano Letters* 11: 5421–5425.

333 Suo, L., Borodin, O., Gao, T. et al. (2015). "Water-in-salt" electrolyte enables high-voltage aqueous lithium-ion chemistries. *Science* 350: 938–943.

334 Jiang, L., Lu, Y., Zhao, C. et al. (2019). Building aqueous K-ion batteries for energy storage. *Nature Energy* 4: 495–503.

335 Zhou, L., Cao, Z., Wahyudi, W. et al. (2020). Electrolyte engineering enables high stability and capacity alloying anodes for sodium and potassium ion batteries. *ACS Energy Letters* 5: 766–776.

336 Huang, J., Lin, X., Tan, H., and Zhang, B. (2018). Bismuth microparticles as advanced anodes for potassium-ion battery. *Advanced Energy Materials* 8: 1703496.

337 Fan, L., Ma, R., Zhang, Q. et al. (2019). Graphite anode for a potassium-ion battery with unprecedented performance. *Angewandte Chemie International Edition* 131: 10610–10615.

338 Fan, L., Ma, R., Wang, J. et al. (2018). An ultrafast and highly stable potassium–organic battery. *Advanced Materials* 30: 1805486.

339 Bie, X., Kubota, K., Hosaka, T. et al. (2017). A novel K-ion battery: hexacyanoferrate (II)/graphite cell. *Journal of Materials Chemistry A* 5: 4325–4330.

340 Yoshii, K., Masese, T., Kato, M. et al. (2019). Sulfonylamide-based ionic liquids for high-voltage potassium-ion batteries with honeycomb layered cathode oxides. *ChemElectroChem* 6: 3901–3910.

341 Arnaiz, M., Bothe, A., Dsoke, S. et al. (2019). Aprotic and protic ionic liquids combined with olive pits derived hard carbon for potassium-ion batteries. *Journal of The Electrochemical Society* 166: A3504.

342 Lu, X., Bowden, M.E., Sprenkle, V.L., and Liu, J. (2015). A low cost, high energy density, and long cycle life potassium-sulfur battery for grid-scale energy storage. *Advanced Materials* 27: 5915–5922.

343 Yoshinari, N., Yamashita, S., Fukuda, Y. et al. (2019). Mobility of hydrated alkali metal ions in metallosupramolecular ionic crystals. *Chemical Science* 10: 587–593.

344 Xiao, R., Li, H., and Chen, L. (2020). High-throughput computational discovery of K_2 CdO_2 as an ion conductor for solid-state potassium-ion batteries. *Journal of Materials Chemistry A* 8: 5157–5162.

345 Burmakin, E., Voronin, V., Akhtyamova, L. et al. (2004). Potassium aluminate crystalline structure and electroconduction. *Russian Journal of Electrochemistry* 40: 619–626.

346 Lei, Y.J., Yang, H.L., Liang, Y. et al. (2022). Progress and prospects of emerging potassium–sulfur batteries. *Advanced Energy Materials* 12: 2202523.

347 Zhao, X., Lu, Y., Qian, Z. et al. (2020). Potassium-sulfur batteries: status and perspectives. *EcoMat* 2: e12038.

348 Ding, J., Zhang, H., Fan, W. et al. (2020). Review of emerging potassium–sulfur batteries. *Advanced Materials* 32: 1908007.

349 Zhang, S., Kong, Y., Gu, Y. et al. (2024). Strong d–π orbital coupling of Co–C_4 atomic sites on graphdiyne boosts potassium–sulfur battery electrocatalysis. *Journal of the American Chemical Society* 146: 4433–4443.

350 Yuan, X., Zhu, B., Feng, J. et al. (2020). Electrochemical insights, developing strategies, and perspectives toward advanced potassium–sulfur batteries. *Small* 16: 2003386.

351 Yao, W., Liao, K., Lai, T. et al. (2024). Rechargeable metal-sulfur batteries: key materials to mechanisms. *Chemical Reviews* 124: 4935–5118.

352 Zhao, Q., Hu, Y., Zhang, K., and Chen, J. (2014). Potassium–sulfur batteries: a new member of room-temperature rechargeable metal–sulfur batteries. *Inorganic Chemistry* 53: 9000–9005.

353 Xiong, P., Han, X., Zhao, X. et al. (2019). Room-temperature potassium–sulfur batteries enabled by microporous carbon stabilized small-molecule sulfur cathodes. *ACS Nano* 13: 2536–2543.

354 Ye, C., Shan, J., Chao, D. et al. (2021). Catalytic oxidation of K_2S via atomic Co and pyridinic N synergy in potassium–sulfur batteries. *Journal of the American Chemical Society* 143: 16902–16907.

355 Song, W., Yang, X., Zhang, T. et al. (2024). Optimizing potassium polysulfides for high performance potassium-sulfur batteries. *Nature Communications* 15: 1005.

356 Qin, L., Lei, Y., Wang, H. et al. (2019). Capillary encapsulation of metallic potassium in aligned carbon nanotubes for use as stable potassium metal anodes. *Advanced Energy Materials* 9: 1901427.

357 Wang, H., Hu, J., Dong, J. et al. (2019). Artificial solid-electrolyte interphase enabled high-capacity and stable cycling potassium metal batteries. *Advanced Energy Materials* 9: 1902697.

358 Lee, S., Park, H., Rizell, J. et al. (2022). High-energy and long-lifespan potassium–sulfur batteries enabled by concentrated electrolyte. *Advanced Functional Materials* 32: 2209145.

359 Gao, H., Xue, L., Xin, S., and Goodenough, J.B. (2018). A high-energy-density potassium battery with a polymer-gel electrolyte and a polyaniline cathode. *Angewandte Chemie International Edition* 130: 5547–5551.

360 Gu, S., Xiao, N., Wu, F. et al. (2018). Chemical synthesis of K_2S_2 and K_2S_3 for probing electrochemical mechanisms in K–S batteries. *ACS Energy Letters* 3: 2858–2864.

361 Ye, C., Shan, J., Li, H. et al. (2023). Reducing overpotential of solid-state sulfide conversion in potassium-sulfur batteries. *Angewandte Chemie International Edition* 135: e202301681.

4

The Electrocatalyst Design for Lithium–Sulfur Battery

Yueyue Kong, Bin Wang, Lu Wang, and Liqiang Xu

Key Laboratory of Colloid and Interface Chemistry (Ministry of Education), School of Chemistry and Chemical Engineering, Shandong University, Jinan, 250100, China

4.1 Background

Li–S batteries stand out among emerging secondary battery technologies, attributing to their high theoretical energy density, ample sulfur reserves, affordability, and eco-friendliness. They have been considered as one kind of top contenders for future energy storage. Their high energy density (2500 Wh kg^{-1} or 2800 Wh L^{-1}) and theoretical specific capacity (1675 mAh g^{-1}) demonstrate extremely broad application prospects, attracting extensive research and attention from scientific researchers (Figure 4.1a) [1]. Li–S batteries far exceed Li-ion batteries in both gravimetric energy density and volumetric energy density (Figure 4.1b) [2].

4.2 Brief Introduction of Li–S Batteries

4.2.1 The Development History of Li–S Batteries

In the 1960s, Herbet and Ulam first proposed the primary Li–S batteries, utilizing elemental sulfur as cathode material [3]. In the 1990s, under the background of the mature Li–ion battery technology and commercialization, research on Li–S batteries experienced a downturn due to inferior cycle stability and safety issues of lithium dendrites. With the increasingly mature technology of Li–ion batteries, their inherent theoretical energy density limitations have gradually become prominent, posing challenges in satisfying the increasing need for high-performance energy storage solutions in society. Therefore, researchers are focusing on Li–S batteries

Electrode Materials in Energy Storage Technologies: Applications in Lithium-, Sodium-, Potassium-, Sulfur- and Zinc-Based Rechargeable Batteries, First Edition. Edited by Liqiang Xu.

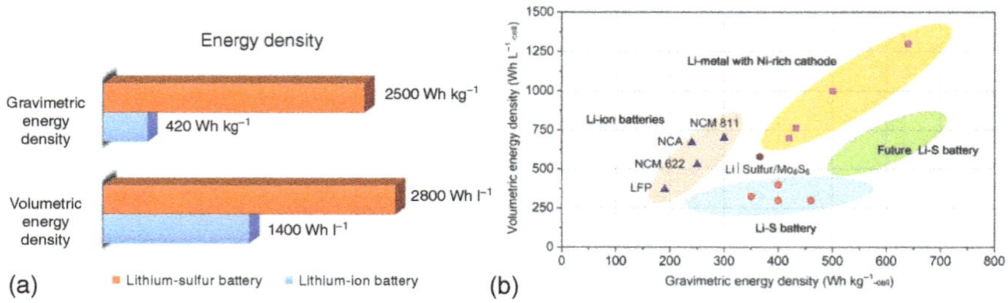

Figure 4.1 (a) Energy density comparison between Li–S batteries and lithium-ion batteries. Source: Seh et al. [1]/The Royal Society of Chemistry. (b) Comparison between volumetric energy density and gravimetric energy density. Source: Liu et al. [2]/John Wiley & Sons.

for their potential high energy density and theoretical capacity. Initially, in the first stage, numerous electrolyte systems were tested to enable battery cycling at room temperature, marking the transformation from a primary battery to a secondary battery [4]. Subsequently, in the second stage, researcher gradually explored how to elevate the capacity retention rate of Li–S batteries during cycling. In 2002, Professor Wang and his team took the lead in proposing an innovative method of combining nanocarbon and sulfur to build cathode materials and adopted a gel electrolyte system. This breakthrough design enabled Li–S batteries to achieve a significant long-lasting capacity during the electrochemical cycling process [5]. In addition, they also introduced sulfurized polyacrylonitrile (S-PAN) as a novel cathode composite material [6]. In 2008, lithium nitrate ($LiNO_3$) was incorporated as an electrolyte additive of Li–S batteries, enabling them to attain higher Coulombic efficiency during cycling [7]. In 2009, Professor Nazar and her team put forward the ordered mesoporous carbon CMK-3 with sulfur composite [8], which enabled the battery to reach a 1320 mAh g^{-1} high specific capacity. They also emphasized the enormous potential of nanoporous carbon materials in the cathode materials. Since then, research work and related reports on Li–S batteries have surged and gained the progressive understanding as basic science research deepens. Different concepts and design strategies have been proposed, such as electrocatalytic effect, catalytic oxidation of Li_2S, single-atom catalyst, and selective catalysis [9]. Several universities and institutions have developed Ah-grade pouch-type cells with an energy density exceeding 300 Wh kg^{-1}, which can achieve more than 50 stable cycles. We have also gradually witnessed the development of Li–S batteries in the realm of energy storage applications.

4.2.2 The Overview of Li–S Batteries

In the realm of the Li–S battery, the cathode comprises sulfur, while the anode is made of metallic lithium. Metal lithium has the highest theoretical capacity storage (3860 mAh g^{-1}) and the lowest potential (−3.04 V vs. SHE). The sulfur cathode undergoes different transformations of polysulfides during the charging/discharging process, exhibiting the biggest theoretical capacity storage of 1675 mAh g^{-1} among solid cathode materials. Under the condition of an average battery voltage of 2.15 V, Li–S batteries have a theoretical energy density of up to 2500 Wh kg^{-1}, significantly surpassing current Li–ion batteries. The theoretical specific capacity is calculated using the Faraday formula: $C = n\text{F}/3.6X$, where n represents the number of electrons shifted, F is the Faraday constant, and X is the molar mass of the reactant in the electrochemical reaction process. The theoretical mass energy density and volumetric energy density are calculated according to the Gibbs free energy. The formula is $E = \Delta_r G^\theta/M$ or V, where M is the total molar mass, and V is the total molar volume.

$$C_{Cathode} = \frac{n\text{F}}{3.6M} = \frac{2 \times 96\,500}{3.6 \times 32} = 1675 \text{ mAh g}^{-1}$$

$$C_{Full\ cell} = \frac{n\text{F}}{3.6M} = \frac{2 \times 96\,500}{3.6 \times 45.927} = 1167 \text{ mAh g}^{-1}$$

$$E_M = C \int_{1.7}^{2.8} \text{d}V = 2500 \text{ Wh kg}^{-1}$$

The Coulombic efficiency of battery = discharge-specific capacity/charge-specific capacity × 100%.

The capacity decay rate per cycle of battery =

$$\frac{\text{Initial capacity-final capacity}}{\text{Initial capacity}} \times \frac{1}{\text{Number of battery cycles}} \times 100\%$$

4.2.3 The Working Principle of Li–S Batteries

The Li–S batteries comprise sulfur cathode, metal lithium anode, electrolyte, and separator (Figure 4.2a) [10]. The sulfur cathode material is usually constructed by the traditional melt diffusion routine, which combines the designed host material with sulfur. Metallic lithium is utilized as the anode. 1 M LiTFSI solution with 2 wt% $LiNO_3$ additive (mixed solvent with a 1 : 1 volume ratio) is usually employed as the electrolyte. The polypropylene separator (PP) is employed as the separator.

The Li–S battery completes the charging and discharging process based on an electrochemical reaction between metallic lithium and elemental sulfur. The overall reaction of the battery is as follows:

$$S_8 + 16Li^+ + 16e^- \leftrightarrow 8Li_2S$$

The overall battery reaction appears straightforward, yet the electrochemical processes involved in charging and discharging are intricate. The specific reaction mechanism is illustrated in Figure 4.2b [11]. According to the charging and discharging curve, we can observe that the solid S_8 molecules undergo complex composition and structural changes in the redox process. The two discharge platforms in the discharge process correspond to the transformation of solid S_8 molecules to liquid

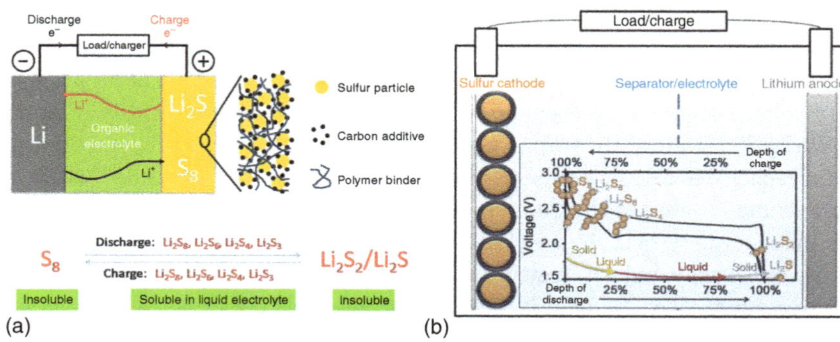

Figure 4.2 (a) Simplified diagram of the composition and operational mechanism of Li–S batteries. Source: Manthiram et al. [10]/American Chemical Society. (b) Setup and a typical charge–discharge curve of Li–S batteries. Source: Chung and Manthiram [11]/John Wiley & Sons.

polysulfides and then to solid Li_2S_2/Li_2S. The discharge process can be segmented into three stages based on the solid–liquid–solid three-phase transformation:

(1) Solid–liquid conversion process from elemental sulfur to liquid Li_2S_8, referring to the first discharge platform (~2.3 V) – solid S_8 gains electrons and combines with Li^+ to form liquid Li_2S_8.

$$S_8 + 2Li^+ + 2e^- \rightarrow Li_2S_8$$

(2) The conversion process from soluble polysulfide Li_2S_8 to Li_2S_6 and then to Li_2S_4 involves a rapid decrease in voltage, corresponding to the descending part of the first platform. As the discharge proceeds, the viscosity of the electrolyte gradually increases, the reaction kinetics gradually slow down, and the polarization increases.

$$3Li_2S_8 + 2Li^+ + 2e^- \rightarrow 4Li_2S_6$$

$$2Li_2S_6 + 2Li^+ + 2e^- \rightarrow 3Li_2S_4$$

(3) The conversion process from liquid-phase Li_2S_4 to solid-phase Li_2S_2/Li_2S, corresponding to the second discharge platform (~2.1 V). Due to the insulating properties of solid Li_2S_2/Li_2S, the reaction kinetics of this process become very slow.

$$Li_2S_4 + 2Li^+ + 2e^- \rightarrow 2Li_2S_2$$

$$Li_2S_2 + 2Li^+ + 2e^- \rightarrow 2Li_2S$$

In summary, the discharge process of the Li–S battery involves a solid–liquid–solid three-phase conversion. In the first and second stages, the conversion process from S_8 to Li_2S_4 correlates with a theoretical capacity of 25% (419 mAh g^{-1}) of the total theoretical capacity. In the third stage, the transformation process from Li_2S_4 to Li_2S_2/Li_2S theoretically provides a capacity of 1256 mAh g^{-1}, contributing to the majority of the capacity of the Li–S battery.

4.2.4 The Problems and Challenges of Li–S Battery

The electrochemical conversion in Li–S batteries entails a complex three-phase transition involving solids and liquids, accompanied by multiple electron transfers [12–15]. However, these transformations to some extent restrict the development and commercialization of the Li–S battery. The specific problems are summarized as follows:

(1) The active material sulfur and its discharge product Li_2S possess inherently significant insulation properties. The insulating properties slow down the kinetics of electron and ion transport, further resulting in sluggish redox kinetics and reduced sulfur usage. Additionally, the insulating nature of Li_2S leads to the extremely slow solid–solid transformation during Li_2S formation, and the formed Li_2S tends to passivate the catalyst surface, reducing the discharge capacity and compromising cycle stability in the battery.
(2) The sulfur cathode undergoes significant volume changes during charge–discharge cycles attributed to its higher density ($2.07\,g\,cm^{-3}$) than that of Li_2S ($1.66\,g\,cm^{-3}$). Moreover, the powdering of the sulfur cathode impairs the contact between sulfur and the designed catalyst, causing deactivation of sulfur and shortening the battery's service life.
(3) The polysulfides shuttle effect aroused due to the liquid-phase polysulfides generated during charging and discharging dissolved in the electrolyte, migrating to the lithium anode side through the separator, influenced by concentration gradients and electric fields. Polysulfides undergo irreversible reactions with the lithium anode, contributing to lithium dendrite formation and endangering battery safety. Polysulfides at the lithium anode side also increase the resistance to Li^+ ion transport, reducing the reaction kinetics of the entire battery system. Furthermore, severe shuttle effects cause the loss of sulfur, which causes swift capacity degradation of the battery.
(4) Lithium dendrites appeared due to uneven plating and stripping of the Li^+ ion; the lithium anode surface becomes uneven and prone to the formation of lithium dendrites. Uncontrolled development of lithium dendrites may puncture the separator, causing short circuits for batteries and posing a severe threat to the safety of Li–S batteries.

The problems in the research of Li–S batteries hinder their development and commercialization [16]. Regarding sulfur-related issues, such as sluggish reaction kinetics and shuttle effects, researchers mainly focus on the design of electrocatalysts for sulfur host or separator modification. Developing highly conductive sulfur host materials with effective physical confinement effects and chemical adsorption properties is extremely important [17, 18]. Modified separators with additional interlayers can adsorb and accelerate the polysulfides conversion, significantly facilitating the performance of Li–S batteries [19, 20]. Regarding lithium-related problems like lithium dendrite growth, the primary research focus includes electrolyte regulation and construction of an artificial modification layer for lithium anode, which significantly promotes uniform plating and stripping of lithium, enhancing the electrochemical performance of the lithium anode. Various electrolyte improvement strategies, including employing additives, high-concentration solutions, and solid-state electrolytes, are gradually being researched and developed [21]. Designing coatings on lithium surfaces is a straightforward method to protect the lithium anode. Next, we will summarize the recent research progress of Li–S batteries from the perspectives of electrocatalyst design strategies and lithium anode protection.

4.3 The Electrocatalyst Design Strategies

Designing effective electrocatalysts for Li–S batteries needs to consider multiple aspects, including conductivity, specific surface area, adsorption capacity, electrocatalytic ability, structural stability, ion transport performance, electrochemical activity, mechanical and chemical stability [22]. First, the high electronic conductivity is essential for the electrocatalyst to achieve rapid migration and transformation of lithium polysulfides (LiPSs) on the electrocatalyst's surface. Good conductivity can reduce the internal resistance of batteries. Second, electrocatalysts should also have specific electrocatalytic capabilities to promote polysulfides conversion, enhancing electrode reaction kinetics, and further boost the performance of batteries. Specifically, the sulfur host should require a high specific surface area and sufficient space to fix active sulfur and alleviate its volume expansion during cycling. This characterization helps to facilitate the utilization efficiency of sulfur, minimize active substance loss, and prolong battery cycle life. In addition, the host material should be able to adsorb effectively and anchor polysulfides, relieve the shuttle effect, and reduce their dissolution and diffusion in the electrolyte, expanding the battery's cycling stability and Coulombic efficiency. Separator modification materials should have good ion transport performance while suppressing polysulfide shuttle. In addition, the separator modification materials should have good mechanical strength and chemical stability to resist various stresses and corrosion during battery operation, ensuring the safety of the battery.

4.3.1 Micro/Nanostructure Design

Through micro/nanostructure design, novel electrocatalysts with excellent electrocatalytic activity can be synthesized. The catalyst can be engineered into nanoscale dimensions by adopting a nanoscale design, enhancing the accessible surface area, providing more reaction sites, and simultaneously shortening the transmission paths for electrons and ions. Catalysts with rich pore structures can be designed to accommodate more sulfur and its redox intermediates while facilitating their diffusion and conversion within the matrix. In addition, by controlling the morphology of the catalyst, such as spherical, sheet-like, and rod-shaped, its surface energy and reaction activity can be optimized.

Metal-organic frameworks (MOFs) and their derivatives offer notable advantages as electrocatalysts for Li–S batteries, including open-pored structure and large specific surface area, plentiful metal active sites, tunable chemical and physical properties, and excellent catalytic performance [23]. Therefore, researchers utilize MOFs as a template, introduce other metal active sites, or further calcine to prepare MOF derivatives that act as electrocatalytic centers for Li–S batteries. MOFs and their derivatives have a porous structure, which provides abundant physical confinement spaces for sulfur and polysulfides, effectively mitigating the shuttle effect and accelerating the battery's cycling stability and capacity retention. The exposed metal ions in derivatives become active sites for catalytic reactions. By adjusting the metal ions, organic ligands, and reaction conditions during the synthesis of MOFs, the structure and properties of MOFs and their derivatives can be precisely controlled. This adjustability enables MOF derivatives to be optimized for the specific needs of the Li–S battery. Zhu and co-workers prepared Co/CoSe junction utilizing ZIF-67 as a template [24]. The charge rearrangement in the heterojunction enhances catalytic activity for sulfur redox reaction. Using Ni-MOF as a precursor, Cabot and coworkers fabricated $NiS_2/NiSe_2$ heterostructures [25], featuring a hollow interior and encapsulated within nitrogen-doped carbon spheres. The resultant porous and hollow architecture effectively encapsulates sulfur within the structure, enhancing sulfur confinement and simultaneously providing ample void spaces that serve as buffers during electrochemical reactions. In addition, the $NiS_2/NiSe_2$ heterostructure facilitates the transition of Ni^{3+} ions from a low-spin to a high-spin state. This high-spin configuration leads to enhanced binding strength and catalytic efficiency toward LiPSs.

Liang and coworkers designed yolk-shelled cubic structure Fe_3O_4/FeP@C nanoreactors (Figure 4.3a) [26], which leverage a unique interfacial built-in electric field (BIEF) to enhance exceptionally efficient electrocatalyst. The achievement of a synergistic effect arises from the adsorption capacity of Fe_3O_4, and the eminent catalytic activity of FeP relies on the BIEF serving as a pivotal "bridge". This unique mechanism significantly mitigates the shuttle effect, profoundly accelerates the conversion kinetics, and facilitates efficient sulfur species reutilization, enhancing overall electrochemical performance. Xu and coworkers reported a novel cascaded dual-cavity nanoreactor (NiCo-B), achieved through precise control of reaction kinetics (Figure 4.3b) [27]. This innovative design offers a promising avenue for realizing hierarchical catalytic behavior, wherein its multilevel architecture proficiently mitigates polysulfide dissolution while expediting sulfur conversion processes. The Ni–B shell possesses a robust capability to adsorb sulfur via the formation of B–S bonds. Concurrently, the Co–B core generates an even stronger catalytic activity by the Co–S bond, which promotes the swift conversion of sulfur species and prevents catalyst passivation. This dual-action mechanism creates optimal adsorption–catalysis conditions for the efficient transformation of sulfur. Yu and coworkers devised porous, hollow, and necklace-like Fe_3C/N-codoped carbon nanoboxes (Fe_3C/NC) [28], as illustrated in Figure 4.3c. These unique nanostructures serve as multifunctional

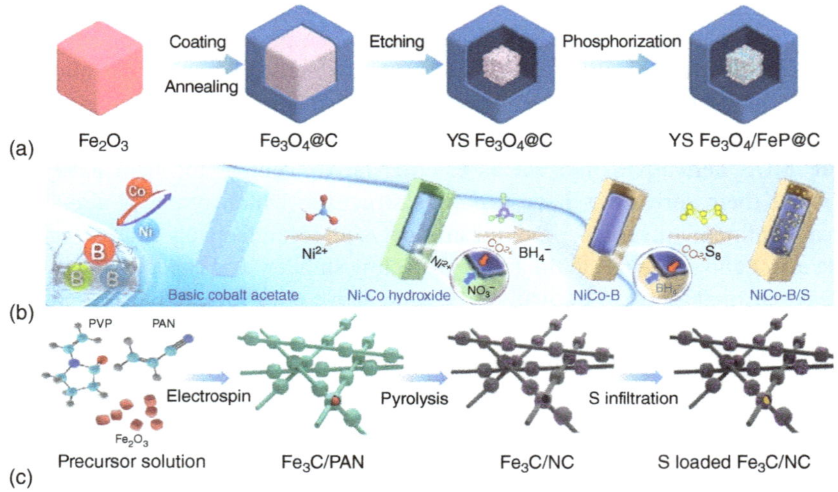

Figure 4.3 (a) Synthesis route for yolk-shelled Fe_3O_4/FeP@C heterostructure nanoreactors. Source: Li et al. [26]/John Wiley & Sons. (b) Synthesis process of the NiCo-B schematic. Source: Wang et al. [27]/John Wiley & Sons. (c) Construction of the Fe_3C/nitrogen-rich carbon (NC) membrane. Source: Zhou et al. [28]/John Wiley & Sons.

sulfur hosts, exhibiting exceptional anchoring capabilities and robust catalytic activities. The uniformly dispersed Fe_3C and N components within the nanoboxes effectively mitigate the shuttle effect. During discharge, the rapid electron transfer via NC nanofibers and enhanced Li^+ ion diffusion via nanobox shells catalyze the conversion of LiPSs into Li_2S. The distinctive "boxes in fibers" architecture of these nanoboxes enhances their sulfur loading capacity and resilience toward volume variations of LiPSs, which promotes a cooperative catalysis mechanism during redox reactions, enhancing their overall performance. Xu and coworkers synthesized molybdenum carbide-adorned N-doped carbon with a mesoporous and double-shelled structure [29]. This design boosts sulfur loading, facilitates ion/electrolyte penetration, and accelerates charge transfer. Furthermore, the microporous outer shell reduces LiPS shuttling and minimizes charge/substance diffusion distances, while the double-shelled architecture accommodates volumetric expansion, ensuring structural integrity and performance stability. Zhang and coworkers constructed a Fe-Co diatomic electrocatalyst anchored on hollow carbon spheres [30], aiming to attain highly efficient catalysis for both the polysulfides conversion and the decomposition of Li_2S. Specifically, the Fe site serves as a catalyst center to expedite reaction kinetics during the discharge process, while the Co atom facilitates the charging process. Therefore, the unique Fe-Co dual-site structure enhances the overall performance and efficiency of the batteries. Li and coworkers introduced a chip-inspired design, reduced graphene oxide (rGO) to embed Mo_2C nanoparticles and nanosulfur [31]. This special design fosters the creation of isolated sulfur nanoreactors organized in a serial arrangement within the rGO. The spatial confinement and protective environment, along with the inherent concentration gradient within the sulfur nanoreactor, proficiently impede the dissolution, diffusion, and loss of LiPSs, consequently bolstering sulfur usage and boosting reaction kinetics. As a result, the Li–S batteries with the devised cathode achieved a remarkable discharge capacity of 1392 mAh g^{-1} at 0.1 C, coupled with a minimal capacity dissipation rate over 1500 cycles at 0.5 C and an exceptional rate capacity. Mai and coworkers introduced Fe_3O_4-doped carbon cubosomes featuring a complex, maze-like structure as sulfur hosts, aiming at crafting cathodes with exceptional high-rate performance and extended cycle durability for Li–S batteries [32]. Their interconnected mesochannels and robust carbon architecture, coupled with dispersed Fe_3O_4 particles, facilitate mass and electron transport, robust polysulfide capture, and rapid LiPS reaction kinetics, which lead to their superior electrochemical behavior. Remarkably, the Li–S batteries using maze-like structure Fe_3O_4-C-based cathode achieved an exceptional reversible capacity of 1303.4 mAh g^{-1} at 0.2 C, alongside robust performance of 691.8 mAh g^{-1} even at 5 C. Furthermore, it delivered outstanding cycling durability, enduring over 1200 cycles at 1 C with a 0.027% minimal dissipation per cycle. Xu and coworkers have successfully embedded NiCoP nanoparticles within N, S-codoped porous carbon matrix derived from petroleum coke (PCPC) to construct PCPC/NiCoP composites as electrocatalysts. Taking advantage of the cooperation effect between PCPC and NiCoP, the Li–S batteries with the PCPC/NiCoP cathode achieve terrific specific capacity and cycling stability [33]. Song and coworkers designed a dual-atom copper electrocatalyst with chlorine bridge bonds [34]. The two copper atoms in this catalyst are connected by a pair of symmetrical chlorine bridge bonds, with a spacing of only 3.5 Å. This dual-core copper catalyst design can fully utilize the proximity effect of two copper atoms to guide the nucleation and decomposition reactions of Li_2S and regulate the lithium stripping/plating behavior. In addition, they observed the deposition behavior of Li_2S through the synchrotron radiation X-ray 3D nano-CT technology. In the presence of a dual-core copper catalyst, Li_2S deposition exhibits a more uniform spatial distribution, demonstrating an excellent catalytic efficiency of the dual-core copper catalyst. To catalyze the rate-determining electrochemical reaction, Wang and coworkers employed cobalt phthalocyanine (CoPc) molecules loaded on carbon surfaces to catalyze the conversion of Li_2S_4 to Li_5S_3 solid products (a 1 : 4 mixture of Li_2S_2 and Li_2S solids) [35]. After initial deposition, the CoPc active sites are covered by polysulfide, and the resulting polysulfides can self-catalyze the reduction of Li_2S_4, continuing to promote the discharge reaction. Zhang and coworkers explored different three-dimensional (3D) dopants (Mn^{2+}, Fe^{2+}, Co^{2+}, Ni^{2+}, or Cu^{2+}) to replace the cations in the parent ZnS lattice and adjust the interaction between ZnS and LiPSs [36]. They revealed the volcanic correlation between the adsorption ability and catalytic characteristic of LiPSs, and developed a $Co_{0.125}Zn_{0.875}S$ composite catalyst based on theoretical analysis. The $Co_{0.125}Zn_{0.875}S$ catalyst exhibits superior performance than simple binary compounds.

4.3.2 Defect Engineering

Defect engineering significantly impacts the design of electrocatalysts for Li–S batteries. Defect engineering can introduce more coordination of unsaturated sites, notably enhancing the capture and transformation ability of electrocatalysts toward LiPSs. In addition, the introduction of defects can alter the electronic and geometric structure of electrocatalysts, which helps to propel the catalytic activity of electrocatalysts [37, 38].

Defect engineering is crucial for designing electrocatalysts in Li–S batteries. Professor Chen and coworkers from the Dalian Institute of Chemical Physics have conducted many research works in exploring various defect-engineered catalysts. They proposed an amorphous niobium pentoxide nanocluster with oxygen defects embedded in microporous carbon nanospheres (A-Nb_2O_{5-x}@MCS) [39]. Its unique structure serving as a nanoreactor achieves uniform sulfur distribution, enhancing LiPS affinity and catalytic activity. As a result, the batteries with A-Nb_2O_{5-x}@MCS cathode exhibit outstanding extended durability, high rate capacity, and favorable areal capacity (6.62 mAh cm^{-2}) even at high sulfur loadings and lean E/S ratios. Then, they introduced anion defects in Sb_2Se_{3-x} to elevate the intrinsic conductance of the electrocatalyst and strengthen sulfur redox conversion efficiency (Figure 4.4a) [40]. Incorporating defects in Sb_2Se_{3-x}/rGO improves conductivity, LiPS adsorption, and catalytic activity, creating an effective barrier that constrains polysulfide shuttling and accelerates sulfur redox in Li–S batteries. The battery employing Sb_2Se_{3-x}/rGO exhibits remarkable long-cycling stability and desirable rate performance. The commonly reported single-atom catalysts always rely on a structure with M–N_4 bonds on a plane, which might not be optimal for host–guest interplay and corresponding catalytic activity. Based on this problem, they adjusted the electronic structure and catalytic activity of the electrocatalyst through defect engineering. They prepared monodisperse iron single atoms with different coordination states on nitrogen-doped carbon by tailoring the ligand structure in the iron precursor (represented as FeN_2-NC and FeN_4-NC) [42]. The FeN_2-NC design with the coordinative deficiency enhances sulfur immobilization and catalytic activity. Recently, as illustrated in Figure 4.4b, they reported novel MOFs featuring dual defects (missing linker and cluster defects) as an innovative catalyst again, designed to harness synergistic catalysis for the intricate multistep LiPS redox reaction [41]. The missing linker defect primarily drives the solid–liquid conversion, while the missing cluster defect primarily facilitates the liquid–solid conversion. Consequently, Li–S batteries utilizing UiO-66D2-modified separators showcase outstanding long-lasting performance, retaining a capacity of 785 mAh g^{-1} even after 500 cyclings.

Sun and coworkers proposed oxygen-defect-abundant Ti_nO_{2n-1} quantum dots onto porous carbon nanosheets to achieve efficient polysulfides immobilization and catalytic conversion [43]. The incorporation of oxygen defects significantly enhances the capture and transition of polysulfides, as it reduces the adsorption barrier and shortens active site–polysulfide bond lengths. Sun et al. employed a selective dual-defect engineering approach to a conventional $MoSe_2$ electrocatalyst [44], involving the strategic introduction of both N-doped and Se-defects. This

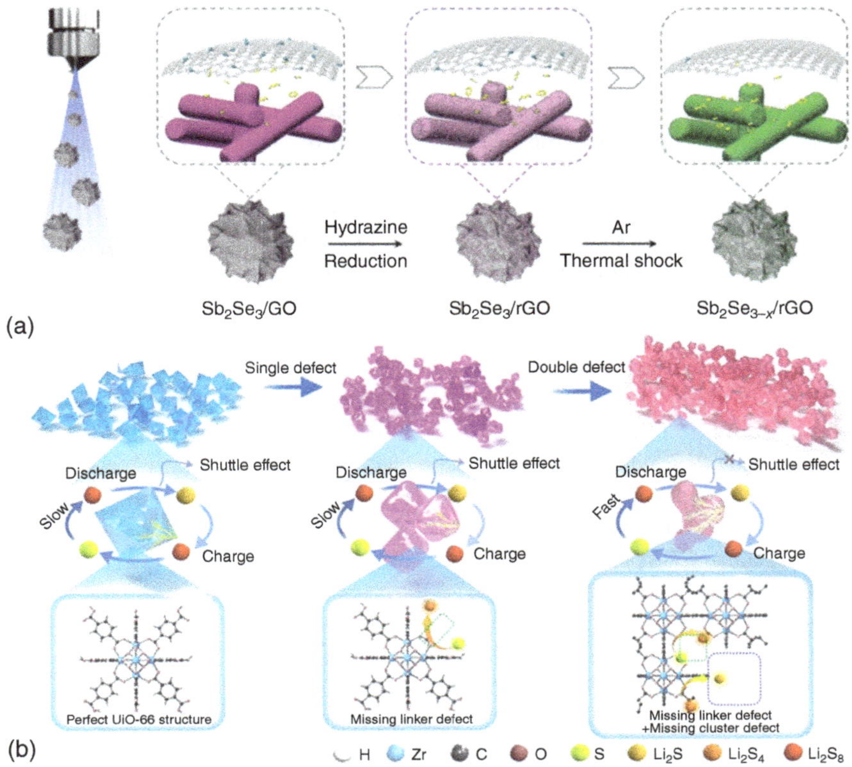

Figure 4.4 (a) Preparation process of Sb_2Se_{3-x}/rGO microspheres. Source: Tian et al. [40]/John Wiley & Sons. (b) Schematic diagram of sample synthesis. Source: Wang et al. [41]/John Wiley & Sons.

manipulation strategy selectively expedites the nucleation and dissociation processes of Li_2S, respectively. Because of this optimal electrocatalytic manipulation of Li_2S, the S@N-$MoSe_{2-x}$/C cathode displays exceptional capacity and remarkable capacity retention. Defect engineering, known for tuning electrocatalyst electronic structures to influence surface adsorption and catalysis, is an admirable method to address challenges among Li–S batteries. Sun et al. designed CoP with phosphorus vacancies to elucidate their improvement mechanisms in Li–S batteries [45]. The phosphorus vacancy engineering enhances the chemical affinity toward LiPSs, effectively mitigating the shuttle effect. Additionally, this modification lowers the reaction energy barrier and facilitates Li^+ ion diffusion, accelerating redox kinetics during LiPS liquid-phase conversions and Li_2S nucleation processes. Furthermore, they explored the direct correlation between the degree of defects and the electrode's capability for adsorption and catalysis, aiming to quantify this relationship [46]. They utilized perovskites $Sr_{0.9}Ti_{1-x}Mn_xO_{3-\delta}$ ($STMn_x$) (x = 0.1–0.3) with various oxygen-defect contents as research models. They summarized a quantitative relationship analysis between the concentration of oxygen vacancies and the performance of Li–S batteries. As the oxygen-vacancy contents enhance, notable enhancements are observed in rate performance, high-loading electrochemical performance, and stability. Similarly, Shao et al. revealed a volcanic-shaped correlation between the defect level of molybdenum sulfide and its adsorption–catalytic ability [47]. The results indicate that sulfur defects optimize LiPS interaction, electron transfer at catalytic interfaces, and Li–ion diffusion during sulfur redox reaction. An optimal vacancy level minimizes Li_2S decomposition potential and accelerates LiPS redox conversion from Li_2S_4 to Li_2S/Li_2S_2. Jordi et al. regulated the electron spin state of the electrocatalyst through a tailored defect engineering method that involves strategically introducing cobalt (Co) vacancies onto the surface of CoSe nanosheets [48]. This manipulation alters the distribution of electron spin states, thereby enhancing the count of unpaired electrons. This localized electronic reconfiguration strengthens the LiPS adsorption ability, decreasing the activation energy required for the sulfur redox reactions. Recently, to simultaneously address both the sulfur cathode and lithium anode challenges, Zhan and coworkers devised a bimetallic telluride electrocatalyst featuring a dense array of heterointerfaces and abundant defects embedded within hollow carbon polyhedron ($N \subset CoTe_{1-x}/ZnTe_{1-y}$@NC, NCZTC) [49]. Both experimental and theoretical analyses reveal that the integration of dense heterointerfaces and abundant defects in NCZTC synergistically tunes the electronic structure, boosts conductivity, accelerates Li^+ ion diffusion, enhances polysulfide adsorption, and promotes catalytic performance. The Li–S batteries with NCZTC-modified separators exhibit remarkable discharge capacity (1385.5 mAh g^{-1} at 0.1 C) and superior rate capability (748.0 mAh g^{-1} at 5 C). Notably, they maintain an impressive areal capacity of 4.31 mAh cm^{-2} over 100 cycles even under 6.0 mg cm^{-2} sulfur loading and limited electrolyte usage.

4.3.3 Composition and Structural Manipulation

By optimizing the microstructure of the catalyst, including its crystallinity, crystal phase, crystal plane, and lattice stress, the catalytic activity, stability, and selectivity, reaction kinetics of the electrocatalyst can be enhanced, thereby promoting the performance of Li–S batteries.

The mixture of crystalline and amorphous phases can provide a larger surface area, increase catalytic active sites, and thus boost the electrocatalytic reaction rate. In addition, crystalline phases usually exhibit better stability and mechanical properties, while amorphous phases possess higher catalytic activity. The combination of the two can coordinate their respective advantages and optimize the electrocatalytic performance. Wang and coworkers reported an innovative amorphous–crystalline heterostructured Nb_2O_5 as a dual-purpose host material, tailored specifically for both sulfur and lithium electrodes [53]. This unique heterostructure significantly improves overall conductivity while effectively integrating the robust sulfur affinity of amorphous Nb_2O_5 with the swift Li^+ ion transfer capabilities of crystalline Nb_2O_5. This strategic combination effectively mitigates the shuttle effect and accelerates conversion kinetics, fostering rapid and stable sulfur reactions. Furthermore, the lithiophilicity of heterostructured Nb_2O_5 facilitates optimal regulation of Li plating

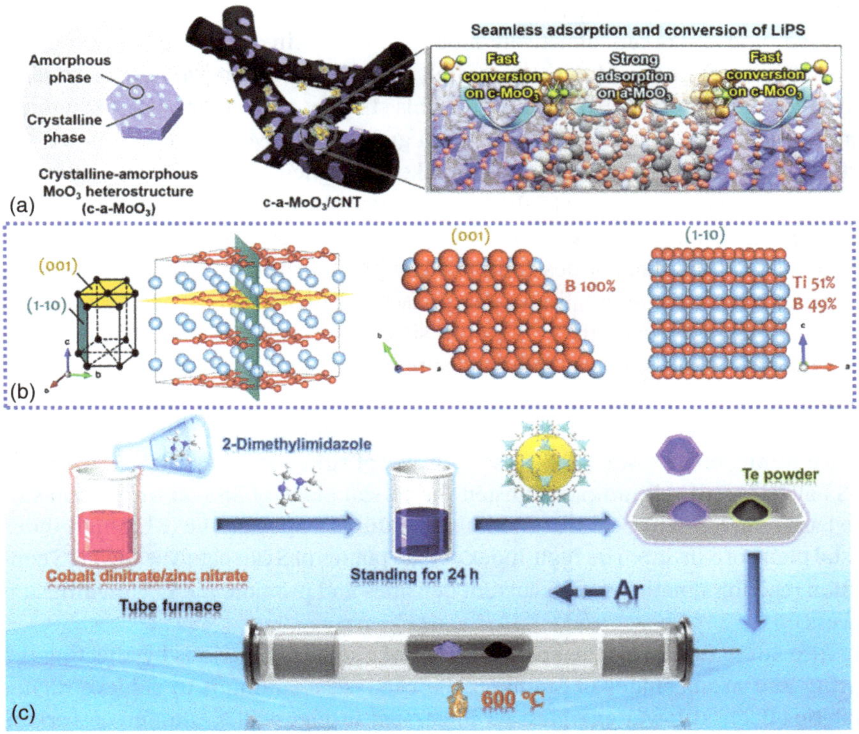

Figure 4.5 (a) Study on the mechanism of crystalline–amorphous MoO_3 heterostructure and polysulfide interaction. Source: Lee et al. [50]/with permission of Elsevier. (b) Crystal structure models of TiB_2 with various crystal facets. Source: Liu et al. [51]/John Wiley & Sons. (c) Schematic diagram of electrocatalyst construction. Source: Wang et al. [52]/John Wiley & Sons.

and stripping processes, thereby inhibiting dendrite formation and bolstering redox reversibility. Recently, Moon and coworkers designed a crystalline–amorphous MoO_3 heterostructure (c-a-MoO_3) (Figure 4.5a) [50]. This distinctive structure, achieved by inhibiting a-MoO_3 phase transition with metal ions, demonstrates remarkable LiPS adsorption and conversion capabilities. Its rapid LiPS adsorption on a-MoO_3 stems from polarization in unsaturated coordination and atomic disorder. Swift LiPS conversion on c-MoO_3 is evidenced by CV peak current intensity and polarization, supported by theoretical calculations testifying low barriers for Li^+ ion diffusion and Li_2S formation.

By regulating the crystal phase and crystal plane, the arrangement of atoms in the catalyst can be changed, thereby altering its electronic structure and band structure, generating strain and electronic effects, and thus regulating the catalytic performance. Hong and coworkers enhanced polysulfides adsorption in cobaltous oxide (CoO) by regulating surface electronic properties via amorphization [54]. Employing amorphous CoO nanosheets as additives, the exceptional electrochemical capacity (1248.2 mAh g^{-1} at 1 C) and remarkable capacity maintenance (1037.3 mAh g^{-1}

over 500 cyclings) are reached. Based on the moderate NH_3 plasma N-doping strategy, Chen and coworkers effectively modulate the desired phase transition (from 2H to 1T phase) of the MoS_2 electrocatalyst, thereby further enhancing its electrocatalytic activity [55]. Benefiting from this strategy, the battery electrochemical characteristic with electrocatalyst has been greatly improved. Through the crystal surface engineering strategy, He and coworkers designed and synthesized ultrathin TiB_2 nanosheets with specific (1–10) exposed crystal faces, achieving highly equivalent exposure of active sites of Ti and B atoms, as illustrated in Figure 4.5b [51]. Compared to traditional borides dominated by (001) surfaces that only expose a single type of atomic active site, this (1–10) surface-dominated TiB_2 nanosheet can not only anchor LiPSs through the dual atomic sites of Ti–S and B–S bonds but also maximize the adsorption of polysulfides and suppress shuttle effects. Meanwhile, it can effectively promote polysulfides conversion, significantly minimize the Li_2S decomposition barrier, and accelerate the redox kinetics. Wang et al. studied the SnS crystal surface tuning effect toward the crystal surface of sulfur and the catalytic ability of Li–S batteries [56]. The author adjusted the crystal plane of SnS on the carbon surface based on different heat treatment temperatures, exposing the (311) high-index crystal plane of S in SnS. The high-index crystal planes of S can catalyze the LiPS conversion reaction kinetics at high current densities and promote the Li^+ ion diffusion process.

Lattice stress can lower the energy barrier of chemical reactions by affecting the spacing and arrangement of atoms in the catalyst, enabling it to achieve higher catalytic effects at lower energies. As illustrated in Figure 4.5c, Xu and coworkers constructed $Co_{0.9}Zn_{0.1}Te_2$@NC (the $CoTe_2$ with uniformly doped zinc atoms) [52]. Experimental and theoretical findings demonstrate that lattice strain modulates the coordination environment of cobalt atoms by shifting the d-band center, thereby increasing electron occupancy in antibonding orbitals, thereby balancing the capture and inherent catalysis of polysulfides. Lattice strain enhances catalysis by lowering the energy barrier for redox reactions, facilitating the cleavage of S–S in polysulfides and S–Li bonds in Li_2S, and enhancing the redox kinetics process. Based on the phosphorus atom-induced phase engineering strategy, Wang et al. induced a phase transition from a mixture of cubic $CoSe_2$ and orthorhombic $CoSe_2$ to a pure orthorhombic $CoSe_2$ structure, resulting in lattice tensile strain [57]. The stretching of Co–Se bonds leads to lattice stretching strain and stronger charge delocalization, causing the d-band center of Co to shift upward and closer to the Fermi level, enhancing the interaction between $CoSe_2$ and LiPSs, thereby reducing the conversion energy barrier.

4.3.4 Heterojunction Construction

Heterogeneous structures play a crucial role in enhancing catalytic sites by regulating the interfaces of different materials, which can more effectively anchor and catalyze polysulfides, thereby boosting LiPS redox kinetics. Various components can play different roles in heterostructure design, synergistically boosting the performance of Li–S batteries. Components with excellent anchoring ability in heterostructures can effectively adsorb polysulfides, reduce their dissolution, and shuttle in electrolytes, thereby improving the actual active sulfur utilization. Some components in heterostructures have excellent conductivity, which helps to enhance the electrocatalyst conductivity and promote rapid electron shift in the battery. In addition, through reasonable heterostructure design, channels conducive to ion transport can be constructed to facilitate the ion migration rate inside batteries, thereby further enhancing battery performance. The above advantages in heterostructure design work together to propel the overall battery capacity and accelerate the functional application of the Li–S battery [18, 20].

The BIEF in heterogeneous electrocatalysts enhances the separation of photogenerated electrons and holes, thereby decreasing their recombination rate. This separation allows electrons and holes to migrate to different regions of the catalyst and participate in different catalytic reactions to further improve catalytic efficiency. The BIEF can also accelerate charge transfer at the interface, allowing electrons and holes to reach the reaction site faster, thus participating in catalytic reactions. This acceleration effect helps to enhance the overall activity of the catalyst. The synergistic impact generated by different components in heterostructures at the interface can improve the overall performance of the catalyst. As part of this synergistic effect, the BIEF can further enhance the catalytic activity of the catalyst. Zhang et al. designed and successfully prepared Co_3O_4/TiO_2 *p-n* heterojunction with selective catalytic conversion, achieving sequential catalysis of two-step continuous reactions of LiPSs (Figure 4.6a) [58]. Among them, *p*-type Co_3O_4 nanosheets and *n*-type TiO_2 nanoparticles promote the conversion of S_8 to Li_2S_4 and Li_2S_4 to Li_2S, respectively. In addition, due to the BIEF at the *p-n* junction interface, Li_2S_4 is induced to undergo directional transfer at the interface, thereby achieving continuous sequential conversion of LiPSs between different components. By utilizing the BIEF, the composition and structural heterogeneity are cleverly coupled with the sequential transformation of LiPSs, greatly suppressing the LiPS shuttle and achieving efficient sulfur usage.

In response to the problems of polysulfide shuttle and instability of metal lithium anode in Li–S batteries, Chen et al. proposed an ultrathin two-dimensional graphene-based mesoporous $SnO_2/SnSe_2$ composite material (G-m$SnO_2/SnSe_2$) [15], a new type of mesoporous material, which combines the dual functions of polysulfide immobilizer and lithium regulator. The BIEF enhances charge transfer and LiPS migration from SnO_2 to $SnO_2/SnSe_2$ interfaces, accelerating diffusion and effectively mitigating the shuttle effect. SnO_2 in the designed G-m$SnO_2/SnSe_2$ material demonstrates high conductivity and enhanced affinity. $SnSe_2$ exhibits a dynamic conversion process (Li_xSnSe_2), which suppresses shuttle behavior, provides fast lithium intercalation transport kinetics, accelerates LiPS conversion, and reduces the Li_2S decomposition barrier. In addition, G-m$SnO_2/SnSe_2$ with lithium affinity promotes uniform Li^+ ion dynamics and prevents Li dendrite formation. Consequently, Li–S batteries with modified separators can realize ideal sulfur utilization and rate performance. Li et al. designed a novel high-conductivity NbB_2-MXene heterostructure with spontaneous BIEF (Figure 4.6b) [59]. Experimental and theoretical results demonstrate that Nb and B atoms can chemically bind with polysulfides, enriching chemical anchoring and catalytic active sites. At

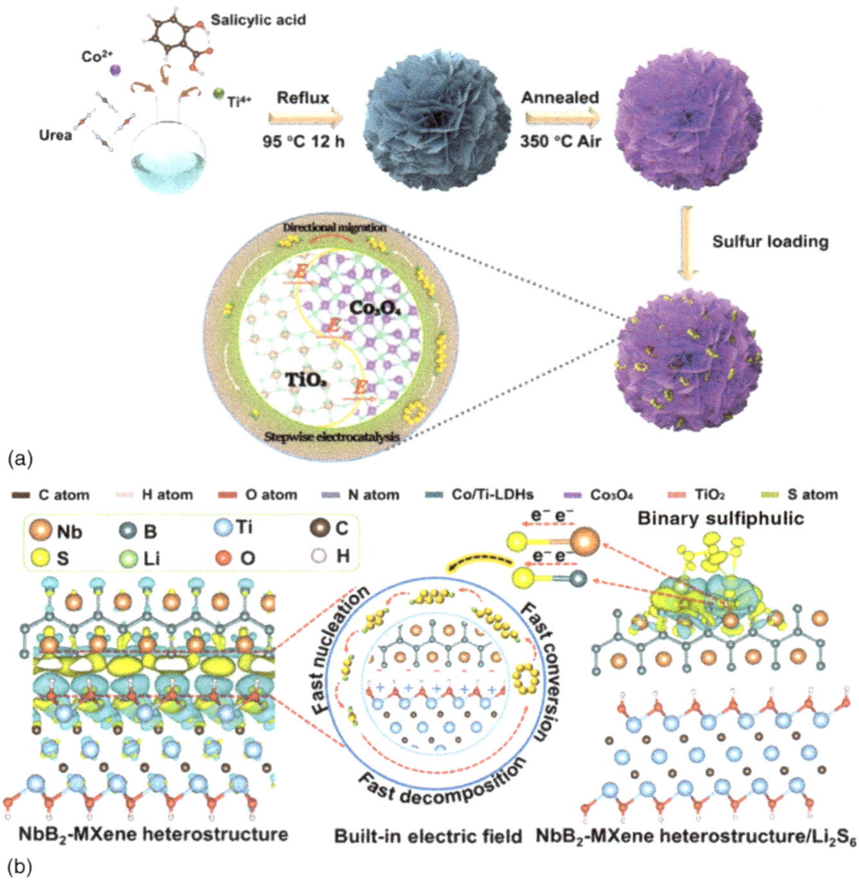

Figure 4.6 (a) Preparation process of hierarchical p-Co_3O_4/n-TiO_2-HPs and the sequential catalytic transformation of LiPSs. Source: Li et al. [58]/John Wiley & Sons. (b) Interfacial charge density difference and LiPS adsorption–catalytic mechanism of the NbB_2-MXene heterostructure. Source: Lu et al. [59]/John Wiley & Sons.

the same time, spontaneous BIEF induces the redistribution of interface charges, allowing more electrons to transfer to the surface of NbB_2, thereby weakening its strong adsorption performance. However, it also accelerates the electron transfer and diffusion of polysulfides on heterogeneous interfaces, providing moderate polysulfide adsorption capacity while reducing energy barriers for sulfur species conversion. This further enhances the intrinsic catalytic activity of NbB_2-MXene and accelerates the bidirectional conversion of sulfur. Then, they proposed a strategy combining cation doping engineering and interface BIEF effect to regulate the p-band center of the catalytic reaction sites and the electron shift rate at the heterojunction interface, thereby enhancing the bidirectional sulfur conversion kinetics in Li–S batteries. They obtained Mo-VS_2/MXene heterostructures by *in situ* growing Mo-doped VS_2 nanosheets on the surface of MXene [60]. The BIEF effect of the interface promotes spontaneous electron rearrangement at heterogeneous interfaces, accelerates electron transfer, and enhances the thermodynamic and kinetic processes of sulfur reduction reactions while improving reaction efficiency and alleviating the shuttle effect.

In the design of heterostructures, various components can fulfill distinct functions, collectively amplifying the electrochemical capacity through synergistic effects. Yang and coworkers proposed an *in situ*-constructed TiO_2-TiN heterostructure [61], which combines TiO_2 with ultrafast anchoring ability and TiN with electronic conductivity to achieve efficient trapping–diffusion–conversion process of LiPSs (Figure 4.7a), thereby enhancing the electrochemical behavior. Specifically, the rapid diffusion characteristics of LiPSs from TiO_2 to TiN provide an effective pathway for efficient polysulfide capture and rapid transformation. Loading the heterostructure onto graphene serves as a barrier for coating the separator, and the shuttle effect is greatly alleviated. With its unique heterostructure design and simple preparation process, this heterostructure demonstrates eminent potential in boosting the cycling life and practical application of Li–S batteries. As schematically illuminated in Figure 4.7b, Xu et al. designed Co/MoN catalysts that fully utilized the synergistic effect of different functional microregions, taking advantage of the catalytic effect of Co and the adsorption effect of MoN. Compared to the original MoN, the introduction of Co significantly improves the electrochemical performance of Li–S batteries [62]. Li et al. reported a macroporous three-dimensional Ni_2P/CoP heterostructure. The macroporous structure of this heterostructure is conducive to accommodating abundant sulfur, facilitating rapid electron transfer and Li^+ ion diffusion. The excellent adsorption and catalytic activity of Ni_2P/CoP heterostructure greatly accelerates the electrochemical reaction kinetics. Li et al. prepared highly dispersed Co/Mo_2C heterostructure nanoparticles [63]. The Co side in Co/Mo_2C heterostructures catalyzes the growth of CNTs, which results in the cross-linked conductive network maintaining the overall morphology of the host and avoiding the coarsening of heterostructure particles. The Mo_2C side of the heterostructure is mainly responsible for the adsorption–catalytic function of LiPSs. The batteries with Co/Mo_2C@CNT-modified separator achieve superb rate capacity and high areal capacity. Wang et al. prepared NiS_2-WS_2 flower-shaped nanospheres composed of ultrathin nanosheets ($\leq$10 nm), enhancing the rapid mass and charge transfer [64]. The flower-shaped nanosphere structure shortens the ion transport pathway, reduces the activation energy barrier, and thus improves the kinetics of redox reactions. NiS_2 regulates the morphology and structure of WS_2, and optimizes its Fermi level, improving its conductivity. NiS_2 regulates the electronic structure of WS_2 that boosts the bidirectional reaction rate of polysulfides and effectively suppresses their shuttle effect. Based on the puzzle-style catalyst design strategy, Zhang and coworkers assembled NbN-NbC coherent nano-heterostructure groups through the *in situ* method to stabilize highly active crystal faces and reduce related reaction energy

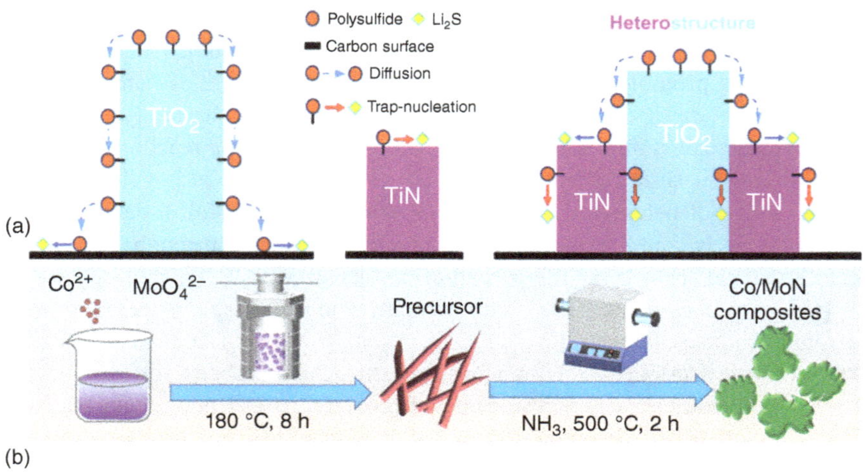

Figure 4.7 (a) Schematic illustration of LiPS conversion mechanisms on TiN, TiO_2, and TiO_2–TiN heterostructure surfaces. Source: Zhou et al. [61]/Royal Society of Chemistry. (b) Fabrication process of Co/MoN. Source: Kong et al. [62]/John Wiley & Sons.

barriers [65]. The stable and highly active crystal surface promotes the aggregation of polysulfides. At the same time, the catalytic activity of the surrounding crystal planes is improved, promoting the Li_2S deposition and the diffusion of Li^+ ions. These advantages synergistically promote continuous and efficient sulfur redox reactions, improving the durability and energy density of the battery.

4.3.5 Alloy Electrocatalyst

Different elements in alloys can form various chemical bonds and active sites, which help enhance the trap and electrocatalytic activity. High-entropy alloys contain multiple elements, and the synergistic effect between these elements can significantly foster the activity of the catalyst. The nanoscale characteristics of high-entropy alloys endow them with a larger specific surface area and numerous adsorption sites, which helps to efficiently adsorb polysulfides, thereby suppressing shuttle effects and improving sulfur utilization efficiency. Additionally, both alloys and high-entropy alloys exhibit relatively stable structures, which help to reduce the loss and failure of catalysts during battery cycling, thus improving the stability and capacity of the battery.

For the first time, Gao et al. introduced a concave nanocube (CNC) nickel–platinum (Ni–Pt) alloy with high-index crystal faces (HIFs) as the electrocatalyst for LiPS conversion [68]. Compared with the low-index facets (LIFs) of traditional nanocube Ni–Pt alloys, HIFs offer moderate chemical coupling for LiPSs and more effectively lower the conversion barrier, thereby strongly enhancing LiPS conversion. Li and coworkers uniformly mixed Ni-MOF with cobalt salt and reduced them to obtain cobalt–nickel alloy [69]. The unique feature is that nickel forms a hexagonal sheet-like structure, while cobalt forms a "border" structure at its edges. The edge

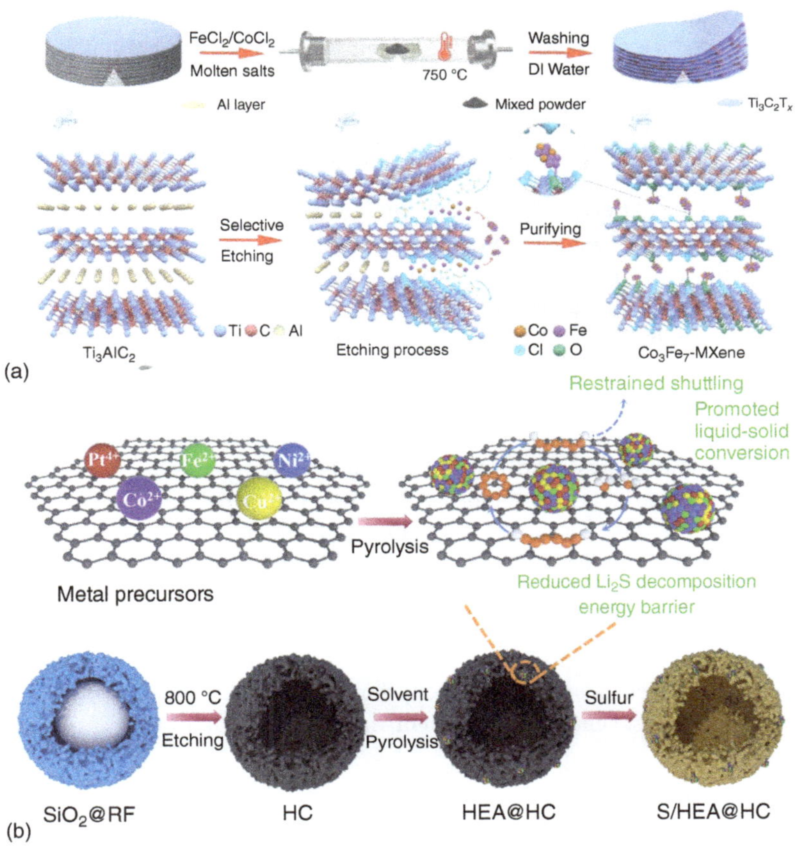

Figure 4.8 (a) Depiction of the etching process mechanism involved in Co_3Fe_7-MXene hybrid materials. Source: Wang et al. [66]/John Wiley & Sons. (b) Synthesis methodology for the S/HEA@HC composite material. Source: Han et al. [67]/American Chemical Society.

geometric effect of cobalt–nickel alloy can cause Li_2S to deposit into a loose sheet-like structure, preventing an early passivation of the electrode surface from dense Li_2S deposition, thereby improving the sulfur utilization efficiency. According to the alloying strategy, Zhang and coworkers regulated and optimized the electronic structure of FeCoNi catalyst [70]. The specific energy band and electronic configuration of iron (Fe) catalysts are regulated by constructing a unique charge environment around iron (Fe) atoms and doping cobalt (Co) and nickel (Ni) atoms with different electron-filling states and atomic radii. The battery equipped with a FeCoNi alloy-based separator delivers a 1000-cycle life at 3 C. Feng et al. prepared a novel MXene-supported ultrasmall Co_3Fe_7 bimetallic alloy hybrid material using a one-pot molten salt etching strategy (Figure 4.8a) [66]. Ultrasmall Co_3Fe_7 alloy particles are formed *in situ* and tightly anchored on MXene substrates. The battery with a Co_3Fe_7-MXene exhibits high areal capacity, outstanding rate performance, and favorable cycling stability.

A large amount of uniformly mixed and distributed metals in high-entropy materials will produce a "cocktail effect," which leads to the diversity of chemical and electronic adsorption sites, thereby catalyzing the redox reaction of LiPSs. Abruña and coworkers designed $Zn_{0.30}Co_{0.31}Cu_{0.19}In_{0.13}Ga_{0.06}S$ high-entropy sulfide material [71], which significantly improved the rate of polysulfide redox reactions. When 2% $Zn_{0.30}Co_{0.31}Cu_{0.19}In_{0.13}Ga_{0.06}S$ additive is added to the Li–S batteries, the initial capacity of the battery is significantly improved. As illustrated in Figure 4.8b, Wu and coworkers utilized the synthesized nano $Pt_{0.25}Cu_{0.25}Fe_{0.15}Co_{0.15}Ni_{0.2}$ high-entropy alloy (HEA) as a sulfur host [67]. HEA nanoparticles demonstrate excellent conductivity and abundant catalytic sites, effectively anchoring LiPSs and accelerating bidirectional liquid–solid conversion ($Li_2S_n \leftrightarrow Li_2S$). In addition, density functional theory (DFT) calculations reveal synergistic modulation related to electronic properties in HEAs. Platinum (Pt) serves as an electron reservoir, while copper (Cu) and iron (Fe) dominate the conduction band and act as electron-consuming sites. The remaining components endow a smooth and continuously regulated charge, promoting the multi-electron reaction. Yang and coworkers added vanadium (V) and palladium (Pd) elements to the traditional transition metal alloy CoNiFe [72]. They developed CoNiFePdV HEA to accelerate the stepwise sulfur redox reaction of Li–S batteries by ultrafast Joule heating synthesis on carbon nanofibers (CNFs). Furthermore, the addition of vanadium (V) significantly increases the electrochemical active surface area of HEA nanocatalysts, thereby enhancing their adsorption capacity for polysulfides. The Li–S batteries containing HEA catalyst achieve up to 1364 $mAh\,g^{-1}$ capacity at 0.1 C.

4.3.6 Other Electrocatalysts

4.3.6.1 MOF Electrocatalysts

MOFs have highly ordered porous structures, enabling them to effectively adsorb and accommodate LiPSs. The structural diversity of MOFs allows researchers to optimize their catalytic performance by designing different metal ions and organic ligands. The chemical modularity of MOFs offers flexibility to researchers for constructing efficient electrocatalysts by introducing metal sites or functional groups with catalytic activity [23]. Multiple catalytic active centers can be formed in MOFs by introducing multiple metal ions or constructing composite structures. The catalytic sites synergistically enhance LiPS conversion and battery performance [73].

Huang and coworkers introduced a strategy for constructing hierarchical porous catalytic MOFs (HPC MOFs) to balance the relationship between porous structures and high-density adsorption–catalytic sites to achieve better catalytic activity (Figure 4.9a) [74]. These HPC MOFs with moderate porosity and abundant catalytic sites were designed through precise pore structure and chemical engineering. Li–S batteries equipped with superior HPC-MOF interlayers exhibit faster redox reaction kinetics, good rate performance, and capacity maintenance. Chen and coworkers further advanced this approach by applying under coordination chemical modification of Zn–Co bimetallic MOF (D-ZIF L) to remove organic ligands from the active center [76]. This process can reduce steric hindrance, promote sufficient

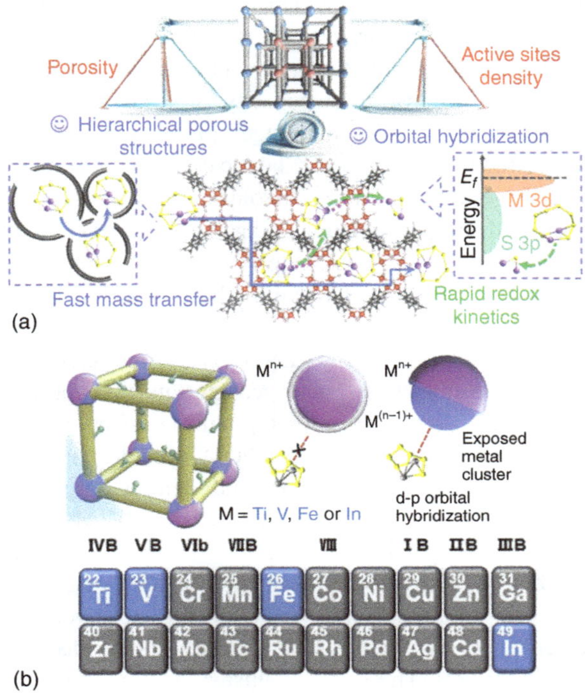

Figure 4.9 (a) HPC-MOF design strategy. Source: Xie et al. [74]/American Chemical Society. (b) Schematic illustration of MSC-MOFs-M with various metal catalytic sites. Source: Lu et al. [75]/John Wiley & Sons.

contact between sulfur species and metal active centers, and thus improve the catalytic efficiency of MOFs. In addition, electron redistribution can be induced under the coordination treatment of metal active centers. The electron density of the Fermi level of metal elements can be increased, and inherent conductivity has been improved. Huang et al. introduced a new type of multisite catalytic MOFs (MSC-MOFs-M, where M = Ti, Fe, V, or In) (Figure 4.9b), characterized by the formation of a synergistically specific adsorption–catalytic interface through multimode designed chemical sites [75]. Specifically, nanocages formed within MOFs, combined with neighboring $-NH_2$ groups, create a synergistic catalytic interface. The $-NH_2$ groups efficiently adsorb polysulfides into the nanocages, concentrating LiPSs around the activity centers via chemical driving force, not just concentration diffusion. Additionally, the co-adsorption of single metal sites with neighboring $-NH_2$ groups boosts the electrocatalytic activity over single-metal-site MOFs.

4.3.6.2 Polyoxometalate (POM) Electrocatalysts

Polyoxometalates offer unique advantages in the electrocatalyst design of Li–S battery electrocatalysts, including rich chemical composition and diverse structures, excellent redox activity, strong adsorption capacity, and adjustable electronic structure. POMs are formed by connecting various transition metal ions (e.g. Mo, W, V, Nb, Ta) with oxygen atoms through co-angle, co-edge, or coplanar connections, and their chemical compositions and structures are diverse. POMs possess multiple variable redox states, which facilitates rich electron transfer pathways during the charging/discharging process, effectively catalyzing the LiPS redox reactions. POMs have abundant negative charges and polar groups on their surface and exhibit strong adsorption capacity for LiPSs. Wei et al. introduced rGO–CNT/PW_{12} catalysts with "electron sponge" property into Li–S battery systems [77], which can achieve bidirectional catalytic conversion from S to Li_2S species. However, the challenges arise from the dissolution of the reduced "hetero-poly blue" substance formed during lithiation, leading to the loss of POM catalysts in the sulfur cathode. Therefore, the key challenge is to find suitable POM structures and simultaneously achieve their effective anchoring within the battery system. Constructing supramolecular/polyoxometalate organic–inorganic hybrid self-assembly framework materials not only effectively immobilizes POM catalysts, but also utilizes supramolecular host–guest recognition to capture polysulfide and electrolyte molecules, thereby suppressing shuttle effects and enhancing Li^+ ion transport. Therefore, they prepared the POM–cyclodextrin–cluster–organic supramolecular framework (POM-CD-COSF) and demonstrated excellent electrochemical performance when using the POM-CD-COSF as separator modification materials (Figure 4.10a) [78].

Chen and coworkers developed a monodisperse Co_4W_{18} cluster catalyst on reduced graphene oxide (rGO) using positively charged polyethylenimine as a linker to synthesize Co_4W_{18}/rGO composite materials through electrostatic assembly [80]. The Co_4W_{18} cluster features unsaturated cobalt atoms that boost polysulfide adsorption and lower the activation energy for polysulfide conversion

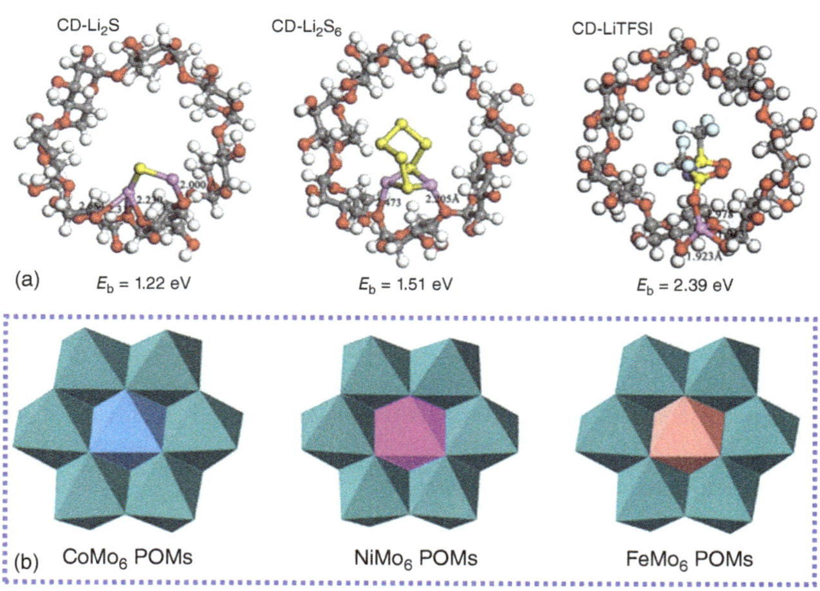

Figure 4.10 (a) Optimal structural model of CD, polysulfides, and LiTFSI. Source: Ni et al. [78]/John Wiley & Sons. (b) Schematic illustration of different POM structures. Source: Li et al. [79]/John Wiley & Sons.

from 0.55 to 0.3 eV in Co_4W_{18}/rGO. In addition, the Co_4W_{18}/rGO composite material exhibits rapid interface charge shift and abundant exposure of active sites, which can attain more uniform Li_2S deposition and reduce the reaction barrier of Li_2S. The capacity of the Co_4W_{18}/rGO/S cathode decreased by only 0.015% per cycle after 1000 cycles at 2 C. Polysulfide catalytic reactions require catalysts with large surface areas, efficient atomic utilization, high conductivity, small sizes, and good stability. As illustrated in Figure 4.10b, He and coworkers used polyoxometalates ($[TMMo_6O_{24}]^{n-}$, TM = Co, Ni, Fe, $TMMo_6$ POMs) as modified materials for Li–S battery separators [79]. They uncovered a layered electron transfer mechanism between $TMMo_6$ units and LiPSs, akin to the collision effect of bowling balls. Therefore, the battery with $CoMo_6$@graphene-based separator has an initial capacity of up to 1588.6 mAh g^{-1} at 0.2 C, and uniform Li^+ transport within 1900 hours.

4.3.6.3 Single-Atom (SA) Electrocatalysts

Each atom of a single-atom catalyst directly participates in the catalytic reaction, achieving 100% atomic utilization efficiency. Due to the high density of surface active sites of SA catalysts, they can more effectively adsorb and catalyze LiPS conversion, thereby accelerating the progress of electrochemical conversion reactions. SA electrocatalysts are anchored on various carriers, including carbon materials, oxides. By tuning carrier properties, their electrocatalytic performance can be further upgraded. The precise design of SA catalyst through coordination atom selection and configuration helps control their electronic structure and geometric configuration, leading to tailored catalytic performance. Wan et al. embedded monodisperse Co atoms into nitrogen-doped graphene as a sulfur host and used it in the Li–S batteries [81]. By combining *in situ* X-ray absorption spectroscopy (XAS) and theoretical calculations, the Co-N-C coordination centers effectively catalyzed both the formation and decomposition of Li_2S during discharge/charge processes. Recently, Mai and coworkers studied a novel SA catalyst (CoSA-N_3PS) [82], which significantly improved the catalytic activity of Co center atoms and promoted d-p orbital hybridization with sulfur atoms by precisely designing the asymmetric co-coordination structure of phosphorus and sulfur. This effectively curbed polysulfide shuttling and sped up the bidirectional redox process of sulfur species. The catalyst allowed the Li–S batteries to have an ultralow capacity degradation rate of 0.027% per cycle at 5 C, maintaining stability over 2000 cycles. At the same time, it demonstrated excellent rate performance of 619 mAh g^{-1} at an ultrahigh rate of 10 C.

Lou et al. utilized the NH_4Cl-assisted pyrolysis method to prepare high mesoporous (NC(p)) with a thin shell and a porous internal network, anchoring Ni single atoms to form an effective sulfur host (Ni NC(p)) for Li–S batteries. (Figure 4.11a) [83]. During the pyrolysis process, HCl generated by NH_4Cl pyrolysis within ZIF-8 will etch ZIF-8, resulting in mesoporous Ni NC(p). The mesoporous Ni NC(p) exhibits superb electron/ion transport, high sulfur loading, and efficient confinement, while the catalytic effect of Ni boosts the reaction kinetics. Many studies have demonstrated that the SA elements in the d-block of the fourth period

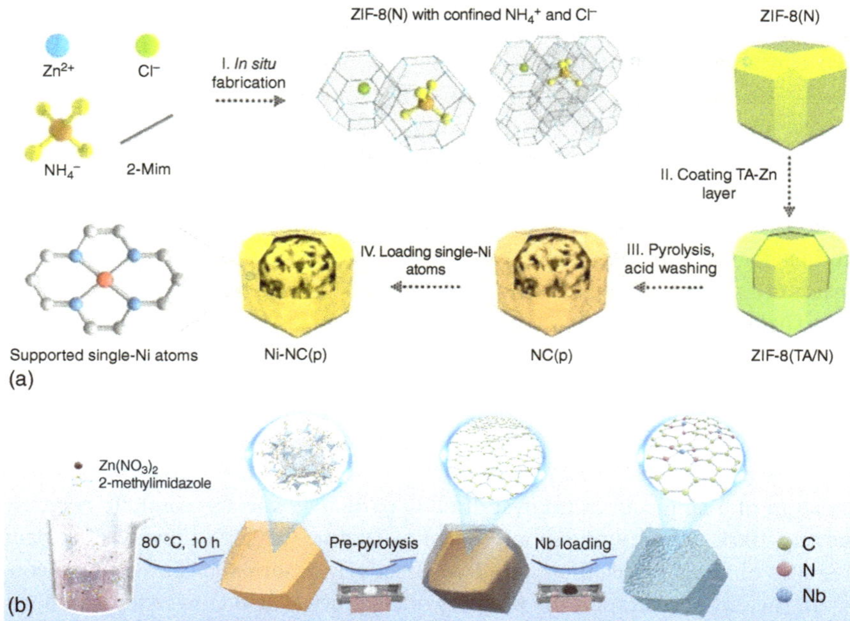

Figure 4.11 Fabrication process of (a) Ni SAC. Source: Li et al. [83]/John Wiley & Sons and (b) Nb SAC. Source: Zhang et al. [84]/Royal Society of Chemistry.

can undergo chemical reactions with the local environment, exhibiting commendable anchoring and electrocatalytic ability toward LiPSs. Guided by theoretical predictions, Yin and coworkers designed a novel SA Nb catalyst for the first time to improve sulfur immobilization and catalytic performance (Figure 4.11b) [84]. The calculation results illuminate the fact that the Nb-N_4 reactive group has abundant unfilled antibonding orbitals, promoting d-p hybridization through the "trapping–coupling–conversion" mechanism, enhancing the anchoring capacity for LiPSs. Most reported SACs for Li–S batteries are based on d-band transition metals with active valence bands formed by their d-orbitals, which are prone to passivation. Recently, Gu and coworkers designed and synthesized the rare metal Sm SACs, which possess electron-rich 4f valence orbitals [85]. Theoretical calculations indicate that the transition between the unique $4f^6 5d^0$ and $4f^5 5d^1$ valence electron configurations of Sm-N_3C_3 leads to enhanced f-d-p hybridization, thereby improving polysulfides conversion and decreasing the Li_2S decomposition barrier. Based on theoretical calculations, Li and coworkers identified that single-atom copper (SA-Cu) endows adsorbed Li_2S with the transition from insulator to metal via perspective of electronic structure [86]. The Li_2S deposition on nitrogen-doped carbon fiber foam modified by SA-Cu is not a conventional two-dimensional layered morphology, but rather three-dimensional spherical clusters that facilitate electron transfer through three-dimensional nucleation, greatly improving the Li_2S deposition ability and catalytic efficiency of catalytic sites.

4.4 Lithium Anode Protection

The lithium anode significantly influences the overall performance, longevity, and safety of Li–S batteries. The exceptional solubility of LiPSs within ether-based electrolytes significantly compromises the stability of lithium metal anode, rendering them highly susceptible to corrosion by the dissolved polysulfide species. This ultimately leads to reduced Coulombic efficiency. Lithium anode faces multiple challenges, such as lithium dendrite growth, severe lithium metal powdering, and volume expansion. Therefore, research on lithium anode protection is essential to enhance the electrochemical capacity and durability of Li–S batteries. The following is a summary of the current research status of lithium anode protection for Li–S batteries.

4.4.1 Electrolyte Regulation

Liquid electrolytes are crucial for Li plating/stripping, mainly due to their impact on the Li–ion solvation sheath and the properties of the SEI layers on Li anode. Therefore, rational regulation of the electrolyte is an important approach to modulate Li plating/stripping and promote anode performance.

$LiNO_3$ was led into electrolyte for Li–S batteries in 2008, enabling them to achieve higher Coulombic efficiency during charging/discharging. Due to the effectiveness of $LiNO_3$ in protecting Li anode, $LiNO_3$ additives are still used in most research on Li–S batteries [7]. Lin and coworkers proposed a P_2S_5 additive and found that P_2S_5 has dual functionality in Li–S batteries [89]. From one perspective, it can promote the decomposition of Li_2S, and from another perspective, it can passivate the lithium anode surface and generate a stable SEI film, inhibiting the formation of lithium dendrites. Zhang et al. proposed FEC additive [90], forming a stable, LiF-rich SEI film on the lithium anode surface, which improved the electrochemical performance of the lithium anode. Wang and coworkers proposed a kind of localized high-concentration electrolyte (LHCE) by incorporating fluoroalkyl ether, specifically 1H,1H,5H-octafluoropentyl-1,1,2,2-tetrafluoroethyl ether [91]. The designed LHCE LiTFSI/DME system can efficiently inhibit Li dendrite growth and alleviate LiPS solubility in electrolytes. Xie and coworkers demonstrated a modified diluted high-concentration electrolyte containing a specific quantity of LiTFSI and DOL [92], which was applied to Li–S batteries. LiTFSI and DOL play crucial roles in the generation of LiF and organic polymers that compose the SEI and cathode–electrolyte interface (CEI), respectively. Owing to the dual advantages, the actual Li–S pouch cells employed within this framework exhibit remarkable cycling stability, maintaining a consistently high Coulombic efficiency of 99.6%. Forming a stable SEI to prevent polysulfide–lithium metal side reactions is key for the stability and efficiency of the Li–S battery. Huang and coworkers introduced 3,5-bis(trifluoromethyl)thiophenol (BTB) to electrolyte to design a SEI (Figure 4.12a) [87]. This SEI containing organic sulfur can shield LiPSs and safeguard the lithium anode. As a result, this design strategy improves the durability and longevity of Li–S batteries under practical conditions. Based on visualization

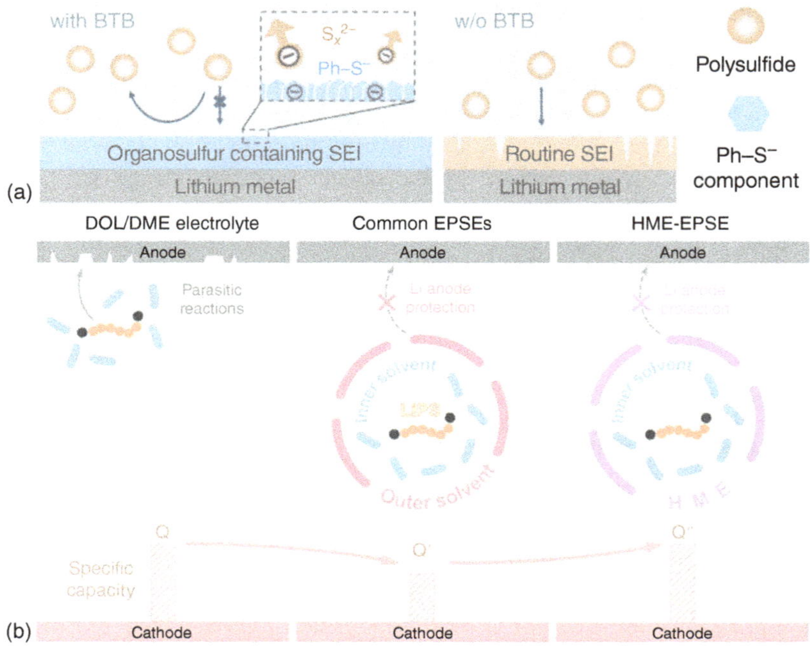

Figure 4.12 (a) Formation process of SEI with and without BTB additive and the barrier effect of SEI toward LiPS corrosion of the lithium anode. Source: Wei et al. [87]/John Wiley & Sons. (b) Working principle diagram of different electrolytes. Source: Su et al. [88]/John Wiley & Sons.

experiments and XPS measurements, it was verified that SEI containing organic sulfur regulated by BTB can effectively suppress the side effects, and its components stably exist in the SEI. The continuous side reactions are an important reason for the short cycle life of Li–S batteries. Adjusting the solvation structure of LiPSs can effectively suppress the side reactions between LiPSs and lithium anode. In the encapsulated lithium polysulfide electrolyte (EPSE), a double-layer solvent is constructed with LiPSs as the center. The solvent in the outer solvation layer has the characteristics of weak solvation ability and high reduction stability, which can dynamically suppress the side reactions between LiPSs and lithium anode, effectively extending the life of Li–S batteries. However, attributed to the introduction of the external solvation layer, the LiPS reaction kinetics are inevitably sacrificed. Huang et al. proposed using hexyl methyl ether (HME) as the solvent for the external solvent layer of EPSE to construct HME-EPSE, boosting the rate capacity of EPSE-based Li–S batteries and achieve the goal of balancing the rate performance and cycling performance of practical Li–S batteries (Figure 4.12b) [88]. Fu and coworkers reported that among three isomeric benzene dithiols (BDTs) used as electrolyte additives, 1,4-BDTs have the ability to bond a greater number of sulfur atoms compared to the other two isomers, greatly improving the capacity of Li–S batteries [93]. Research has shown that the functionalization of 1,4-BDT on the cathode and anode involves chemical reactions of thiol groups. 1,4-BDT reacts with sulfur through oligomerization to form S-S bonds, altering the original redox pathway of sulfur and inhibiting the LiPS shuttling. At the same time, 1,4-BDT can produce a stable SEI, which helps Li//Li symmetric batteries to cycle stably at ultralow overpotential of 0.08 V for more than 300 hours at 5 mA cm^{-2}. Zhang et al. first proposed and systematically studied lithium bonds in Li–S batteries [94]. Lithium bonds can regulate the interactions between lithium ions and solvent molecules or anions in electrolytes. The strength of lithium bonds influences the solvation and desolvation processes of lithium ions, which in turn affects the bulk transport properties of lithium ions and the interfacial reaction properties of electrolyte components. The batteries are severely troubled by the instability of the SEI formed in conventional ether electrolytes. Here, Zhang et al. reported an electrolyte with 1,3,5-trioxane (TO) and 1,2-dimethoxyethane (DME) as auxiliary solvents to construct a highly mechanically stable SEI by enriching the organic components in Li–S batteries [95]. As illustrated in Figure 4.13, TO with high polymerization ability can preferentially decompose and form SEI containing abundant organic matter, enhancing the mechanical stability of SEI and alleviating SEI rupture and regeneration. This emerging electrolyte design can construct a highly mechanically stable SEI on lithium anodes, thereby decreasing the consumption of active lithium, LiPSs, and electrolytes, ultimately improving the stability of batteries. The LiTFSI could exacerbate the surface gel of MoS_2 electrocatalyst. The trifluoromethanesulfonyl group in LiTFSI interacts with the Lewis acidic site on the MoS_2 electrocatalyst, forming an electron-deficient center with strong Lewis acidity. This center triggers the cationic polymerization of the 1,3-dioxolane solvent, resulting in a surface gel layer that diminishes the electrocatalytic activity. In order to solve the above problems, Zhang and coworkers introduced Lewis base lithium

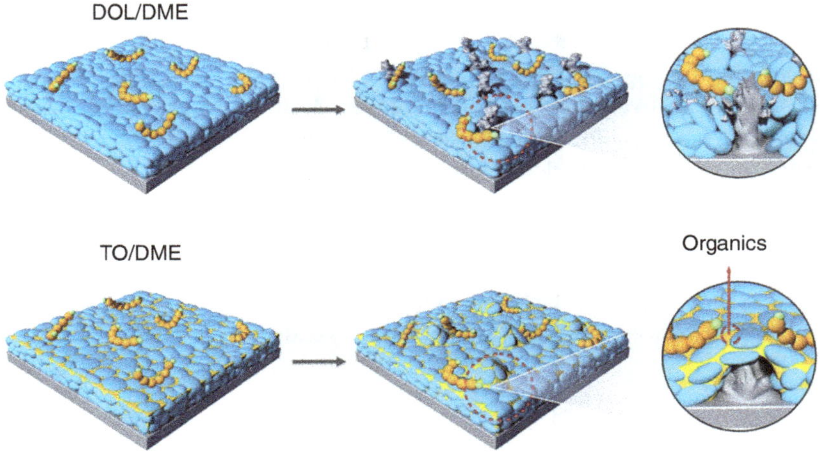

Figure 4.13 Schematic illustration for the evolution process of the SEI during lithium deposition in two distinct electrolyte systems: DOL/DME electrolyte and TO/DME electrolyte. Source: Hou et al. [95]/John Wiley & Sons.

iodide (LiI) to prevent the interaction between LiTFSI and MoS_2 and suppress the formation of surface gel [96]. The Li–S batteries with LiI additives and MoS_2 electrocatalysts exhibit superior rate performance, higher discharge capacity, and ultrahigh energy density of 416 Wh kg^{-1}.

In summary, the research on electrolyte additives can protect lithium anode, restrain the lithium dendrite growth, and improve the safety of batteries. Constructing a stable and efficient SEI can promote the interface Li^+ transport, ultimately promoting the Coulombic efficiency and longevity of the battery.

4.4.2 Artificial Modification Layer on Lithium Anode

In response to the problem of lithium dendrite, Choi and coworkers deposited 10-nm-thick MoS_2 on the lithium anode surface as a protective layer [97], effectively boosting the electrochemical behavior of Li–S batteries. The lithium metal coated with MoS_2 maintains a low overpotential at 10 mA cm^{-2}, and its cycle life is three times compared to that of the bare lithium metal anode. The batteries assembled a lithium anode coated with MoS_2, and a three-dimensional carbon nanotube sulfur cathode exhibit an energy density of 589 Wh kg^{-1} and a power density of 295 W kg^{-1}. An artificial mixed SEI layer of Li_3Sb and LiF was constructed on the Li metal anode to study Li^+ ion diffusion behavior [98]. Experimental and theoretical studies indicate Li_3Sb facilitates Li^+ ion diffusion in the SEI, and LiF provides electron shielding at the SEI/Li interface. The synergy of these two components enables homogeneous Li plating at the Li/SEI interface. Furthermore, the artificial hybrid SEI layer offers high mechanical strength and electrochemical stability, preventing Li dendrite growth during cycling. Lv and coworkers fabricated a dense, stable, and ultrathin protective layer composed of octadecylamine (ODA) and Al-Li alloy (ALA) (represented as ALA-ODA) [99]. When lithium metal is immersed in a mixed solution of $AlCl_3$-ODA-tetrahydrofuran (THF), the -NH_2 group reacts with lithium metal to form Li_3N, resulting in the rapid, directional, and self-limiting assembly of ODA on the lithium anode surface, forming a thin and dense layer that effectively prevents polysulfide ions from entering. Al and Li form an ion-conductive Al-Li alloy on the surface of lithium anode, providing a uniform Li^+ ion transport path for stable lithium deposition. Lee and coworkers proposed a surface adaptive double-layer protective layer (APL) strategy, which reduces the interface energy in contact with the lithium metal anode through an inner layer of high surface free energy polymer [100]. The outer layer of high ductility polymer acting as a shield effectively blocks the penetration of electrolytes and LiPSs. This design notably enhances the cycling stability of Li–S batteries.

4.5 *In situ* Characterization Method for Li–S Batteries

The *in situ* characterization of Li–S batteries primarily examines the physical and chemical transformations within the sulfur cathode, lithium anode, and electrolyte throughout the charge/discharge cycle, to obtain a better insight into the working mechanism, capacity degradation mechanism, and performance improvement strategies of the battery. Here are some main *in situ* characterization methods for batteries.

In summary, the purpose of *in situ* characterization methods in batteries is to disclose the structural changes and electrochemical reaction mechanisms during the charging/discharging process, optimize battery performance, and promote the development and application of related technologies.

4.5.1 *In situ* Spectroscopy Characterization

4.5.1.1 *In situ* Raman Characterization

In situ Raman characterization technology can monitor the real-time fabrication and transformation of polysulfides during the charging/discharging process for Li–S batteries, helping to unravel the mechanisms of these complex chemical reactions. This helps researchers better understand the internal electrochemical mechanism of batteries and improve their performance and stability. Zhang and coworkers provide an overall toolbox for Li–S battery research. When processing *in situ* Raman testing data, we can refer to the Raman shift positions corresponding to different polysulfides proposed in this article [101]. Duan and coworkers utilized *in situ* Raman spectroscopy to detect the inhibition effect of the designed separator modification material toward polysulfides [102]. The author tested the Raman signal of polysulfides on the lithium anode side using a coin cell with a quartz window. Compared with the original separator, the test results demonstrate that the modified separator with reduced graphene oxide (rGO)/sodium lignosulfonate composite significantly suppressed polysulfides shuttling effect. Based on *in situ* Raman testing, the research results indicate that the cathode electrode containing CoP with phosphorus vacancy exhibits more pronounced polysulfide signals and regular polysulfides conversion. As illustrated in Figure 4.14, Sun and coworkers confirmed that introducing phosphorus vacancies in CoP significantly improved sulfur utilization efficiency and polysulfide conversion reaction kinetics [45]. Abruña and coworkers employed *in situ* confocal Raman microscopy technology to study and quantify the kinetics of Li–S redox reactions, offering insights into the mechanisms of polysulfide formation/evolution and sulfur deposition [103]. The author used confocal Raman scanning imaging to examine the transformation of polysulfides in a certain area. The Raman images captured on the surface of the sulfur cathode and inside the electrolyte during the constant discharge and charging process depict the spatial distribution and intensity variations of polysulfides under different discharge/charge states. The research results indicate that the discharge process is a gradual reduction process, but the charging process is a parallel oxidation mechanism. Wan and coworkers detect Raman signal from separator/lithium interface to evaluate the shuttle degree of polysulfides. The results indicated that the Raman spectrum signals of LiPSs are almost inapparent on the usage of modified separator with Bi_2S_3-rGO [104]. Based on *in situ* Raman analysis results, Xu and coworkers demonstrated Co/MoN-modified separators improved the LiPS

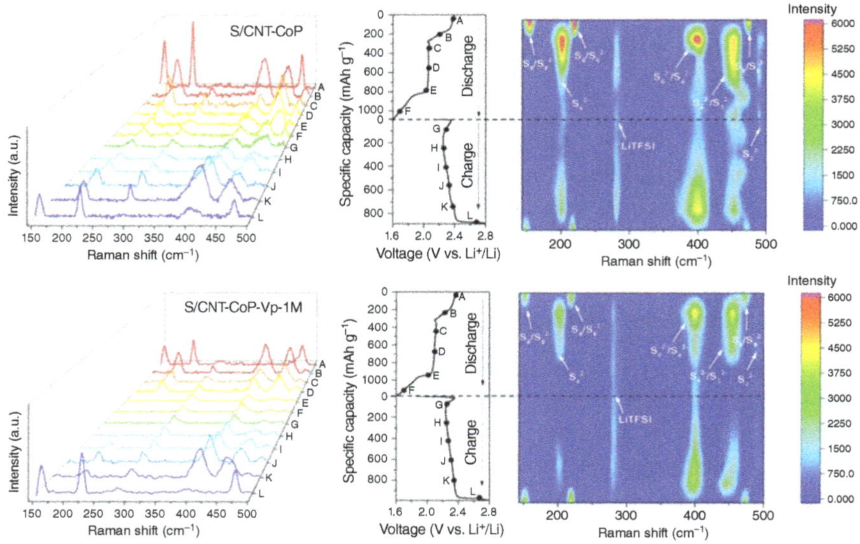

Figure 4.14 *In situ* Raman spectra and the corresponding contour profiles of the S/CNT-CoP cathode and the S/CNT-CoP-Vp-1M cathode. Source: Sun et al. [45]/John Wiley & Sons.

conversion kinetics and reversibility of sulfur species [62]. Zhu and coworkers found the characteristic signal peak of Mn–S and regular polysulfides variations, indicating that MnP–Mn_2P/C have great affinity toward LiPSs and enhance sulfur redox reversibility through the *in situ* Raman test [105].

4.5.1.2 *In situ* Ultraviolet–Visible Spectroscopy (UV–vis) Characterization

UV–vis spectroscopy represents a form of absorption spectroscopy characterized by a distinctive array of absorption bands within the UV–vis region. In 2013, Dominko and his team pioneered the design of a reflection-based setup to capture *in situ* UV–vis absorption spectra from an operational Li–S battery for the first time [106]. This innovative *in situ* UV–vis cell adopts a pouch-type configuration. Yan et al. synthesized a novel type of binder featuring amino functional groups (AFGs), achieved through the polymerization reaction between hexamethylene diisocyanate (HDI) and the polymer polyethylenimine (PEI) [107]. The strong interaction between the binder and the discharged polysulfides was verified through *in situ* UV–vis spectroscopy measurements. Yang and coworkers designed VC@INFeD composite inspired by the integration of iron dextran (INFeD) and ascorbic acid (VC), which displayed a remarkable acceleration in the conversion of LiPSs and effective catalytic role in facilitating the decomposition of Li_2S [108]. Based on *in situ* UV–vis measurements, the author affirmed that INFeD additives expedited the transformation of long-chain LiPSs, while VC promotes the conversion of short-chain LiPSs. Wang and coworkers directly detected Li_2S_2 evolution through *in situ* UV–vis measurements (Figure 4.15) [109]. During operando measurements conducted on the pouch cell level, the evolution of Li_2S_2 was closely monitored. The results reveal

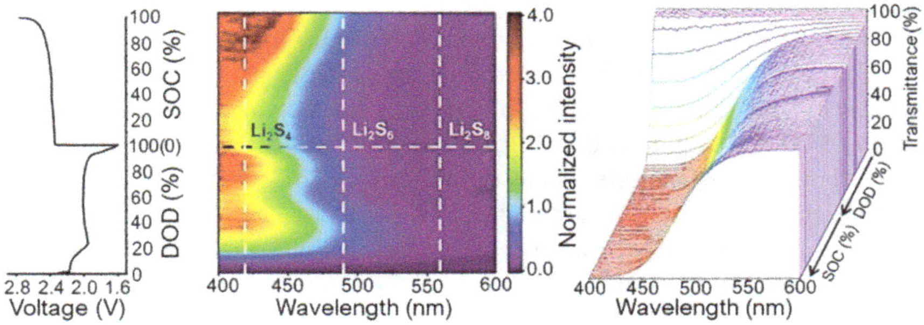

Figure 4.15 *In situ* UV–vis measurement of the fresh Li–S pouch cell. Source: Luo et al. [109]/John Wiley & Sons.

that Li_2S undergoes a concurrent process of sluggish electrochemical reduction and chemical disproportionation during discharging. This dual mechanism results in the dissolution of additional LiPSs and the formation of Li_2S, yet these transformations do not contribute to the overall capacity. Notably, even after the charging process is completed, Li_2S_2 remains present in significant amounts, contrasting with the fully oxidized state of Li_2S. The sluggish Li_2S_2 conversion kinetics are the primary factors constraining sulfur utilization. This ultimately corresponds to the observed limited reversible capacity in Li–S batteries. Qiao and coworkers used *in situ* UV–vis to monitor the concentration changes of LiPSs during the sulfur reduction reaction (SRR) process in real time and established the kinetic trend of SRR [110]. Research dug out the fact that the conversion efficiency and kinetic current of each step of LiPSs increase with the increase in initial LiPSs species concentration. As a result, it is crucial to increase the concentration of LiPSs to achieve rapid kinetics of SRR.

4.5.1.3 *In situ* Fourier Transform Infrared Spectroscopy (FT-IR) Characterization

In situ infrared spectroscopy is a powerful tool to directly monitor the molecular-level changes of electrode materials during charge and discharge processes, exposing reaction mechanisms and interface dynamics, which are crucial for optimizing the performance of Li–S batteries. *In situ* infrared spectroscopy monitors the generation and transformation of the intermediate products in real time, helping to elucidate the reaction mechanism of sulfur cathodes. The anode electrode of Li–S batteries is metallic lithium, which interacts with the electrolyte during charging/discharging to form a SEI. *In situ* infrared spectroscopy can study the formation process and composition of SEI, revealing its impact on battery performance.

Porter and coworkers quantitatively assessed the order and concentration of polysulfides through *in situ* FT-IR spectroscopy [111]. This technique precisely quantifies the dissolved polysulfides as they vary with the state of charge in a Li–S battery. Across multiple charge–discharge cycles, a discernible capacity fade emerges, manifesting alterations in both the order and concentration of polysulfides. Notably, in subsequent cycles, crystalline sulfur fails to be fully recovered during charging, resulting in the depletion of active material as insoluble Li_2S. Then, they developed a method that combines theoretical calculations with *in situ* infrared measurements for detecting polysulfides [112]. Based on FT-IR spectroscopy, experimental analysis of LiPS solutions, synthesized through the reaction of lithium sulfide with sulfur, has corroborated a theoretical prediction. The simulated spectra indicate that as the polysulfide order escalates, a distinctive S–S vibrational mode shifts toward higher wavenumbers. This diagnostic approach is used as a foundational tool for *in situ* spectroscopic investigations of Li–S cells. Yang and coworkers revealed for the first time the influence mechanism of $LiNO_3$ on the dynamic evolution of the SEI layer on the lithium anode, using *in situ* FT-IR and *ab initio* molecular dynamics (AIMD) techniques [113]. The FT-IR and AIMD results directly indicate that $LiNO_3$ accelerates the reduction of LiTFSI to small molecule substances rich in Li_2O on the lithium anode, thereby forming a dense and stable SEI after immersion in the electrolyte containing $LiNO_3$. The SEI's stability prevents LiTFSI and solvent decomposition during lithium deposition and dissolution, enhancing Coulombic efficiency and long-term cycling stability.

4.5.2 *In situ* X-Ray Characterization

4.5.2.1 *In situ* X-Ray Diffraction (XRD) Characterization

Through *in situ* XRD characterization techniques, the phase transition process of electrode materials directly monitored in Li–S batteries during charge/discharge. For example, XRD reveals the phase transition reaction of sulfur cathode from α-S_8 to LiPSs, then to Li_2S, back to LiPSs, and finally to β-S_8. *In situ* XRD demonstrates detailed changes in crystal structure changes of electrode materials, thereby helping researchers to gain deeper insight into the mechanics. Based on *in situ* XRD measurements, using TiS_2 as a host material, Ke and coworkers first presented the concept of dynamic sulfur electrocatalyst [114]. The *in situ* XRD study exhibits that during discharging, Li^+ ions initially intercalate into the TiS_2 layers to form Li_xTiS_2 ($0 < x < 1$), followed by simultaneous intercalation of Li^+ ions into the TiS_2 layers and lithiation reactions with sulfur. Upon discharge completion, $LiTiS_2$ and Li_2S phases emerged. During the charging process, Li^+ ions are removed from the $LiTiS_2$ phase, and after charging, the final product is Li_xTiS_2 ($0 < x < 1$) rather than the original host material TiS_2, indicating that some Li^+ ions remain intercalated within the TiS_2 layers. Therefore, it is Li_xTiS_2 ($0 < x < 1$) that interacts with the redox products of sulfur during the charging/discharging process, rather than the initial host material TiS_2. Chen and coworkers gain insight into the phase transition of $SnSe_2$ through *in situ* XRD testing. At the start of the discharge stage, the electrode material features (001) and (101) planes of $SnSe_2$ at 14.4° and 30.7°, respectively [15]. The battery is further discharged to 2.1 V; compared with standard $SnSe_2$, the characteristic peak of $SnSe_2$ shifts toward a lower angle position, indicating an increase in lattice parameters after Li embedding into Li_1SnSe_2. At the end of continuous discharge to a voltage of 1.7 V, the signal strength of the $SnSe_2$ slowly weakens and disappears, which is due to the formation of the Li_2SnSe_2/Li_3SnSe_2 phase. After charging, the $SnSe_2$ materials achieve high reversibility in the intercalation transformation evolution process.

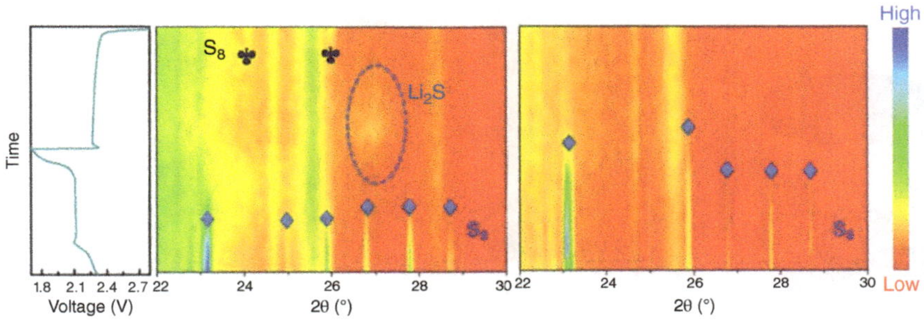

Figure 4.16 *In situ* XRD analysis of HfB_2/rGO/S (left) and rGO/S (right) cathodes. Source: Wang et al. [116]/John Wiley & Sons.

Based on the *in situ* XRD technology, Xiong and coworkers monitor Li_2S evolution during the LiPSs redox process. The assembled batteries exhibit prominent diffraction patterns characteristic of α-S_8 [115]. Upon initiating the discharge process, the intense Li_2S signals appear at the onset of the second discharge platform (~2.1 V). Subsequently, during the charging phase, these Li_2S signals undergo a complete disappearance at approximately 2.4 V, which is concomitant with the formation of the β-S_8 phase. Compared to Zn_8 clusters, the acetylene black electrode exhibits a sluggish and incomplete LiPS conversion process. Xu and coworkers confirmed that HfB_2 promoted the LiPS transformation reversibility through *in situ* XRD measurement (Figure 4.16) [116]. At the initial stage of the discharge process, S_8 diffraction peaks can be clearly observed. As the discharge proceeds, the signal peak of Li_2S emerges. During subsequent discharge and charge cycles, the diffraction peaks of S_8 and Li_2S exhibit reversible changes. In contrast, the lack of the Li_2S signal in the rGO-based electrode indicates that HfB_2 greatly improves LiPS conversion efficiency.

4.5.2.2 *In situ* X-Ray Absorption Spectroscopy (XAS) Characterization

In situ XAS can track the migration and transformation process of polysulfides in batteries, revealing the specific mechanism of the shuttle effect. In addition, it can also analyze the mechanism of catalyst action in batteries, revealing the catalyst–sulfur species interaction. Nazar and coworkers first utilized operando XAS on the S K-edge to monitor the sulfur species [117]. The findings present the first comprehensive evidence of sulfur redox chemistry mechanisms during cycling. These revelations offer vital fundamental insights and perspectives that are pivotal for the development of the Li–S batteries. To elucidate the inhibition mechanism of $LiNO_3$ toward LiPS shuttling, Guo et al. studied its impact on SEI formation using *in situ* XAS [118]. The improved cycling capacity results from the synergistic effect of $LiNO_3$ with intermediate LiPSs, where $LiNO_3$ oxidizes shuttled polysulfides to Li_2SO_3/Li_2SO_4 while reducing to $LiNO_2$, forming a compact, stable layer that mitigates the side reactions between polysulfide and lithium, alleviating the shuttling and enhancing electrochemical property. As shown in Figure 4.17, Zhang and coworkers proposed a shuttle-suppression strategy by incorporating an electrocatalytic layer composed of twinborn bismuth sulfide/bismuth oxide

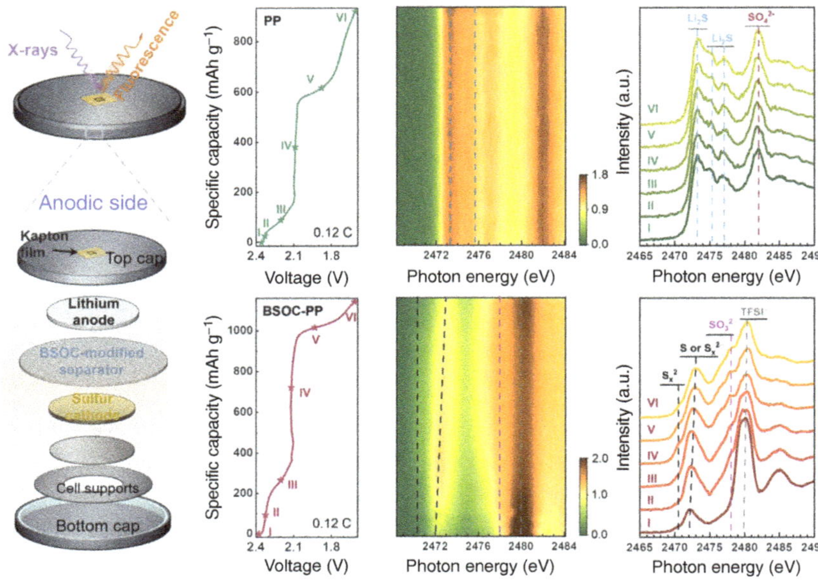

Figure 4.17 *In situ* XAS study at the anodic side of Li–S cells with/without BSOC layer. Source: Jia et al. [119]/John Wiley & Sons.

nanoclusters embedded in a carbon matrix (denoted as BSOC) [119]. Utilizing *in situ* XAS, they traced polysulfide migration in Li–S cells by real-time monitoring sulfur species at cathode–electrolyte and anode–electrolyte interfaces. Their S K-edge XAS findings demonstrated BSOC modification to be dual-functional, both trapping polysulfides and catalyzing sulfur species conversion. Utilizing *in situ* XAS, Abruña et al. conducted a comprehensive investigation on the cobalt single-atom electrocatalyst (Co-SAs/NC) in catalyzing Li–S redox processes [120]. Based on *in situ* XAS, they monitored the real-time sulfur species transformation and cobalt (Co) SA catalytic activity in Li–S batteries. Furthermore, *in situ* XAS analysis of the cobalt (Co) K-edge captured the dynamic alterations in Co catalytic sites, showing the formation of Co-S bonds during electrocatalysis.

4.5.3 *In situ* Electrochemical Impedance Spectroscopy (EIS) Characterization

According to *in situ* EIS analysis of Li-S batteries, it can be observed that the impedance spectrum changes at different charging/discharging states for the battery. Typically, impedance spectra consist of multiple semicircles and diffusion lines in the low-frequency region, corresponding to different electrochemical processes. By combining the distribution of relaxation times (DRT) technology, subtle changes in EIS data can be more accurately analyzed, revealing the impact of different electrode processes on battery performance. The different regions in the DRT curve correspond to different electrode reaction processes, and the characteristic time constants and peak areas of these processes reflect their contributions to the

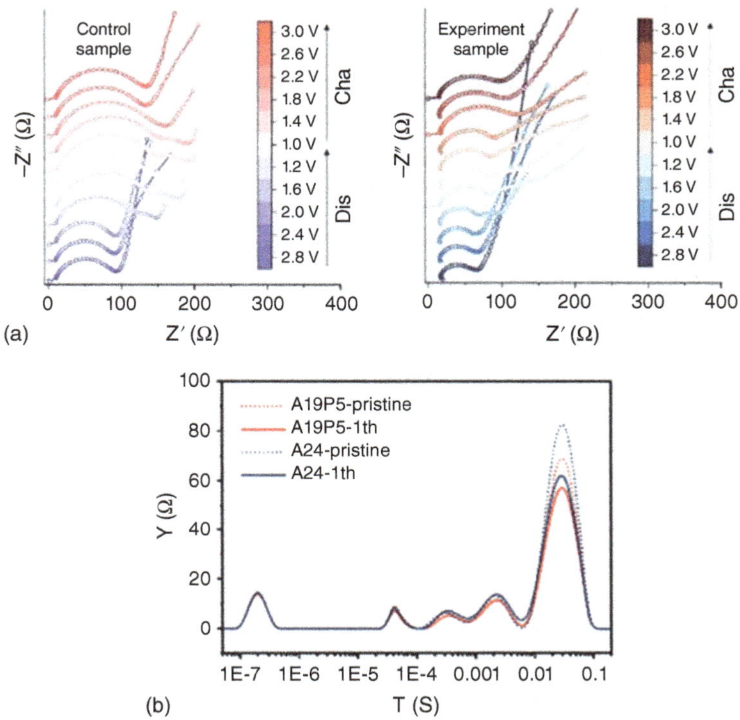

Figure 4.18 (a) *In situ* EIS test for control and experiment sample. Source: Nie et al. [122]/John Wiley & Sons. (b) DRT profiles. Source: Ma et al. [123]/John Wiley & Sons.

total polarization impedance of the battery. Zhou and coworkers employed *in situ* impedance to observe the kinetic characteristics of sulfur redox [121]. Through *in situ* impedance measurement, the author observed a trend of decreasing and then increasing charge transfer impedance during the electrochemical reaction. Compared to the $CoPS_3$/NC, $FeCoPS_3$/NC batteries demonstrate lower charge transfer resistance, indicating enhanced electron/ion transportation. In addition, they executed *in situ* EIS to evaluate the contact ability of the electrode/electrolyte interface for solid-state Li–S batteries (Figure 4.18a) [122]. Yin and coworkers testified the reversible conversion of polysulfides during charging and discharging processes through *in situ* impedance analysis [84]. Furthermore, the author carried out the DRT as a semiquantitative analysis method to identify the durability of the Li–S batteries equipped with Nb–SAs@NC. Miller and coworkers displayed the utilization of the DRT method for a quantitative decomposition of EIS spectra obtained from Li–S batteries. This approach unveiled the individual contributions of eight distinct electrode processes that collectively contributed to the overall cell polarization [124]. According to DRT analysis, Yang and coworkers presented us with clearer methods for distinguishing different electrochemical processes during LiPS conversion (Figure 4.18b) [124].

4.5.4 *In situ* Transmission Electron Microscopy (TEM) Characterization

In situ TEM technology permits real-time observation of interface reactions in Li–S batteries, enhancing our understanding of the interface reaction and reaction mechanism of Li–S batteries. Yushin and colleagues first captured the electrochemical lithiation of sulfur within carbon nanotube pores using *in situ* TEM studies [125]. When subjected to a constant potential, the lithiation process progressed uniformly and linearly along the direction of sulfur encapsulated within the CNTs. This observed lithiation behavior strongly suggests that the interfacial region between Li_2S and sulfur possesses significant electrical conductivity. Zhang and coworkers developed a solid-state Li–S nanobattery for *in situ* TEM analysis [126]. Comprehensive investigations were conducted to deepen our understanding of the electrochemical lithiation process that occurs in carbon-coated sulfur materials. For the first time, the remarkable S/Li_2S phase separation phenomenon has been directly observed in solid-state Li–S batteries. This nanoscale phenomenon not only dramatically decreases the diffusion distances but also creates a dense network of S/Li_2S interfaces. This interface network facilitates Li^+ ion and electron diffusion during the lithiation. Khalil Amine et al. delved into the operational mechanisms and lithiation dynamics of a sulfur cathode featuring a graphene nanocage structure integrated with sulfur (S-GNC) [127]. Their findings revealed that this nanocage architecture not only ensures robust structural integrity, effectively mitigating volumetric fluctuations during the lithiation process, but also fosters an electronically conductive circumstance that optimizes the performance of the encapsulated active sulfur species. Huang and coworkers employed an *in situ* TEM technique, integrated with a microelectromechanical systems (MEMS) heating device, to investigate the precipitation and decomposition processes of Li_2S at elevated temperatures [128]. Their findings reveal that during electrochemical lithiation at room temperature, Li_2S undergoes a transformation from an amorphous–nanocrystalline state to a polycrystalline form. Notably, the precipitation of Li_2S is more thorough at higher temperatures compared to room temperature due to the enhanced Li^+ ion conduction at high temperatures. These observations confirm that Li^+ ion diffusion in the electrochemical decomposition of Li_2S governs its reversibility within solid-state Li–S batteries. Liao et al. utilized *in situ* liquid battery electrochemical TEM technology to directly observe the atomic-scale transformation of LiPSs on the electrode surface (Figure 4.19) [129]. The incorporation of active centers in metal nanoclusters promotes the aggregation and charge storage of LiPSs on the surface, resulting in an instantaneous shift from the LiPS-enriched phase to nonequilibrium Li_2S nanocrystals. This reaction pathway uncovers a new conversion mechanism, distinct from the classical pathway of LiPSs gradually converting to Li_2S_2 and Li_2S in traditional electrochemical reaction processes, contributing to deeper understanding of Li–S batteries.

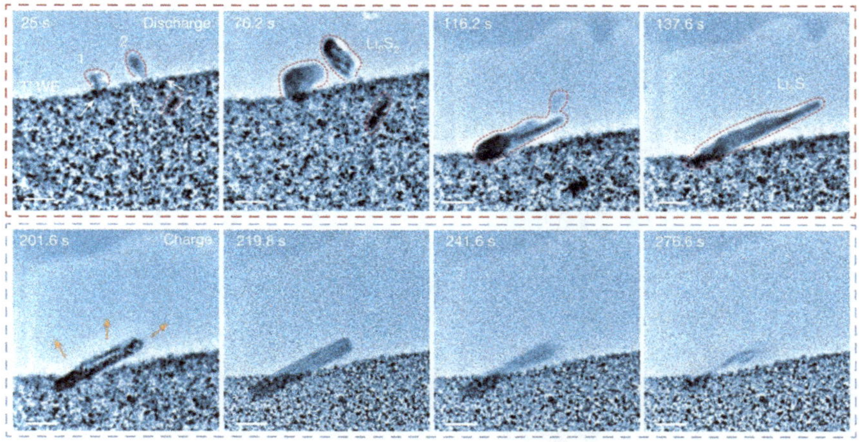

Figure 4.19 *In situ* TEM images of Li_2S deposition and dissolution. Source: Zhou et al. [129]/Springer Nature.

4.5.5 *In situ* Stress Monitoring Characterization

In situ stress detection technology actively monitors the stress state of the internal electrodes of the battery in real time, providing an important means for a deeper knowledge of the reaction mechanism of Li–S batteries and optimizing battery design. During the charging/discharging, the cathode material undergoes complex chemical and physical changes, resulting in volume expansion and contraction, thereby generating stress. By monitoring the evolution process of cathode electrode stress in real time, the relationship between these changes and battery performance can be revealed. This helps optimize battery design and improve battery performance. Huang and coworkers monitored the chemical mechanical stress evolution inside the sulfur-based cathode under three different mechanisms, and the obtained signal changes were consistent with the structural evolution and volume changes of the electrode [130]. By comparing the stress changes of the sulfur cathode under different electrochemical conditions, it was found that the stress changes of the sulfur cathode electrode were the largest under the solid–solid conversion mechanism during the charging and discharging process. The research results indicate that the severity of stress changes is related not only to the degree of volume change but also to the properties of the material. Recently, Zhou and coworkers employed an *in situ*-partitioned stress monitoring system to probe the stress distribution in Li–S soft pack batteries (Figure 4.20) [131], to decouple local Li plating/stripping information, and to obtain a more profound insight into the coupling relationship between cathode and anode electrodes at the soft pack battery level. The results indicate that the sponge-like structure cathode design can not only buffer volume change but also adjust the local current density to optimize the Li^+ ion plating.

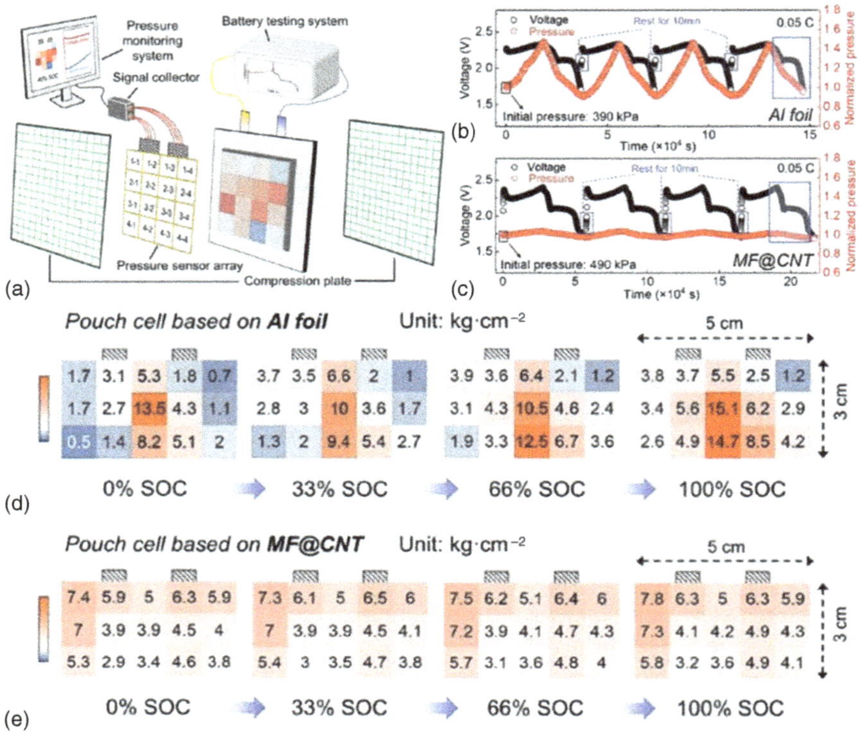

Figure 4.20 (a) *In situ* stress monitoring device for pouch cells. GDC curves and the pressure change for (b) the S-Al foil and (c) the S-MF@CNT cell. The local stress distribution of (d) S-Al foil and (e) S-MF@CNT at various states of charge. Source: Gao et al. [131]/John Wiley & Sons.

4.6 Practical Application Research of Li–S Batteries

The practical research of Li–S batteries is an indispensable part after basic research, directly influencing their successful commercialization. During the fundamental research stage, researchers have proposed multiple solutions to the major challenges faced by Li–S batteries, such as poor conductivity of the sulfur cathode, significant volume changes during charging and discharging, shuttle effects of polysulfides, and instability of the lithium anode. These solutions have been initially validated in coin cell batteries, successfully improving the cycling life and Coulombic efficiency of Li–S batteries. However, most of these performance improvements were achieved under ideal experimental conditions, such as low sulfur surface loading, low sulfur mass fraction, higher electrolyte-to-sulfur ratio, low discharge rate, and excessive use of the lithium metal anode. Under such conditions, the potential problems that Li–S batteries may encounter in practical applications have not been fully exposed. Therefore, the guiding significance of these research results for enhancing the practicality of Li–S batteries is relatively restricted. Therefore, it is particularly important to conduct in-depth practical research on Li–S batteries. By simulating conditions that are closer to real-world application scenarios, we can expose and mitigate new problems associated with Li–S batteries, which may encounter on the commercialization path, to truly promote the progress of Li–S battery technology and facilitate its transition from the laboratory to the market.

4.6.1 Pouch-Type Li–S Battery

Pouch-type Li–S batteries are increasingly recognized for their flexible design according to requirements and can be assembled on a small scale in the laboratory to meet the practical research needs of batteries. As the most representative high specific energy secondary battery, pouch-type cells are the most promising application prospects in the field of consumer electronics. Currently, the practical devices for Li–S batteries are mostly pouch cells.

Sion Power company integrated Li–S batteries into drones in 2010, which can fly continuously for 14 days, providing guidance for the development of Li–S batteries in special fields. OXIS company reported Li–S polymer batteries with an energy density exceeding 300 Wh kg^{-1}, making outstanding contributions to solving key challenges in Li–S batteries. Researchers from the Dalian Institute of Chemical Physics, Chinese Academy of Sciences (CAS), led by Chen and his team, have successfully developed a Li–S battery with a rated capacity of 15 Ah, and have formed a small batch preparation capacity. The test results of the third party (Institute of Physics, Chinese Academy of Sciences) demonstrate that the specific energy of the battery is more than 430 Wh kg^{-1}, which exceeds the technical indicators of the Li–S battery developed by Sion Power company. Afterward, they developed high specific energy battery, which was tested via a third party according to the safety requirements of the national military standards. From the past 520 and 570 Wh kg^{-1} to the current 609 Wh kg^{-1}, it again breaks the leading position of secondary batteries in terms of specific energy in the same field. The battery endures excellent environmental adaptability, with a discharge-specific energy of 400 Wh kg^{-1} at −20 °C, and can operate in extremely cold environments of −60 °C, demonstrating significantly better low-temperature performance than lithium-ion batteries. Dr. Mahdokht Shaibani from Monash University led an international research team to develop a high-capacity Li–S battery. This kind of high-energy battery delivers continuous power to smartphones for five days or allows electric vehicles to travel over 1000 km without needing to be recharged. Wang and coworkers from the Institute of Chemical Defense of the Academy of Military Sciences fabricated a defect-coordinated SA indium electrocatalyst with an optimized electronic structure, integrated with carbon nanotubes, to enhance multiphase redox kinetics in Li–S batteries [132]. Indium is uniformly integrated onto carbon nanotubes (SAIn@CNT) through π–π self-assembly, overcoming the challenges of aggregation effects and atomic size mismatch during pyrolysis. Therefore, the pouch-type batteries equipped with SAIn@CNT achieved an energy density of 495 Wh kg^{-1} under high sulfur loading and low electrolyte conditions. As illustrated in Figure 4.21, Suo and coworkers from the Institute of Physics, Chinese

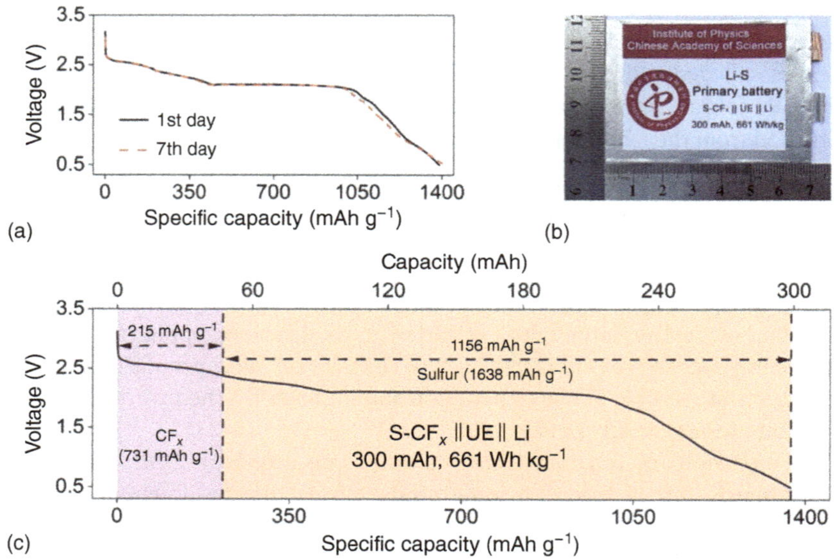

Figure 4.21 (a) Discharge curves of pouch-type cell for 1st day and 7th day. (b) Digital photo and (c) the discharge curve of the 661 Wh kg^{-1} pouch cell. Source: With permission of [133]/American Chemical Society.

Academy of Sciences, have successfully designed and experimentally verified an ultralight electrolyte with a protective encapsulation solvation structure, as well as an innovative solution using fluorinated graphite composite sulfur-based cathodes [133]. Based on the above innovation, the pouch-type cell achieved a high energy density of 661 Wh kg^{-1}. Guangdong Mengshi New Energy Technology Company has signed a “Cooperation Research Agreement” with the Singapore Institute of Materials Research and Engineering to jointly develop Li–S battery technology. This cross-border technology cooperation not only timely delivers cutting-edge scientific theories to Mengshi Technology but also provides strong technical support and development direction for scientific research and innovation of the company. Huang et al. designed a Li–S pouch-type cell with sulfurized polyacrylonitrile (S-PAN) as the cathode and graphite (Gr) as the anode [134]. This pouch cell design can avoid shuttle effects, dendrite, and side reaction problems caused by using lithium metal as the anode, ultimately achieving Ah-level soft pack batteries with 1000 stable cycles and 99% capacity retention.

At present, pouch-type Li–S cells are still in the initial phase of research. To attain higher energy density, ongoing optimization of battery structure, electrode composition, and electrolyte–sulfur ratio is essential. In addition, the longevity, Coulombic efficiency, and rate performance of the battery also need to be further improved. By further improving the above aspects, pouch-type Li–S batteries with high specific energy and long cycle stability will be designed.

4.6.2 High- and Low-Temperature Electrochemical Performance

The performance of Li–S batteries is highly sensitive to temperature. Through research on their high- and low-temperature performance, understanding deepens for battery working characteristics, which provides a theoretical basis for optimizing battery design. In high-temperature environments, the chemical reaction rate within the battery will accelerate, which may lead to safety issues such as overheating, short circuits, and even explosions. Studying the high-temperature performance of Li–S batteries can help identify potential safety hazards and take corresponding protective measures. Different application scenarios in various fields demand various battery performance levels, with some requiring operation under extreme temperature conditions. Studying the high- and low-temperature performance of Li–S batteries can broaden their application range and meet the needs of more diverse fields.

Dong and coworkers reported a composite material using graphene-supported BN nanosheets, exhibiting a strong adsorption effect on polysulfides [135]. While alleviating the shuttling effect, it enables Li–S batteries to deliver excellent electrochemical performance over a −40 to −70 °C temperature range. Sun and coworkers proposed Fe_3C electrocatalyst for improving the electrochemical capacity of batteries [136]. The Fe_3C electrocatalyst facilitates the stable functioning of Li–S batteries with exceptional areal capacities exceeding 6.0 mAh cm^{-2} across a broad temperature, ranging from −10 to 40 °C. Jin et al. designed a multibranched vanadium nitride (MB-VN) electrocatalytic material for separator modification of Li–S batteries [137]. The Li–S batteries equipped with MB-VN-modified separators exhibit stable performance from −20 to 60 °C. The electrochemical enhancement can be attributed primarily to the decrease in electrolyte viscosity, a reduction in charge transfer resistance, and an overall improvement in reaction kinetics. Wang and coworkers constructed Ni nanoparticles embedded within carbon nanotubes (Ni@C/CNT) [138]. The Li–S cells modified with Ni@C/CNT demonstrate remarkable stability during cycling across an extensive temperature range from −50 to 70 °C. Li and coworkers successfully synthesized a series of lithium thiophosphate (mP_2S_5–nLi_2S_x) complexes, showcasing their potential as promising catholytes for Li–S batteries [139]. Through meticulous optimization efforts encompassing sulfur chain length adjustment, precise control of complexation stoichiometry, and solvent selection, they have crafted complexes that exhibit remarkable solubility and desirable electrochemical characteristics. Furthermore, the all-liquid nature of the electrochemical conversion process within these complexes facilitates exceptional battery performance at low temperatures, delivering outstanding capacities of over 400 mAh g^{-1} at −40 °C and above 200 mAh g^{-1} even at −60 °C. The phenomenon known as the shuttle effect tends to intensify as temperatures rise, while the reaction kinetics undergo a significant decline as temperatures drop. Regarding the above problems, Wu and coworkers introduced porous bismuth vanadate ($BiVO_4$) spheres as sulfur host [140]. The cell demonstrates an impressive discharge capacity exceeding 400 mAh g^{-1}, nearly 100 cycles even at a low temperature of −40 °C. Upon returning the temperature to the standard 25 °C, the capacity remains at approximately 1000 mAh g^{-1}, equal to electrochemical performance at room temperature. This underscores the exceptional electrochemical stability and reversibility of the cell equipped with $BiVO_4$ spheres, capable of maintaining superior performance across varying temperature conditions during the charging/discharging process.

4.7 Conclusion and Prospect

This chapter offers an in-depth overview of Li–S batteries and summarizes research efforts to enhance their electrochemical performance and cycle life. The discussion emphasizes on the design of sulfur host material, separator modification, and lithium anode protection. Next, the application of *in situ* characterization methods and the practicality exploration of Li–S batteries were addressed. At present, great progress has been made in the commercialization of Li–S batteries, several challenges remain that hinder their development including poor sulfur conductivity, polysulfide shuttling, lithium dendrite formation, and side reactions on the anode hinder Li–S battery advancement.

In future research, more diverse testing methods and *in situ* characterization technologies are essential to further study and analyze the mechanism of polysulfide conversion reaction during electrocatalysis, optimize battery composition and structure in a targeted manner, and continuously improve the electrochemical property of batteries. In addition, the combination of machine learning and DFT calculations has become a trend in the research of Li–S batteries. They not only improve the screening efficiency of catalysts and materials but also contribute to a better understanding of the internal reaction mechanism of batteries, providing strong support for the rapid development of the Li–S battery technology.

References

1 Seh, Z.W., Sun, Y., Zhang, Q., and Cui, Y. (2016). Designing high-energy lithium–sulfur batteries. *Chemical Society Reviews* 45 (20): 5605–5634.

2 Liu, Y., Liu, S., Li, G., and Gao, X. (2020). Strategy of enhancing the volumetric energy density for lithium–sulfur batteries. *Advanced Materials* 33 (8): 2003955.

3 Herbert, D. and Ulam, J. (1962). Electric dry cells and storage batteries. *United States Patent Office*, 3043896.

4 Yamin, H., Penciner, J., Gorenshtain, A. et al. (1985). The electrochemical behavior of polysulfides in tetrahydrofuran. *Journal of Power Sources* 14: 129–134.

5 Wang, J., Yang, J., Xie, J. et al. (2002). Sulfur–carbon nano-composite as cathode for rechargeable lithium battery based on gel electrolyte. *Electrochemistry Communications* 4: 499–502.

6 Wang, J., Yang, J., Xie, J., and Xu, N. (2002). A novel conductive polymer-sulfur composite cathode material for rechargeable lithium batteries. *Advanced Materials* 14 (13–14): 963–965.

7 Aurbach, D., Pollak, E., Elazari, R. et al. (2009). On the surface chemical aspects of very high energy density, rechargeable Li–sulfur batteries. *Journal of the Electrochemical Society* 156 (8): A694–A702.

8 Ji, X., Lee, K.T., and Nazar, L.F. (2009). A highly ordered nanostructured carbon–sulphur cathode for lithium–sulphur batteries. *Nature Materials* 8 (6): 500–506.

9 Song, Y., Wei, X., Song, L. et al. (2024). Advanced detections for deciphering the catalytic reaction behaviors of lithium–sulfur batteries. *Energy Storage Materials* 70: 1034914.

10 Manthiram, A., Fu, Y., and Su, Y. (2013). Challenges and prospects of lithium–sulfur batteries. *Accounts of Chemical Research* 46 (5): 1125–1134.

11 Chung, S.H. and Manthiram, A. (2019). Current status and future prospects of metal–sulfur batteries. *Advanced Materials* 31 (27): 1901125.

12 Li, M., Chen, H., Guo, C. et al. (2023). Interfacial engineering on cathode and anode with iminated polyaniline@rGO-CNTs for robust and high-rate full lithium–sulfur batteries. *Advanced Energy Materials* 13 (25): 2300646.

13 Zhao, Z., Yi, Z., Li, H. et al. (2021). Synergetic effect of spatially separated dual co-catalyst for accelerating multiple conversion reaction in advanced lithium sulfur batteries. *Nano Energy* 81: 105621.

14 Yao, W., Zheng, W., Xu, J. et al. (2021). ZnS-SnS@NC heterostructure as robust lithiophilicity and sulfiphilicity mediator toward high-rate and long-life lithium–sulfur batteries. *ACS Nano* 15 (4): 7114–7130.

15 Yao, W., Xu, J., Cao, Y. et al. (2022). Dynamic intercalation–conversion site supported ultrathin 2D mesoporous $SnO_2/SnSe_2$ hybrid as bifunctional polysulfide immobilizer and lithium regulator for lithium–sulfur chemistry. *ACS Nano* 16 (7): 10783–10797.

16 Yao, W., Xu, J., Ma, L. et al. (2023). Recent progress for concurrent realization of shuttle-inhibition and dendrite-free lithium–sulfur batteries. *Advanced Materials* 35 (32): 2212116.

17 Feng, T., Zhao, T., Zhang, N. et al. (2022). 2D amorphous Mo-doped CoB for bidirectional sulfur catalysis in lithium sulfur batteries. *Advanced Functional Materials* 32 (30): 2202766.

18 Wang, T., He, J., Zhu, Z. et al. (2023). Heterostructures regulating lithium polysulfides for advanced lithium–sulfur batteries. *Advanced Materials* 35 (47): 2303520.

19 Zhou, C., Li, M., Hu, N. et al. (2022). Single-atom-regulated heterostructure of binary nanosheets to enable dendrite-free and kinetics-enhanced Li–S batteries. *Advanced Functional Materials* 32 (33): 2204635.

20 Pu, J., Wang, T., Tan, Y. et al. (2023). Effect of heterostructure-modified separator in lithium–sulfur batteries. *Small* 19 (42): 2303266.

21 Liu, T., Shi, Z., Li, H. et al. (2021). Low-density fluorinated silane solvent enhancing deep cycle lithium–sulfur batteries' lifetime. *Advanced Materials* 33 (38): 2102034.

22 Yao, W., Liao, K., Lai, T. et al. (2024). Rechargeable metal-sulfur batteries: key materials to mechanisms. *Chemical Reviews* 124 (8): 4935–5118.

23 Qi, F., Sun, Z., Fan, X. et al. (2021). Tunable interaction between metal-organic frameworks and electroactive components in lithium–sulfur batteries: status and perspectives. *Advanced Energy Materials* 11 (20): 2100387.

24 Wen, Y., Shen, Z., Hui, J. et al. (2023). Co/CoSe junctions enable efficient and durable electrocatalytic conversion of polysulfides for high-performance Li–S batteries. *Advanced Energy Materials* 13 (20): 2204345.

25 Huang, C., Yu, J., Zhang, C.Y. et al. (2024). Electronic spin alignment within homologous $NiS_2/NiSe_2$ heterostructures to promote sulfur redox kinetics in lithium–sulfur batteries. *Advanced Materials* 36 (25): 2400810.

26 Li, J., Wang, Z., Shi, K. et al. (2023). Nanoreactors encapsulating built-in electric field as a "bridge" for Li–S batteries: directional migration and rapid conversion of polysulfides. *Advanced Energy Materials* 14: 2303546.

27 Wang, B., Wang, L., Mamoor, M. et al. (2024). Manipulating atomic-coupling in dual-cavity boride nanoreactor to achieve hierarchical catalytic engineering for sulfur cathode. *Angewandte Chemie International Edition* 63: e202406065.

28 Zhou, C., Li, X., Jiang, H. et al. (2021). Pulverizing Fe_2O_3 nanoparticles for developing Fe_3C/N-codoped carbon nanoboxes with multiple polysulfide anchoring and converting activity in Li–S batteries. *Advanced Functional Materials* 31 (18): 2011249.

29 Li, W., Chen, K., Xu, Q. et al. (2021). Mo_2C/C hierarchical double-shelled hollow spheres as sulfur host for advanced Li–S batteries. *Angewandte Chemie International Edition* 60 (39): 21512–21520.

30 Sun, X., Qiu, Y., Jiang, B. et al. (2023). Isolated Fe-Co heteronuclear diatomic sites as efficient bifunctional catalysts for high-performance lithium–sulfur batteries. *Nature Communications* 14 (1): 291.

31 Lan, Y., Wang, Y., Wang, Y. et al. (2024). Chip-inspired design of high-performance lithium–sulfur batteries by integrating monodisperse sulfur nanoreactors on graphene. *ACS Nano* 18 (24): 15638–15650.

32 Zhang, H., Zhang, M., Liu, R. et al. (2024). Fe_3O_4-doped mesoporous carbon cathode with a plumber's nightmare structure for high-performance Li-S batteries. *Nature Communications* 15 (1): 5451.

33 Zhang, B., Wang, L., Wang, B. et al. (2022). Petroleum coke derived porous carbon/NiCoP with efficient reviving catalytic and adsorptive activity as sulfur host for high performance lithium–sulfur batteries. *Nano Research* 15 (5): 4058–4067.

34 Yang, Q., Cai, J., Li, G. et al. (2024). Chlorine bridge bond-enabled binuclear copper complex for electrocatalyzing lithium–sulfur reactions. *Nature Communications* 15 (1): 3231.

35 Zhong, Y., Wang, Q., Bak, S.-M. et al. (2023). Identification and catalysis of the potential-limiting step in lithium–sulfur batteries. *Journal of the American Chemical Society* 145 (13): 7390–7396.

36 Shen, Z., Jin, X., Tian, J. et al. (2022). Cation-doped ZnS catalysts for polysulfide conversion in lithium–sulfur batteries. *Nature Catalysis* 5 (6): 555–563.

37 Zhang, Y., Tao, L., Xie, C. et al. (2020). Defect engineering on electrode materials for rechargeable batteries. *Advanced Materials* 32 (7): 1905923.

38 Shi, Z., Li, M., Sun, J., and Chen, Z. (2021). Defect engineering for expediting Li–S chemistry: strategies, mechanisms, and perspectives. *Advanced Energy Materials* 11 (23): 2100332.

39 Luo, D., Zhang, Z., Li, G. et al. (2020). Revealing the rapid Electrocatalytic behavior of ultrafine amorphous defective Nb_2O_{5-x} nanocluster toward superior Li–S performance. *ACS Nano* 14 (4): 4849–4860.

40 Tian, Y., Li, G., Zhang, Y. et al. (2019). Low-bandgap Se-deficient antimony selenide as a multifunctional polysulfide barrier toward high-performance lithium–sulfur batteries. *Advanced Materials* 32 (4): 1904876.

41 Wang, X., Zhang, X., Zhao, Y. et al. (2023). Accelerated multi-step sulfur redox reactions in lithium–sulfur batteries enabled by dual defects in metal-organic framework-based catalysts. *Angewandte Chemie International Edition* 62 (42): e202306901.

42 Wang, J., Qiu, W., Li, G. et al. (2022). Coordinatively deficient single-atom Fe-N-C electrocatalyst with optimized electronic structure for high-performance lithium–sulfur batteries. *Energy Storage Materials* 46: 269–277.

43 Zhang, H., Yang, L., Zhang, P. et al. (2021). MXene-derived Ti_nO_{2n-1} quantum dots distributed on porous carbon Nanosheets for stable and Long-life Li–S batteries: enhanced polysulfide mediation via defect engineering. *Advanced Materials* 33 (21): 2008447.

44 Shi, Z., Sun, Z., Cai, J. et al. (2021). Manipulating electrocatalytic Li_2S redox via selective dual-defect engineering for Li–S batteries. *Advanced Materials* 33 (43): 2103050.

45 Sun, R., Bai, Y., Bai, Z. et al. (2022). Phosphorus vacancies as effective polysulfide promoter for high-energy-density lithium–sulfur batteries. *Advanced Energy Materials* 12 (12): 2102739.

46 Hou, W., Feng, P., Guo, X. et al. (2022). Catalytic mechanism of oxygen vacancies in perovskite oxides for lithium–sulfur batteries. *Advanced Materials* 34 (26): 2202222.

47 Guo, Y., Li, J., Yuan, G. et al. (2023). Elucidating the volcanic-type catalytic behavior in lithium–sulfur batteries via defect engineering. *ACS Nano* 17 (18): 18253–18265.

48 Yu, J., Huang, C., Usoltsev, O. et al. (2024). Promoting polysulfide redox reactions through electronic spin manipulation. *ACS Nano* 18 (29): 19268–19282.

49 Wu, X., Xie, R., Cai, D. et al. (2024). Engineering defect-rich bimetallic telluride with dense heterointerfaces for high-performance lithium–sulfur batteries. *Advanced Functional Materials* 34 (26): 2315012.

50 Lee, H., Nam, H., and Moon, J.H. (2024). Seamless integration of nanoscale crystalline-amorphous MoO_3 domains for high-performance lithium–sulfur batteries. *Energy Storage Materials* 70: 103551.

51 Liu, K., Feng, J., Guo, J. et al. (2023). (1–10) facet-dominated TiB_2 nanosheets with high exposure of dual-atom-sites for enhanced polysulfide conversion in Li-S batteries. *Advanced Functional Materials* 34 (17): 2314657.

52 Wang, B., Wang, L., Ding, D. et al. (2022). Zinc-assisted cobalt ditelluride polyhedra inducing lattice strain to endow efficient adsorption-catalysis for high-energy lithium–sulfur batteries. *Advanced Materials* 34 (50): 2204403.

53 Wang, J., Li, G., Luo, D. et al. (2021). Amorphous–crystalline-heterostructured niobium oxide as two-in-one host matrix for high-performance lithium–sulfur batteries. *Journal of Materials Chemistry A* 9 (18): 11160–11167.

54 Li, R., Rao, D., Zhou, J. et al. (2021). Amorphization-induced surface electronic states modulation of cobaltous oxide nanosheets for lithium–sulfur batteries. *Nature Communications* 12 (1): 3102.

55 Liang, X., Cai, J., Qiu, S. et al. (2022). Bidirectional catalyst design for lithium–sulfur batteries: phase regulation cooperates with N-doping. *Journal of Materials Chemistry A* 10 (44): 23780–23789.

56 Wang, C., Hao, Y., Xiao, Q. et al. (2024). Exposed high-order crystal planes of sulfur to catalyze polysulfide conversion process by regulating crystal planes of stannous sulfide. *Materials Today Chemistry* 35: 101839.

57 Wang, J., Xu, Y., Zhuang, Y. et al. (2024). Lattice strain and charge localization dual regulation of phosphorus-doped $CoSe_2$/MXene catalysts enable kinetics-enhanced and dendrite-free lithium–sulfur batteries. *Advanced Energy Materials* 14: 2401630.

58 Li, H., Chen, C., Yan, Y. et al. (2021). Utilizing the built-in electric field of *p–n* junctions to spatially propel the stepwise polysulfide conversion in lithium–sulfur batteries. *Advanced Materials* 33 (51): 2105067.

59 Lu, D., Wang, X., Hu, Y. et al. (2023). Expediting stepwise sulfur conversion via spontaneous built-in electric field and binary sulfiphilic effect of conductive NbB_2-MXene heterostructure in lithium–sulfur batteries. *Advanced Functional Materials* 33 (15): 2212689.

60 Wang, X., Chen, L., Yu, Y. et al. (2024). Tuning p-band centers and interfacial built-in electric field of heterostructure catalysts to expedite bidirectional sulfur redox for high-performance Li–S batteries. *Advanced Functional Materials* 34: 2406290.

61 Zhou, T., Lv, W., Li, J. et al. (2017). Twinborn TiO_2–TiN heterostructures enabling smooth trapping–diffusion–conversion of polysulfides towards ultralong life lithium–sulfur batteries. *Energy & Environmental Science* 10 (7): 1694–1703.

62 Kong, Y., Wang, L., Mamoor, M. et al. (2023). Co/MoN invigorated bilateral kinetics modulation for advanced lithium–sulfur batteries. *Advanced Materials* 36 (13): 2310143.

63 Liu, W., Lei, M., Zhou, X., and Li, C. (2023). Heterojunction interlocked catalysis-conduction network in monolithic porous-pipe scaffold for endurable Li–S batteries. *Energy Storage Materials* 58: 74–84.

64 Wang, J., Zhou, L., Guo, D. et al. (2023). Flower-like NiS_2/WS_2 heterojunction as polysulfide/sulfide bidirectional catalytic layer for high-performance lithium–Sulfur batteries. *Small* 19 (11): 2206926.

65 Zhao, Z., Pan, Y., Yi, S. et al. (2023). Enhanced electron delocalization within coherent nano-heterocrystal ensembles for optimizing polysulfide conversion in high-energy-density Li-S batteries. *Advanced Materials* 36 (13): 2310052.

66 Wang, Z., Jiang, H., Wei, C. et al. (2024). Ultrasmall CoFe bimetallic alloy anchored on fluoride-free MXene by one-pot etching strategy for the barrier-adsorption-catalyst functions of polysulfides in lithium–sulfur batteries. *Advanced Functional Materials* 34 (23): 2315178.

67 Han, F., Wang, Z., Jin, Q. et al. (2024). High-entropy alloy electrocatalysts bidirectionally promote lithium polysulfide conversions for long-cycle-life lithium–sulfur batteries. *ACS Nano* 18 (23): 15167–15176.

68 Wang, Z., Zhang, B., Liu, S. et al. (2022). Nickel–platinum alloy nanocrystallites with high-index facets as highly effective core catalyst for lithium–sulfur batteries. *Advanced Functional Materials* 32 (27): 2200893.

69 Jiang, Y., Liu, S., Gao, X., and Li, G. (2023). Morphology control of Li_2S deposition via geometrical effect of cobalt-edged nickel alloy to improve performance of lithium–sulfur batteries. *Advanced Functional Materials* 33 (45): 2304965.

70 Zuo, Y., Jiao, X., Huang, Z. et al. (2024). Surface electron reconstruction of catalyst through alloying strategy for accelerating sulfur conversion in lithium–sulfur batteries. *Advanced Functional Materials* 34: 2405853.

71 Theibault, M.J., McCormick, C.R., Lang, S. et al. (2023). High entropy sulfide nanoparticles as lithium polysulfide redox catalysts. *ACS Nano* 17 (18): 18402–18410.

72 Xu, Y., Yuan, W., Geng, C. et al. (2024). High-entropy catalysis accelerating stepwise sulfur redox reactions for lithium–sulfur batteries. *Advanced Science* 11 (31): 2402497.

73 Li, Z., Wang, J., Yuan, H. et al. (2024). Recent progress and challenge in metal–organic frameworks for lithium–sulfur battery separators. *Advanced Functional Materials* 2405890.

74 Xie, L., Xiao, Y., Zeng, Q. et al. (2024). Balanced mass transfer and active sites density in hierarchical porous catalytic metal–organic framework for enhancing redox reaction in lithium–sulfur batteries. *ACS Nano* 18 (20): 12820–12829.

75 Lu, H., Zeng, Q., Xu, L. et al. (2024). Multimodal engineering of catalytic interfaces confers multi-site metal-organic framework for internal preconcentration and accelerating redox kinetics in lithium–sulfur batteries. *Angewandte Chemie International Edition* 63: e202318859.

76 Wang, J., Zhang, X., Wang, X. et al. (2024). Activation of MOF catalysts with low steric hindrance via undercoordination chemistry for efficient polysulfide conversion in lithium–sulfur battery. *Advanced Energy Materials* 2402072.

77 Ni, L., Yang, G., Liu, Y. et al. (2021). Self-assembled polyoxometalate nanodots as bidirectional cluster catalysts for polysulfide/sulfide redox conversion in lithium–sulfur batteries. *ACS Nano* 15 (7): 12222–12236.

78 Ni, L., Gu, J., Jiang, X. et al. (2023). Polyoxometalate-cyclodextrin-based cluster-organic supramolecular framework for polysulfide conversion and guest–host recognition in lithium–sulfur batteries. *Angewandte Chemie International Edition* 62 (36): e202306528.

79 Li, Y., Sun, K., Fu, Y. et al. (2024). "Bowling collision effect" of $CoMo_6$ polyoxometalate units enables wide temperature range from −20 to 60 C and dendrite mitigation Li-S batteries. *Advanced Materials* 2406343.

80 Lei, J., Fan, X., Liu, T. et al. (2022). Single-dispersed polyoxometalate clusters embedded on multilayer graphene as a bifunctional electrocatalyst for efficient Li-S batteries. *Nature Communications* 13 (1): 202.

81 Du, Z., Chen, X., Hu, W. et al. (2019). Cobalt in nitrogen-doped graphene as single-atom catalyst for high-sulfur content lithium–sulfur batteries. *Journal of the American Chemical Society* 141 (9): 3977–3985.

82 Dong, C., Ma, C., Zhou, C. et al. (2024). Engineering d-p orbital hybridization with P, S Co-coordination asymmetric configuration of single atoms toward high-rate and long-cycling lithium–sulfur battery. *Advanced Materials* 36: 2407070.

83 Li, Y., Zeng, Y., Chen, Y. et al. (2022). Mesoporous N-rich carbon with single-Ni atoms as a multifunctional Sulfur host for Li-S batteries. *Angewandte Chemie International Edition* 61 (47): e202212680.

84 Zhang, Y., Kang, C., Zhao, W. et al. (2023). d-p hybridization-induced "trapping–coupling–conversion" enables high-efficiency Nb single-atom catalysis for Li–S batteries. *Journal of the American Chemical Society* 145 (3): 1728–1739.

85 Zhou, R., Ren, Y., Li, W. et al. (2024). Rare earth single-atom catalysis for high-performance Li–S full battery with ultrahigh capacity. *Angewandte Chemie International Edition* 63 (31): e202405417.

86 Xiao, R., Yu, T., Yang, S. et al. (2022). Electronic structure adjustment of lithium sulfide by a single-atom copper catalyst toward high-rate lithium–sulfur batteries. *Energy Storage Materials* 51: 890–899.

87 Wei, J.Y., Zhang, X., Hou, L. et al. (2020). Shielding polysulfide intermediates by an organosulfur-containing solid electrolyte interphase on the lithium anode in lithium–sulfur batteries. *Advanced Materials* 32 (37): 2003012.

88 Su, L., Yao, N., Li, Z. et al. (2024). Improving rate performance of encapsulating lithium-polysulfide electrolytes for practical lithium–sulfur batteries. *Angewandte Chemie International Edition* 63 (10): e202318785.

89 Lin, Z., Liu, Z., Fu, W. et al. (2012). Phosphorous pentasulfide as a novel additive for high-performance lithium–sulfur batteries. *Advanced Functional Materials* 23 (8): 1064–1069.

90 Zhang, X., Cheng, X., Chen, X. et al. (2017). Fluoroethylene carbonate additives to render uniform Li deposits in lithium metal batteries. *Advanced Functional Materials* 27 (10): 1605989.

91 Zheng, J., Ji, G., Fan, X. et al. (2019). High-fluorinated electrolytes for Li–S batteries. *Advanced Energy Materials* 9 (16): 1803774.

92 Jiang, Z., Zeng, Z., Hu, W. et al. (2021). Diluted high concentration electrolyte with dual effects for practical lithium–sulfur batteries. *Energy Storage Materials* 36: 333–340.

93 Lian, J., Guo, W., and Fu, Y. (2021). Isomeric organodithiol additives for improving interfacial chemistry in rechargeable Li–S batteries. *Journal of the American Chemical Society* 143 (29): 11063–11071.

94 Hou, T., Xu, W., Chen, X. et al. (2017). Lithium bond chemistry in lithium–sulfur batteries. *Angewandte Chemie International Edition* 56 (28): 8178–8182.

95 Hou, L., Li, Y., Li, Z. et al. (2023). Electrolyte design for improving mechanical stability of solid electrolyte interphase in lithium–sulfur batteries. *Angewandte Chemie International Edition* 62 (32): e202305466.

96 Li, X., Feng, S., Zhao, C. et al. (2022). Regulating lithium salt to inhibit surface gelation on an electrocatalyst for high-energy-density lithium–sulfur batteries. *Journal of the American Chemical Society* 144 (32): 14638–14646.

97 Cha, E., Patel, M.D., Park, J. et al. (2018). 2D MoS_2 as an efficient protective layer for lithium metal anodes in high-performance Li–S batteries. *Nature Nanotechnology* 13 (4): 337–344.

98 Hu, A., Chen, W., Du, X. et al. (2021). An artificial hybrid interphase for an ultrahigh-rate and practical lithium metal anode. *Energy & Environmental Science* 14 (7): 4115–4124.

99 Han, Z., Ren, H., Huang, Z. et al. (2023). A permselective coating protects lithium anode toward a practical lithium–sulfur battery. *ACS Nano* 17 (5): 4453–4462.

100 Choi, B., Kim, K.G., Lim, M. et al. (2024). Surface adaptive dual-layer protection of Li-metal anode for extending cycle-life of Li–Sulfur batteries with lean electrolyte. *Advanced Functional Materials* 34 (28): 2316838.

101 Zhang, G., Zhang, Z., Peng, H. et al. (2017). A toolbox for lithium–sulfur battery research: methods and protocols. *Small Methods* 1 (7): 1700134.

102 Lei, T., Chen, W., Lv, W. et al. (2018). Inhibiting polysulfide shuttling with a graphene composite separator for highly robust lithium–sulfur batteries. *Joule* 2 (10): 2091–2104.

103 Lang, S., Yu, S.-H., Feng, X. et al. (2022). Understanding the lithium–sulfur battery redox reactions via operando confocal Raman microscopy. *Nature Communications* 13 (1): 4811.

104 Hua, W., Shang, T., Li, H. et al. (2023). Optimizing the p charge of S in p-block metal sulfides for sulfur reduction electrocatalysis. *Nature Catalysis* 6 (2): 174–184.

105 Deng, Q., Dong, X., Shen, P.K., and Zhu, J. (2023). Li–S chemistry of manganese phosphides nanoparticles with optimized phase. *Advanced Science* 10 (9): 2207470.

106 Patel, M.U.M. and Dominko, R. (2014). Application of in operando UV/Vis spectroscopy in lithium–sulfur batteries. *ChemSusChem* 7 (8): 2167–2175.

107 Chen, W., Qian, T., Xiong, J. et al. (2017). A new type of multifunctional polar binder: toward practical application of high energy lithium sulfur batteries. *Advanced Materials* 29 (12): 1605160.

108 Li, T., Cai, D., Yang, S. et al. (2022). Desolvation synergy of multiple H/Li-bonds on an iron-dextran-based catalyst stimulates lithium–sulfur cascade catalysis. *Advanced Materials* 34 (51): 2207074.

109 Luo, Y., Fang, Z., Duan, S. et al. (2023). Direct monitoring of Li_2S_2 evolution and its influence on the reversible capacities of lithium–sulfur batteries. *Angewandte Chemie International Edition* 62 (11): e202215802.

110 Li, H., Meng, R., Ye, C. et al. (2024). Developing high-power Li||S batteries via transition metal/carbon nanocomposite electrocatalyst engineering. *Nature Nanotechnology* 19 (6): 792–799.

111 Saqib, N., Ohlhausen, G.M., and Porter, J.M. (2017). In operando infrared spectroscopy of lithium polysulfides using a novel spectro-electrochemical cell. *Journal of Power Sources* 364: 266–271.

112 Saqib, N., Silva, C.J., Maupin, C.M., and Porter, J.M. (2017). A novel optical diagnostic for in situ measurements of lithium polysulfides in battery electrolytes. *Applied Spectroscopy* 71 (7): 1593–1599.

113 Wu, L., Hu, J., Chen, S. et al. (2023). Lithium nitrate mediated dynamic formation of solid electrolyte interphase revealed by in situ Fourier transform infrared spectroscopy. *Electrochimica Acta* 466: 142973.

114 Liu, X.-C., Yang, Y., Wu, J. et al. (2018). Dynamic hosts for high-performance Li–S batteries studied by cryogenic transmission electron microscopy and in situ X-ray diffraction. *ACS Energy Letters* 3 (6): 1325–1330.

115 Wang, P., Xu, T., Xi, B. et al. (2022). A Zn8 double-cavity Metallacalix[8]arene as molecular sieve to realize self-cleaning intramolecular tandem transformation of Li–S chemistry. *Advanced Materials* 34 (51): 2207689.

116 Wang, B., Wang, L., Kong, Y. et al. (2023). Hafnium diboride spherical superstructure born of 5d-metal Hf-MOF-induced p orbital activity of B atom and enhanced kinetics of sulfur cathode reaction. *Advanced Energy Materials* 13 (22): 2300590.

117 Cuisinier, M., Cabelguen, P.-E., Evers, S. et al. (2013). Sulfur speciation in Li–S batteries determined by operando X-ray absorption spectroscopy. *The Journal of Physical Chemistry Letters* 4 (19): 3227–3232.

118 Zhang, L., Ling, M., Feng, J. et al. (2018). The synergetic interaction between $LiNO_3$ and lithium polysulfides for suppressing shuttle effect of lithium–sulfur batteries. *Energy Storage Materials* 11: 24–29.

119 Jia, L., Wang, J., Ren, S. et al. (2020). Unraveling shuttle effect and suppression strategy in lithium/sulfur cells by in situ/operando X-ray absorption spectroscopic characterization. *Energy & Environmental Materials* 4 (2): 222–228.

120 Xu, W., Lang, S., Wang, K. et al. (2023). Fundamental mechanistic insights into the catalytic reactions of Li–S redox by Co single-atom electrocatalysts via operando methods. *Science Advanced* 9 (33): eadi5108.

121 Li, H., Chuai, M., Xiao, X. et al. (2023). Regulating the spin state configuration in bimetallic phosphorus trisulfides for promoting sulfur redox kinetics. *Journal of the American Chemical Society* 145 (41): 22516–22526.

122 Nie, L., Zhu, J., Wu, X. et al. (2024). A large-scale fabrication of flexible, ultrathin, and robust solid electrolyte for solid-state lithium–sulfur batteries. *Advanced Materials* 36 (29): 2400115.

123 Ma, R., Fan, Y., Jin, Y. et al. (2024). Construction of elastic and conductive channels for high-rate and high-areal-capacity sulfur cathodes in all-solid-state lithium–sulfur batteries. *Advanced Energy Materials* 14 (19): 2304412.

124 Soni, R., Robinson, J.B., Shearing, P.R. et al. (2022). Lithium–sulfur battery diagnostics through distribution of relaxation times analysis. *Energy Storage Materials* 51: 97–107.

125 Kim, H., Lee, J.T., Magasinski, A. et al. (2015). In situ TEM observation of electrochemical lithiation of sulfur confined within inner cylindrical pores of carbon nanotubes. *Advanced Energy Materials* 5 (24): 1501306.

126 Yang, Z., Zhu, Z., Ma, J. et al. (2016). Phase separation of Li_2S/S at nanoscale during electrochemical lithiation of the solid-state lithium–sulfur battery using in situ TEM. *Advanced Energy Materials* 6 (20): 1600806.

127 Yuan, Y., Tan, G., Wen, J. et al. (2018). Encapsulating various sulfur allotropes within graphene nanocages for Long-lasting lithium storage. *Advanced Functional Materials* 28 (38): 1706443.

128 Wang, Z., Tang, Y., Zhang, L. et al. (2020). In situ TEM observations of discharging/charging of solid-state lithium–sulfur batteries at high temperatures. *Small* 16 (28): 2001899.

129 Zhou, S., Shi, J., Liu, S. et al. (2023). Visualizing interfacial collective reaction behaviour of Li–S batteries. *Nature* 621 (7977): 75–81.

130 Miao, Z., Li, Y., Xiao, X. et al. (2022). Direct optical fiber monitor on stress evolution of the sulfur-based cathodes for lithium–sulfur batteries. *Energy & Environmental Science* 15 (5): 2029–2038.

131 Gao, R., Zhang, M., Han, Z. et al. (2023). Unraveling the coupling effect between cathode and anode toward practical lithium–sulfur batteries. *Advanced Materials* 36 (1): 2303610.

132 Guo, Y., Jin, Z., Lu, J. et al. (2023). Engineering a deficient-coordinated single-atom indium electrocatalyst for fast redox conversion in practical 500 Wh kg^{-1}-level pouch lithium–sulfur batteries. *Energy & Environmental Science* 16 (11): 5274–5283.

133 Feng, J., Liu, T., Li, H. et al. (2024). Ultralight electrolyte with protective encapsulation solvation structure enables hybrid Sulfur-based primary batteries exceeding 660 Wh/kg. *Journal of the American Chemical Society* 146 (6): 3755–3763.

134 Zhang, H., Zhang, Y., Cao, C. et al. (2024). Lithium–sulfur pouch cells with 99% capacity retention for 1000 cycles. *Energy & Environmental Science* https://doi.org/10.1039/d4ee02149e.

135 Deng, D., Xue, F., Bai, C. et al. (2018). Enhanced adsorptions to polysulfides on graphene-supported BN nanosheets with excellent Li–S battery performance in a wide temperature range. *ACS Nano* 12 (11): 11120–11129.

136 Zeng, P., Yuan, C., An, J. et al. (2022). Achieving reversible precipitation-decomposition of reactive Li_2S towards high-areal-capacity lithium–sulfur batteries with a wide-temperature range. *Energy Storage Materials* 44: 425–432.

137 Ma, L., Wang, Y., Wang, Z. et al. (2023). Wide-temperature operation of lithium–sulfur batteries enabled by multi-branched vanadium nitride electrocatalyst. *ACS Nano* 17 (12): 11527–11536.

138 Zhang, H., Chen, J., Li, Z. et al. (2023). Operating lithium–sulfur batteries in an ultrawide temperature range from −50 °C to 70 °C. *Advanced Functional Materials* 33 (48): 2304433.

139 Wang, P., Kateris, N., Li, B. et al. (2023). High-performance lithium–sulfur batteries via molecular complexation. *Journal of the American Chemical Society* 145 (34): 18865–18876.

140 Deng, D., Xiong, H., Luo, Y. et al. (2024). Accelerating the rate-determining steps of sulfur conversion reaction for lithium–sulfur batteries working at an ultrawide temperature range. *Advanced Materials* 36: 2406135.

5

Room Temperature Sodium–Sulfur Batteries: Challenges and Progress

Mingyue Wang[2], Nana Wang[3], and Zhongchao Bai[1]

[1] *Institute of Energy Materials Science, University of Shanghai for Science and Technology, 516 Jungong Road, Shanghai 200093, China*
[2] *Institute for Superconducting and Electronic Materials, Faculty of Engineering and Information Sciences, University of Wollongong, North Wollongong, NSW 2500, Australia*
[3] *Centre for Clean Energy Technology, School of Mathematical and Physical Sciences, Faculty of Science, University of Technology Sydney, Sydney, NSW 2007, Australia*

5.1 Introduction

Over recent decades, the global increase in energy consumption across both industrial sectors and everyday life has emphasized the urgent need to harness renewable energy efficiently for sustainable growth. While lithium-ion batteries are the standard for powering portable devices and electric vehicles, their application remains constrained by low energy density and high material costs in large-scale energy storage. [1] This has driven the search for next-generation battery technologies that either offer higher energy densities or are made from more cost-effective materials. [2] Room temperature sodium–sulfur (RT Na–S) batteries are an up-and-coming alternative within this field. The widespread availability of Na and S, combined with their high theoretical capacities (1166 and 1672 mAh g^{-1} for Na and S, respectively), thus leading to a potential gravimetric energy density of 1274 Wh kg^{-1} for RT Na–S batteries. [3] This impressive energy capacity positions this battery system as a scalable and cost-effective technology, capable of meeting the increasing energy storage demands in a future powered by renewable energy solutions.

While RT Na–S batteries are valued for their affordability and environmental advantages, technical challenges persist, such as rapid capacity decay, low Coulombic efficiency (CE), and stability issues. These problems stem primarily from the low conductivity of S, significant volume expansion during discharging and charging, and the formation of Na polysulfides (NaPSs), which shuttle between electrodes and influence during charge–discharge cycles. [4] Additionally, the buildup of insoluble short-chain NaPSs, such as Na_2S_2 and Na_2S, hampers the electrochemical reactions, further complicating matters. [5] Furthermore, the unchecked Na dendrites growth during prolonged cycling can degrade the Na metal surface, reducing CE and posing safety risks. [6] To overcome these obstacles, researchers are investing in the exploration of several strategies, including the development of advanced S hosts to better

Electrode Materials in Energy Storage Technologies: Applications in Lithium-, Sodium-, Potassium-, Sulfur- and Zinc-Based Rechargeable Batteries, First Edition. Edited by Liqiang Xu.

accommodate the active material, enhancing separators to prevent polysulfides migration, optimizing electrolyte compositions to reduce undesirable reactions, and refining metal anode designs to curb dendritic growth. Such comprehensive approaches are crucial to enhance the safety and electrochemical performance, ensuring more stable S conversion reactions and broader viability of RT Na–S battery technology.

This chapter provides an in-depth overview of the exploration of Na–S batteries, which is structured to first discuss the historical development and the underlying reaction mechanisms and then provide a detailed analysis of the various challenges. We delve into recent advancements in key materials such as S cathodes, metal anodes, modified separators, and electrolytes. These developments showcase the strategies employed to address critical issues that impede the electrochemical performance of the batteries. Finally, this chapter presents forward-looking insights, offering perspectives on potential advancements of RT Na–S batteries and suggesting paths for future research and technology optimization.

5.2 History of Na–S Batteries

The development of Na–S batteries can be traced back to the 1960s when Kummer and his team discovered sodium beta-alumina as a high-temperature (HT) solid-state Na-ion conductor. [9] Typically, as illustrated in Figure 5.1a, HT Na–S batteries operate above 300 °C to keep the S cathode and Na anode in a molten state, with a Na β-alumina tube serving as both electrolyte and separator.[1a] This high temperature is essential for beta-alumina electrolytes to sustain high ion conductivity, which is essential for achieving adequate energy densities and power. The electrode and overall Na–S cell reactions are detailed in Eqs. (5.1)–(5.3)

Cathode:

$$xS + e^- \leftrightarrow S_x^{2-} \tag{5.1}$$

Anode:

$$2Na \leftrightarrow 2Na^+ + 2e^- \tag{5.2}$$

Overall:

$$2Na + xS \leftrightarrow Na_2S_x \tag{5.3}$$

Specifically, during discharge, Na is oxidized at the surface of β-alumina, and Na ions transfer through it to the S cathode, interacting with S and forming Na_2S_5 at 2.075 V. Initially, a two-phase liquid forms because Na_2S_5 is immiscible with S at the operating temperature. As the discharge continues, single-phase reactions take place from Na_2S_5 to Na_2S_4 and further to Na_2S_3 at 1.74 V (Figure 5.1b). [10] With continued discharge, the voltage remains 1.74 V, causing solid Na_2S_2 and Na_2S to precipitate inside the molten cathode and on the solid-electrolyte surface. However, these non-electrochemically active precipitates are detrimental to the cycling life,

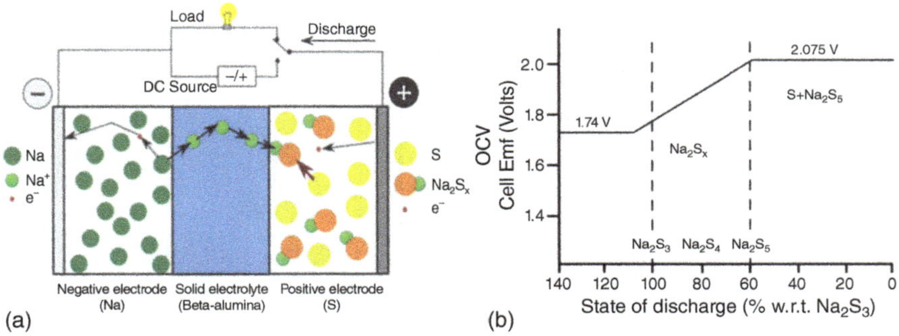

Figure 5.1 (a) Operating principle of HT Na–S batteries during discharge. Source: [7]/John Wiley & Sons. (b) Voltage profile of HT Na–S cells with the different phases during charging. Source: [8]/John Wiley & Sons.

so discharging at 1.74 V should be avoided. This significantly lowers the practical capacity to 557 mAh g^{-1}, limiting the reaction to S and Na_2S_3.

Besides the reduced energy density, high operating temperatures in Na–S systems pose significant safety risks and increase operating costs. Lowering Na–S batteries' operating temperature can enhance their safety, energy density, and cycle life while reducing costs associated with thermal management and cell packaging. Consequently, there is significant ongoing research and development focused on RT Na–S battery systems.

5.3 Reaction Mechanism of Room Temperature Na–S Batteries

Recent breakthroughs in novel organic liquid electrolytes and materials have enabled Na–S redox reactions to take place at RT. The transition from S to Na_2S involves a complex reaction process of these batteries. S (cyclo-S_8 molecules) undergoes ring opening during discharge, followed by the formation of long-chain and short-chain NaPSs intermediates. [11] Additionally, they demonstrate distinct electrochemical behaviors in two primary nonaqueous electrolytes: ether- and carbonate-based electrolytes, which are discussed in the following part separately.

5.3.1 Reaction Mechanisms in Ether-Based Electrolytes

Figure 5.2a–b displays the cyclic voltammetry (CV) and galvanostatic charge–discharge (GCD) curves of an ether-based (1.5 M $NaClO_4$ + 0.3 M $NaNO_3$ + TEGDME) RT Na–S battery. [12] The transformation of cyclo-S_8 to Na_2S (the final discharge product) occurs in four distinct regions (Eqs. (5.4)–(5.9)). [12] In Region I, a voltage plateau around 2.2 V (*vs.* Na/Na^+) is observed, where cyclo-S_8 transitions from a solid to a liquid phase, converting into soluble Na_2S_8:

$$S_8 + 2Na^+ + 2e^- \rightarrow Na_2S_8 \tag{5.4}$$

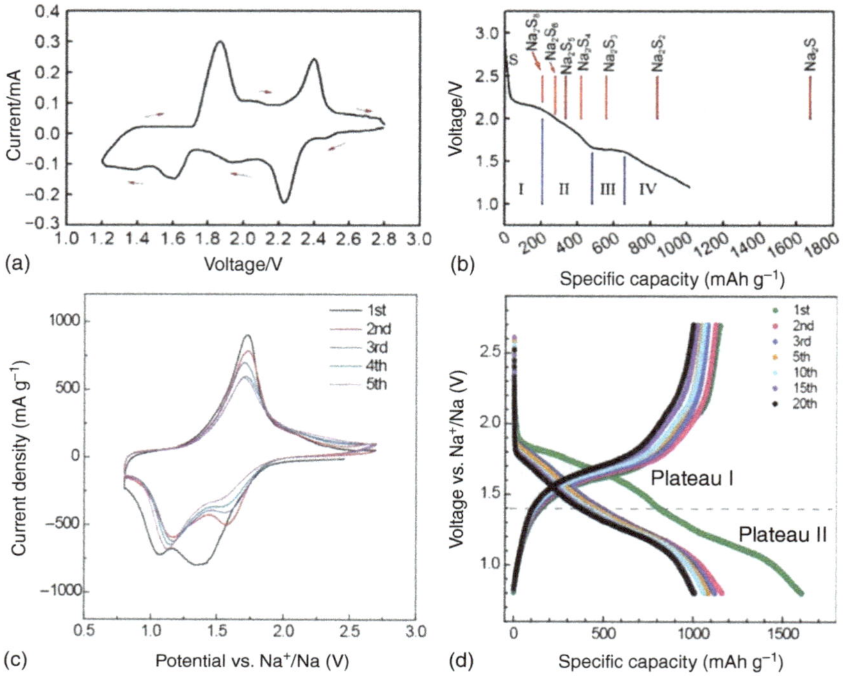

Figure 5.2 (a) CV curve and (b) GCD profile with ether-based electrolyte of RT Na–S batteries. Source: Reproduced from [12]/with permission of John Wiley & Sons. (c) CV curves and (d) GCD profile with carbonate-based electrolyte of RT Na–S batteries. Source: Reproduced from [13]/with permission of John Wiley & Sons.

Region II features a sloping voltage range from 2.2 V to 1.65 V, where a liquid–liquid transition occurs, converting Na_2S_8 to Na_2S_4:

$$Na_2S_8 + 2Na^+ + 2e^- \rightarrow 2Na_2S_4 \tag{5.5}$$

Region III, a low-voltage plateau at approximately 1.65 V, indicates a liquid–solid transition where dissolved Na_2S_4 converts to insoluble Na_2S_3, Na_2S_2, and Na_2S:

$$Na_2S_4 + 2/3Na^+ + 2/3e^- \rightarrow 4/3Na_2S_3 \tag{5.6}$$

$$Na_2S_4 + 2Na^+ + 2e^- \rightarrow 2Na_2S_2 \tag{5.7}$$

$$Na_2S_4 + 6Na^+ + 6e^- \rightarrow 4Na_2S \tag{5.8}$$

Region IV, ranging from 1.65 to 1.2 V, represents a solid–solid reaction, converting Na_2S_2 to Na_2S:

$$Na_2S_2 + 2Na^+ + 2e^- \rightarrow 2Na_2S \tag{5.9}$$

Ether-based electrolytes are known for their high solubility of high-order NaPSs. This S redox process, transitioning from S_8 to soluble NaPSs, and subsequently to Na_2S_2 and Na_2S, enhances rapid polysulfides conversion kinetics in the liquid phase. Nevertheless, the high solubility of NaPSs results in significant side reactions with Na metal anodes, leading to poor cycling performance and reduced capacity in RT Na–S batteries.

5.3.2 Reaction Mechanisms in Carbonate-Based Electrolytes

Carbonate-based electrolytes complicate electrochemical reactions in RT Na–S batteries due to significant side reactions between the solvent and the S cathode. Recent research identifies two typical electrochemical mechanisms. The first mechanism suggests a solid–liquid–solid reaction pathway at the S cathode, which is similar yet distinct from that observed with ether-based electrolytes: S initially converts to NaPSs, before being reduced to Na_2S. As displayed in Figure 5.2c, CV curves of carbonate-based electrolytes in Na–S batteries exhibit two distinct peaks; however, the rest of the area is minimal. [14] Because the solubility of NaPSs in carbonate-based electrolytes is low, this mechanism is often referred to as a quasi-solid-state reaction. The reaction equation is:

$$S_8 + Na^+ \rightarrow Na_2S_x \rightarrow Na_2S_2/Na_2S \tag{5.10}$$

Another perspective suggests that in carbonate-based electrolytes, small molecular sulfur undergoes a solid-state transformation, directly converting to Na_2S_2/Na_2S without forming intermediate NaPSs. [13, 15] The reaction equation is as follows:

$$S_n + Na^+ \rightarrow Na_2S_2/Na_2S\,(n = 1, 2, 3, 8) \tag{5.11}$$

As depicted in Figure 5.2d, the reaction between S and Na initially yields Na_2S_2 molecules (Plateau I) and subsequently produces Na_2S at a lower discharge voltage (Plateau II), marking the completion of the reaction. [13] The considerably lower solubility of NaPSs in carbonate-based electrolytes, compared to ether-based ones, helps mitigate the polysulfides shuttle effect, thereby improving S utilization and enhancing cycle stability. Consequently, the majority of reported Na–S batteries have predominantly utilized carbonate-based electrolytes.

5.4 Challenges of Room Temperature Na–S Batteries

The practical application of Na–S batteries remains challenging due to inherent issues related to the chemistry at both the S cathode and Na anode, as illustrated in Figure 5.3. [16] These challenges result in low reversible capacity, significant self-discharging, and reduced cycle stability. The main challenges include:

(i) The inherently low conductivity of S and its discharge products (Na_2S_2/Na_2S) results in substantial resistance within the S cathode, which directly impedes electrochemical reaction kinetics and leads to poor S utilization. This sluggishness in redox reactions results in significant polarization and reduced discharge capacity. The generation of an insulating Na_2S_2/Na_2S passivation film on the cathode surface during discharge hinders dissociation and greatly degrades Na–S battery performance.

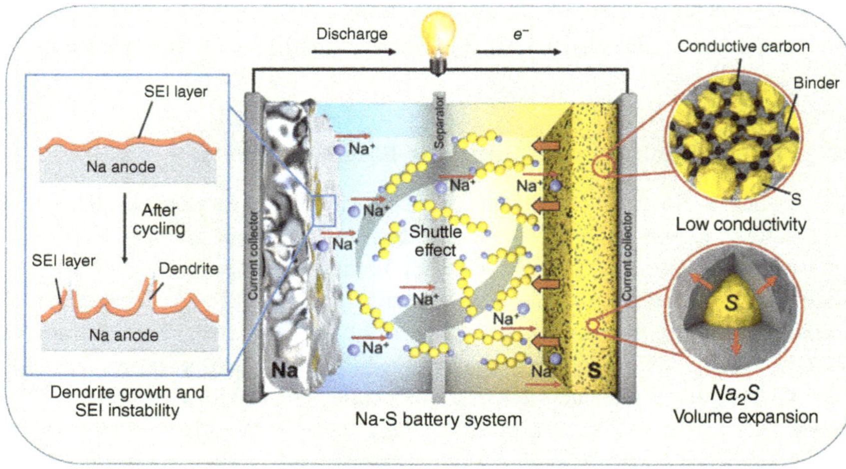

Figure 5.3 Schematic representation of RT Na–S battery, highlighting practical challenges. Source: Reproduced with permission from [16]/John Wiley & Sons.

(ii) During discharge, S undergoes a substantial volume expansion (about 170%), leading to pulverization and structural breakdown of the cathode. [17] Consequently, the insulating S may lose contact with conductive agents and the current collector, which results in sluggish electrochemical kinetics and loss of active materials.

(iii) The generation of NaPSs and the subsequent polysulfide shuttle effect pose significant challenges in Na–S batteries. During cycling, soluble high-order NaPSs form and dissolve in the electrolyte, leading to continuous accumulation. This accumulation prompts the diffusion of NaPSs toward the Na anode, driven by concentration gradients, where they are reduced to form insoluble Na_2S. Known as the "shuttle effect", this continuous diffusion and migration during charging cause deterioration of Na metal anodes. These irreversible reactions, along with the low active materials utilization in electrodes, significantly accelerate capacity decay and reduce CE during cycling.

(iv) The sluggish redox kinetics of S is another significant challenge that leads to low CE in Na–S batteries. The transformation process of S involves various phases, including solid S_8, liquid NaPSs, and solid Na_2S/Na_2S_2. During discharge, the conversion of soluble NaPSs to insoluble Na_2S presents a higher activation barrier, resulting in the accumulation of soluble NaPSs. Similarly, during charging, the oxidation of Na_2S back to Na_2S_x requires substantial activation energy due to the electrochemically inert and insulating nature of Na_2S, contributing to the sluggish kinetics of Na–S chemistry.

(v) The nonuniform solid-electrolyte interphase (SEI) layer on the Na metal anode fails to protect the surface adequately, leading to ongoing detrimental side reactions with the electrolyte. Uneven Na plating/stripping continuously breaks and reforms the SEI layer, causing irreversible loss of active Na and electrolytes during cycling. This contributes to cell failure due to high electrochemical internal resistance across the thickened SEI layer. Furthermore, the diffusion of NaPSs between the Na anode and the S cathode forms an insulating Na_2S passivation layer on the anode, significantly reducing cell reversibility and stability.

The uncontrolled growth of Na dendrites compromises contact with the Na anode, further reducing CE. Additionally, these dendrites can pierce separators, resulting in internal short circuits and significant safety risks. Managing dendrite growth is a critical challenge in the practical application of the Na anode in this battery system.

5.5 Progress on Room Temperature Na–S Batteries

Significant efforts have concentrated on optimizing RT Na–S battery designs to improve cycling stability. This section summarizes recent progress on cathodes, anodes, electrolytes, and separators for RT Na–S batteries.

5.5.1 Cathodes

Rationally designing cathode materials has proven highly effective in enhancing the conductivity and activity of S-based cathodes, buffering volume expansion, and reducing polysulfide dissolution. Current conventional strategies involve incorporating S into carbonaceous matrices, covalently bonding S to polymers, and embedding polar metal-based sulfidic sites within the cathode to immobilize polysulfides through strong polar interactions.

5.5.1.1 Carbon-S Cathodes

Due to their superior conductivity, structural and electrochemical stability, and adjustable pore structures, an increasing number of carbon materials have recently been developed as S-host materials. Wang and coworkers designed an interconnected mesoporous hollow carbon nanosphere as S hosts, which delivered excellent electrochemical performance. The hollow interior accommodates a high S content and buffers volume changes, while the outer carbon nanoshells trap NaPSs. [18] However, the physical space confinement effect of NaPSs, achieved solely through morphology, is limited and insufficient to mitigate the shuttle effect. To mitigate this challenge, researchers have developed heteroatom-doped carbon materials as a solution. Li and coworkers prepared N/O-codoped multichambered carbon nanoboxes (MCCBs) by etching and carbonizing monodisperse ZIF-8 nanocubes. The N/O-doped carbon chemically anchors polysulfides through strong adsorption, while the porous shell and internal carbon grids enhance electron and ion transport, thereby improving electrode kinetics. [19] Hence, the S@MCCBs cathode achieves outstanding electrochemical performance.

In addition to adjusting morphology and doping with heteroatoms, the pore structure is also crucial for S hosts. Some studies have focused on molecularly confining S within micropores. Xin and colleagues prepared a S-microporous carbon composite (S/(CNT@MPC)), featuring a carbon nanotube (CNT) core and a

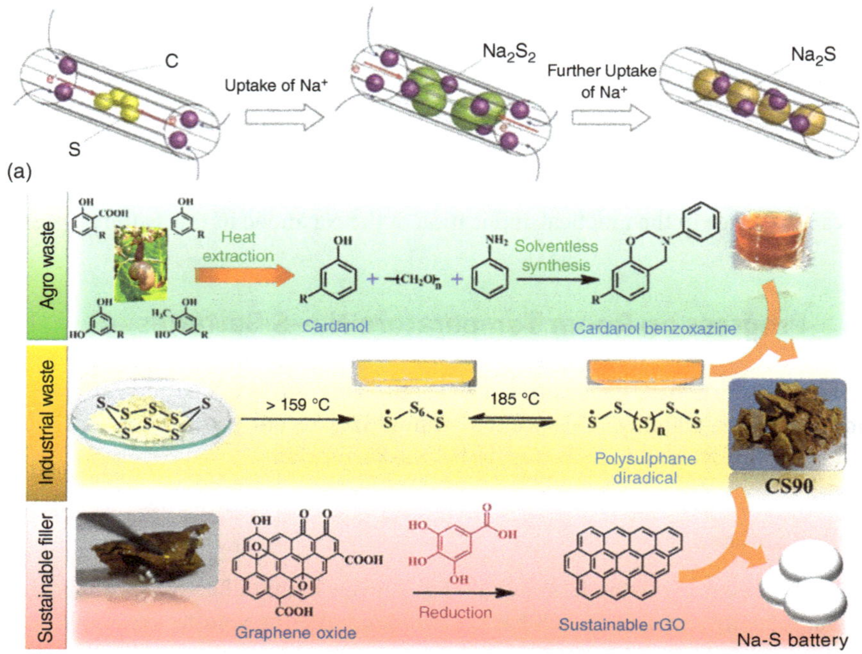

Figure 5.4 (a) Reactions between S and Na^+ in S/(CNT@MPC). Source: Reproduced from [13]/with permission of John Wiley & Sons. (b) Strategies utilized to prepare the CS_{90}-rGO. Source: Reproduced with permission from [20]/American Chemical Society.

S-containing microporous carbon sheath. [13] The S is confined within micropores of 0.5 nm as S_{2-4}. As illustrated in Figure 5.4a, this "size effect" results in S reacting with Na via a solid–solid reaction, eliminating the high-order NaPSs formation, thus mitigating the harmful shuttle effect. This allows for the maintenance of high capacity and stable cyclability, achieving a specific capacity of 600 mA h g^{-1} after 200 cycles at 1 C. Guo and coworkers developed a microporous carbon derived from coffee residue, which is characterized by abundant 0.5 nm micropores and a high specific surface area (1100 m^2 g^{-1}), serving as a micropore confinement cathode. This design facilitates a one-step reaction, preventing the formation of polysulfides. The carbon-S composite electrode exhibits neglectable capacity degradation after 2000 cycles at 1 C, maintaining 100% CE, which demonstrates its exceptional reversibility and cycling stability. [21]

5.5.1.2 Polymer-S Cathodes

Designing polymer-S composites that confine NaPSs through chemical bonding represents is a potent approach to developing high-performance cathodes. Sized pyrolyzed polyacrylonitrile (S-pPAN) is particularly noteworthy for its excellent cycling stability with carbonate-based electrolytes. The preparation of covalent S-pPAN composites involves a series of steps, including precursor mixing, cyclization, dehydrogenation, and embedding S at high temperatures to form molecular-level S composites. [22] This method fosters strong covalent interactions between S and polyacrylonitrile (PAN) carbon framework, effectively preventing the formation of free long-chain polysulfides and ensuring high S dispersion at the molecular level to avoid aggregation. Building on this method, Hwang and coworkers reported that PAN-derived C-S composite with one-dimensional (1D) fibers replaced PAN powders through electrospinning. [23] The 1D nanofiber morphology of S-pPAN, as opposed to uncontrolled powder forms, enhances battery performance, reaching a capacity of 153 $mAh\,g^{-1}$ after 500 cycles at a 1 C rate. Although S-pPAN effectively reduces NaPSs shuttling and demonstrates excellent electrochemical performance, its limited S content (around 50 wt%) restricts the overall energy density. Therefore, novel polymer-S composites with the polymer backbone containing S chains have been developed to achieve higher S contents. Ghosh and coworkers created an S-embedded polymer that secures substantial S confinement (~90 wt%) within the cathode structure through thermal ring-opening polymerization of benzoxazine with elemental S (CS_{90}) and then combined with reduced graphene oxide (rGO). The CS_{90} composite exhibits a uniform S distribution within the polymer backbone, which enables maximum S utilization and results in high CE (99%) and robust cycle life, as depicted in Figure 5.4b. [20]

5.5.1.3 Electrocatalyst-Enhanced S Cathode

To enhance the suppression of NaPSs, the introduction of materials with polar effects into the carbon matrix has garnered increasing attention. Currently, various polar materials have been tested in cathode materials, including metal-based nanomaterials, metal compounds, and heterostructure composites. These materials enhance electrochemical performance by immobilizing NaPSs through strong chemical interactions and improving the S redox kinetics and cycle stability.

For metal-based nanomaterials, transition-metal nanoparticles, transition-metal single-atom, or metal nanocluster catalysts are reported to improve cyclic stability and kinetics. Guo and coworkers synthesized a three-dimensional (3D) network (Ni-NCFs) composed by nitrogen-doped carbon fibers (NCFs) and hollow nickel (Ni) units that buffer volume fluctuations of S during cycling and improve the conductivity of the carbon fibers. [24] Additionally, during the S-loading process, a small portion of Ni becomes polarized, forming a polar Ni–S bond. The Ni-NCFs composite effectively absorbs NaPSs due to the presence of NCFs, which further promotes the Ni unit catalysis in the redox reactions of NaPSs. Consequently, the S@Ni-NCFs cathode exhibits excellent electrochemical performance. Gold atoms (Au) are noted for their stability than other transition metals in the electrochemical reaction; therefore, incorporating Au into RT Na–S batteries is anticipated to resolve the sluggish kinetics issue due to their superior catalytic activity and electrical conductivity. Wang and coworkers developed a sulfidic host by decorating gold nanodots on hierarchical N-doped carbon microspheres (CN/Au/S). This configuration facilitates the complete S conversion by electrocatalyzing the transformation of Na_2S_4 into Na_2S during discharge and back to S during charging, thereby overcoming low reaction kinetics. [14] Furthermore, the Au nanodots and N-doped carbon enhance the S cathode conductivity and offer strong polar–polar adsorption to NaPSs, mitigating shuttling effects. As a consequence, the CN/Au/S cathode demonstrates excellent high-rate capability.

Transition-metal single-atom catalysts (SACs) combined with a porous carbon matrix have proven to be an effective strategy for achieving efficient catalysis. Jiang and coworkers developed a 3D nitrogen-doped hierarchical porous carbon framework (termed 3D-PNCV) based on first principles calculations to improve the adsorption, reactivity, and catalytic conversion performance of NaPSs (Figure 5.5a). [25] The 3D-PNCV host owns abundant active vanadium sites, a high electrical conductivity framework, and strong chemical adsorption and conversion capabilities due to vanadium–nitrogen (V–N) bonding. These characteristics help immobilize polysulfides and facilitate the reversible catalytic conversion of polysulfides to Na_2S. Consequently, achieving a high reversible capacity and outstanding rate capability. Building on the concept of enhancing material interactions with NaPSs, Li and coworkers developed a novel asymmetrical manganese-nitrogen (Mn-N_2) structure embedded into nitrogen-doped carbon nanofibers (Mn-N_2/CNs) using thermal NH_3 etching of a symmetrical Mn-N_2O_2 structure. Due to the increased hybridization of *d*-*p* orbitals between *d* orbitals of Mn and *p* orbitals of S in NaPSs, the Mn-N_2 structure enhances the bonding affinity and catalytic conversion rate of NaPSs. As a result, the Mn-N_2/CNs@S composite demonstrates a high specific capacity of 458 mAh g^{-1} at 3 C and outstanding stability with a capacity fade of only 0.23% over 2300 cycles. [30]

In addition to metal-based electrocatalysts, a variety of metal compounds, such as metal oxides, sulfides, carbides, and nitrides, have been successfully employed to enhance the redox kinetics of NaPSs. Zhou and coworkers employed amorphous silica (AS) with appropriate binding ability as the S host to develop advanced S cathodes. [31] The flexibility and high pore volume of the AS favor high S loading and accommodate the large volume fluctuation of S during charging and discharging. In addition, due to the unique physical structure of AS and the Na–O chemical bond formed with NaPSs, it exhibited excellent physical and chemical limiting ability to NaPSs. The S electrodes achieved a high S content (73.48 wt%), exhibited remarkable cycling stability, and maintained a high capacity of 1180 mAh g^{-1} at 1.0 A g^{-1}. Yan and coworkers prepared a novel sulfidic host by *in situ* growing of FeS_2 nanoparticles in nitrogen-doped multilevel carbon microspheres (FeS_2@NCMS) with large pore volumes. After loading S into the host, a composite material (FeS_2@NCMS/S) is formed with high S-loading content, which is expected to improve polysulfides immobilization and conversion (Figure 5.5b). [26] Thanks to its optimized micro-/nano-architectures and powerful electrocatalytic properties, the prepared FeS_2@NCMS/S cathode achieves a specific capacity of 1471 mAh g^{-1}.

Mo_2C, a typical transition-metal carbide, is extensively utilized in the battery field due to its high electronic conductivity (approximately 1.02×10^2 S cm^{-1}) and effective adsorption of alkali metal ions. [32] Zhou and coworkers designed graded porous carbon porous carbon polyhedrons with uniformly embedded Mo_2C nanoparticles for S loading (Figure 5.5c). [27] The micro and mesoporous structure of porous carbon significantly accommodates the volume changes of the S cathode during redox processes. Additionally, the Mo_2C nanoparticles exhibit strong chemical effects on

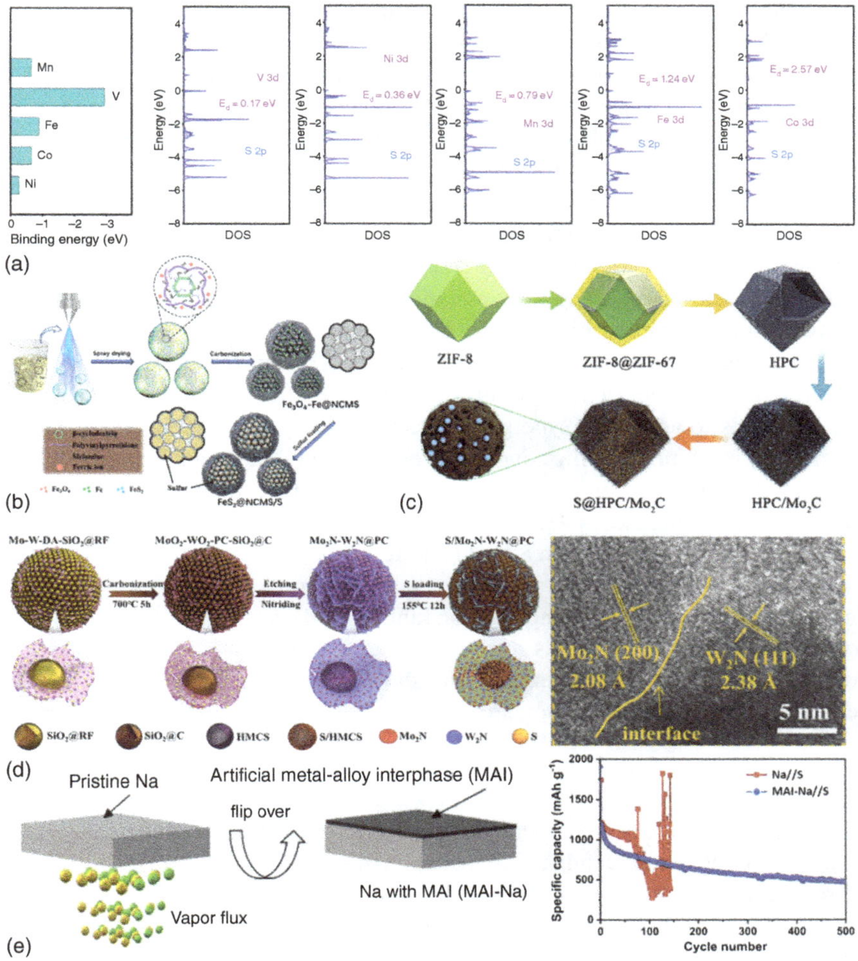

Figure 5.5 (a) Binding energy of Na_2S_4 adsorption on SAC-X (X = V, Ni, Mn, Co, and Fe); density of state for p-bands and d-bands of Na_2S_4 in SAC-X-Na_2S_4 (X = V, Ni, Mn, Fe, and Co) systems. Source: Reproduced from [25]/with permission of American Chemical Society. (b) Schematic illustration of FeS_2@NCMS/S composite fabrication process. Source: Reproduced from [26]/with permission of John Wiley & Sons. (c) Schematic illustration of the preparation process of the S@HPC/Mo_2C. Source: Reproduced from [27]/with permission of John Wiley & Sons. (d) Schematic illustration of the preparation of the spherical S/Mo_2N-W_2N@PC superstructure and HRTEM images of the spherical Mo_2N-W_2N@PC superstructure. Source: Reproduced with permission from [28]/John Wiley & Sons. (e) Graphical representation of the MAI interphase formation process and cycle stability of RT Na–S battery with and without MAI layer. Source: Reproduced from [29]/with permission of Elsevier.

polysulfides and catalyze their transformation, effectively inhibiting the polysulfides dissolution and promoting reaction kinetics. Consequently, the prepared S cathode demonstrates a high specific capacity and superior rate performance. Qi and coworkers developed a Fe_3N catalyst on N-doped multilayered carbon networks (NMCN) that offers high adsorption of NaPSs in batteries. [33] The NMCN features a layered pore structure that can accommodate a substantial amount of S and inhibit volume expansion. Fe_3N exhibits a strong affinity for NaPSs through Na–N and Fe–S bonding, facilitating NaPSs dissociation. The S@Fe_3N-NMCN cathode can be used directly as an electrode without the need for binders, conductive additives, or current collectors, reducing the time and cost of battery assembly. The battery achieved a high S loading of 85 wt% (2.6 mg cm^{-2}). Notably, the S@Fe_3N-NMCN cathode exhibits impressive performance in RT Na–S batteries.

Heterostructures represent a promising strategy that involves coupling two or more materials to achieve a synergistic effect. This approach enhances RT Na–S performance by combining the advantages of each material. Zhang and coworkers developed Mo_2N-W_2N heterostructures, as evidenced by high-resolution transmission electron microscopy (HRTEM), inserted into spherical carbon structures (Mo_2N-W_2N@PC) to serve as an S host (Figure 5.5d). [28] The hierarchical porous spherical superstructures boost excellent electrical conductivity and an abundant microporous structure, improving reaction kinetics and accommodating the volume variation of S cathode materials during cycling. The Mo_2N-W_2N heterostructure with strong adsorption ability inhibits the shuttle effect, enhances ion diffusivity, and facilitates rapid NaPSs conversion.

5.5.2 Anodes

Similar to Na-ion batteries, Na metal anodes can form dendrites in RT Na–S batteries during cycling. These dendrites are needle-like structures that can grow through electrolytes, potentially piercing the separator and causing short circuits, which can lead to battery failure or even safety hazards such as fires and explosions. Moreover, due to the notorious polysulfides, the interface reaction of Na anodes in RT Na–S batteries is more complex than in traditional Na-ion batteries. Polysulfides increase resistance and the activation energy required for the SEI layer formation. It is believed that some polysulfides can penetrate the SEI layer and corrode fresh Na metal, resulting in capacity loss. To combat these challenges, Kumar and coworkers developed an artificial metal-alloy interphase (MAI) through a straightforward solid-vapor reaction (Figure 5.5e). [29] The newly formed MAI exhibited a combination of four critical attributes: strong electrode adhesion, high Na-ion conductivity, elevated Young's modulus, and inferior electrolyte permeability. Consequently, RT Na–S batteries utilizing this anode technology were constructed, demonstrating impressive long-term cycling performance.

5.5.3 Separators

Polypropylene (PP) and glass fiber are commonly used separators in RT Na–S batteries. These separators possess a rich porosity that not only facilitates ion transport

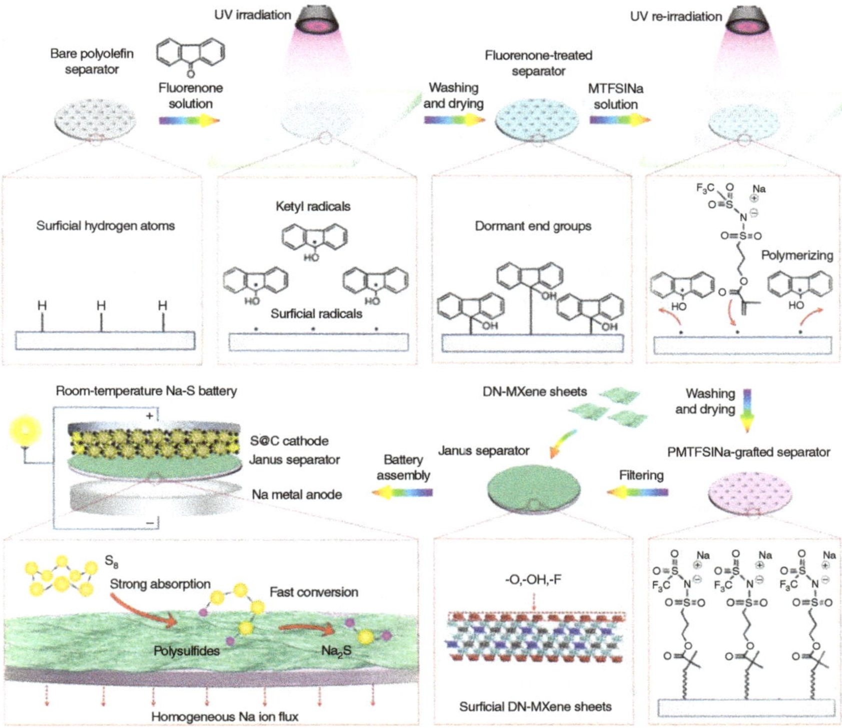

Figure 5.6 Schematic illustration of PMTFSINa-grafted and DN-MXene-coated separators preparation process. Source: Reproduced from [34]/with permission of John Wiley & Sons.

but also inadvertently promotes the shuttling of polysulfides. To mitigate this issue, researchers are exploring modifications to the separators that aim to obstruct polysulfide shuttling and enhance the conversion efficiency of NaPSs. Zhou and coworkers devised a sophisticated grafting-filtering technique to modify commercial PP separators (Figure 5.6). [34] This led to the creation of single-ion-conducting polymer-grafted and low-dimensional material-coated separators. The grafted layer of poly(1-[3-(methacryloyloxy)propylsulfonyl]-1-(trifluoromethanesulfonyl)imide Na) (PMTFSINa), serves as a gel polymer electrolyte (GPE), which effectively blocks the diffusion of NaPSs through electrostatic repulsion and concurrently inhibits the Na dendrite growth. Additionally, the rich in defects and nitrogen-containing $Ti_{3-x}CNT_y$ MXene (labeled as DN-MXene) coating layer enhances the kinetics of NaPSs transitions electrocatalytically. Batteries with this Janus separator show improved specific capacity, extended cycle life, and reduced electrolyte requirements compared to conventional glass fiber separators. Xu and coworkers developed a novel approach to enhance battery safety and performance using a functionalized porous ZIF-based membrane (PZM). This membrane incorporates metal ions within its zeolitic imidazolate framework, which effectively binds with polysulfide anions to prevent their harmful migration. Additionally, the specifically engineered porous structure allows for selective and consistent passage of Na ions, significantly reducing uneven deposition and preventing the growth of dendritic Na. Furthermore, the inherently fire-resistant properties of the ZIF-based membrane safeguard against thermal incidents and the decomposition of electrolytes. Employing this advanced, multifunctional PZM in conjunction with S/C cathode results in a robust separator that enhances the safety and reliability of lean electrolyte Na–S batteries. [35] Covalent organic frameworks (COFs), recognized for their tunable structures and intrinsic nanopores, are emerging as a novel class of crystalline polymers utilized in modifying separators. Yin and coworkers synthesized thin films of COFs that incorporate azobenzene side groups attached to the pore walls. [36] These azobenzene modifications serve a dual purpose: they reduce the aperture to sub-nanometer sizes, effectively impeding the shuttle effect, and they function as ion-hopping sites to facilitate the Na-ion transfer. Owing to these innovations, the Na–S batteries utilizing these COFs thin films as separators achieve remarkable electrochemical performance.

5.5.4 Electrolytes

The electrolyte in RT Na–S batteries plays a crucial dual role. It facilitates the ion transport of Na ions while simultaneously serving as a barrier that prevents electron transfer between the anode and cathode, reducing the risk of short circuits. Currently, electrolytes in these batteries are typically categorized into three types: liquid electrolytes (LEs), GPEs, and solid-state electrolytes (SSEs).

5.5.4.1 Liquid Electrolytes (LEs)

Na–S batteries typically employ organic electrolytes, which are often categorized into carbonate-based and ether-based types. The formation of NaPSs during operation can induce Na dendrite growth and trigger side reactions at the cathode. This not only affects the stability of Na–S batteries but also results in S loss when NaPSs migrate toward the anode. Consequently, substantial research efforts are directed toward improving LEs to support the commercial viability of Na–S batteries. He and coworkers demonstrated localized high-concentration electrolytes (LHCEs) with a specific molar ratio of dimethoxyethane (DME) to sodium bis(fluorosulfonyl)imide and 1,1,2,2-tetrafluoroethyl 2,2,3,3-tetrafluoropropyl ether, set at 1 : 1.2 : 1. [40] They discovered that as the DME ratio decreases, the solubility of NaPSs also diminishes. Crucially, they found that the redox reaction of S shifted from the traditional dissolution–precipitation dynamics to a quasi-solid-state process within this specially formulated electrolyte. This transition helps suppress the shuttling of NaPSs and guarantees faster reaction kinetics. Additionally, the electrolyte aids in forming a stable solid SEI on the anode, preventing the growth of dendritic Na. These improvements contribute to enhancing the long-cycle stability of the battery, maintaining integrity over 300 cycles. Guo and coworkers proposed a novel electrolyte concept for Na–S batteries, termed localized saturated electrolyte, which utilizes 2-methyltetrahydrofuran (MeTHF) as the inner solvent. [37] This innovative approach leverages the unique MeTHF to create a saturated inner sheath. This sheath enhances the coordination between Na ions and anions,

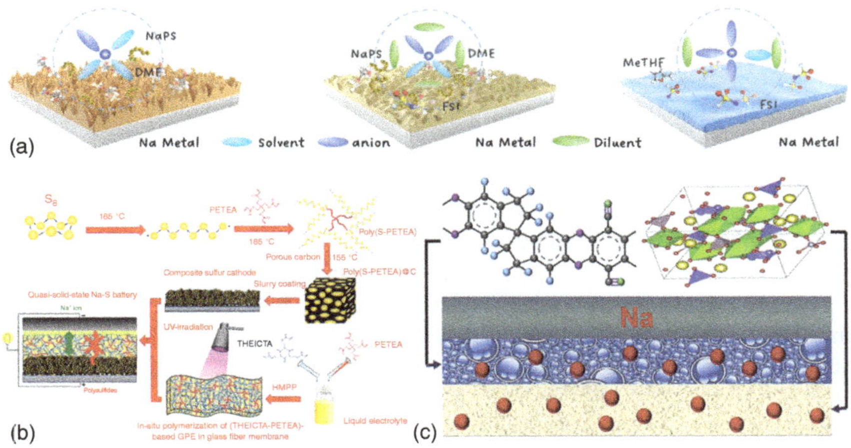

Figure 5.7 (a) The solvation structure diagrams in electrolytes of different concentrations (from left to right: low concentration, LHCE, MeTHF, respectively); the interfacial compatibility between NaPSs or electrolytes and Na anode. Source: Reproduced from [37]/with permission of John Wiley & Sons. (b) Structural diagram of the quasi-solid-state Na–S battery. Source: Reproduced from [38]/with permission of John Wiley & Sons. (c) Schematic illustration of the polymer with intrinsic nanoporosity (PIN) coated $Na_3Zr_2Si_2PO_{12}$. Source: Reproduced from [39]/with permission of John Wiley & Sons.

reducing the interaction between anions and the diluent (Figure 5.7a). Such a configuration is crucial for establishing a continuous and stable interphase, facilitating a quasi-solid-state S charging and discharging process. As a result, it enables the creation of a battery that is free from dendrite formation and polysulfide shuttling. This electrolyte has also demonstrated promising cycling performance in pouch cells under challenging conditions.

Wu and coworkers developed a novel carbonate-based electrolyte, which consists of trimethyl phosphate (TMP), fluoroethylene carbonate (FEC), and Na trifluoromethanesulfonimide (NaTFSI). [41] This innovative electrolyte contributes to highly reversible Na plating and stripping. Employing this NaTFSI/TMP + FEC electrolyte, the dendrite-free battery delivers an impressive specific capacity of 788 mAh g^{-1} at a 1 C rate. Xu and colleagues further advanced the development of the Na–S battery by meticulously optimizing the electrolyte system, which includes propylene carbonate (PC), FEC, NaTFSI, and indium triiodide (InI_3). [42] Through both theoretical and experimental analysis, it has been demonstrated that FEC solvent, combined with the high salt concentration, greatly diminishes the solubility of NaPSs and effectively establishes a durable SEI on the anode during operation. Furthermore, the introduction of InI_3 serves as a redox mediator, accelerating the conversion kinetics of Na sulfide at the cathode while simultaneously forming a protective indium (In) layer on the anode, shielding it from polysulfide erosion. These innovations result in Na–S batteries with enhanced capacity and longer cycle stability.

In addition, ionic liquid-based electrolytes have attracted attention owing to their high ionic conductivity, wide electrochemical window, and flame retardancy.

However, their adoption is limited by high viscosity. While they have not yet been widely used as the primary solvent, their application as additives in organic electrolytes has proven to enhance performance significantly. Wei and coworkers developed a Na–S battery assembled from a microporous carbon/S composite cathode and enhanced the solvation of the electrolyte by adding 1-methyl-3-propylimidazolium-chlorate ionic liquid bound to silica nanoparticles (SiO_2-IL-ClO_4). [15] These additives stabilize the electrodeposition process. The synergistic effect of the cathode carrier and customized electrolyte ensures strong S retention within the cathode, promoting electrochemical reactions limited to all solid states, with Na_2S as the sole product. This design delivers exceptional performance, achieving 100% CE, even at high current densities and with substantial S content in the cathode.

5.5.4.2 Gel Polymer Electrolytes (GPEs)

To address the risks posed by flammable organic LEs in RT Na–S batteries, researchers have developed GPEs and SSEs suitable for use with Na metal anodes and S cathodes. These GPEs are produced by incorporating a specified amount of LE into gel polymers, followed by solvent evaporation, resulting in a flexible matrix. This approach effectively restricts the number of organic solvents within the gel polymer structure, remarkably reducing side reactions and boosting the thermal stability of batteries. Zhou and coworkers developed a Na–S battery, which mainly consists of a cathode of poly(S-pentaerythritol tetra acrylate [PETEA]) and a GPE composed of PETEA and tris[2-(acryloyloxy)ethyl] isocyanurate, as illustrated in Figure 5.7b. [38] The S electrode effectively anchors S through chemical bonding, mitigating the shuttle effect. Simultaneously, the polymer electrolyte, with high Na-ion conductivity, stabilizes the Na anode–electrolyte interface and limits NaPSs dissolution. This innovative quasi-solid-state Na–S cell exhibits a high specific capacity of 877 $mA\,h\,g^{-1}$ at 0.1 C. Murugan and colleagues developed a LE solution that contains 1 M Na tetraphenylborate, PC, and 7.0 wt% FEC. This solution was crosslinked through *in situ* free radical polymerization of pentaerythritol triacrylate, triggered by 1.0 wt% azobisisobutyronitrile. [43] Polymerization occurs on both the Na anode surface and within the porous sulfurized polyacrylonitrile (SPAN) cathode, ensuring excellent interfacial contact. The resulting GPEs demonstrated superior ionic conductivity compared to traditional LEs, especially at elevated temperatures. The polymer matrix within the electrolyte effectively confines discharge products, preventing cathode pore blockage by insulating crystallites and promoting electrochemical conversion. Na-SPAN full cells with these GPEs display exceptional cycling stability.

5.5.4.3 Solid-State Electrolytes (SSEs)

To address the detrimental side reactions between organic solvents and both the Na anode and S cathode in batteries, research has significantly focused on the development of SSEs. These SSEs eliminate the use of organic liquid solvents, thereby reducing the continued side reactions of Na metal with the electrolyte and associated gas evolution. Additionally, the high shear modulus of SSEs helps prevent the penetration of Na metal dendrites, enhancing safety. For S cathodes,

SSEs prevent the dissolution of NaPSs, countering the shuttle effect. However, while SSEs offer numerous advantages, their practical application faces several challenges. The key challenges include poor ionic conductivity and chemical instability when in contact with polysulfides and Na, leading to difficulty in maintaining stable interfacial contact at electrolyte/electrode interfaces. To tackle these challenges, various efforts have been undertaken, achieving notable progress in enhancing the performance and stability of SSEs. Tanibata and coworkers developed all-solid-state cells using Na_3PS_4 ceramic as a solid electrolyte. [44] These all-solid-state Na–S cells leverage the complete convention of S active material to the final discharge product, Na_2S, achieving a high capacity of approximately 1100 mAh g^{-1} of S at RT. To address the key issues of Na dendrite formation and polysulfide shuttling, Yu and his colleagues developed a novel hybrid-electrolyte method. This approach employs a Na-ion SSE membrane ($Na_3Zr_2Si_2PO_{12}$), as a replacement for traditional porous separators first in the field. [39] They coated this membrane with a layer of PIN, which not only facilitates a smoother interface with the Na anode but also serves as an elastic buffer (Figure 5.7c). This buffer effectively mitigates the risk of the solid-electrolyte pellet fracturing during the assembly of the cells, enhancing the structural integrity and operational reliability of the batteries.

5.6 Summary and Outlook

The RT Na–S battery stands out as a highly promising electrochemical system, potentially surpassing the energy densities of existing battery technologies. However, several significant challenges hinder its development: (i) low electronic conductivity of S, (ii) volume expansion of active S, (iii) polysulfide shuttling, (iv) sluggish reaction kinetics, (v) nonuniform SEI, and (vi) the formation of Na dendrites. This chapter has provided an extensive discussion of various strategies to overcome these hurdles, including the modulation of S cathodes, anode design, separator modifications, and electrolyte optimization. For achieving practical energy densities exceeding 300 Wh kg^{-1}, a high S load and minimal electrolyte usage are essential. [45] However, there has been insufficient focus on pursuing a low electrolyte-to-S (E/S) ratio while maintaining robust electrochemical performance, as the intrinsic challenges are exacerbated at lower E/S ratios. Continued research and development are required, and potential future directions for advancing RT Na–S battery technology are outlined below:

1) One prevalent strategy to mitigate the polysulfide shuttle behavior in RT Na–S batteries is the careful design of the S cathode host. This strategy typically involves the use of carbonaceous materials, and carbonaceous materials coupled with polar inorganic components (single metals, metal clusters, metal oxides/sulfides, and metal nitrides). Carbonaceous materials physically confine polysulfides and improve electron transfer, while polar additives chemically adsorb NaPSs intermediates. However, these non-active materials can detract from the battery's overall energy density. To improve the practicality of the S electrode, increasing both the S content and its utilization is critical, as these factors directly influence the areal capacity of cathodes. Therefore, for boosting

the energy density, it is essential to select S-host materials that incorporate a minimal mass of non-active components while maximizing S content.

2) Optimizing catalytic materials is crucial for addressing the shuttle effect and enhancing the sluggish redox kinetics in batteries. Although several studies have focused on designing electrocatalyst-enhanced S cathodes, the chemical properties of RT Na–S batteries are not fully understood. Variations in CV curves and charge/discharge profiles across different studies indicate that the conversion mechanisms of S cathodes are still a subject of debate. Detailed analyses of CV curves and specific S conversion processes remain elusive, and the catalytic mechanisms related to S conversion at the cathode need further clarification. To advance our understanding, there is a pressing need for advanced *in situ* characterization techniques. For example, synchrotron radiation and *in situ* TEM. These methods would provide real-time insights into the behavior of S species during electrochemical conversion. Additionally, integrating theoretical calculations to determine the free energy and binding energy could provide deeper insights into material interactions and facilitate more effective cathode materials development.
3) The design of Na–S batteries requires a focused approach to mitigating metal dendrite formation and their interaction with polysulfides. Employing porous, conductive, and lightweight materials can substantially increase the available surface area for Na deposition, thereby enhancing anode electrochemical performance. Furthermore, incorporating specific electrolyte additives can curb the Na dendrite formation and aid in the SEI layer formation, which is vital for the longevity and operational safety of the battery. Additionally, considering the aim of achieving large-scale battery production in the future, it is essential to ensure that the Na hosts are both abundant and economical, which is key to the feasibility and broad-scale implementation of the technology.
4) Modified separators are pivotal for managing ion flux and minimizing the formation of Na metal dendrites, while ion-selective separators help reduce the shuttling of NaPSs, thereby boosting the reversibility of the cells. Research should continue to explore multifunctional separators that can both prevent Na metal dendrite growth and control NaPSs shuttling. Additionally, these separators must be designed to be lightweight to preserve the gravimetric energy density of the full cells, ensuring that enhancements in functionality do not compromise the overall battery performance.
5) Exploring new electrolyte systems, especially those that are compatible with Na metals, is critical. There is an urgent need to develop electrolytes that are both compatible and cost-effective, supporting reversible metal deposition and improved ion transfer. Simultaneously, formulating electrolyte additives for batteries is essential to meet these requirements. Furthermore, the reaction mechanisms between electrolytes and additives still demand comprehensive study. Future research needs to develop strategies for recovering "dead metal" to improve metal anode utilization. Additionally, the typical use of flammable LEs in RT Na–S batteries poses substantial safety risks. Future designs of SSEs should prioritize not only preventing these issues but also ensuring robust physical contact and compatibility with electrodes to optimize performance and durability.

In the research of Na–S battery systems, several key parameters, including S loading, the negative-to-positive (N/P) ratio, and the E/S ratio, require careful consideration for their practical application. It is crucial to design S composites with a high S content to realize high energy densities. Furthermore, the often-overlooked E/S ratio significantly affects the electrochemical performance of these systems and should be minimized to optimize energy densities. Many studies report extensive cycling with either low S loading or high E/S ratios, potentially misleading the research community. Future research must specify these parameters to effectively guide and enhance ongoing efforts. Additionally, maintaining a balanced N/P ratio without reducing the energy density is essential, especially since current practices often involve using an excess of Na metal anodes, which can diminish the entirety energy density of the cell. Addressing the residual metal in anodes without deteriorating cell performance necessitates improvements in Na metal anodes' stability and reversibility. Exploring anode-free cell configurations could provide valuable insights into the fundamental challenges and guide targeted solutions. Moreover, while most Na–S battery research employs coin-type cells, transitioning to practical applications like increased cell size necessitates the design and testing of more scalable solutions such as pouch cells. This approach will help bridge the gap between laboratory research and real-world application, fostering the development of commercially viable Na–S battery technologies.

References

1 (a) Dunn, B., Kamath, H., and Tarascon, J.-M. (2011). *Science* 334: 928. (b) Larcher, D. and Tarascon, J.M. (2015). *Nature Chemistry* 7: 19.

2 Zhao, C., Wang, Q., Yao, Z. et al. (2020). *Science* 370: 708.

3 (a) Wang, M., Bai, Z., Yang, T. et al. (2022). *Advanced Energy Materials* 12: 2201585. (b) Yao, W., Liao, K., Lai, T. et al. (2024). *Chemical Reviews* 124: 4935. (c) Wang, C., Zheng, Y., Chen, Z.N. et al. (2023). *Advanced Energy Materials* 13: 2204125.

4 (a) Li, Z., Wang, C., Ling, F. et al. (2022). *Advanced Materials* 34: 2204214. (b) Zhao, L., Tao, Y., Zhang, Y. et al. (2024). *Advanced Materials* 36: 2402337.

5 Liu, H., Lai, W.H., Lei, Y. et al. (2022). *Advanced Energy Materials* 12: 2103304.

6 Soni, C.B., Sungjemmenla, S., and Vineeth, V.K. (2022). *Energy Storage* 4: e264.

7 Wang, Y.X., Zhang, B., Lai, W. et al. (2017). *Advanced Energy Materials* 7: 1602829.

8 Oshima, T., Kajita, M., and Okuno, A. (2004). *International Journal of Applied Ceramic Technology* 1: 269.

9 Yao, Y.-F.Y. and Kummer, J. (1967). *Journal of Inorganic and Nuclear Chemistry* 29: 2453.

10 Hueso, K.B., Armand, M., and Rojo, T. (2013). *Energy & Environmental Science* 6: 734.

11 Manthiram, A. and Yu, X. (2015). *Small* 11: 2108.

12 Yu, X. and Manthiram, A. (2014). *ChemElectroChem* 1: 1275.

13 Xin, S., Yin, Y.X., Guo, Y.G., and Wan, L.J. (2014). *Advanced Materials* 26: 1261.
14 Wang, N., Wang, Y., Bai, Z. et al. (2020). *Energy & Environmental Science* 13: 562.
15 Wei, S., Xu, S., Agrawral, A. et al. (2016). *Nature Communications* 7: 11722.
16 Eng, A.Y.S., Kumar, V., Zhang, Y. et al. (2021). *Advanced Energy Materials* 11: 2003493.
17 Zhang, S., Yao, Y., and Yu, Y. (2021). *ACS Energy Letters* 6: 529.
18 Wang, Y.-X., Yang, J., Lai, W. et al. (2016). *Journal of the American Chemical Society* 138: 16576.
19 Li, D., Gong, B., Cheng, X. et al. (2021). *ACS Nano* 15: 20607.
20 Ghosh, A., Shukla, S., Monisha, M. et al. (2017). *ACS Energy Letters* 2: 2478.
21 Guo, Q., Li, S., Liu, X. et al. (2020). *Advancement of Science* 7: 1903246.
22 Wei, S., Ma, L., Hendrickson, K.E. et al. (2015). *Journal of the American Chemical Society* 137: 12143.
23 Hwang, T.H., Jung, D.S., Kim, J.-S. et al. (2013). *Nano Letters* 13: 4532.
24 Guo, B., Du, W., Yang, T. et al. (2020). *Advancement of Science* 7: 1902617.
25 Jiang, Y., Yu, Z., Zhou, X. et al. (2023). *Advanced Materials* 35: 2208873.
26 Yan, Z., Liang, Y., Xiao, J. et al. (2020). *Advanced Materials* 32: 1906700.
27 Zhou, X., Yu, Z., Yao, Y. et al. (2022). *Advanced Materials* 34: 2200479.
28 Zhang, S., Yao, Y., Jiao, X. et al. (2021). *Advanced Materials* 33: 2103846.
29 Kumar, V., Eng, A.Y.S., Wang, Y. et al. (2020). *Energy Storage Materials* 29: 1.
30 Li, Z., Chen, X., Yao, G. et al. (2024). *Advanced Functional Materials* 34: 2400859.
31 Zhou, J., Yang, Y., Zhang, Y. et al. (2021). *Angewandte Chemie, International Edition* 60: 10129.
32 Wan, J., Wu, J., Gao, X. et al. (2017). *Advanced Functional Materials* 27: 1703933.
33 Qi, Y., Li, Q.-J., Wu, Y. et al. (2021). *Nature Communications* 12: 6347.
34 Zhou, D., Tang, X., Guo, X. et al. (2020). *Angewandte Chemie, International Edition* 59: 16725.
35 Xu, H., Xiang, Y., Xu, X. et al. (2024). *Advanced Functional Materials* 34: 2403663.
36 Yin, C., Li, Z., Zhao, D. et al. (2022). *ACS Nano* 16: 14178.
37 Guo, D., Wang, J., Lai, T. et al. (2023). *Advanced Materials* 35: 2300841.
38 Zhou, D., Chen, Y., Li, B. et al. (2018). *Angewandte Chemie, International Edition* 57: 10168.
39 Yu, X. and Manthiram, A. (2019). *Matter* 1: 439.
40 He, J., Bhargav, A., Shin, W., and Manthiram, A. (2021). *Journal of the American Chemical Society* 143: 20241.
41 Wu, J., Liu, J., Lu, Z. et al. (2019). *Energy Storage Materials* 23: 8.
42 Xu, X., Zhou, D., Qin, X. et al. (2018). *Nature Communications* 9: 3870.
43 Murugan, S., Klostermann, S.V., Schützendübe, P. et al. (2022). *Advanced Functional Materials* 32: 2201191.
44 Tanibata, N., Deguchi, M., Hayashi, A., and Tatsumisago, M. (2017). *Chemistry of Materials* 29: 5232.
45 Wang, B., Wang, L., Guo, B. et al. (2024). *Advanced Materials* 36: 2411725.

6

Zinc-Ion Rechargeable Battery

Guangmeng Qu and Liqiang Xu

Key Laboratory of Colloid and Interface Chemistry (Ministry of Education), School of Chemistry and Chemical Engineering, Shandong University, Jinan, Shandong, 250100, P. R. China

6.1 Overview of Aqueous Zinc-Ion Battery

The rapid development of clean energy sources such as ocean, solar, wind, and tidal power presents a significant challenge: effectively storing intermittent energy for consistent use. Achieving this is crucial for the utilization of renewable energy. Therefore, the advancement of high-efficiency energy storage systems holds essential significance for broad coverage of clean renewable energies. Currently, lithium-ion and lead-acid batteries dominate the energy storage market. However, concerns about lithium scarcity and potential safety hazards, as well as the low lifespan and environmental impact of lead-acid batteries, drive the search for new storage technologies that are safer, environmentally friendly, and more resource-abundant.

One of the earliest energy storage systems, which uses zinc metal as the active material in anodes, has a long development history. From the early days of zinc–copper primary batteries to modern aqueous zinc-ion batteries (AZIBs), zinc-based batteries have significantly evolved. Systems like zinc–manganese dry cells and zinc–nickel secondary batteries represent a notable share in the energy storage market. However, their operational voltage and lifespan have not matched mainstream lithium-ion batteries, hindering their entry into power sources and large-scale storage. In recent years, escalating demands for energy storage systems with higher safety and environmental standards have spurred rapid developments in zinc-based energy technologies, notably AZIBs. These batteries offer superior environmental sustainability and safety compared to commercially available lithium-ion and lead-acid batteries, repositioning zinc-ion batteries on the energy stage. Developing high-performance zinc-ion rechargeable batteries offers potential solutions to mitigate issues such as poor safety in lithium-ion batteries and lithium resource scarcity, gradually replacing low-energy-density and toxic lead-acid batteries. Compared to well-commercialized alkaline zinc–nickel and zinc–manganese batteries, near-neutral AZIBs offer higher active material utilization and superior overall performance, yet scaling for commercial applications remains a challenge.

Electrode Materials in Energy Storage Technologies: Applications in Lithium-, Sodium-, Potassium-, Sulfur- and Zinc-Based Rechargeable Batteries, First Edition. Edited by Liqiang Xu.

While the electrode configurations of AZIBs resemble those of the aforementioned systems, their fundamentally different charge carriers and electrode conversion mechanisms mean that the research experience from alkaline zinc-based batteries is of limited relevance. Similarly, the electrode materials and electrolyte systems that have illustrated significant progress in lithium-ion batteries cannot be directly applied to multivalent zinc-ion batteries. Therefore, achieving commercially viable zinc-ion batteries requires comprehensive efforts from developing electrode materials and innovating electrolyte solutions to improving assembly processes and establishing new testing and evaluation frameworks. In recent years, researchers, both domestically and internationally, have made significant strides in exploring zinc-ion battery cathode and anode materials, electrolytes, other non-active components, battery reaction characterization techniques in aqueous zinc-based systems, as well as theoretical calculations and simulation techniques.

As research progresses, the prominent merits of zinc-ion batteries in the energy storage field are gradually being uncovered, showcasing immense commercial potential in the diverse landscape of twenty-first century energy storage. This chapter offers a concise summary of the development history of zinc-ion batteries and categorizes current research on zinc-ion batteries and related process technologies based on energy storage systems and component structures. Key sections include zinc-ion battery cathode materials, anode materials, electrolytes, non-active components, battery reaction characterization techniques in zinc-ion batteries, and theoretical calculations and simulation techniques.

6.2 Introduction

6.2.1 Development History

The earliest zinc-based battery, also the earliest of all batteries, was invented by Alessandro Volta in 1799 using a circular arrangement of silver and zinc plates with a moistened cloth soaked in saltwater in between. Volta created what became known as the Voltaic pile. This setup generated electric current from the potential difference between zinc and silver metals. In 1868, Georges Leclanché introduced the first generation of zinc–manganese dry cell batteries ($Zn|ZnCl_2\text{-}NH_4Cl|MnO_2$), which became commercially viable by 1888. Since then, zinc primary batteries have undergone continual redesign, improvement, and miniaturization, with zinc consistently serving as the primary anode material while the cathode has evolved to include copper, manganese, silver, and even air. However, the inability to recharge primary batteries proved limiting. Thus, in the late nineteenth century, the first rechargeable zinc-based battery based on zinc bromide electrolyte was developed. Thomas Edison patented the first Ni–Zn battery in 1901 and explored its use in inventions such as the phonograph, railway signals, and electric fans. By 1959, commercially viable alkaline zinc-based batteries were established, featuring a zinc powder anode that significantly extended lifespan compared to carbon–zinc dry cells. In 1973, a hybrid zinc flow battery system was proposed, utilizing zinc metal as the anode and a Br/Br^- redox couple in the electrolyte. In 1993, researchers made a technological breakthrough by further improving the alkaline electrolyte and finally commercialized the nickel–zinc battery in 2000 [1].

The prototype of true AZIBs emerged in 1986 when Yamamoto and colleagues successfully substituted neutral $ZnSO_4$ electrolytes for the KOH utilized in rechargeable alkaline zinc batteries. This first battery used MnO_2 with a layered structure as the cathode, zinc metal as the anode, and $ZnSO_4$ aqueous solution as the electrolyte but demonstrated a limited lifespan with fewer than 50 charge cycles. In 2011, Professor Feiyu Kang's team at Tsinghua University introduced the concept of "zinc-ion batteries," reporting reversible intercalation/deintercalation of Zn^{2+} ions in α-MnO_2 [2]. These batteries, utilizing divalent zinc ions as charge carriers, showed substantially improved cycle reversibility and capacity compared to traditional alkaline zinc batteries. Since then, the concept of AZIBs has been firmly established, with a substantial number of researchers directing their efforts toward elucidating reaction mechanisms and developing advanced electrode materials for these batteries.

6.2.2 Structure and Components of AZIBs

A common structural diagram of AZIBs is illustrated in Figure 6.1. Similar to lithium-ion and sodium-ion batteries, their active structure primarily consists of a zinc metal anode (occasionally organic materials or metal compounds capable of storing zinc ions), a separator made of materials like glass fiber or other hydrophilic substances, a mild zinc salt electrolyte solution, and a cathode material that facilitates rapid insertion/extraction or adsorption/desorption of zinc ions.

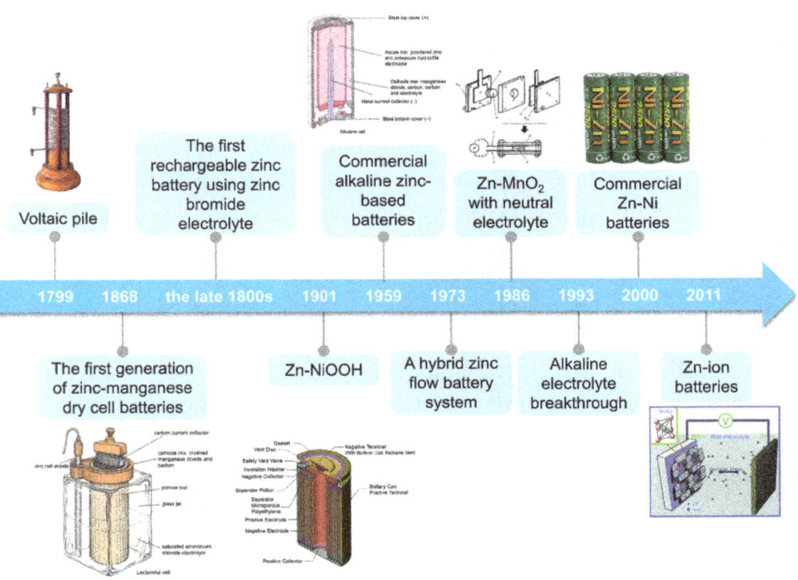

Figure 6.1 History of AZIBs development.

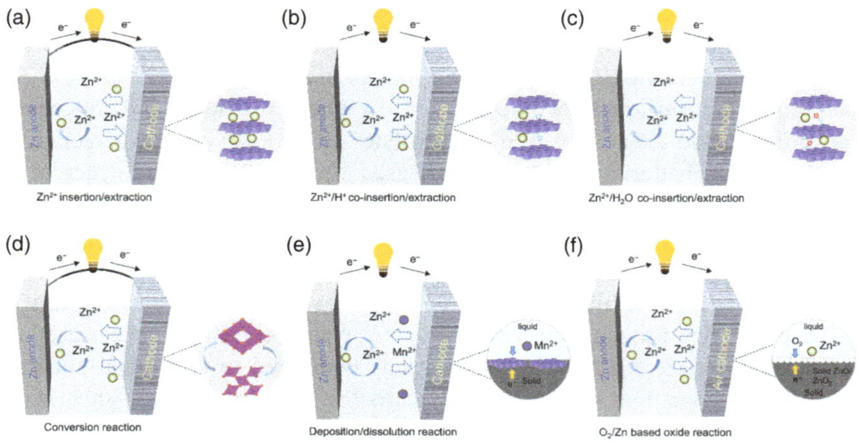

Figure 6.2 Energy storage mechanism of AZIBs. (a) Zn^{2+} insertion/extraction; (b) Zn^{2+}/H^{+} co-insertion/extraction; (c) Zn^{2+}/H_2O co-insertion/extraction; (d) Conversion reaction; (e) Deposition/dissolution reaction.

Unlike alkali metal ion batteries, zinc-ion batteries achieve two-electron transfer through the migration of a single ion, potentially achieving higher energy densities. As depicted in Figure 6.2, under a mild electrolyte system, the anode reaction in AZIBs primarily involves the deposition and dissolution of zinc metal (Eq. (6.1)).

$$Zn^{2+} + 2e^{-} \leftrightarrow Zn \tag{6.1}$$

In specific cases involving intercalation reactions and adsorption/desorption processes, the anode materials mainly include organic materials and low-potential metal compounds. The detailed reaction mechanisms will be discussed in Section 4, focusing on the zinc anode materials.

In AZIBs, the anode typically consists of metallic zinc, meaning that the energy storage mechanism is mainly determined by the cathode material. Confirmed cathode materials for AZIBs include vanadium-based compounds (mainly V_2O_5), manganese-based compounds (mainly MnO_2), Prussian blue and its analogs (PBAs), along with various organic compounds. Unlike other traditional battery systems, due to the application of near-neutral aqueous electrolytes, during the battery cycling process, conventional current collectors such as copper foil, aluminum foil, and nickel foam undergo varying degrees of dissolution. Therefore, the current collectors used for the cathode materials in AZIBs mainly include titanium foil, stainless steel mesh, and hydrophilic carbon cloth. These substrates exhibit both high hydrogen evolution overpotential and stable chemical properties.

The electrolyte, apart from the electrodes, is another critical component of the battery, serving as the medium for ion transport and connecting the cathode and anodes. An ideal electrolyte for AZIBs should meet several requirements: a broad electrochemical range and excellent compatibility with other battery components, environmental friendliness, low toxicity, and cost-effectiveness. Currently, electrolytes for AZIBs mainly include $ZnSO_4$ aqueous solution and $Zn(CF_3SO_3)_2$ and other zinc salts. Due to the dissolution of active materials, the capacity of zinc-ion batteries rapidly deteriorates. Although neutral electrolytes in zinc-ion batteries do not exhibit severe zinc dendrite growth at low current densities, dendritic issues cannot be ignored at high current densities. The presence of significant free water in the aqueous solution leads to unavoidable side reactions, including the dissolution of active material at the cathode, corrosion of the zinc anode, and the formation of additional products on both the cathode and anodes.

$CF_3SO_3^-$ possesses a larger ionic radius and molecular weight, which effectively diminishes the solvation effect of Zn^{2+} ions, facilitating a quicker insertion and extraction of Zn^{2+} ions within the cathode material. Therefore, the overall performance of the $Zn(CF_3SO_3)_2$ electrolyte is better than the $ZnSO_4$ electrolyte. Furthermore, a high $Zn(CF_3SO_3)_2$ solution concentration can achieve more uniform zinc deposition without forming zinc dendrites. However, the widespread application of $Zn(CF_3SO_3)_2$ electrolyte is limited due to its high cost. Therefore, there is an urgent need to explore reasonably priced new zinc salts and additives to adjust the electrolyte concentration, expand the electrochemical stability range, enhance Coulombic efficiency, and prolong battery cycle stability in aqueous electrolytes.

6.2.3 Energy Storage Mechanism

The reaction mechanism of zinc-ion batteries in aqueous solutions, especially involving manganese-based and vanadium-based cathode materials, remains unclear and controversial. Unlike other alkali metal ion batteries, which involve intercalation reactions, zinc-ion batteries primarily define their energy storage mechanism based on cathodic reactions due to the zinc metal deposition/dissolution on the anode side.

Various mechanisms for AZIBs have been suggested, such as the insertion/extraction of Zn^{2+} ions, co-insertion/extraction of H^+/Zn^{2+} ions, co-insertion/extraction of H_2O/Zn, chemical conversion reactions, and deposition/dissolution processes. The energy storage mechanism of zinc-ion batteries is typically classified and illustrated in Figure 6.2 based on cathodic energy storage. Detailed descriptions and representative electrode materials corresponding to these mechanisms will be discussed in Section 4 on cathode materials.

(1) Zn^{2+} Insertion/Extraction Mechanism (Figure 6.2a):

The reversible insertion/extraction of Zn^{2+} ions for charge transfer was initially proposed for mild rechargeable AZIBs. Representative cathode materials based on this reaction mechanism include layered and tunnel-structured manganese oxides, vanadium oxides, as well as polyanion materials with open-framework structures, and PBAs. Kang et al. first reported this mechanism, illustrating the reversible insertion and extraction of Zn^{2+} ions into and from the α-MnO_2 cathode during charge and discharge cycles, while zinc metal undergoes dissolution/deposition on the anode side. However, debates continue over the energy storage mechanism of α-MnO_2 cathodes. Other mechanisms for manganese oxide-based cathodes have also been elucidated. Nazar and coworkers described the Zn ion insertion reaction in the $Zn_{0.25}V_2O_5{\cdot}nH_2O$ cathode [3], describing a rapid and reversible storage process where more than one Zn^{2+} ion can be accommodated per unit lattice (with a capacity of up to 300 mAh g^{-1}). First-principles calculations indicate that around 0.5 mol of Zn^{2+} can be inserted into the VO_2 cathode host material during discharge. The insertion process of Zn^{2+} ion is divided into two parts: initially inserting 0.125 mol of Zn^{2+}, forming $Zn_{0.125}V_2O_5$ at an average potential of 0.78 V, followed by further discharge to 0.55 V to insert an additional 0.375 mol of Zn^{2+} ion, forming $Zn_{0.5}V_2O_5$. Similar mechanisms are observed in other vanadium-based compounds like $Li_xV_2O_5{\cdot}nH_2O$, $Ca_{0.25}V_2O_5{\cdot}nH_2O$, $Na_2V_6O_{16}{\cdot}3H_2O$, and $Zn_3V_2O_7(OH)_2{\cdot}2H_2O$. Vanadium-based compounds adjust to the insertion of Zn^{2+} ions by evolving from V^{5+} to V^{4+} or even V^{3+}, while preserving a stable lattice structure [4].

(2) H^+/Zn^{2+} Co-Insertion/Extraction Mechanism (Figure 6.2b):

The sluggish insertion of Zn^{2+} ions is due to the strong electrostatic repulsion from high-valence charges and large hydrated hydrogen ions. The insertion of ions with faster diffusion kinetics, such as H^+ and Na^+ ions, can increase the utilization of active sites in the main structure, thereby enhancing the performance of zinc-ion batteries. Common AZIBs cathode materials include vanadium-based oxides with tunnel and layered structures, which facilitate charge storage through an H^+/Zn^{2+} co-insertion/extraction mechanism.

Wang et al. first introduced the H^+/Zn^{2+} co-insertion mechanism in Zn/ε-MnO_2 batteries. GITT and cyclic voltammetry (CV) curves exhibited that proton insertion precedes Zn^{2+} insertion, forming MnOOH, which then transforms to $ZnMnO_4$ [5]. Professor Ji and colleagues also confirmed the reversible H^+/Zn^{2+} co-insertion in α-MnO_2 cathodes [6]. Initially, H^+ inserted into α-MnO_2, forming $[Zn(OH)_2]_3$ $ZnSO_4{\cdot}5H_2O$, followed by Zn^{2+} insertion into the host to create $ZnMn_2O_4$, with less efficient Zn^{2+} insertion. Kang et al. observed similar sequential H^+ and Zn^{2+} ions insertion in MOF-derived α-Mn_2O_3, transforming discharge product morphology. A similar mechanism occurs in vanadium-based cathodes [6]. In $NaV_3O_8{\cdot}1.5H_2O$ with 1 M Na_2SO_4/1 M $ZnSO_4$ electrolyte, H^+/Zn^{2+} co-insertion is observed, represented by Eq. (6.2) [7]

$$NaZn_{0.1}V_3O_8 \cdot 1.5H_2O + 3.9H^+ + 0.4Zn^{2+} + 4.7e^- \rightarrow H_{3.9}NaZn_{0.5}V_3O_8 \cdot 1.5H_2O \tag{6.2}$$

The lighter proton co-insertion improves zinc-ion insertion's kinetics and diffusion dynamics. It is noteworthy that H^+ and Zn^{2+} ions co-insert rather than stepwise intercalation. Furthermore, the mechanism by which H^+ affects ion migration requires further study.

(3) H_2O/Zn^{2+} Co-insertion/Extraction Mechanism (Figure 6.2c)

Beyond the proton co-insertion mechanism, vanadium-based oxide cathode materials (V_2O_5, V_6O_{13}, and $H_2V_3O_8$) exhibit similar co-insertion behavior involving both water molecules and zinc ions. Solvated water mitigates Zn^{2+} ion charge density and weakens electrostatic interactions within the framework, promoting Zn^{2+} ion transfer. Mai et al. highlighted the importance of structured water in facilitating Zn^{2+} ions insertion, [8] as the solvated Zn^{2+} ion carries a reduced effective charge, minimizing its electrostatic interaction with the V_2O_5 structure and enhancing ion diffusion. Simultaneously, water molecules, Zn^{2+} ions, and $CF_3SO_3^-$ ions are inserted to maintain charge neutrality, decreasing the interlayer spacing in V_2O_5 from 12.6 to 10.4 Å at 1.3 V, which is attributed to hydrogen bonding between Zn^{2+} ions, $CF_3SO_3^-$, H_2O, and lattice oxygen, pulling the V_2O_5 bilayers closer together. Upon discharge to 0.2 V, the spacing increases to 13.5 Å. Cheng et al. also observed simultaneous zinc and water insertion into V_2O_5 [9] during the initial discharge, forming an open-structured $Zn_{3+x}(OH)_2V_2O_7{\cdot}2H_2O$ matrix, followed by reversible Zn^{2+} ions extraction. Similar to V_2O_5, water molecules can intercalate into V_6O_{13} to form $Zn_aV_6O_{13}{\cdot}mH_2O$ and $Zn_xV_2O_7(OH)_2{\cdot}2H_2O$, with cycling depending on Zn^{2+} insertion/extraction. Huang et al. discovered that in the Zn||$NaCa_{0.6}V_6O_{16}{\cdot}3H_2O$ battery, the structured water layer not only widens the channel spacing for better Zn^{2+} storage but also enhances reaction kinetics. The electrochemical reaction of the cathode is as follows: [10]

$$NaCa_{0.6}V_6O_{16} \cdot 3H_2O + 4.3Zn^{2+} + 8.6e^- \leftrightarrow Zn_{4.3}NaCa_{0.6}V_6O_{13} \cdot 3H_2O \quad (6.3)$$

Additionally, Tao et al. discovered that pre-insertion of Ca^{2+} and water enhances the reversible stability of Zn^{2+} ions insertion/extraction in manganese-based layered materials, with Ca^{2+} and H_2O acting as pillars for layered δ-MnO_2 [11]; Yu and coworkers reported on a sodium-ion and crystallized water co-intercalated $Na_{0.55}Mn_2O_4{\cdot}0.57H_2O$ cathode, where the presence of sodium ions and crystallized water expands the interlayer spacing and facilitates the transport of zinc ions [12].

(4) Conversion Reaction Mechanism (Figure 6.2d)

Compared to ion insertion chemistry, conversion reactions in batteries generally provide higher capacity contributions. Thus, integrating effective conversion reactions represents a promising approach for developing high-performance zinc-ion batteries. Notable electrode materials include manganese-based oxides, tunnel-structured vanadium-based oxides, and halogen- and chalcogen-based cathodes.

Liu et al. illustrated the chemical conversion reaction mechanism within an α-MnO_2 system. Initially, protons from water interact with MnO_2, forming the MnOOH phase in the fully discharged state. The remaining OH^- ions then react with H_2O and $ZnSO_4$ in the solution to maintain the neutrality. The reaction mechanism is as follows [13]:

$$H_2O \leftrightarrow H^+ + OH^- \quad (6.4)$$

$$MnO_2 + H^+ + e^- \leftrightarrow MnOOH \quad (6.5)$$

$$\frac{1}{2}Zn^{2+} + OH^- + \frac{1}{6}ZnSO_4 + OH^- + \frac{x}{6}H_2O \leftrightarrow \frac{1}{6}ZnSO_4[Zn(OH)_2]_3 \cdot xH_2O \quad (6.6)$$

Qiao and coworkers proposed a multiple redox reaction involving both MnO_2 deposition dissolution and zinc-ion intercalation. In the first cycle, water molecules participate in the oxidation of Mn^{2+} ions (with the oxidation product being solid MnO_2), exhibiting poor reversibility. In the subsequent process, hydrogen ions and zinc ions reversibly participate in redox reactions, as shown below [14]:

$$Mn^{2+} + H_2O \rightarrow MnO_2 + 4H^+ + 2e^- \quad (6.7)$$

$$MnO_2 + 4H^+ + 2e^- \leftrightarrow Mn^{2+} + 2H_2O \tag{6.8}$$

$$MnO_2 + H^+ + e^- \leftrightarrow MnOOH \tag{6.9}$$

$$MnO_2 + 0.5Zn^{2+} + e^- \leftrightarrow Zn_{0.5}MnO_2 \tag{6.10}$$

Liu et al. identified both H^+ conversion and Zn^{2+} intercalation occurring in the bulk phase of δ-MnO_2. Meanwhile, Kang et al. suggested a proton-involved conversion mechanism in β-MnO_2 cathodes. During the first discharge cycle, β-MnO_2 transforms into MnOOH and dissolves into Mn^{2+}. In the subsequent cycle, the deposited MnO_2 reacts with protons, resulting in the formation of MnOOH and Mn^{2+}. Thus, this represents a combined mechanism involving manganese deposition/dissolution alongside proton conversion reactions, as illustrated below.

In addition, similar energy storage mechanisms have been observed in β-MnO_2 and δ-MnO_2 electrodes. Liu et al. identified both H^+ conversion and Zn^{2+} intercalation occurring in the bulk phase of δ-MnO_2. Meanwhile, Kang and coworkers suggested a proton-involved conversion mechanism in β-MnO_2 cathodes. During the first discharge cycle, β-MnO_2 transforms into MnOOH and dissolves into Mn^{2+}. In the subsequent cycle, the deposited MnO_2 reacts with protons, forming MnOOH and Mn^{2+}. Thus, this represents a combined mechanism involving manganese deposition/dissolution alongside proton conversion reactions, as illustrated below [15]:

$$MnO_2 + H^+ + e^- \leftrightarrow MnOOH \tag{6.11}$$

$$MnOOH + 3H^+ + e^- \leftrightarrow Mn^{2+} + H_2O \tag{6.12}$$

$$4Zn^{2+} + SO_4^{2-} + 8H_2O \leftrightarrow Zn_4SO_4(OH)_6 \cdot 5H_2O + 6H^+ \tag{6.13}$$

(5) Deposition–Dissolution Reaction Mechanism (Figure 6.2e)

Unlike intercalation or conversion reactions, manganese oxides exhibit reversible MnO_x deposition reactions. The addition of Mn^{2+} dopant ions facilitates the storage of active zinc on the electrode surface. Zn^{2+} ions intercalate into the surface undissolved MnO_x surface layer, as observed by Kim et al. in spinel-type $ZnMn_2O_4$ [16]. The reaction mechanism is as follows:

$$Mn^{2+} + xH_2O \leftrightarrow MnO_x + 2xH^+ + 2e^- \tag{6.14}$$

$$MnO_x + Zn^{2+} + 2e^- \leftrightarrow ZnMnO_x \tag{6.15}$$

Liang et al. also explored the dissolution–deposition reaction mechanism of α and δ-MnO_2 cathodes in $MnSO_4$ and $ZnSO_4$ electrolytes [17]. Tetravalent manganese dissolves into divalent manganese ions. During discharge, water manganese mineral-MnO_2 reversibly deposits on the electrode surface. Zinc ions precipitate on the α-MnO_2 surface as layered zinc hydroxysulfate ($Zn_4(OH)_6SO_4{\cdot}5H_2O$). During discharge, trivalent manganese ions are formed at the cathode, which are unstable and prone to disproportionation, leading to manganese dissolution into the electrolyte. As the pH of the electrolyte rises, precipitation occurs. The pH drops during charging, and the zinc hydroxysulfate dissolves into the electrolyte. The reaction mechanism is as follows:

$$MnO_2 + 2H_2O + 2e^- \leftrightarrow Mn^{2+} + 2OH^- \tag{6.16}$$

$$3MnO_2 + 8Zn^{2+} + 2SO_4{}^{2-} + 16H_2O + 6e^- \leftrightarrow Mn^{2+} + 2Zn_4(OH)_6SO_4 \cdot 5H_2 \tag{6.17}$$

The energy storage mechanisms proposed for zinc-ion batteries are complex and still in development. In vanadium-based compounds, Zn^{2+} insertion/extraction, H^+/Zn^{2+} co-insertion/extraction, and H_2O/Zn co-insertion/extraction are common. Manganese-based compounds, on the other hand, exhibit mechanisms such as Zn^{2+} insertion/extraction, H^+/Zn^{2+} co-intercalation/extraction, chemical conversion reaction, and deposition–dissolution reaction mechanisms. Further clarification and investigation of these mechanisms are essential to designing more efficient zinc-ion battery systems.

6.3 Cathode Materials for AZIBs

As mentioned previously, the electrochemical performance of AZIBs is largely influenced by the cathode materials, which serve as the primary storage sites for zinc ion. The properties of the cathode materials largely determine these energy storage mechanisms. Therefore, developing new cathode materials with enhanced capacity and improved cycling stability is critical for advancing AZIB technology. Cathode materials in ZIBs (zinc-ion batteries) can generally be divided into categories such as manganese-based and vanadium-based compounds, PBAs, and organic materials. The unique crystal structures and morphological characteristics of each material type play a crucial role in shaping their respective energy storage mechanisms, leading to variations in electrochemical performance.

6.3.1 Vanadium-Based Compound Cathode Materials

Vanadium-based compounds are a class of important anode materials for AZIBs, characterized by high theoretical capacity, diverse crystal structures, abundant elemental reserves, and low cost. Currently reported structures of vanadium-based cathode materials typically include layered, tunnel-like, polyoxoanion, and spinel structures (Figure 6.3a–f) [18]. Given that vanadium commonly exists in oxidation states of +2, +3, +4, and +5, it can participate in multi-electron transfer redox reactions, offering the potential for high specific capacity. However, a notable drawback is that the stable redox couple is generally V^{3+}/V^{4+}, leading to a relatively low average discharge voltage (typically 1.1–0.4 V, Figure 6.3g), which affects the energy density of vanadium-based aqueous batteries. Furthermore, due to the solubility of various vanadium species in water (Figure 6.4a), particularly in acidic solutions, vanadium-based cathode materials experience significant

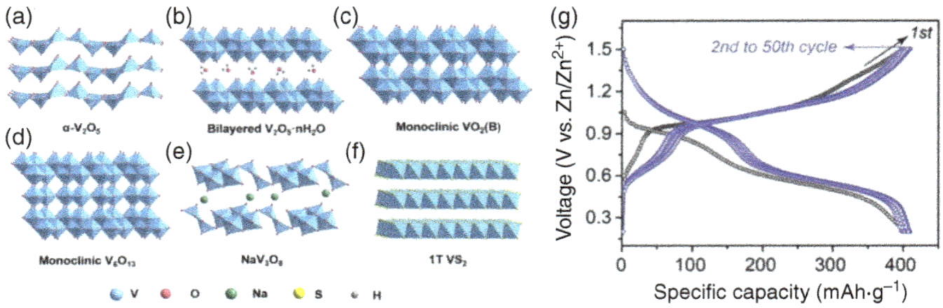

Figure 6.3 (a–f) Common vanadium-based compounds and their corresponding crystal structures. Source: Guo et al. [18]/John Wiley & Sons. (g) Corresponding galvanostatic charge/discharge curves of vanadium-based oxides. Source: Reproduced from Liao et al. [19]/with permission of John Wiley & Sons.

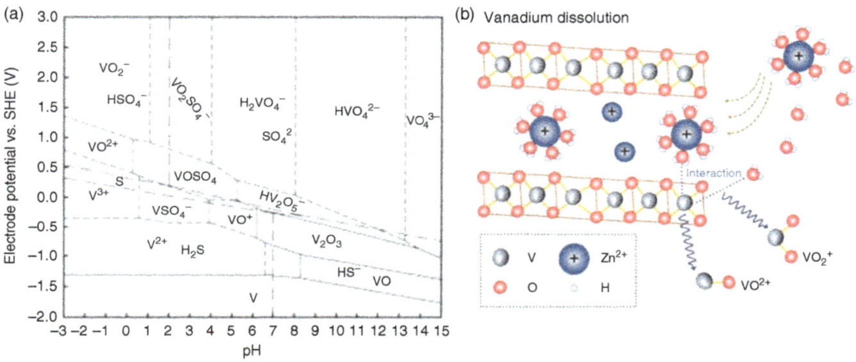

Figure 6.4 Pourbaix diagram of vanadium-based species in sulfate aqueous solution. Source: Deniz et al. [20]/John Wiley & Sons. (b) Dissolution behavior of V_2O_5 during charge storage in aqueous electrolyte. Source: Dai et al. [21]/John Wiley & Sons/CC BY 4.0.

dissolution during charge–discharge cycles in weakly acidic aqueous electrolytes (Figure 6.4b), severely impacting the cycle stability of vanadium-based zinc-ion batteries. Consequently, researchers have extensively optimized the composition and structure of vanadium-based cathode materials to achieve higher energy density and longer cycle life for these batteries.

6.3.1.1 Layered Vanadium-Based Compound

Early research on vanadium-based ZIB cathode materials focused primarily on layered vanadium oxides. The most common of these, V_2O_5, exhibits a typical layered structure. Within the V_2O_5 lattice, strong covalent bonds are present, while the interlayer interactions are weaker, including van der Waals forces and hydrogen bonds, with an interlayer spacing of 0.577 nm. Theoretical capacity derived from the two-electron redox reaction ($V^{5+} \leftrightarrow V^{3+}$) can achieve 589 mAh g^{-1}. However, in practical applications, layered V_2O_5 typically exhibits V^{5+}/V^{3+} redox couple only during initial discharge; after multiple Zn^{2+} ions intercalation/deintercalation cycles, the V^{4+}/V^{3+} couple becomes predominant, resulting in lower capacity and operating

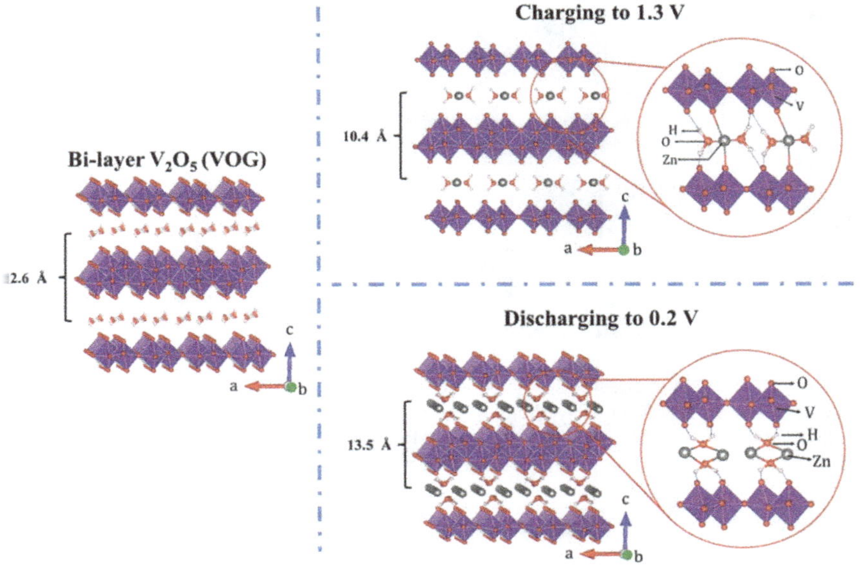

Figure 6.5 Effect of crystal water on zinc ions storage process for V_2O_5 cathode. Source: Yan et al. [8]/John Wiley & Sons.

voltage and leading to unstable structural phase changes. Introducing lattice water or metal/nonmetal ions can stabilize the layered structure of V_2O_5. $V_2O_5{\cdot}nH_2O$ features an increased interlayer spacing and incorporates structural water molecules between the V-O layers. These water molecules function as a buffer, mitigating the strong interactions between divalent Zn^{2+} ions and the oxide layers. As a result, this configuration enhances the diffusion kinetics of Zn^{2+} and improves the rate performance of the complete batteries (Figure 6.5).

In 2016, Nazar and coworkers reported a layered vanadium bronze electrode material ($Zn_{0.25}V_2O_5{\cdot}nH_2O$) in *Nature Energy* [3]. This material was synthesized using microwave hydrothermal techniques, producing ultra-long nanoribbons within hours. TEM images (Figure 6.6a,b) confirmed the anisotropic, planar morphology and single-crystal characteristics of the nanoribbons. High-resolution transmission electron microscopy (HRTEM) images (Figure 6.6c,d) revealed a lattice spacing of 0.537 nm, corresponding to the (200) plane, which is parallel to the longitudinal direction of the $Zn_{0.25}V_2O_5{\cdot}nH_2O$ nanoribbons. The nanoribbons have a width of 150 nm, corresponding to the *a*-axis direction, that is, the Zn^{2+} insertion path length. The detailed crystal structure, as illustrated in Figure 6.6e–h, consists primarily of a double-layer two-dimensional (2D) V_2O_5 framework, similar to the δ-phase vanadium pentoxide-layered structure, extending indefinitely in the a-b plane. Ordered ZnO_6 octahedra act as pillars to stabilize the 2D V_2O_5 layers in a framework, surrounding cross-pathways along [110] and [110] directions in the a-b plane, thereby accommodating additional Zn^{2+} ions during discharge.

Apart from vanadium-based oxides, vanadium sulfides like VS_2 and VS_4 have shown significant promise as cathode materials for ZIBs. VS_2 is a 2D layered

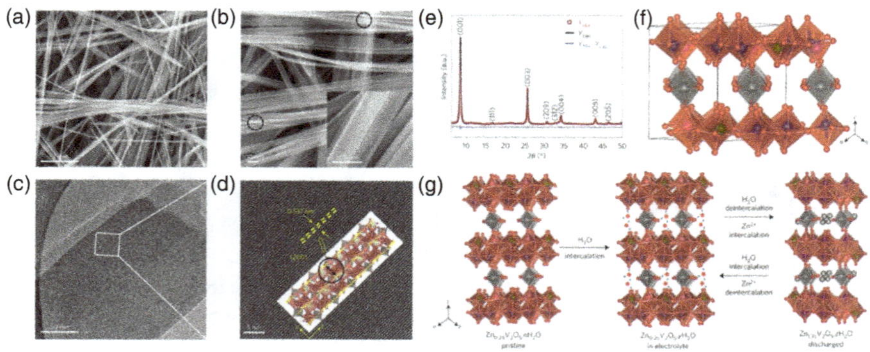

Figure 6.6 (a,b) SEM images of $Zn_{0.25}V_2O_5{\cdot}nH_2O$. (c,d) TEM images of $Zn_{0.25}V_2O_5{\cdot}nH_2O$. (e) XRD pattern of $Zn_{0.25}V_2O_5{\cdot}nH_2O$. (f) Crystalline structure of $Zn_{0.25}V_2O_5{\cdot}nH_2O$. (g) Charge storage mechanism of $Zn_{0.25}V_2O_5{\cdot}nH_2O$. Source: Reproduced with permission from [3]/Springer Nature.

transition metal disulfide characterized by its sandwich-like S–V–S structure, where the layers are held together by weak van der Waals forces. It features a notable interlayer spacing of 5.76 Å and exhibits excellent conductivity. Compared to other transition metal disulfides, such as MoS_2 and TiS_2, VS_2 offers superior multivalent ion charge storage capacity. When utilized as a cathode in ZIBs, layered VS_2 presents a voltage profile akin to that of V_2O_5 cathodes, achieving a specific capacity of approximately 190 mAh g^{-1} [22]. Additionally, it demonstrates impressive cycling stability, maintaining 98% of its initial capacity after 200 cycles at a current density of 0.5 A g^{-1}. *In situ* characterization has revealed a highly reversible two-step Zn^{2+} intercalation/deintercalation mechanism within the voltage range of 0.4–1.0 V (relative to Zn^{2+}/Zn). During discharge, zinc ions intercalate in two steps to form $Zn_{0.99}VS_2$ and $Zn_{0.23}VS_2$ phases, corresponding to two discharge platforms at potentials of 0.82–0.65 and 0.65–0.45 V (vs. Zn^{2+}/Zn). The interlayer spacing of VS_2 exhibits favorable elastic characteristics, adapting to the intercalated zinc ions; when discharged to 0.4 V, the interlayer spacing expands by 1.73%. During charging, zinc ions are deintercalated, allowing reversible restoration of the original VS_2 phase and interlayer spacing. Although this study proposes VS_2 as a viable cathode candidate for AZIBs, its further development is hindered by structural instability. The Zn||VS_2 battery exhibits satisfactory cycling performance within a limited voltage range of 0.4–1.0 V (relative to Zn^{2+}/Zn). However, extending the cycling voltage beyond this range may result in rapid capacity degradation. This degradation is primarily due to substantial volume changes that occur during the insertion of a large number of Zn^{2+} ions, which can lead to delamination and pulverization of the layered VS_2 structure.

6.3.1.2 Vanadium-Based Compounds with Tunnel Structure

Unlike layered structures, tunnel structures provide unique one-dimensional (1D) spaces that enable rapid and directional diffusion of Zn^{2+}. These structures not only enhance structural stability but also offer abundant storage sites for Zn^{2+}, making them more resilient during repeated Zn^{2+} intercalation and deintercalation

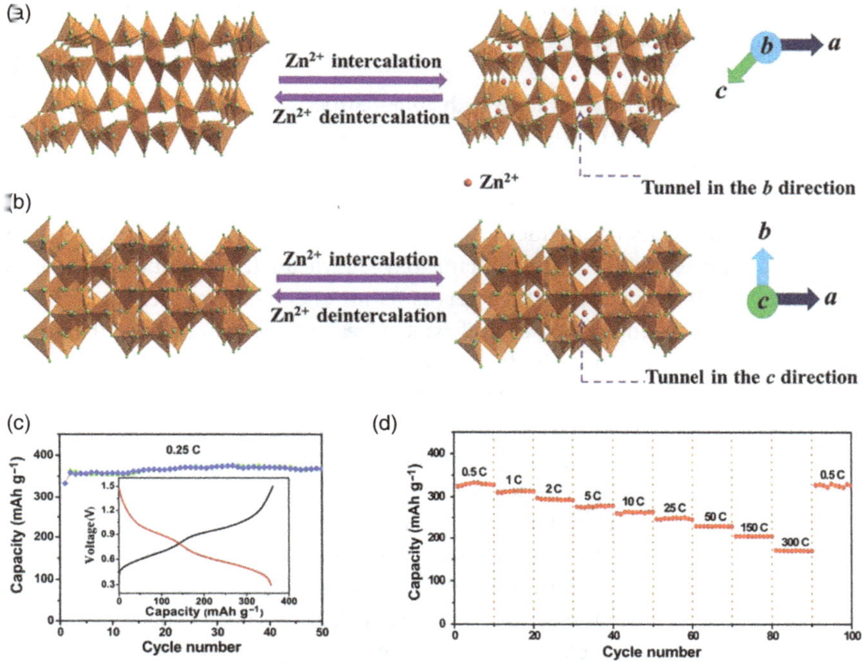

Figure 6.7 Diagram of intercalation/deintercalation of Zn^{2+} ions along (a) b- and (b) c-axes in VO_2(B) nanofiber body phase. (c) Cyclic properties of VO_2 cathode materials. Illustration: GCD profile. (d) Magnification properties of VO_2 cathode materials. Source: Reproduced from Ding et al. [23]/with permission of John Wiley & Sons.

processes. Theoretical considerations suggest that tunnel-structured materials are ideal hosts for zinc-ion storage, with tunnel size and stability playing crucial roles in this capacity. Monoclinic VO_2 is a typical example of a tunnel structure. Its framework, consisting of distorted VO_6 octahedra, offers a large surface area that facilitates rapid Zn^{2+} diffusion and maintains structural stability during the Zn^{2+} insertion and de-insertion process (Figure 6.7a,b). As a result, monoclinic VO_2(B) nanofibers utilized as cathode materials exhibit a high reversible capacity of 357 mAh g^{-1}, excellent rate performance of 171 mAh g^{-1} at 300 C, and high energy density of 297 Wh kg^{-1} at 180 W kg^{-1}, as illustrated in Figure 6.7c,d [23]. However, tunnel-type vanadium-based cathode materials also have drawbacks, such as poor conductivity.

To address this issue, numerous studies have explored methods to improve the zinc storage capability of monoclinic VO_2(B) by integrating it with carbon materials like graphene and reduced graphene oxide. For instance, Cui et al. developed a VO_2(B)/RGO composite using a straightforward hydrothermal technique [24]. This composite demonstrates a high specific capacity of 292 mAh g^{-1} at 5 A g^{-1} and commendable cycling stability. The enhanced electrochemical performance is mainly due to the even distribution of VO_2(B) nanoparticles on the rGO nanosheets, which facilitates rapid electron transport and helps maintain structural integrity during the Zn^{2+} insertion/de-insertion process.

6.3.1.3 Vanadium-Based Compounds with Spinel Structure

Spinel-structured vanadium-based compounds feature stable ionic diffusion channels due to their interconnected network, which aids the movement of zinc ions. However, the closely packed atomic arrangement in the spinel structure requires the reduction of repulsive interactions between the existing metal ions and the intercalated zinc ions. Thus, introducing oxygen vacancies or cation vacancies and defects within the spinel lattice serves as an effective approach to mitigate lattice repulsion challenges. Wu et al. first synthesized spinel $Zn_3V_3O_8$ using the sol–gel method [25]. As a cathode material for AZIBs, $Zn_3V_3O_8$ demonstrates remarkable cycling performance, primarily due to a dual-electron reaction mechanism involving zinc ions and vanadium ($V^{5+} \leftrightarrow V^{3+}$). The crystal structure of $Zn_3V_3O_8$ was elucidated through X-ray diffraction (XRD) data refinement, as illustrated in Figure 6.8a,b. The XRD patterns reveal a relatively pure phase of $Zn_3V_3O_8$, indicating good crystallinity for the sample synthesized via the sol–gel method. The structure comprises a three-dimensional (3D) framework of VZn_2O_6 octahedra, facilitating the diffusion of zinc ions through its tunnels. The cycling performance of the $Zn_3V_3O_8$ electrode was assessed, delivering a capacity retention of 192 mAh g^{-1} after 200 cycles and 72.6% after 2000 cycles. Additionally, *in situ* XRD was employed to examine the structural changes of the $Zn_3V_3O_8$ electrode during cycling (Figure 6.8c). During the initial charge–discharge cycle, the intensity of two diffraction peaks associated with $Zn_3V_3O_8$ diminished. Upon reaching a voltage of 1.6 V during charging, these peaks completely disappeared, indicating structural collapse and phase transition. As the characteristic diffraction peaks of $Zn_3V_3O_8$ faded, a new set of peaks emerged, potentially corresponding to $ZnyV_2O_{5-x}{\cdot}nH_2O$.

6.3.1.4 Polyanion-Type Vanadium-Based Compounds

Polyanion vanadium-based compounds are constructed from polyhedral polyanions linked by strong covalent bonds to transition metal polyhedra, resulting in robust and open 3D network structures. A classic example of a polyanion-based vanadium compound is the sodium superionic conductor type (NASICON) $Na_3V_2(PO_4)_3$. The strong P–O covalent bonds in polyanionic compounds, driven by the "inductive effect," enhance the covalency of V–O bonds, resulting in elevated redox potentials. Consequently, $Na_3V_2(PO_4)_3$-based ZIBs exhibit high voltage characteristics, with an average operating voltage (~1.15 V) significantly higher than that of vanadium oxides and vanadates (~0.8 V), as illustrated in Figure 6.9. Additionally, the strong P–O covalent bonds stabilize the oxygen in the lattice, giving the NASICON-type $Na_3V_2(PO_4)_3$ good theoretical stability.

When $Na_3V_2(PO_4)_3$/graphene composites were initially employed as cathodes in AZIBs, their zinc storage performance was proved inadequate (initial capacity of 97 mAh g^{-1}, with a capacity retention of 74% after 100 cycles at 0.5 C). The rapid capacity decay in NASICON-type $Na_3V_2(PO_4)_3$ was primarily attributed to

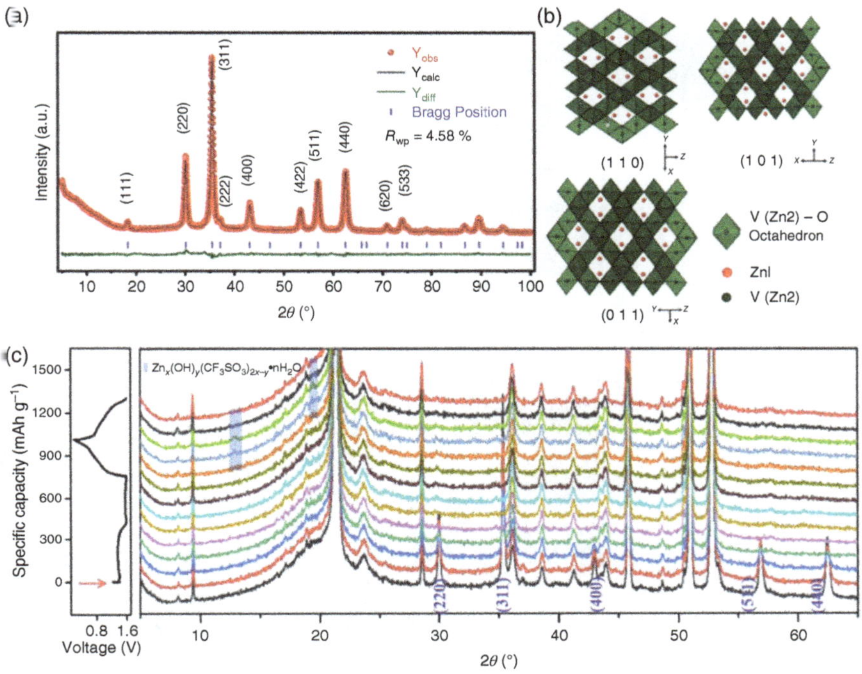

Figure 6.8 (a) XRD pattern of $Zn_3V_3O_8$ with spinel-structure. (b) The structure diagrams of spinel-$Zn_3V_3O_8$. (c) *In situ* XRD of $Zn_3V_3O_8$ during zinc-ion storage. Source: Reproduced from Wu et al. [25]/with permission of Elsevier.

lattice distortion induced by zinc-ion insertion. Cui et al. found that introducing a mix of Na^+/Zn^{2+} at M1 and M2 sites in $Na_3V_2(PO_4)_3$ significantly enhanced its structural stability when used as a cathode in AZIBs. Compared to $Na_3V_2(PO_4)_3$, the Zn^{2+}-intercalated $ZnNaV_2(PO_4)_3$ exhibits enhanced electrical conductivity and mechanical properties, opening up a new avenue for improving the performance of NASICON cathode materials.

6.3.1.5 Modification Strategy of Vanadium-Based Cathode Materials

As previously mentioned, vanadium-based cathode materials hold great potential for application in AZIBs. However, significant issues persist, such as electrode dissolution, structural collapse, limited electronic conductivity, and slow kinetics, which impede the advancement of vanadium-based materials. To overcome these challenges, researchers have implemented a range of modification strategies (Figure 6.10a), including constructing micro/nanostructures, designing composite materials, ion/molecule pre-intercalation, dopant incorporation, and defect introduction.

(1) Constructing Micro/Nano Structures

V_2O_5, serving as an ion intercalation material, leverages its graphite-like layered structure to enable fast and reversible ion migration. However, it suffers from

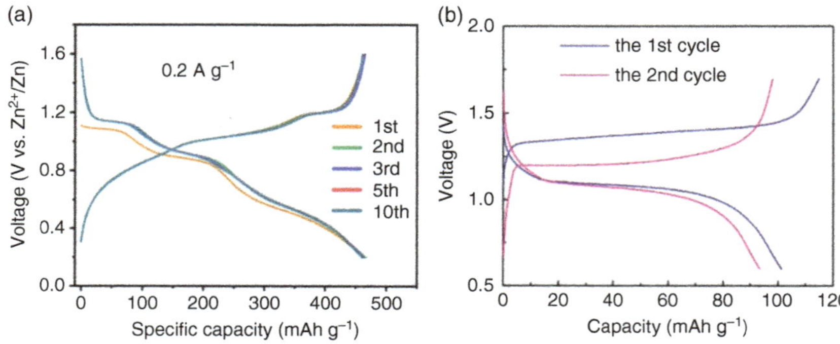

Figure 6.9 Comparison of charge and discharge curves of V_2O_5 and $Na_3V_2(PO_4)_3$ as cathodes of zinc-ion batteries. (a) V_2O_5. Source: Wang et al. [26]/with permission of Elsevier; (b) $Na_3V_2(PO_4)_3$. Source: Reproduced from Hu et al. [27]/with permission of John Wiley & Sons.

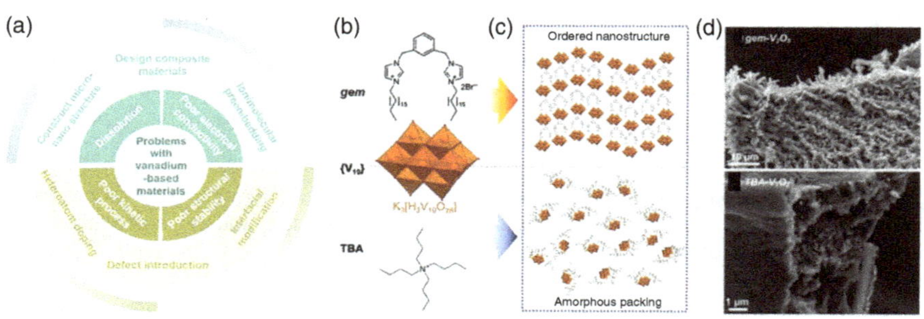

Figure 6.10 (a) The problems and improvement strategies of vanadium-based compounds as AZIBs cathodes. (b,c) Assembly diagrams of TBA-V_2O_5 and gem-V_2O_5 precursors. (d) SEM images of TBA-V_2O_5 and gem-V_2O_5. Source: [28]/John Wiley & Sons/CC BY 4.0.

diffusion limitations and low electronic mobility. Nanostructuring can modify the electronic structure of the cathode material and regulate diffusion path lengths, enhancing performance. Early synthesis methods for V_2O_5 often relied on processing simple vanadate precursors under various conditions (e.g., hydro/solvothermal methods, pyrolysis, electrospinning, exfoliation); however, these methods typically lacked precise control over the product's structure and morphology, rather than targeted synthetic design. When $[V_{10}O_{28}]_6^-$ decavanadate anions are mixed with Gemini surfactants, the self-assembly of surfactant-activated decavanadate cluster nanostructures takes place (Figure 6.10b,c). Air calcination of this material results in the formation of highly crystalline, sponge-like V_2O_5 (gem-V_2O_5), while calcining amorphous decavanadate tetra-*n*-butylammonium yields microcrystalline V_2O_5 (TBA-V_2O_5), which tends to aggregate. The larger grain size and long-range microstructure (Figure 6.10d) of gem-V_2O_5 contribute to higher initial capacity and improved capacity retention compared to TBA-V_2O_5. However, TBA-V_2O_5, with its smaller grain size and increased surface area, demonstrates faster lithium insertion kinetics and enhanced rate performance [28].

Designing porous and multidimensional structures enhances electrolyte wetting and alleviates volumetric changes during charge–discharge cycles, allowing the material to maintain good cycling stability while delivering higher capacity. For instance, Chen et al. fabricated a 3D porous V_2O_5 (3D-NRAs-V_2O_5) by assembling V_2O_5 nanorods through a sol–gel reaction followed by an annealing process [29]. The 3D porous structure increases the effective surface area, enhances electrolyte penetration, and greatly improves zinc-ion diffusion throughout the material, resulting in a high reversible capacity of 336 mAh g^{-1} at a current rate of 0.05 A g^{-1}.

(2) Phase/Crystal Plane Regulation

Amorphous materials experience less structural change during ion diffusion compared to crystalline structures, which aids in cycle stability. Additionally, amorphous materials have more isotropic ion diffusion pathways and abundant vacancies, promoting rapid ion diffusion and offering additional active sites. Thus, designing V_2O_5 with an amorphous structure offers more isotropic Zn^{2+} transport and higher specific capacity due to these isotropic diffusion paths and enhanced reaction sites.

For tunnel-type structure materials, despite similarly exposed crystal planes, rate performance can still be influenced by the electrode's crystal plane orientation. By synthesizing VO_2 (B) nanobelts with either dispersed or aggregated forms and dominated by (00l) crystal planes, it has been demonstrated that optimizing the tunnel orientation in vanadium oxides can fully leverage the potential of tunnel structure materials (Figure 6.11) [30], thus significantly improving ion transport kinetics. Compared to the isotropic diffusion behavior of aggregated VO_2-A, the dispersed VO_2-D tends to lay flat on the electrode, forming ion transport channels along the *c*-axis direction, ensuring rapid zinc-ion diffusion. Additionally, crystal engineering successfully designed the preferentially oriented $Na_2V_6O_{16}\cdot 3H_2O$ cathode material. Based on different zinc-ion pathways in the $Na_2V_6O_{16}\cdot 3H_2O$ cathode, it was found that the linear tunnel has a lower migration barrier, indicating that samples with more active surfaces and $b \neq 0$ possess higher diffusion coefficients. Multiphysics simulations also showed that during zinc-ion insertion, the preferentially oriented $Na_2V_6O_{16}\cdot 3H_2O$ maintains a more uniform zinc-ion concentration distribution internally and demonstrates excellent rate performance and cycling stability, offering promising improvements in ZIB performance [31].

(3) Designing Composite Materials

Constructing composite materials is a common approach in designing battery electrode materials. By delving into molecular to micro-nano scale composites and adjustments, synergistic effects between materials can be achieved, compensating for deficiencies and enhancing overall electrode performance. For the design of vanadium oxide electrodes in ZIBs, composites of different vanadium oxides, carbon materials, and conductive polymers are primarily used to improve the electrical conductivity of vanadium-based compounds and suppress dissolution behavior during extended cycling. The solvothermal method is an effective approach for

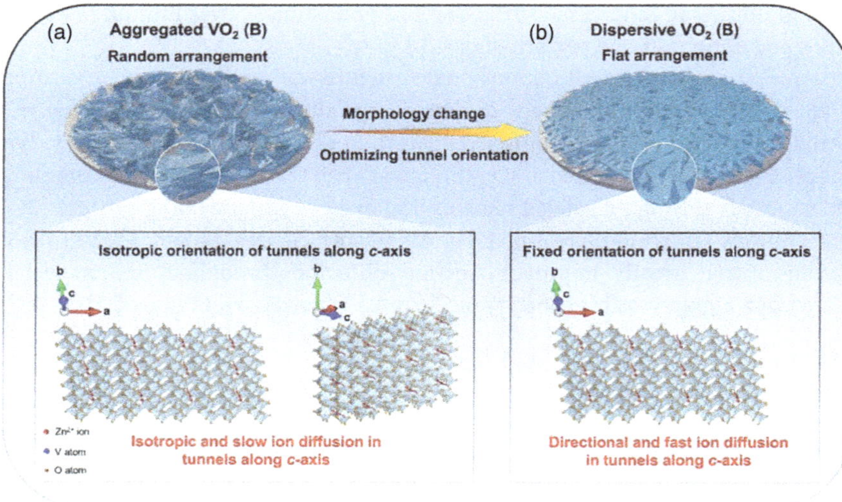

Figure 6.11 Schematic illustration of the effect of material morphology-related electrode arrangement on the tunnel orientation and ion diffusion behavior. Source: Reproduced from He et al. [30]/with permission of John Wiley & Sons.

synthesizing $H_2V_3O_8$/graphene composites. Graphene's 2D nanosheet structure enables efficient face-to-face or edge-to-edge contact with active materials, promoting swift electron and ion transport [32]. Additionally, graphene nanosheets are highly flexible, making it easier to form encapsulating structures that limit dissolution. Besides graphene, MXenes (2D transition metal carbides and nitrides) are also widely used in composites. Their rich composition, 2D atomic layer structure, metallic conductivity, and active surfaces endow the composite materials with excellent electronic conductivity and structural stability, while also serving as a uniform dispersion substrate to reduce agglomeration of active materials and accelerate zinc-ion diffusion. Using an *in situ* growth strategy with MXene materials, introducing 5.25% MXene into the nucleation and growth process of vanadium oxide provides heterogeneous nucleation sites and acts as a seed to regulate the morphology and structure of vanadium oxide (as shown in Figure 6.12). The synthesized cathode material greatly improves both the capacity and rate performance of ZIBs [33].

(4) Ion/Molecule Pre-intercalation

Taking the traditional orthorhombic V_2O_5 as an example, this material features a typical 2D layered structure composed of VO_5 square pyramids. However, the interlayer spacing is limited to 0.44 nm because of van der Waals forces, which hinders structural stability and high electron transfer rates during ion intercalation and deintercalation, leading to unsatisfactory cycling performance. Zinc ions, compared to monovalent lithium, sodium, and potassium ions, exhibit stronger electrostatic interactions during intercalation, impeding the rapid insertion and extraction of Zn^{2+}, causing structural collapse and rapid capacity degradation.

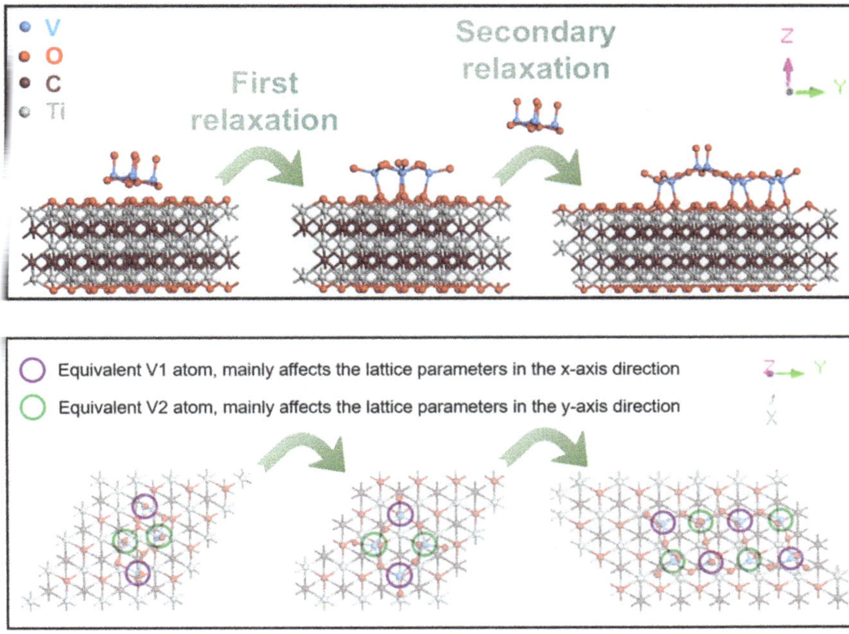

Figure 6.12 DFT simulation for the growth of vanadium oxide units on the surface of MXene. Source: Reproduced from Chen et al. [33]/with permission of John Wiley & Sons.

Pre-intercalation of cations such as Na^+, NH^{4+}, Sn^{4+}, V^{4+}, Ba^{2+}, Ni^{2+}, Mn^{2+}, H^+, Zn^{2+}, and Ca^{2+} can effectively modify the crystal structure of vanadium-based cathode materials, enhancing the rapid diffusion of zinc ions within the layers of the vanadium compounds. Pre-intercalating zinc ions into V_2O_5 layers results in the formation of ZnO_6 octahedra along with VO_6 octahedra and VO_5 triangular bipyramids, ensuring a more stable structure throughout the charge–discharge cycles. $Zn_{0.25}V_2O_5 \cdot nH_2O$ (ZVO) demonstrates reversible structural changes, ensuring stability over 1000 cycles [3]. Xia et al. experimented with pre-intercalating calcium ions into V_2O_5 layers [34]. Compared to ZVO, $Ca_{0.25}V_2O_5 \cdot nH_2O$ (CVO) has a larger interlayer distance due to the larger volume of CaO_7 polyhedra compared to ZnO_6 octahedra (Figure 6.13). The increased interlayer spacing promotes the easier diffusion of zinc ions, enabling a greater number of zinc ions to intercalate and thereby leading to a higher reversible capacity. Additionally, the stronger interaction between calcium ions in CaO_7 polyhedra and the three top oxygens and four co-planar H_2O molecules in V_2O_5 further stabilizes the structure. As a result, the CVO electrode can sustain up to 3000 stable cycles while retaining 96% of its capacity. The pre-insertion of NH_4^+ also broadens the V_2O_5 layer spacing, accelerating electron transfer and Zn ion migration [35].

(5) Structural Water Pre-intercalation

In layered oxides, interlayer structural water molecules can form solvation complexes with embedded Zn^{2+} ions, significantly reducing the effective surface

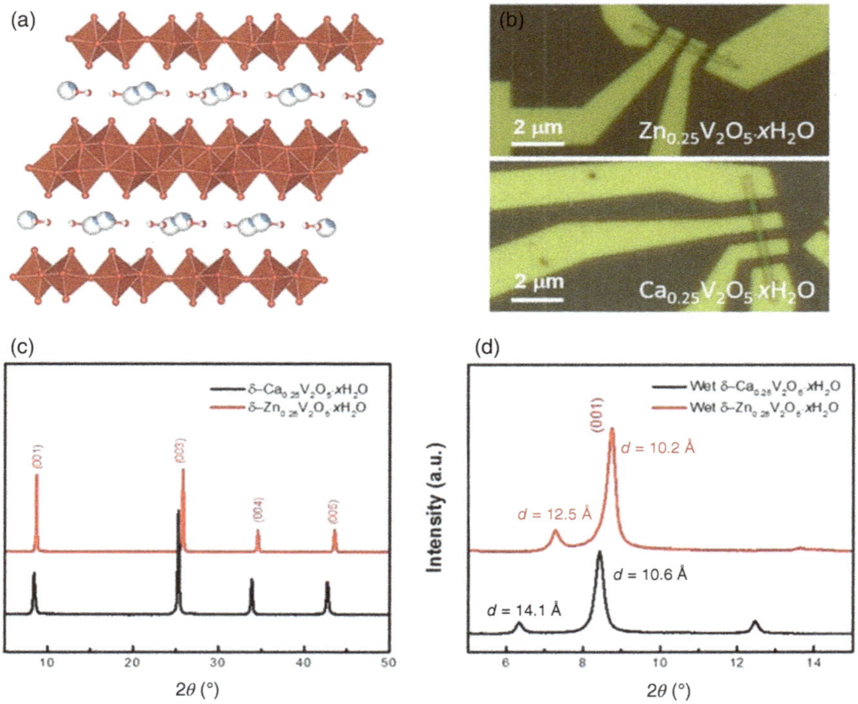

Figure 6.13 (a) View of the CVO crystal structure along the *b*-axis, illustrating a double-layered configuration with an open framework. (b) Individual nanodevices based on CVO and $Zn_{0.25}V_2O_5 \cdot nH_2O$. (c,d) Characteristic XRD patterns of the synthesized CVO. Source: Reproduced with permission from [34]/John Wiley & Sons.

charge and diminishing electrostatic repulsion with the oxide layers, thereby enhancing the diffusion kinetics of Zn^{2+} ion and acting as a "lubricant" for ion diffusion. Similarly, in tunnel-structured V_6O_{13} materials, Zn^{2+} ion intercalates with some solvation water molecules during discharge. The presence of water molecules enhances the electrostatic shielding effect, facilitating the diffusion of Zn^{2+} ions at the electrode–electrolyte interface and throughout the cathode material, while also preserving structural stability. This hydration intercalation technique greatly enhances both the rate performance and cycling stability of vanadium-based oxide cathodes. Besides electrostatic shielding, Huang's team found through experiments and density functional theory (DFT) calculations that structural water in $NaCa_{0.6}V_6O_{16} \cdot 3H_2O$ (NaCaVO) expands the interlayer spacing, providing wider diffusion channels for Zn^{2+} and increasing active sites to accommodate more Zn^{2+} ions (Figure 6.14), thereby enhancing diffusion kinetics and storage capacity. Electrochemical tests reveal that NaCaVO electrodes achieve a specific capacity of 347 mAh g^{-1} at 0.1 A g^{-1}, maintaining an impressive capacity retention of 83% after 10 000 cycles at 5 A g^{-1}, which is superior to that of dry NaCaVO electrodes [36].

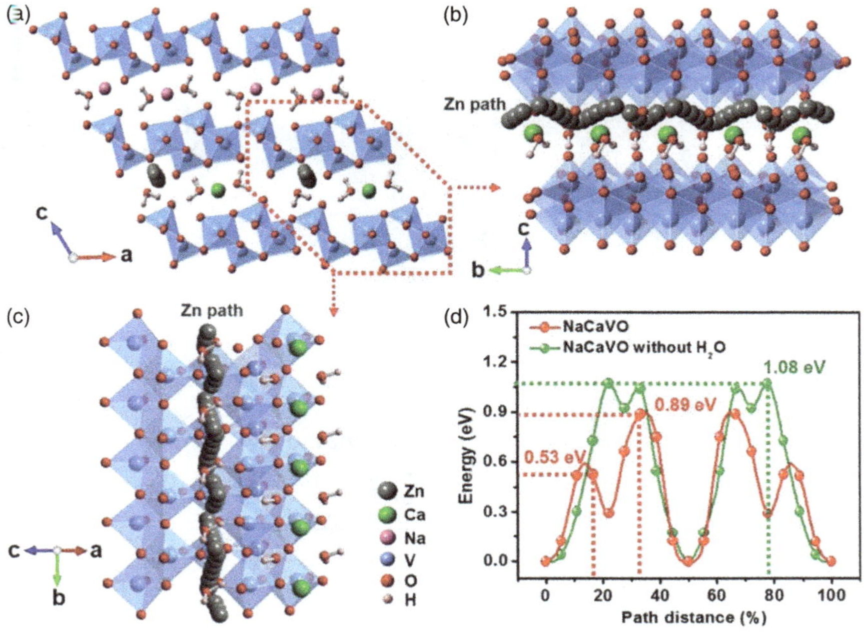

Figure 6.14 (a) Potential Zn^{2+} ion migration pathways in NaCaVO as viewed along the *b*-axis; (b) bottom-up perspective and (c) side perspective along the *a*-axis of the highlighted red area in (a); (d) energy barriers associated with Zn^{2+} ion migration pathways in NaCaVO, with and without structural water. Source: Reproduced from Zhu et al. [36]/with permission of John Wiley & Sons.

(6) Polymer/Small Molecule Pre-intercalation

In addition to cation and water intercalation, *in situ* insertion of conductive organic polymer molecules between oxide layers is an emerging strategy. The incorporation of these polymers can increase the crystal lattice spacing, creating efficient diffusion pathways for Zn^{2+}. Furthermore, the distinctive π-conjugated structure of conductive polymers can significantly reduce the electrostatic interactions between the host framework and the intercalated Zn^{2+}, thereby improving the kinetics of ion diffusion. Polyaniline (PANI) is a commonly used organic intercalant; for instance, PANI-intercalated V_2O_5 shows an exceptionally large lattice spacing of 13.9 Å (Figure 6.15a,b) [37]. This ZIB exhibits remarkable rate performance, providing a capacity of 197.1 mAh g^{-1} at a current density of 20 A g^{-1}, along with exceptional cycling stability. Poly(3,4-ethylenedioxythiophene) (PEDOT) can also serve as a polymer intercalant (Figure 6.15c). Xia's team prepared PEDOT-*in situ* intercalated $NH_4V_3O_8$ [38], which increased the interlayer spacing from 7.8 to 10.8 Å, effectively improving diffusion kinetics and cycling stability.

(7) Defect Introduction

The electronic structure of electrode materials, which is closely linked to electrochemical reactions, governs the rates of carrier diffusion. Defects within the crystal

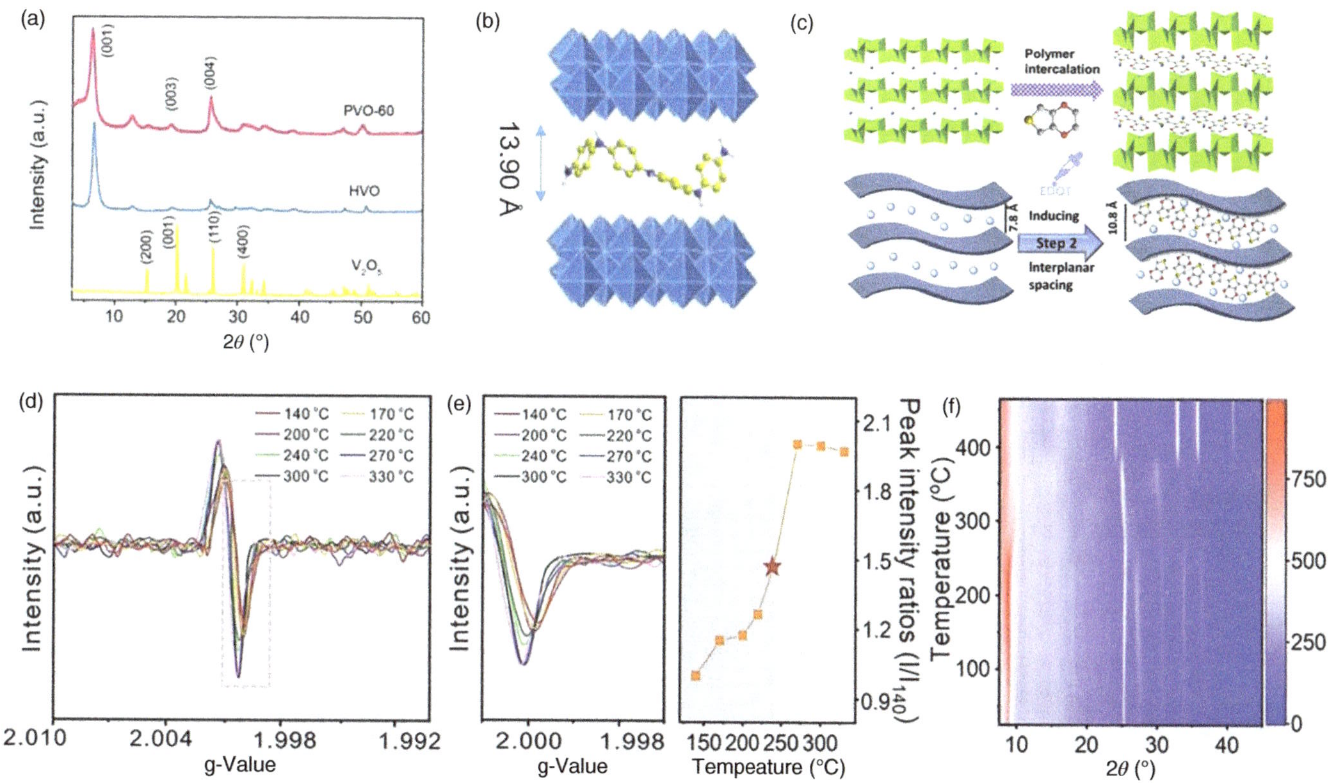

Figure 6.15 (a) XRD profile of vanadium-based oxide intercalated with PANI. (b) Schematic representation of the structure of PANI-intercalated vanadium-based oxide. Source: Liu et al. [37]. (c) Synthesis process of PEDOT-intercalated $NH_4V_3O_8$-layered material. Source: Reproduced from Bin et al. [38]/with permission of IOP Publishing. (d) *In situ* EPR spectra of $Zn_{0.125}V_2O_5{\cdot}0.95H_2O$ calcined at temperatures ranging from 140 to 330 °C. (e) Enlarged partial EPR spectra (left) and illustration of peak intensity ratios at various temperatures (right). (f) *In situ* XRD patterns of $Zn_{0.125}V_2O_5{\cdot}0.95H_2O$ throughout the calcination process. Source: Reproduced from Ye et al. [39]/with permission of The Royal Society of Chemistry.

structure influence the distribution of this electronic structure. These defects can be categorized into various types, including point defects (such as vacancies), line defects (like dislocations), planar defects (e.g. grain boundaries), and bulk defects (which arise from lattice disorder caused by impurities). Defects provide additional storage/adsorption/active sites for carriers, reduce electrostatic repulsion between ions, overcome migration and diffusion barriers, and lower reaction energy barriers, thus accelerating ion intercalation/deintercalation.

Dou's team introduced oxygen vacancies to design $Zn_{0.125}V_2O_5 \cdot 0.95H_2O$ nanomaterials [39]. Characterization using *in situ* electron paramagnetic resonance (EPR, Figure 6.15d,e) and XRD (Figure 6.15f) revealed the generation of multiple oxygen vacancies caused by the partial escape of lattice water at relatively low reaction temperatures. Additionally, *in situ* spectroscopy and DFT calculations showed that these oxygen vacancies reduce the Zn^{2+} diffusion barrier and diminish the interactions between Zn and O atoms, thereby enhancing electrochemical performance.

6.3.2 Manganese-Based Oxide Cathode Materials

Manganese-based oxides are considered some of the most promising materials for Zn^{2+} storage in AZIBs, thanks to their affordability, eco-friendliness, low toxicity, and ability to exist in multiple valence states (Mn^{x+}, where x can be 0, 2, 3, 4, or 7). Additionally, manganese-based oxide cathodes exhibit a variety of crystal structures, with manganese dioxide (MnO_2) having the most diverse structures (Figure 6.16), including $\alpha/\beta/\gamma/\varepsilon/\delta$-$MnO_2$ [40]. Manganese dioxide consists of a fundamental crystal framework of octahedral units with six oxygen atoms and one manganese atom. These MnO_2 octahedral units are connected to form corresponding layered or tunnel structures, depending on the specific number of MnO_6 octahedra. The following sections will discuss different structures of MnO_2 individually.

6.3.2.1 α-MnO_2

α-MnO_2 exhibits a tunnel structure of [2×2, approximately 0.46×0.46 nm], classified under the body-centered tetragonal crystal system with the space group I4/m. Its fundamental octahedral MnO_6 units form a double-chain structure through corner sharing. The comparatively spacious tunnel structure of α-MnO_2 facilitates the swift intercalation and deintercalation of Zn^{2+} and other metal ions along the z-axis. In 2009, Kang et al. were the first to document the intercalation and deintercalation behavior of Zn^{2+} in α-MnO_2 when using a mildly acidic $Zn(NO_3)_2$ electrolyte Alfaruqi et al. [41]. They later showed that during discharge, Zn^{2+} intercalates into the α-MnO_2 cathode, resulting in the formation of discharge to form $ZnMn_2O_4$. The charging process then reverses this reaction, leading to a high capacity of 210 mAh g^{-1} at a current density of 200 mA g^{-1}, demonstrating a significant degree of reversibility in the intercalation and deintercalation of Zn^{2+} within α-MnO_2 [2]. However, early reports on α-MnO_2 cathodes indicated poor cycling performance due to significant Mn^{2+} dissolution and John-Taylor distortion.

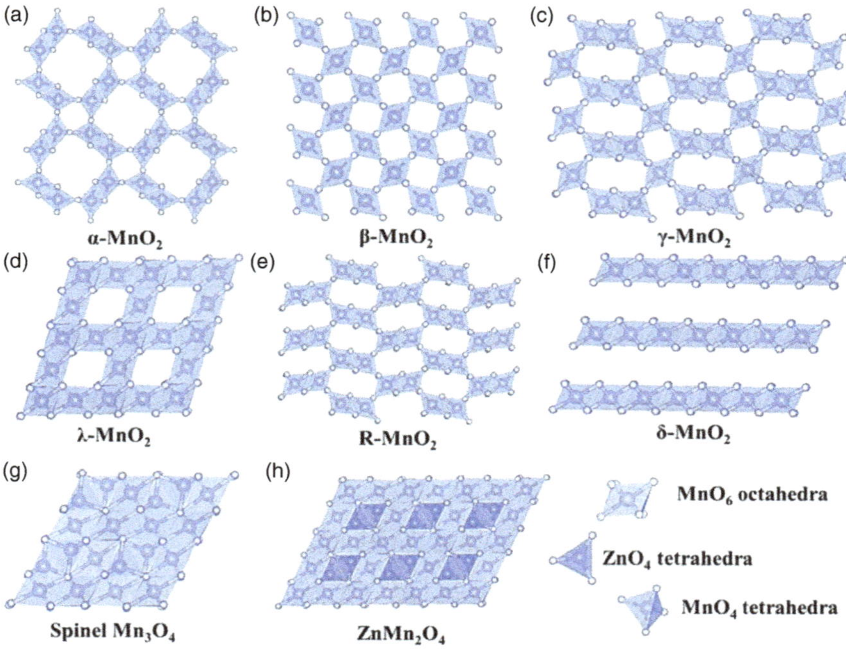

Figure 6.16 Crystal structures of various manganese-based oxides. (a) α-MnO_2. (b) β-MnO_2. (c) γ-MnO_2. (d) λ-MnO_2. (e) R-MnO_2. (f) δ-MnO_2. (g) Mn_3O_4. (h) $ZnMn_2O_4$ (spinel-type). Source: Reproduced from Xiong et al. [40]/with permission of John Wiley & Sons.

6.3.2.2 β-MnO_2

β-MnO_2 possesses a [1×1, ~0.23×0.23 nm] tunnel structure formed by corner-sharing MnO_6 octahedra. It is classified within the tetragonal crystal system and exhibits a rutile-type structure, where oxygen atoms are tightly packed in a hexagonal arrangement, and manganese atoms occupy the centers of octahedra, with oxygen atoms at the octahedral vertices. The relatively narrow tunnel structure of β-MnO_2 theoretically hinders the reversible intercalation of Zn^{2+}. Islam et al. were the first to introduce microwave-assisted hydrothermal synthesis for producing β-MnO_2 nanorods as cathode materials in AZIBs. This material exhibited a reversible capacity of 270 mAh g^{-1} at a current density of 100 mA g^{-1}. *In situ* XRD analysis showed that during the initial intercalation of Zn^{2+}, the peak corresponding to the (101) plane of β-MnO_2 experienced a slight shift to a lower angle; however, it returned to its original position upon full charging, indicating minimal structural distortion [42].

6.3.2.3 γ-MnO_2

γ-MnO_2 is composed of alternating [1×1, ~0.23×0.23 nm, pyrolusite] and [1×2, ~0.23×0.46 nm, birnessite] tunnels and belongs to the hexagonal crystal system. The disordered nature of γ-MnO_2's tunnel structure results in lower crystallinity and a higher presence of defects, such as twin planes (Tw or Mt), Mn(III) substituting for Mn(IV), and Mn vacancies. γ-MnO_2 was first employed as a cathode material in weakly acidic AZIBs in 1986, prompting investigations into the insertion and extraction behavior of Zn^{2+} in γ-MnO_2. Despite these efforts, the detailed mechanism of this process has not yet been fully elucidated [43]. It was not until 2015 that Alfaruqi et al. demonstrated that during discharge, Zn^{2+} insertion into γ-MnO_2 leads to phase changes, forming layered L-Zn_yMnO_2, $ZnMn_2O_4$, and γ-Zn_xMnO_4, which then revert to γ-MnO_2 during recharging [44]. This study elucidated the mechanisms of charge storage and the structural changes of γ-MnO_2 throughout the charge–discharge cycles. Additionally, reports suggest that temperature adjustments can control the structural transformation of γ-MnO_2 to γ/β-MnO_2, which exhibits superior electrochemical performance compared to γ-MnO_2, providing insights for designing high-performance manganese dioxide materials.

6.3.2.4 δ-MnO_2

δ-MnO_2 is a layered electrode material characterized by a large interlayer spacing of approximately 0.7 nm, formed by corner-sharing MnO_6 octahedra, and belongs to the monoclinic crystal system with the P2/m space group. Alfaruqi et al. investigated the insertion and extraction behavior of Zn^{2+} in layered nano δ-MnO_2, finding that Zn^{2+} insertion causes δ-MnO_2 to transform into spinel-type $ZnMn_2O_4$ and layered δ-Zn_xMnO_2, confirming phase transitions during cycling [45]. Later, Jin et al. demonstrated that during the initial rapid charging in $Zn(TFSI)_2$ electrolyte, Zn^{2+} inserts into δ-MnO_2 without significant phase change [46]. The subsequent reaction process is dominated by H^+-involved conversion reactions, leading to stable high capacity with 93% capacity retention after 4000 cycles. Furthermore, the low proton adsorption energy of δ-MnO_2 enables the co-insertion of H^+ and Zn^{2+} into the electrode structure. Its substantial interlayer spacing promotes rapid diffusion and storage of Zn^{2+}, contributing to its widespread application as an electrode material [47]. Unlike other tunnel-structured MnO_2 (such as α-, β-, γ-, and λ-MnO_2), which undergo transitions from tunnel to layered structures affecting battery stability, the "layered to layered" transition in δ-MnO_2 is more favorable for Zn^{2+} storage kinetically.

6.3.2.5 λ-MnO_2

λ-MnO_2 features a [1 × 3] tunnel structure and is a classic spinel phase, where manganese and oxygen ions occupy the octahedral 16d and 32e sites, respectively. In 1981, Hunter successfully synthesized λ-MnO_2 by initially preparing lithium manganese oxide via a precursor method, followed by treatment with acidic solution to induce oxidation and delithiation, ultimately yielding λ-MnO_2. Due to the limited 3D tunnel size of λ-MnO_2, its ability to intercalate/de-intercalate zinc ions was initially considered inadequate, making it less favored as a zinc-ion cathode material. However, Yuan et al. discovered that λ-MnO_2 could be synthesized by treating $LiMn_2O_4$ with H_2SO_4 [48]. When employed as an electrode material in rechargeable aqueous batteries, λ-MnO_2 exhibited remarkable zinc-ion storage performance, achieving an impressive discharge capacity of 545.6 mAh g^{-1}.

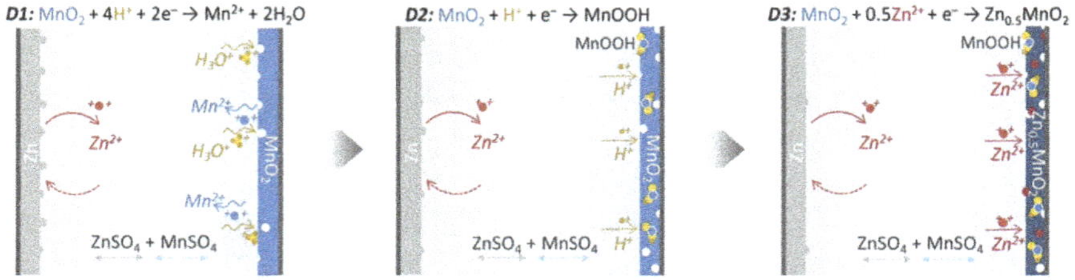

Figure 6.17 Schematic illustration and charge storage mechanism for the MnO_2–Zn ion battery operating in 1 M $ZnSO_4$ and 1 M $MnSO_4$ electrolyte. Source: Reproduced from Chao et al. [14]/with permission of John Wiley & Sons.

6.3.2.6 ε-MnO_2

Hexagonal ε-MnO_2 contains densely packed [MnO_6] octahedra and does not inherently possess a void structure, which restricts the deintercalation and diffusion of zinc ions. However, it can achieve charge storage and transfer through deposition-dissolution mechanisms, enabling the Mn^{4+}/Mn^{2+} dual-electron redox reaction, with a theoretical capacity reaching as high as 616 mAh g^{-1}. The Zn/MnO_2 full-cell theoretical voltage exceeds 2 V (vs. Zn/Zn^{2+}). This system functions in an electrolyte composed of 1 M $ZnSO_4$ and 1 M $MnSO_4$. During the initial constant current charging phase, Zn^{2+} and Mn^{2+} cations from the electrolyte are deposited onto the current collector through reduction and oxidation reactions, resulting in the formation of Zn metal and ε-MnO_2, respectively. The subsequent reversible constant current discharge process consists of three stages: the deposition–dissolution reaction of MnO_2/Mn^{2+} and the intercalation of H^+ and Zn^{2+} ions, with the former contributing most to the capacity, as illustrated in Figure 6.17. Additionally, the voltage window of this system can be extended to 2.4 V by further acidifying the electrolyte.

The acidified electrolyte leads to the formation of ε-MnO_2 with a high density of Mn vacancies, which improves the kinetics of proton and electron transfer during the redox reaction. Utilizing this reaction mechanism, the engineered Zn–MnO_2 battery can achieve a discharge voltage of up to 1.95 V, along with a high specific capacity of 570 mAh g^{-1} [14].

6.3.2.7 Other Manganese-Based Compounds

R-MnO_2 features a [1 × 2] tunnel framework constructed from a double-chain structure with shared corners, while this double-chain structure comprises edge-sharing MnO_6 octahedra, classifying it as a type of sapphire structure. R-MnO_2 possesses a relatively wide tunnel structure, facilitating rapid intercalation and deintercalation of charge carriers, thus achieving a high battery capacity.

Manganese monoxide (MnO), characterized by a rock-salt structure, has been previously identified as a potential anode material for lithium-ion batteries and supercapacitors. However, due to its lack of ionic intercalation channels (either interlayer or tunnel), its charge storage mechanism was initially considered to be dominated by surface charge adsorption/desorption, leading to its classification as an ineffective anode material in AZIBs. Recent studies have revealed that MnO can be activated into layered MnO_2 during the initial charging process through a phase transition, enabling charge storage in conventional weakly acidic electrolytes. The XPS spectra of Mn 3*s* reveal a decrease in the binding energy gap from 5.8 to 4.7 eV after three cycles, indicating the oxidation of manganese from +2 to +3.9. This observation confirms the conversion of MnO into MnO_2. In the following cycles, the newly generated MnO_2 serves as an active host for the sequential intercalation of H^+ and Zn^{2+} ions, leading to the formation of $Zn_4(OH)_6SO_4{\cdot}nH_2O$ on the electrode surface. Another study indicates that the electrochemical activation of MnO occurs through the introduction of manganese defects ($Mn_{0.61}\square_{0.39}O$, where □ denotes manganese vacancies) during the initial charging phase. These resulting defects generate accessible migration pathways and active sites, enabling the insertion and extraction of Zn^{2+} without leading to structural collapse. Comparisons between defect-containing and defect-free MnO indicate that the differences in charge distribution demonstrate electron accumulation around defects, generating a strong electric field that attracts Zn^{2+} ions. The CV curves and Mn 3*s* binding energy shifts after initial charging are consistent between defect-activated MnO and p hase-activated MnO. However, through morphological imaging and XRD results, this research attributes the increase in manganese oxidation state to the extraction of Mn^{2+} from the MnO host rather than the oxidation of MnO to MnO_2. However, based on morphological imaging and XRD results, this study attributes the rise in manganese oxidation state to the extraction of Mn^{2+} from the MnO host instead of the oxidation of MnO to MnO_2.

Mn_3O_4, also known as hausmannite, is a multivalent manganese oxide with a spinel-type structure, where Mn^{2+}ions are located in tetrahedral sites, and Mn^{3+} ions occupy the octahedral sites. The operation of Mn_3O_4 involves an initial phase transition followed by subsequent reversible cycling. As depicted in Figure 6.18, during the initial charging process, Mn_3O_4 undergoes transformation into layered birnessite, with the formation of Mn_5O_8 as an intermediate product, a reaction that does not involve zinc ions [49]. As charging progresses, Mn_5O_8 converts to birnessite, aided by the dissolution of Mn^{2+} and the intercalation of H_2O. During discharge, zinc ions intercalate into the birnessite layers, forming Zn-birnessite. This intercalation and subsequent Zn^{2+} extraction during charge–discharge cycles play a key role in boosting battery capacity and maintaining structural stability, highlighting a distinct electrochemical reaction mechanism compared to MnO_2.

6.3.2.8 Modification Strategy of Mn-Based Oxide Cathode Materials

(1) Micro/Nano Structural Design and Phase Regulation

Nanoscale electrode materials exhibit a high surface area and remarkable size-related effects (both small-size and real-size effects). These properties increase the contact area with the electrolyte, facilitating easier carrier diffusion across the nanostructured surface, which greatly improves the battery's rate performance. Additionally, nanomaterials can alleviate volume expansion issues and maintain structural stability, leading to prolonged cycling life. According to the relationship between ionic diffusion and material nanosize (Eq. (6.18)), synthesizing nanometer-sized electrode materials (i.e., reducing *L*) can effectively reduce the ion

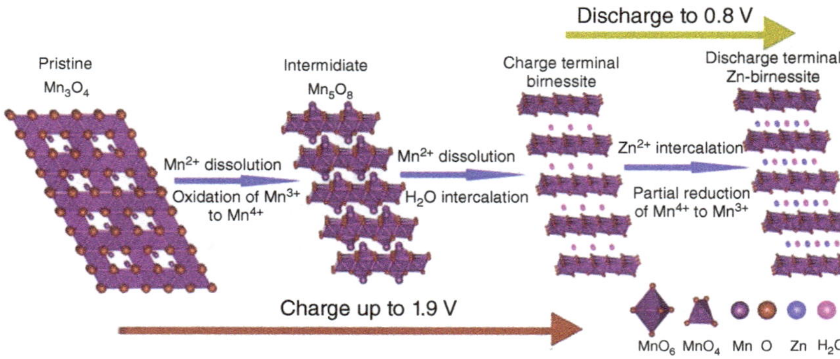

Figure 6.18 Reaction pathway schematic for the Mn_3O_4 within the first cycle. Source: Reproduced from Jin et al. [49]/with permission of Elsevier.

diffusion time τ, thereby improving battery charge and discharge rates:

$$\tau = L2/2D \tag{6.18}$$

where τ represents the diffusion time, L is the material size, and D is the diffusion coefficient.

The crystal morphology of MnO_2 significantly influences Zn^{2+} storage performance. Various nanostructures and crystal phases of MnO_2 have been extensively studied for aqueous zinc-manganese secondary batteries. Kim et al. reported that 3D-layered mesoporous γ-MnO_2 nanospheres (UW-MO) exhibit superior cycle stability and charge capacity compared to bulk γ-MnO_2 [50]. After 50 cycles, UW-MO maintains a specific capacity of 400 mAh g^{-1}, whereas bulk γ-MnO_2 decreases to 180 mAh g^{-1}. Additionally, β-MnO_2, with its narrower tunnel structure, is less favorable for Zn^{2+} storage compared to other tunnel-type MnO_2 crystal forms. Islam et al. employed a microwave-assisted hydrothermal method to synthesize β-MnO_2 nanorods that were exposed on the (101) plane, which served as cathode materials for aqueous zinc-manganese batteries (Figure 6.19a,b) [42]. The (101) peak shifts to a lower angle upon Zn^{2+} insertion, attributed to manganese reduction and elongation of the Mn–O bond (Figure 6.19c). After Zn^{2+} insertion and extraction, the β-MnO_2 nanorods display distinct Fourier Transform (FT) peaks associated with Mn–O, shared-edge Mn–Mn, and shared-angle Mn–Mn interactions. Rotating the β-MnO_2 nanorods along the a-axis uncovers $[1 \times 1]$ open tunnels, which enhance the electrode's contact area with the electrolyte, thus facilitating Zn^{2+} insertion and extraction. The constructed battery reached a discharge capacity of 270 mAh g^{-1} at a current density of 100 mA g^{-1}, maintaining a capacity retention of 75% and achieving 100% Coulombic efficiency after 200 cycles at 200 mA g^{-1}. The excellence of nanostructures is also observed in other crystal forms. For instance, Liu et al. proposed δ-MnO_2 nanowires with a (1D–3D) mixed network structure, enhancing contact area with the electrolyte, providing more active sites, and facilitating rapid ion transport [52]. Additionally, Alfaruqi et al. reported a nano α-MnO_2 cathode with exceptional structural stability, significantly improving the reversibility of Zn^{2+} insertion [53].

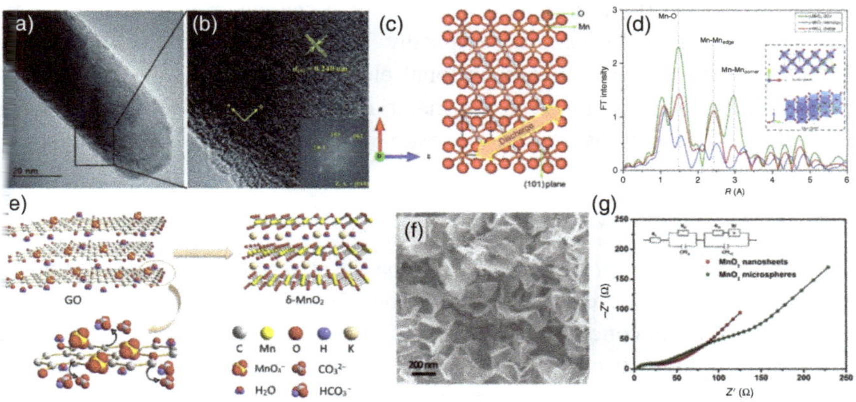

Figure 6.19 (a) TEM image depicting the β-MnO_2 nanorod. (b) HR-TEM image showcasing the β-MnO_2 nanorod. (c) Schematic representation of (101) lattice planes in the β-MnO_2 structure. (d) EXAFS spectra of the β-MnO_2 nanorod cathode obtained after discharging/charging in the Zn test cell, including an inset of corner-shared and edge-shared [MnO_6] in the β-MnO_2 structure. Source: Reproduced with permission from [42]/Royal Society of Chemistry. (e) Illustration of the δ-MnO_2@GO nanosheets design. (f) SEM image of δ-MnO_2 nanosheets. (g) EIS comparison between δ-MnO_2 nanosheets and δ-MnO_2 nanospheres used as cathodes. Source: Reproduced with permission from [51]/Elsevier.

Nanometer-sized electrode materials reduce ion diffusion distance, shorten diffusion time, and improve electrochemical reaction kinetics. Guo et al. synthesized ultrathin δ-MnO_2 nanosheets and found via electrochemical impedance spectroscopy (EIS) that their charge transfer resistance (46 Ω) is significantly lower than that of δ-MnO_2 nanospheres (223 Ω). This difference is likely attributed to the shorter diffusion path in the 2–4 nm thick nanosheets compared to the 50 nm thick nanospheres (Figure 6.19e–g) [51]. Wang et al. reported a 2D single-atom-layer MnO_2 providing interconnected ion transport channels both horizontally and vertically, resulting in faster electron/ion transport compared to multilayer nanomaterials. This indicates that smaller nanostructures lead to faster electrochemical reaction rates [54]. Moreover, nanometer-sized active materials are better equipped to handle volume changes during ion insertion and extraction, significantly reducing the likelihood of structural collapse from phase transitions and resulting in outstanding cycling stability. Liu et al. utilized an electrochemical deposition method to prepare α-MnO_2/CNT hybrid materials, where the unique composite structure enhances electrochemical stability [55]. The dense network formed by α-MnO_2 nanofibers and CNTs creates faster charge transfer pathways and offers numerous active sites, greatly enhancing charge transfer dynamics and the battery's rate performance. Despite these advantages, nanomaterials also face challenges such as reduced thermal stability and increased surface side reactions. Nanoparticles, characterized by their large specific surface area and surface energy, are prone to aggregation. This leads to increased contact resistance and can result in electrochemical agglomeration during cycling, ultimately diminishing battery capacity and raising the risk of side reactions with the electrolyte. Furthermore, nanostructured electrode materials typically have a narrower electrochemical stability window compared to conventional electrodes. Therefore, when designing modified nanostructured electrode materials, it is essential to consider both the advantages and disadvantages to create high-performance zinc-manganese secondary battery cathodes.

(2) Design of Composite Materials

Mn-based oxides are similar to vanadium-based materials, and their composite material modification strategies primarily involve combining carbon-based materials, 2D materials, and conductive polymers. The primary goals include improving the electrical conductivity of electrode materials, increasing the specific surface area and active sites, and effectively preventing structural collapse from repeated intercalation/deintercalation or phase transitions. These enhancements aim to boost the electrochemical performance of aqueous zinc-manganese secondary batteries.

Commonly used carbon-based materials for composites include carbon fibers, graphene, and carbon nanotubes. Carbon fibers have unique ion/electron transport channels and direct electron collection functions, and they exhibit low electrochemical activity and a wide electrochemical window in aqueous media. Carbon fibers also provide structural support to prevent collapse, enhancing the cycle stability of the battery. Carbon cloth, a typical carbon fiber material, features light weight, strong bend resistance, high electrical conductivity, and excellent chemical stability. Electroplating MnO_2 on its surface can create high-performance electrode materials, effectively improving ion/electron transport rates and exhibiting good conductivity and electrochemical stability.

In addition to carbon cloth, optimized carbon materials can significantly improve their physicochemical properties and are applicable in enhancing battery performance. For instance, nitrogen-doped carbon nanosheets (N-CNSs) networks, formed by interconnected N-CNS with rich nanopores, increase the material's specific surface area and provide more ion/electron transport channels. The defects created by nitrogen doping effectively enhance reaction activity, making them excellent conductive materials. Chen et al. synthesized a vertically aligned MnO_2 nanosheet structure grown on carbon nanofibers (CNFs) [56]. As shown in Figure 6.20a, the conductive CNFs and the closely connected MnO_2 form an interconnected network that significantly enhances the ion/electron transfer kinetics of the positive electrode material, while interlayer crystallization water acts as a structural scaffold, maintaining the stability of the active material's internal structure. Additionally, the ultrathin MnO_2 nanosheets enhance the interface contact area between the electrode and electrolyte, significantly boosting the number of reaction active sites.

Yu and coworkers synthesized layered manganese dioxide on nitrogen-doped carbon nanowires, resulting in a 3D core-shell structure for aqueous ZIB cathodes – MnO_2@NC (Figure 6.20b–d) [57]. The Zn-MnO_2@NC battery exhibited a high reversible capacity of 325 mAh g^{-1} at a current density of 0.1 A g^{-1}, alongside excellent rate performance and stable cycling durability. Mechanistic studies attributed the enhanced performance of the MnO_2 composite electrode to nitrogen

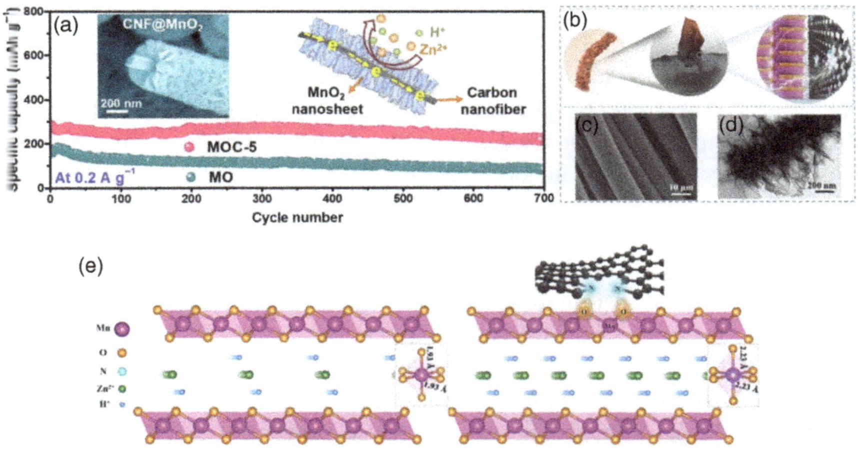

Figure 6.20 (a) Cyclic stability and SEM image of MnO_2 nanosheet structure grown on carbon nanofibers. Source: [56]/with permission of Elsevier. (b) Illustration of the interface connections between MnO_2 nanosheets and N-doped carbon. (c) SEM image of MnO_2@NC. (d) TEM image of MnO_2@NC. (e) Mechanism of Zn^{2+}/H^+ ion co-insertion into MnO_2 and MnO_2@NC. Source: [57]/John Wiley & Sons.

doping, which reduces the Mn^{4+} valence state in MnO_2, thereby decreasing electrostatic repulsion of Zn^{2+}/H^+ during intercalation. This modification facilitates greater cation intercalation and promotes faster ion migration (Figure 6.20e).

Graphene, a classic 2D carbon-based material, is widely used in composite material design. Mai used α-MnO_2/rGO scrolls (MGS) as highly reversible aqueous zinc-ion cathode materials [58]. The rGO scrolls not only enhance the conductivity of MnO_2 but also mitigate its dissolution, leading to improve electrochemical performance of the battery. Compared to bare MnO_2 (MNW), the rGO-modified cathode demonstrates impressive capacity (362.2 mAh g^{-1} at a current density of 0.3 A g^{-1}), a long cycle life (87.4 mAh g^{-1} after 800 cycles at 7 A g^{-1}, and 145.3 mAh g^{-1} at 3 A g^{-1}), and excellent rate capability. Additionally, the authors further explored the electrochemical reaction mechanism of the MGS electrode, confirming a two-step intercalation process of zinc ions – first into the MnO_2 framework layers, followed by insertion into the tunnels.

3D porous foam scaffolds can also be used to enhance electron transfer and ion diffusion rates. For example, a thin-film material composed of α-MnO_2 and 3D porous carbon nanotubes (CNTs) (α-MnO_2@CNT) effectively improves ion/electron diffusion kinetics, providing the electrode material with faster ion diffusion rates and superior conductivity, delivering a discharge specific capacity of 308.5 mAh g^{-1} at a rate of 0.97 C (with 1 C equivalent to 308 mA g^{-1}) [59].

Conductive polymers not only offer high conductivity, excellent chemical stability, and strong charge storage capabilities but also have environmentally friendly and cost-effective characteristics, making them ideal protective coatings for active materials. Zeng et al. reported a PEDOT coating used to modify the MnO_2 interface [60]. The PEDOT buffering layer suppresses manganese dissolution and structural collapse while maintaining high carrier transport rates. Moreover, MnO_2@PEDOT can be directly synthesized onto flexible carbon cloth, eliminating the need for binders. When paired with a PVA/$ZnCl_2$/$MnSO_4$ gel electrolyte, the resulting flexible quasi-solid-state Zn||MnO_2@PEDOT battery exhibits excellent cycling stability and Coulombic efficiency. Additionally, other conductive polymers such as polypyrrole (PPy) and polyaniline (PANI) can also be used as coatings on MnO_2 to mitigate the dissolution of active materials during charge and discharge cycles.

Chen and coworkers developed a novel carbon-based poly(4,4″-oxydiphenylamine)/MnO_2 organic–inorganic hybrid cathode (C@PODA/MnO_2) using a simple two-step electrodeposition technique [61]. The nanoporous carbon framework facilitates the deposition of nanoparticle-shaped PODA and MnO_2. Through *in situ*/non-*in situ* analysis and DFT calculations, it was revealed that the C=N groups within the PODA chains offer more active surfaces and interfaces for ion/electron migration and Zn ion storage. Notably, the newly formed Mn-N interfacial bonds significantly enhance ion diffusion, prevent Mn atom dissolution, and improve the redox kinetics and structural stability of MnO_2. As a result, the C@PODA/MnO_2 cathode delivers enhanced capacity (321 $mAh\,g^{-1}$ or 1.7 $mAh\,cm^{-2}$ at 0.1 $A\,g^{-1}$), excellent rate performance (88 $mAh\,g^{-1}$ at 10 $A\,g^{-1}$), and outstanding cycling stability.

Another effective approach to enhance the electrochemical performance of MnO_2 is by compositing it with MXenes. A 3D high-density MXene-MnO_2 composite cathode material was synthesized using a spray-drying technique [62]. MnO_2 nanoparticles are uniformly encapsulated within the wrinkled MXene nanosheets, forming a stable, conductive 3D nanoflower architecture. This design promotes rapid ion and electron transport while ensuring high structural stability.

(3) Defect Engineering

The electrochemical reaction is intrinsically linked to the electronic structure of the electrode material, as it dictates the carrier diffusion rate. Defects in the crystal lattice can alter the electronic structure distribution. These defects may take various forms, including point defects (e.g. vacancies), line defects (e.g. dislocations), surface defects (e.g. grain boundaries), and bulk defects (e.g. lattice disorder induced by impurities). The presence of defects can provide additional sites for carrier storage/adsorption/activation, effectively reduce electrostatic repulsion between ions, overcome migration and diffusion barriers, and lower reaction energy barriers, thus enhancing ion insertion/extraction rates.

Doping heterogeneous elements into MnO_2 can generate new defect states (such as vacancies) in the original bandgap. The altered electronic structure at defect sites reduces the energy barrier for zinc-ion insertion, resulting in improved electrochemical performance. For example, Lian et al. reported Ti-doped α-MnO_2, where Ti doping causes layer contraction but generates oxygen vacancies to compensate for the decrease in manganese valence [63]. Oxygen vacancies create openings in the [MnO_6] octahedral walls, leading to uneven charge distribution and local electric field imbalances within the crystal structure. This facilitates enhanced ion and electron diffusion rates, contributing to improved rate performance and extended cycle stability. Additionally, Fang et al. introduced a composite material consisting of K^+-doped and oxygen-defect-containing α-MnO_2 ($K_{0.8}Mn_8O_{16}$, KMO) [64]. Compared to the original α-MnO_2, KMO has a faster H^+ diffusion rate, effectively improving the electrochemical reaction kinetics (Figure 6.21). Oxygen defects reduce the energy barriers for electron transport and charge transfer during redox reactions, while also opening the walls of the [MnO_6] polyhedra to facilitate H^+ diffusion (Figure 6.21b). The Zn||$K_{0.8}Mn_8O_{16}$ battery exhibited an impressive energy density of 398 Wh kg^{-1} and maintained stable capacity without significant degradation after 1000 charge–discharge cycles, highlighting the promise of defect engineering in the advancement of high-performance cathode materials. Moreover, Kim et al. reported a β-MnO_2 cathode material with fluorine (F) doping in an intercalated ball-shaped nanosheet structure [65]. This F-doping strategy effectively expanded the lattice size of β-MnO_2, providing more electrochemical reaction active sites. The introduced oxygen defects not only improve electrical conductivity by altering the local electronic structure but also lower the energy barrier for zinc insertion, thereby promoting zinc-ion insertion/extraction behavior.

(4) Guest Pre-intercalation Strategies

Irreversible nonequilibrium phase transitions and the instability of intercalation layers can cause capacity loss and adversely affect battery performance. Pre-intercalating guest species, such as water molecules, organic compounds, and inorganic metal ions, into layered MnO_2 can create a stable internal framework for the insertion and extraction of Zn^{2+}/H^+ ions. This strategy mitigates structural collapse during the intercalation–deintercalation process, optimizes the crystal structure, reduces the electrostatic interactions between Zn^{2+} ions and MnO_2, and enhances carrier diffusion rates. Among various MnO_2 polymorphs with tunnel structures, δ-MnO_2 – with its broader interlayer spacing – is particularly suited for guest species pre-intercalation, making it the preferred choice in many pre-intercalation methods.

Huang et al. presented a layered MnO_2 cathode material that has been pre-intercalated with polyaniline. The polymer intercalation significantly increased the interlayer spacing of MnO_2 (~10 nm) and enhanced the stability of the crystal structure. Combined with the advantages of the nanolayered structure, this approach effectively facilitates charge transfer, thereby improving the battery's long-cycle stability. At a current density of 280 mAh g^{-1}, the material can undergo stable cycling for 200 cycles. Additionally, interlayer water molecules function as "lubricants," facilitating the rapid insertion and extraction of zinc ions. They also act as structural pillars, preventing collapse during charge–discharge cycles and helping to maintain the integrity of the main framework. Nam et al. reported a crystalline water pre-intercalated layered MnO_2 (cw-MnO_2), which features a wider interlayer spacing of 0.725 nm that is conducive to Zn^{2+} ion insertion. Discrete Fourier transform (DFT) calculations and EXAFS analysis revealed that upon insertion, Zn^{2+} ion initially forms a spherical structure with water molecules before evolving into a stable Zn-Mn dumbbell configuration. This stable configuration enhances long-cycle stability and remarkable rate performance.

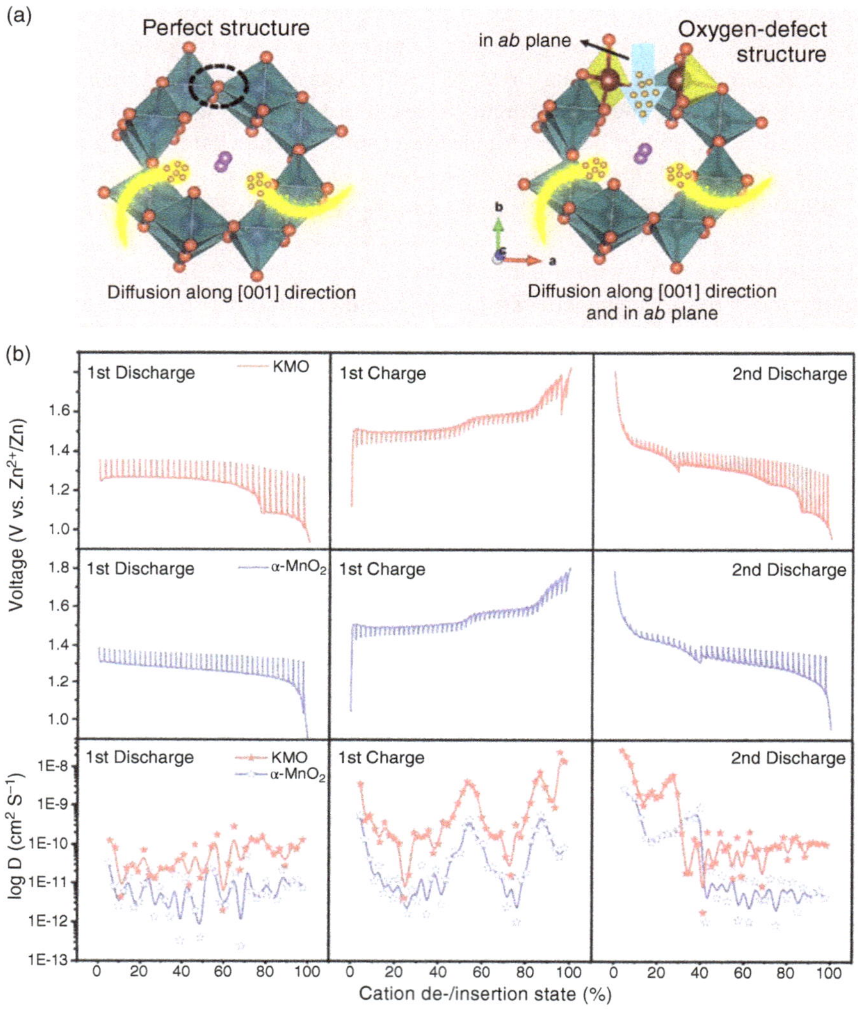

Figure 6.21 (a) Schematic representation of H^+ diffusion in KMO with both ideal and oxygen-defect structures; (b) GITT curves alongside H^+ diffusion coefficients at various discharge/charge state for KMO and α-MnO_2 electrodes. Source: Reproduced from Fang et al. [64]/with permission of John Wiley & SonsWiley Online Library.

Additionally, Wang et al. introduced a δ-MnO_2 pre-intercalated with Na^+ ion and water molecules, referred to as δ-NMOH. Non-*in situ* XRD testing showed that the XRD diffraction peaks of δ-NMOH remained unchanged before and after cycling, with no formation of spinel-type $ZnMn_2O_4$ and MnOOH by-products, indicating a stable electrochemical process. After 10 000 cycles, the cathode material retained up to 98% of its capacity. Furthermore, a quasi-solid-state self-healing Zn-δ-NMOH battery, constructed using carboxylated polyurethane (CPU) as a substrate, demonstrated high charge–discharge capacity even after numerous bending and recovery cycles. It retained 81.2% of its capacity after four bending/recovery cycles at 10 C, achieving a specific capacity of 100 mAh g^{-1} (with 1 C = 380 mA g^{-1}). This provides a valuable reference for the advancement of flexible zinc batteries.

Pre-intercalating metal cations such as K^+, Ba^{2+}, Na^+, and $Co^{2+/3+}$ into MnO_2 can occupy specific lattice sites or increase the lattice spacing of the electrode, interacting with Zn^{2+} ion and altering the conductivity of the electrode. Zhang et al. proposed a La^{3+} ion pre-intercalated nano δ-MnO_2 electrode material (LMO). The pre-intercalated La^{3+} ion effectively reduced the interaction between Zn^{2+} and MnO_2, improving the kinetics of Zn^{2+} ion insertion and extraction, and resulted in a higher zinc storage capacity, faster reaction rate, and reversible redox reactions. Additionally, La^{3+} ion supports the crystal structure, preventing structural collapse during cycling, and the Zn||MnO_2 full battery exhibited an extremely high capacity density of 375.9 Wh kg^{-1} (at a current density of 100 mA g^{-1}). Although cation pre-intercalation in MnO_2 is typically regarded as beneficial for expanding interlayer spacing, promoting carrier insertion/extraction, and improving structural stability, some studies indicate that pre-introduced metal cations may hinder carrier diffusion. This can occur due to physical blocking or electrostatic repulsion when the tunnels are filled with nonremovable metal cations, significantly reducing the rate of carrier insertion/extraction. Therefore, in designing cation pre-intercalated MnO_2 cathodes, it is crucial to balance the hindrance of carrier transport by foreign ions with their contribution to material stability, to develop superior electrode materials.

(5) Element Doping

Hetero-elements have different electronic configurations compared to MnO_2 and can adjust the material's surface morphology, inducing structural transformations that significantly affect the electrode material's structure and electrochemical performance. Additionally, the bonding interactions between Mn and hetero-elements can enhance the stability of the crystal structure, preventing collapse during phase transitions and improving the charge storage performance of MnO_2 cathodes. Rare earth elements, with their unique 4f electron configurations, can interact with MnO_2 to modify its surface morphology. By integrating the benefits of nanostructures with metal doping, Wang et al. developed a Ce-doped MnO_2 nanorod cathode material through a hydrothermal synthesis method. The introduction of cerium (Ce) led to a phase transformation in MnO_2, shifting from the initial β-phase to a novel α-phase characterized by a [2 × 2] tunnel structure. CV analysis revealed that the voltage difference between the oxidation and reduction peaks for Ce-doped β-MnO_2 is notably smaller compared to that of undoped β-MnO_2. Furthermore, Ce-doped β-MnO_2 exhibits significantly lower charge transfer resistance than pristine β-MnO_2, indicating enhanced conductivity. The Zn||MnO_2 batteries utilizing this modified β-MnO_2 demonstrated a remarkable capacity of 134 mAh g^{-1} at a current density of 5 C (1 C = 308 mA g^{-1}). In addition to rare earth elements, Alfaruqi et al. introduced a V-doped MnO_2 nanocathode fabricated via a redox reaction. This V-doped MnO_2 features a smaller average grain size compared to the original MnO_2, which increases the specific surface area and provides additional reactive sites, thereby effectively enhancing conductivity. Compared to the original

MnO_2, it shows a higher discharge capacity (266 vs. 213 mAh g^{-1}). Additionally, doping with trace elements (such as N, La, Ca, etc.) can also reduce charge transfer resistance, thereby enhancing the conductivity of MnO_2 and improving battery electrochemical performance. Fluorine doping is a common method to enhance electrode electrochemical storage. It not only induces defect formation and provides more electrochemical active sites for Zn^{2+} ion but also effectively adjusts electrode performance due to its extremely high electronegativity and ion adsorption capacity, promoting Zn^{2+} ion insertion/extraction behavior. Liu et al. also proposed a "pinning effect" caused by Mn-F bonds, which constructs a highly active and stable lattice framework and provides rapid ion transport channels, achieving a remarkable specific capacity of 311.6 mAh g^{-1} at a current density of 0.6 C (where 1 C = 308 mA g^{-1}), with no significant capacity loss observed after 1200 cycles at a current density of 5C. It is noteworthy that oxygen has high electronegativity and replacing it with low electronegativity elements such as sulfur (S) and carbon (C) could potentially reduce the diffusion barrier of Zn^{2+} ion in the crystal structure. However, there is currently limited research on anion doping, which requires further investigation.

6.3.3 Prussian Blue Analogs (PBAs) Cathode Materials in Zinc-Ion Batteries

Beyond the conventional Mn-based and V-based materials, PBAs present notable potential for application in ZIBs. These materials feature open-framework face-centered cubic lattices that belong to the fm3m space group. Their 3D structure not only facilitates efficient ion transport and storage but also positions them favorably within the realm of electrochemical energy storage technologies. PBAs are characterized by the general formula $A_xM[M'(CN)_6]_{1-y}\square_y{\cdot}wH_2O$, where A denotes alkali metal ions (such as Na or K), M and M′ refer to transition metal elements (including Fe, Co, Ni, Cu, and Zn), and □ indicates vacancies within the hexacyanometalate structure. The crystal architecture of PBAs, depicted in Figure 6.22, shows that M′ forms bonds with carbon atoms while M bonds with nitrogen atoms, generating octahedra $M'C_6$ and MN_6. These octahedra are interconnected by C ≡ N bonds, resulting in an expansive 3D framework. Previous studies have revealed several advantages of PBAs: (1) the 3D structure and ample ion insertion sites enable rapid insertion and extraction of Zn^{2+} ions, ensuring robust structural stability. (2) PBAs exhibit high charge/discharge plateaus, excellent ionic conductivity, and substantial specific capacity. (3) PBAs are cost-effective, simple to synthesize, and environmentally friendly. Nonetheless, conventional synthesis techniques for PBAs often result in considerable amounts of $[Fe(CN)_6]^{4-}$ vacancies and lattice water. The random distribution of these vacancies can worsen lattice deformation, leading to the breaking of M-CN-M′ bridges during ion insertion and extraction, thereby compromising structural stability. Furthermore, lattice water can limit the availability of guest ion sites and hinder ion diffusion pathways. To improve the quality of PBA compounds as efficient zinc-ion insertion materials, significant efforts are underway to enhance their crystallinity.

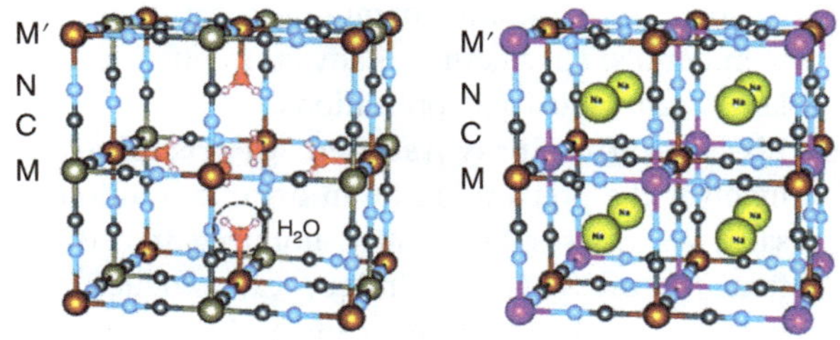

Figure 6.22 Typical PBA crystal structure.

6.3.3.1 Synthesis Methods for PBA Cathode Materials

1) **Coprecipitation Method**: The coprecipitation method is prone to rapid precipitation rates, which can lead to defects in PBAs when no additives are used. These defects significantly impact ion storage performance and result in several issues: (i) An abundance of water molecules can coordinate with unsaturated bonds at defect sites, which makes these bonds vulnerable to oxidative decomposition at high potentials, consequently diminishing the capacity for guest ion insertion. (ii) Water molecules may occupy some of the interlayer positions, obstructing ion transport pathways and slowing down ionic kinetics; (iii) repeated ion delamination processes can lead to lattice distortion or bond rupture, severely degrading cycling performance. Recent research indicates that incorporating sodium citrate or potassium citrate can effectively yield low-defect PB/PBAs. This suggests that optimized coprecipitation could be a viable approach for large-scale production.
2) **Single-Source Method**: PB/PBAs can be synthesized via the decomposition and recombination of the iron cyanide complex itself. In acidic solutions, $[Fe(CN)_6]^{4-}$ gradually, releasing Fe^{2+} ion, which subsequently reacts with the remaining $[Fe(CN)_6]^{4-}$ to form PB/PBAs with uniform morphology. Materials produced through this method typically exhibit high crystallinity and low defect content; however, the yield of PB-derived materials is often relatively low. Additionally, this process frequently generates toxic CN^- as a byproduct, which poses challenges for its application in large-scale industrial production.
3) **Electrochemical Deposition**: PBAs can be synthesized using a three-electrode system through electrochemical deposition. For example, immersing a working electrode in a solution containing $Fe^{3+}/K_3Fe(CN)_6$ and performing CV within a specified potential range allows for the continuous formation of PBAs through repeated scanning. Several factors can affect the reaction, including the concentration of the solution, stirring speed, temperature, and the use of surfactants. While this method is versatile and suitable for producing PB/PBA composite materials on various substrates, it requires complex equipment and operational procedures.
4) **Ball Milling Method**: This solid-state synthesis technique involves the reaction of transition metal salts with $Na_4Fe(CN)_6{\cdot}10H_2O$ in an inert gas environment. Throughout the milling process, the crystallization water is expelled, causing the reactants to dissolve and produce polybrominated biphenyls. Key parameters that influence the reaction include the water content of the starting materials,

rotation speed, duration of milling, atmospheric conditions, and the ratio of balls to materials. This method is advantageous due to its high yield and suitability for large-scale industrial applications.

5) **Electrostatic Spray-Assisted Coprecipitation (ESAC)**: This technique involves the preparation of an organic solution that contains transition metal salts, which is subsequently atomized into droplets using high-voltage electrostatic forces. This technique can produce low-defect crystals; however, it involves complex equipment and high costs, making it less suitable for industrial scale-up.

6.3.3.2 Optimization Strategy of PBA Cathode Materials

To improve the electrode performance of PBAs, researchers have suggested various optimization strategies focusing on material composition, microstructure, and morphology. The specific approaches include:

(1) Structural Water Regulation

The vacancy and coordination water content in PBAs can be diminished by augmenting the amount of alkali metal ions during synthesis to decrease the adsorption sites for zeolitic water, or by utilizing vacuum drying to eliminate interstitial water. This approach enhances the capacity delivery of PBA electrodes. In 2015, Goodenough et al. demonstrated that vacuum-dried $Na_2MnFe(CN)_6$, which was dehydrated (Figure 6.23a,b) [66], displayed a flatter and more reversible capacity of 150 mAh g^{-1} in a 3.5 V sodium half-cell, in contrast to hydrated $Na_{2-\delta}$MnHCF. This finding highlights the substantial influence of interstitial water on the electrochemical voltage distribution of the electrode.

(2) Composite Material Regulation

Combining PBAs with various materials, such as carbon-based substances and conductive polymers, through physical or chemical methods can improve conductivity and reversibility. For instance, in 2015, Yang et al. introduced a reduced graphene oxide-PB composite material (GOPC) designed as a cathode for sodium-ion batteries. This composite underwent thermal treatment, facilitating electron exchange between the cubic PB particles and graphene oxide (GO), which helped eliminate coordination water from the PB structure (Figure 6.23b) [67]. The resulting treated PB was entirely covered by reduced GO, leading to enhanced conductivity and minimized side reactions. Consequently, the Prussian blue/rGO cathode with less coordinated water demonstrated remarkable rate performance and cycling stability, achieving a capacity of 163.3 mAh g^{-1} at 30 mA g^{-1}, with a capacity retention rate of 91.9% after 500 cycles.

(3) New Morphology Regulation

Through promoting ion transport through new morphologies of active materials, the electrochemical performance of PBAs can be improved. Ren et al. enhanced Na^+ ion storage capacity in $Ni[Fe(CN)_6]$ by developing a distinctive NiHCF nanoflower structure through a surface etching process [68]. This treatment increased the BET

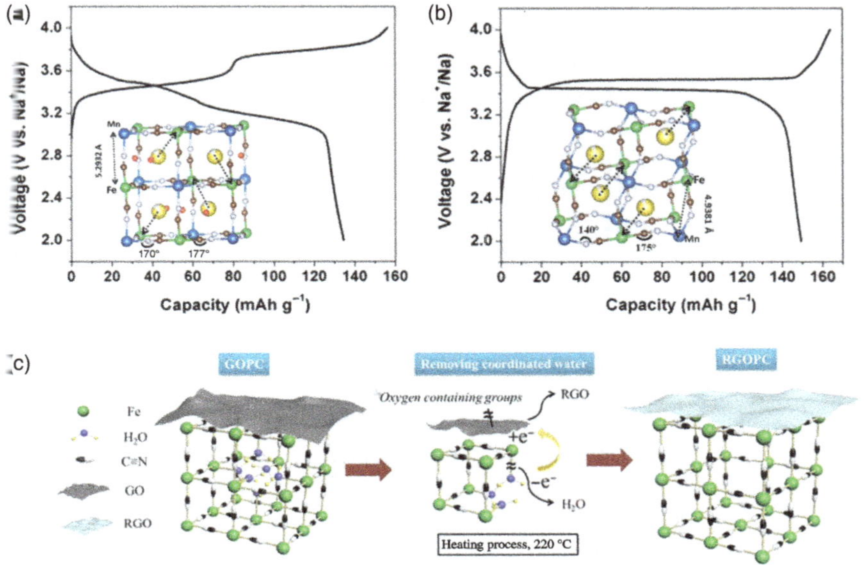

Figure 6.23 Galvanostatic initial charge and discharge profiles of (a) air-dried and (b) vacuum-dried $Na_{2-\delta}MnHCF$ at a current of 0.1 C. Insets illustrate the local structures of M-$Na_{2-\delta}MnHCF$ and R-$Na_{2-\delta}MnHCF$, highlighting the Na^+ displacements and framework distortions. Source: Reproduced from Song et al. [66]/with permission of American Chemical Society. (c) A schematic representation of the mechanism for removing coordinated water from rGO/PB cathode. Source: Reproduced from Yang et al. [67]/with permission of The Royal Society of Chemistry.

surface area of the etched material from 12.7 to 38.0 $m^2\ g^{-1}$, which enhanced the electrode/electrolyte interface and facilitated ion and charge transfer. Additionally, compared to the original NiHCF cubes, the etched NiHCF electrode exhibited more pronounced redox peaks, indicating higher specific capacity and faster Na^+ ion reaction kinetics.

(4) Metal Substitution Regulation

Metal substitution is a widely employed technique for modifying the electrochemical performance of electrodes by replacing alkali metal cations with different metal ions or by doping transition metal ions with other elements. For instance, Wang et al. investigated how the incorporation of low levels of Ni in Na-rich $NaFeFe(CN)_6$ (NaFeHCF) cathodes affected their electrochemical properties [69]. Through citric acid-assisted controlled crystallization coprecipitation, the Ni-substituted NaFeHCF not only stabilized its crystal structure but also enhanced its conductivity. Consequently, the $NaNi_{0.23}Fe_{0.77}HCF$ exhibited a remarkable discharge capacity of 105.9 mAh g^{-1} at a current density of 200 mA g^{-1}, along with a capacity retention of 73.1% when tested at 1000 mA g^{-1}.

6.3.4 Halogen-Based (Iodine/Bromine) Cathode Materials for AZIBs

In recent decades, intercalation-type cathode materials have been extensively utilized in zinc-ion batteries. However, their fundamental limitations tied to the intercalation mechanism led to less-than-ideal capacity and cycling performance. Conversion-type cathode materials, being diverse and cost-effective, can achieve high specific capacities through multi-electron transfer. AZIBs with conversion-type cathodes have seen remarkable advancements. Notably, iodine- and bromine-based halogen cathodes are particularly promising because of their high redox potentials and theoretical capacities. The rapid phase transition and charge storage in halogen-based electrodes offer a flat voltage plateau and high energy density, which are unparalleled advantages compared to intercalation-type cathodes.

6.3.4.1 Iodine-Based Cathode Materials for AZIBs

Zinc-iodine batteries have seen rapid advancements recently, primarily due to their high theoretical capacity and operational voltage. The I^-/I^0 redox couple, which involves single-electron transfer, has a theoretical specific capacity of 211 mAh g^{-1} and a redox potential of 0.54 V vs. SHE. Experimental findings indicate that by activating additional $I^0/I^+/IO^{3+}$ redox couples, capacities can achieve up to 422 mAh g^{-1} (I^-/I^+), and 1266 mAh g^{-1} (I^-/IO^{3+}), significantly exceeding the capacities of traditional insertion-type cathodes, which are around 200 mAh g^{-1}. However, inherent physical properties of iodine and the non-flowing storage mechanism of aqueous zinc-iodine batteries lead to several issues: (i) The poor conductivity of iodine leads to slow redox kinetics and limits iodine utilization; (ii) the natural sublimation of iodine results in considerable energy loss; (iii) low Coulombic efficiency, pronounced self-discharge behavior, and the shuttling effect of polyiodides (I^-, I_2, I_3^-) contribute to rapid capacity degradation.

Physical Properties of Iodine **Low Conductivity:** Iodine exhibits an electronic conductivity of approximately 10^{-6} S cm^{-1}, which is superior to that of sulfur (~5×10^{-30} S cm^{-1}) and certain insertion-type materials such as $LiFePO_4$ (~10^{-9} S cm^{-1}) and $Li_4Ti_5O_{12}$ (~10^{-13} S cm^{-1}). However, this conductivity is still two orders of magnitude lower than that of layered and spinel-type oxides (e.g. $LiCoO_2$ and $LiMn_2O_4$, ~10^{-4} S cm^{-1}). Consequently, most iodine-based cathodes necessitate the incorporation of a conductive network for optimal functionality.

Sublimation: Although iodine has a boiling point of 184.4 °C, its elevated vapor pressure encourages spontaneous sublimation, resulting in undesirable thermal instability. This pronounced sublimation causes a continuous loss of active iodine, which in turn diminishes the overall performance of the battery.

Solubility: Based on the electronic donor–acceptor theory, elemental iodine, acting as an electron acceptor, readily dissolves in various organic solvents (electron donors) such as chloroform, esters, alcohols, benzene, ethers, and even room-temperature ionic liquids. However, its solubility in water is quite limited at 0.034 g per 100 ml, and it does not form hydrates due to its higher formation heat of −4.5 kcal mol^{-1}, which makes aqueous electrolytes more appealing than nonaqueous ones. However, common transformation products like iodide (I_3^-) have greater solubility. This leads to reactions between iodine and I^- to form I_3^- and other polyiodides, which dissolve in both aqueous and nonaqueous solvents. The shuttling effect of these polyiodides results in a loss of active material and unwanted self-discharge behavior. Moreover, iodine can react directly with certain metals (such as Mg, Ca, Al, Zn, Sn, Ni, and Fe), forming insulating layers that increase internal resistance, impede ion transport, and passivate the metal anode.

(1) Modification Strategies for Iodine-Based Cathode Materials

Physical Adsorption: Developing appropriate hosts for the physical adsorption of iodine can effectively mitigate the aforementioned challenges. By anchoring I_2 within a host material, it is possible to enhance thermal stability and reduce the migration of polyiodides. High-conductivity hosts are essential to fulfill the kinetic demands associated with insulating iodine. Commonly used materials for this purpose include carbon-based substances with layered porous structures and extensive surface areas, such as carbon black, carbon nanotubes, and activated carbon. However, the interaction between nonpolar carbon and polar polyiodides is inadequate to prevent their dissolution and shuttling behavior. To overcome this limitation, doping carbon hosts with heteroatoms (such as nitrogen and phosphorus) presents a promising strategy to enhance both stability and iodine loading capacity.

Chemical Interactions: Enhancing iodine composites through chemical interactions represents a more sophisticated approach than mere physical adsorption. By establishing chemical bonds between iodine species and host materials, the affinity between them can be significantly improved, resulting in greater long-term stability. For instance, polymers like β-cyclodextrin (β-CD), polyvinylpyrrolidone (PVP), and linear starch have demonstrated the ability to form charge transfer complexes with iodine. This occurs when these polymers are immersed in iodine solutions or exposed to iodine vapor, effectively increasing their interaction with iodine and enhancing the overall stability of the composite. Among these, PVP-I_2 exhibits the highest stability due to hydrogen bonding. PVP and I_2 combinations improve thermal stability and suppress shuttling effects by reducing vapor pressure and solubility of iodine species [70]. Conductive polymers like polyaniline (PANI) offer rapid charge transfer and chemical fixation of iodine species, showing rapid rate capability and long-term stability [71]. Organic groups such as methylammonium iodide (MAI) and tetrabutylammonium iodide ($TBAI_3$) also exhibit strong interactions and have shown extremely long lifespans (10 000 cycles) [72]. In general, similar to pure I_2, most polymer-I_2 complexes and organic iodide salts still depend on conductive networks to improve their naturally low conductivity. Common porous carbon hosts, including activated carbon, graphene, and carbon hollow spheres, can aid in electron transport while also capturing dissolved iodine ions, thus enhancing stability.

Constructing conductive hosts derived from metal-organic frameworks (MOFs) is another method that leverages chemical interactions while addressing sluggish electrochemical redox kinetics. 2D MOFs featuring extended π-conjugation facilitate the delocalization of π-electrons, which enables the effective transport of iodine ions within the π-d conjugated planes. This structural characteristic leads to enhanced conductivity. Moreover, MOFs exhibit strong chemical adsorption of iodine ions, and their excellent charge transfer capabilities ensure rapid electrochemical kinetics. In addition to their strong interactions with iodine ions, the metal ions present in the framework can serve as electrocatalytic sites, enhancing the kinetics of the reactions. For example, PBAs, $Co[Co_{1/4}Fe_{3/4}(CN)_6]$) have been employed as electrocatalytic hosts, where iodine is completely converted to I^-, as shown by UV–visible absorption spectroscopy (Figure 6.24a) [73]. DFT calculations further suggest that the low energy barrier for the reaction arises from the synergistic effect of Co and Fe (Figure 6.24b). As a result of the rapid reaction kinetics, even at a high current density of 20 A g^{-1} (~94 C), the rechargeable Zn-PBA/I_2 battery demonstrates an impressive specific capacity of 151.4 mAh g^{-1}. Through an efficient interhalogen activation strategy and interlayer confinement design, we successfully stimulated and stabilized a highly reversible four-electron redox reaction from I^- to I^+ in zinc-iodine batteries. This advancement has led to significant improvements in the specific capacity, electrode potential, and cycling performance of the zinc-iodine battery. Using Br^- as an activator, the interhalogen interactions facilitate the four-electron transfer reaction from I^- to I^+. Nickel-iron-layered double hydroxide (Ni–Fe LDH) was synthesized and used as the iodine host, which, due to its large interlayer spacing and stability, ensures that the I^-/I^+ redox reactions occur within the layers, effectively suppressing the shuttling effect of iodides [74]. Consequently, the zinc-iodine full battery based on the I_2@Ni-Fe LDH and routine $ZnSO_4$ electrolyte containing $ZnBr_2$ demonstrated significant improvements in output voltage, cycling stability – with a capacity retention of 94.6% after 10 000 cycles – and a specific capacity of 350 mAh g^{-1} at a current density of 1 A g^{-1}.

(2) Separator Design

Incorporating an interlayer between the cathode and separator presents an effective approach to confine soluble polyiodides within the cathode region. Functional separators that facilitate selective ion transport have been successful in reducing the shuttle effect seen in lithium-sulfur batteries and iodine-based redox flow batteries. Cation exchange membranes (CEMs), which prevent iodine ion migration, have been extensively researched for use in iodine-based redox flow batteries. A MOF known as Zn-BTC has been developed as a multifunctional separator that effectively inhibits the migration of I_3^- ions in the electrolyte, regulates the solvation process, and subsequently mitigates both the shuttle effect and side reactions involving zinc anodes [75]. Researchers also reported that low-cost zeolite molecular sieves with a porous structure exhibited similar functionality to the Zn-BTC MOF, highlighting the potential for functional separators in large-scale applications. Overall, the introduction of iodine hosts through physical adsorption and chemical interactions can stabilize iodine species, contributing to the long-term stability of zinc-iodine batteries.

(3) Electrolyte Improvements

Soluble iodine in the electrolyte can migrate to the anode, where it may react with the metal, leading to anode passivation and significant self-discharge issues. To

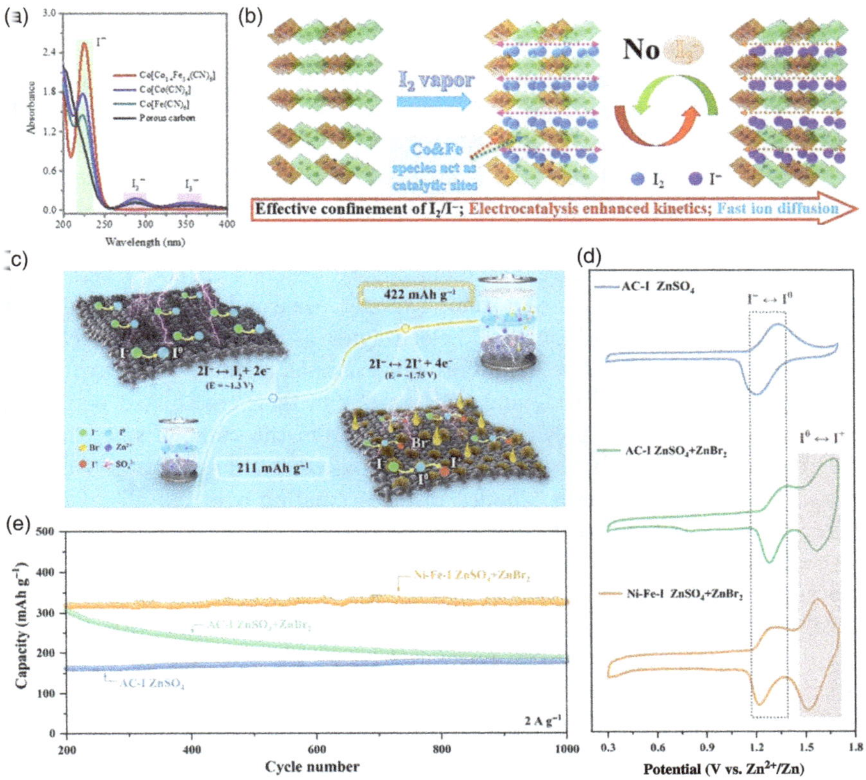

Figure 6.24 (a) UV–vis absorption spectra of electrolytes for $Co[Co_{1/4}Fe_{3/4}(CN)_6]$, $Co[Co(CN)_6]$, $Co[Fe(CN)_6]$, and porous carbon after sufficient IRR. (b) Schematic illustration of the fabrication of I_2-in-PBAs frameworks (denoted as PBAs/I_2) and electrochemical procedure with electrocatalytically active sites of Co and Fe species. Source: Ma et al. [73]/John Wiley & Sons. (c) Schematic illustration of energy storage mechanism in a Zn-Iodine battery with or without the Br^- activator; (d) CV curves at 1 mV s^{-1} of AC–I electrode in $ZnSO_4$ and $ZnSO_4 + ZnBr_2$ electrolytes, and Ni–Fe–I electrode in $ZnSO_4 + ZnBr_2$ electrolyte; (e) corresponding cycling performance at 2 A g^{-1}. Source: Wang et al. [74]/John Wiley & Sons.

tackle these challenges, optimizing the electrolyte to prevent iodine dissolution and protect the anode from corrosion is crucial. The electrolyte-to-iodine ratio (E/I) plays a vital role in influencing solid–liquid reactions, as the host's limitations on iodine compete with the solvation of iodine in the electrolyte. Reducing the electrolyte volume or increasing the iodine loading can help strike a balance and minimize iodine loss, thereby alleviating capacity fade and enhancing Coulombic efficiency. Additionally, the choice of anions in the electrolyte significantly impacts the electrochemical performance of aqueous zinc-iodine batteries. For instance, in diluted $ZnSO_4$ solutions and zinc trifluoromethanesulfonate ($Zn(CF_3SO_3)_2$) electrolytes, the binding energy between iodine and SO_4^{2-} is considerably stronger than that between iodine and $CF_3SO_3^-$. This suggests that the low utilization of iodine may result from the SO_4^{2-} capture effect.

The salt-in-water electrolyte strategy can also address the shuttle issue. In dilute electrolytes, I_2 reacts with I^- to form soluble I_3^-, whereas in 15 M $ZnCl_2$ electrolytes, I^- ions tend to be captured by forming $[ZnI_x(OH_2)_{4-x}]^{2-x}$ superhalide ions, thereby limiting iodide dissolution. Besides optimizing liquid electrolytes, employing all-solid electrolytes can effectively solve the polyiodide shuttle problem in aqueous zinc-iodine batteries. The solid-phase conversion of iodine into insoluble polyiodides in solid electrolytes can result in sluggish kinetics and poor reversibility. To address this issue, quasi-solid-state hydrogel electrolytes have emerged as effective alternatives to traditional solid electrolytes, as they can enhance reaction kinetics and inhibit polyiodide migration. For example, alginate-based hydrogels, which feature abundant –COO– and –OH groups in their crosslinked networks, are capable of preventing polyiodide migration from the cathode side, thereby providing quasi-solid Zn-I_2 batteries with enhanced cycling stability. Another approach to improve reaction kinetics involves creating a mixed all-solid-state electrolyte structure with additional dispersive layers. While significant efforts have been made to stabilize the I^-/I^0 redox reaction, researchers have only recently achieved success in activating and stabilizing the I^0/I^+ redox couple using chloride electrolytes [76]. This double-electron transfer mechanism, along with a higher redox potential of 0.99 V vs. SHE, presents a more attractive option for energy density compared to the single-electron transfer mechanism of the I^-/I^0 conversion reaction, which has a lower redox potential of 0.54 V vs. SHE.

6.3.4.2 Bromine-Based Cathode Materials for AZIBs

In comparison to zinc-iodine batteries, bromine exhibits a more favorable theoretical energy density of 419 Wh kg^{-1}, attributed to its higher redox potential (1.84 V for Br^0/Br^- vs. Zn/Zn^{2+}). Current research primarily emphasizes zinc-bromine flow batteries, which employ circulating aqueous solutions (like $ZnBr_2$) and auxiliary components such as microporous membranes. This system, along with electrolyte flow mechanisms (including tanks, pumps, and pipelines), effectively separates and circulates the catholyte and anolyte in distinct compartments, thereby reducing the diffusion rate of bromine species and improving the current density of zinc deposition. However, these systems face long-term challenges, including high battery resistance, low energy efficiency (typically around 80%), and significant system scale and complexity, all of which contribute to increased overall system costs. Additionally, the flow of bromine species, which are both mobile and corrosive, further undermines system reliability, with battery failure often attributed to the corrosion of flow battery components rather than the degradation of the internal redox reactions.

In this context, non-flow zinc-bromine batteries, similar to lithium-ion battery configurations, avoid the aforementioned issues. This streamlined design eliminates the need for auxiliary components, significantly improving energy density, lowering costs, preventing electrolyte leakage, and enhancing battery safety. However, the formation of polybromide species (Br_n^-, where $n = 2x + 1$) during the charging and discharging processes leads to uncontrolled diffusion of bromine components, resulting in substantial self-discharge and capacity decline. During charging, the cathode generates Br_2, which reacts with Br^- in the aqueous electrolyte to form highly soluble bromides. The dynamic equilibrium of Br_n^- contributes to poor utilization of active bromine. Additionally, the migration of negatively charged polybromide ions toward the anode, driven by electric fields and concentration gradients, leads to capacity loss, decreased Coulombic efficiency, and zinc anode corrosion.

To prevent polybromide crossover, conventional methods employ physical adsorption to confine liquid bromides within a porous carbon matrix. Hee-Tak Kim and colleagues proposed a membrane-free non-flow zinc-bromine battery using protonated pyridine nitrogen-doped microporous graphite felt (NGF) as the positive electrode material [77]. As shown in Figure 6.25a, the abundant micropores and protonated nitrogen-doped sites in NGF effectively capture bromine atoms and polybromide anions, promoting the transformation of bromine atoms into polybromide anions through electrochemical–chemical growth mechanisms. This effectively slows the diffusion of elemental bromine and polybromide anions, suppresses self-discharge, and allows the membraneless and flowless Zn-Br battery with NGF to maintain stable cycling for 1000 cycles, achieving an energy efficiency of 80%.

Introducing complexing agents that can form stable solid-phase complexes effectively mitigates the shuttle issues of Br^- and Br_3^-. Liang et al. utilized a highly effective complexing agent, tetrabutylammonium bromide (TPABr) [79], to reversibly convert Br_2/Br_3^- species into a solid phase. The solid complex is confined within porous carbon with excellent interface contact, offering physical constraints that prevent shuttle effects and self-discharge. This design mitigates the high self-discharge rate and low energy efficiency commonly associated with zinc-bromine batteries, while retaining the benefits of the zinc-bromine redox couple. The team developed a practical zinc-bromine battery that harnesses the combined effects of complexation chemistry at the electrode and the salting-out phenomenon in the aqueous electrolyte. The isotropic $ZnSO_4$ electrolyte, with its strong salting-out effects due to structural water formation, effectively suppresses the dissolution and hydrolysis of the isotropic pyridinium–polybromide complex in the electrode. This exclusion–complexation mechanism facilitates the creation of quasi-solid cathodes, enhancing redox kinetics while efficiently inhibiting unwanted oxygen and chlorine evolution reactions. The highly reversible $Br^-/Br_0/Br^+$ redox couple achieves a near-theoretical specific capacity of 215 mAh g^{-1}, with an average Coulombic efficiency of 99.8%. The pouch cell delivers an energy density of up to 106 Wh kg^{-1}, an energy efficiency of 87.8%, and a low material cost of $28 per kilowatt-hour.

Compared to complexation mechanisms, chemical adsorption is also an effective method for confining polybromides, though research in this area is still limited. Liu and coworkers developed an aqueous zinc-bromine battery based on covalent organic framework (COF), using Br_2-exCOF as the cathode. They employed *in situ* electrochemical Raman spectroscopy to observe the chemical structural changes in the Br_2-exCOF electrode during bromine storage, providing insights into the bromine storage mechanisms [78]. UV–VIS (Figure 6.25b) was used to detect the adsorption performance of exCOF for polybromides, while DFT calculations

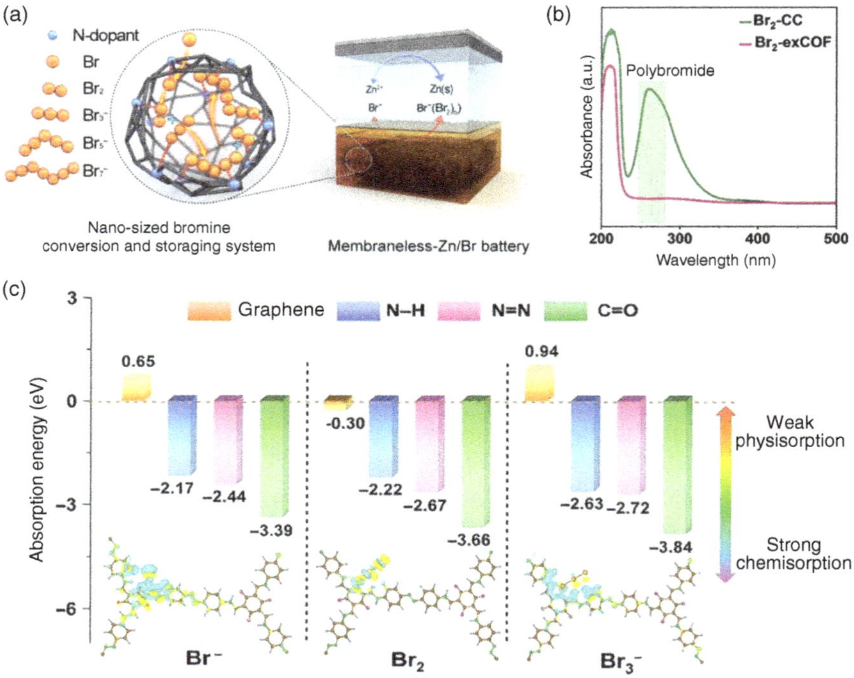

Figure 6.25 (a) Schematics of proposed membraneless and flowless Zn-Br battery employing NGF electrode and its Br conversion/storage mechanism. Source: Reproduced with permission from [77]/John Wiley & Sons. (b) UV–vis spectra of electrolyte with Br_2-CC and Br_2-exCOF as cathode. (c) The summarized adsorption energy on C=O, N=N, and N—H for various bromine species and the charge-density patterns of Br species around C=O adsorption sites. Source: Reproduced from Zhang et al. [78]/with permission of Elsevier.

explored the interaction mechanisms between exCOF (C=O, N=N, and N—H) and graphene with Br^-, Br_2, and Br_3^- (Figure 6.25c). Results indicated that exCOF exhibits strong chemical adsorption toward Br species, facilitating effective conversion between elemental bromine and polybromides. Through comprehensive characterization of electrodes based on the 1,8-octanediamine (ODA) framework ($ODABr_2$, $ODAI_2$, and ODABrI), Zhi et al. uncovered their unique electrochemical properties [80]. DFT simulations, alongside spectroscopic analyses, revealed that the improved chemical adsorption between the double ammonium host and halide ions is key to the battery's excellent redox kinetics and cycling stability. This strong host–guest interaction significantly boosts electron shuttle efficiency and effectively suppresses polybromide cross-diffusion. As a result, the designed AZIBs achieve near-theoretical capacities of 235 mAh g^{-1} for $ODAI_2$ and 312 mAh g^{-1} for $ODABr_2$, while demonstrating extended cycle lives, with $ODAI_2$ lasting 10 000 cycles and $ODABr_2$ lasting 3000 cycles.

Although these strategies effectively suppress the shuttle issues of polybromides (Br^-, Br_2, Br_3^-) and improve the cycling life of zinc-bromine batteries, the inherent insulating properties of bromine and polybromides hinder electron transfer, reducing redox kinetics and leading to poor active bromine utilization and rate performance. Wei et al. designed a high-performance aqueous Zn||Br_2 battery by incorporating 2D conjugated nickel phthalocyanine (NiPPc) as the adsorptive-catalytic cathode host [81]. This innovative approach tackles key issues in conventional Zn||Br_2 batteries, including shuttle effects, low bromine utilization, and low discharge voltage. Experimental data and theoretical simulations revealed that atomically dispersed Ni-N_4 sites in NiPPc serve as efficient active sites for bromine redox reactions and exhibit strong polarity, enhancing bromine and polybromide adsorption. This synergistic mechanism significantly improves the battery's overall electrochemical performance. The resulting Zn||Br_2 battery delivers a high specific capacity of 265 mAh g^{-1} at a discharge current density of 2 A g^{-1}, with a discharge voltage of up to 1.82 V, close to theoretical values. Even at a current density of 5 A g^{-1}, it maintains a specific capacity of 240 mAh g^{-1} and retains approximately 95% capacity after 3000 charge–discharge cycles. The NiPPc host, leveraging its unique adsorption-catalytic mechanism, overcomes the limitations of traditional NF-ZBB systems and holds promise as a candidate material for high-performance aqueous batteries.

6.3.5 Organic Compounds Cathode Materials for AZIBs

Organic compounds can be classified into *p*-type, *n*-type, and bipolar types according to their redox mechanisms. As illustrated in Figure 6.26a, *p*-type organic materials donate electrons and become positively charged upon oxidation. Examples of these include nitrogen-oxide-free radicals and triphenylamine derivatives. On the other hand, *n*-type organic materials gain electrons and are reduced to a negatively charged state, such as carbonyl and imine compounds. Bipolar organic compounds exhibit both *p*-type and *n*-type characteristics, primarily as conductive polymers. *P*-type organic materials generally show faster reaction kinetics, whereas *n*-type materials have a larger theoretical specific capacity. Bipolar materials exhibit higher electron conductivity, which facilitates electron transfer in the cathode. Organic cathode materials for AZIBs can be broadly classified based on their active groups into conductive polymers, carbonyl compounds, imine compounds, nitrogen-oxide free radical compounds, COFs, MOFs, and triphenylamine derivatives. Figure 6.26b presents a comparison of the electrochemical performance indicators for various types of organic cathode materials. The subsequent discussion will delve into the reaction mechanisms and energy storage behaviors of each type of organic compound cathode material in detail.

6.3.5.1 Conductive Polymers

The conductive polymers, such as PT, PPy, and PANI, have been widely explored as cathode materials for secondary ion batteries due to their long-range π-electron conjugation, which imparts high electrical conductivity. Notably, PANI and PPy have been directly employed as cathode materials in AZIBs, frequently exhibiting bipolar

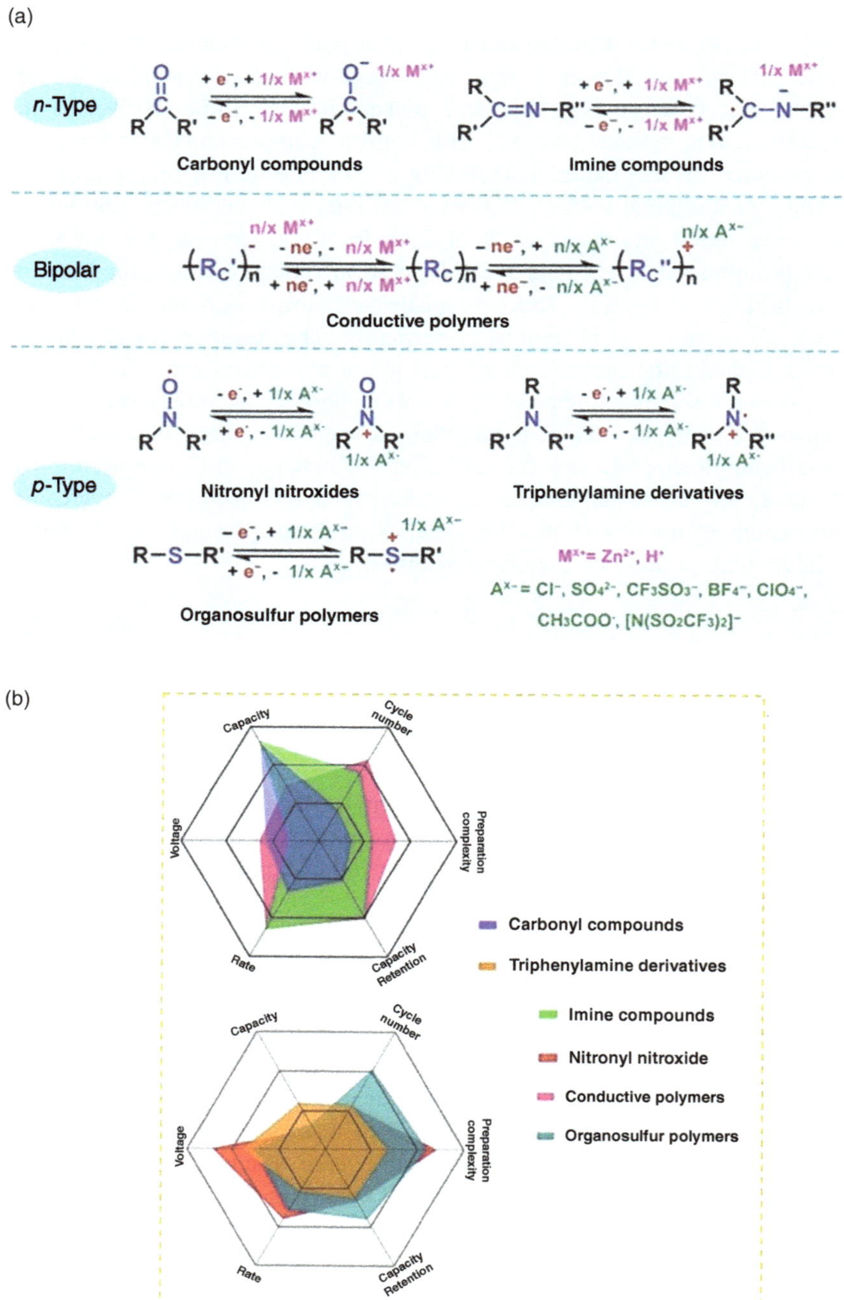

Figure 6.26 (a) Charge storage mechanisms of various organic electrode materials (*n*-type, *p*-type, and bipolar). (b) Performance comparison of different organic compounds as cathodes for AZIBs. Source: Reproduced from Tie et al. [82]/with permission of John Wiley & Sons.

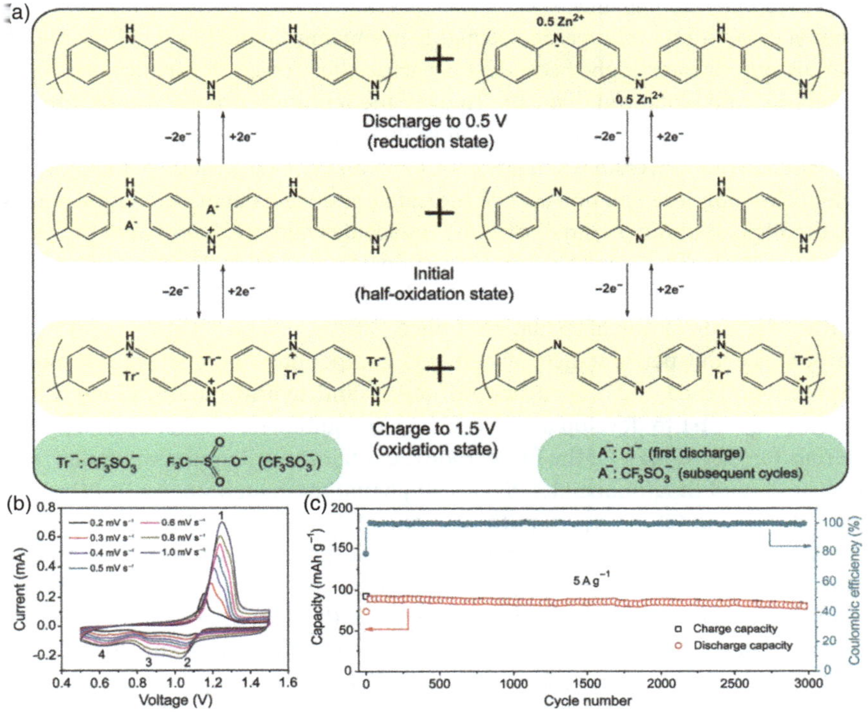

Figure 6.27 (a) The redox mechanism of PANI/CFs. (b) CV curves of PANI/CFs (vs. Zn^{2+}/Zn). (c) Cycling performance of PANI/CFs at 5 A g^{-1}. Source: Reproduced from Li et al. [80]/with permission of John Wiley & Sons.

redox mechanisms in aqueous electrolytes. Wan et al. studied the Zn^{2+} insertion/extraction process and the bipolar mechanism of the Zn||PANI battery during charging and discharging (Figure 6.27a) [83]. The corresponding charge–discharge behavior at each stage is illustrated in the Figure 6.27b. The battery achieves a high capacity retention of 92% after 3000 cycles at a current density of 5 A g^{-1}, as shown in Figure 6.27c. Furthermore, the high conductivity and flexibility of PANI/CFs cathodes make them suitable for flexible battery electrodes. Wang et al. deposited Zn anodes and PPy cathodes on PET film, creating a flexible electrochromic Zn||PPy battery with excellent flexibility [84]. However, the low specific capacity and energy density of conductive polymers often limit their ability to meet the energy storage demands of batteries when used alone. Nonetheless, their high conductivity allows them to replace conductive additives in cathode materials. Typically, conductive polymers are combined with inorganic materials to boost their overall performance. This modification enhances the electrical conductivity of the inorganic components while also stabilizing their crystal structure, leading to improved reaction kinetics and significantly better cycle stability for the battery (refer to Section 6.3.1.5 for detailed modification strategies).

6.3.5.2 Carbonyl Compounds

Carbonyl compounds are regarded as highly promising candidates for ZIB cathode materials due to their high theoretical capacities, fast reaction kinetics, excellent reversibility, and structural diversity. To date, the reported carbonyl-based cathode materials include quinones, ketones, and acylamide compounds. Their functionality is based on the reversible activation of C=O and C—O$^-$ groups, which enable efficient Zn^{2+} ion intercalation and deintercalation during battery operation. However, the high solubility of small carbonyl molecules and their discharge products in aqueous electrolytes inevitably results in poor cycle stability. Unlike monovalent ions (Li^+, Na^+, and K^+), zinc ions need to interact with two adjacent C=O groups, requiring the overcoming of spatial hindrance.

To enhance the performance of carbonyl compounds, various improvements have been proposed, such as structural design and hybrid modifications. Chen et al. were the first to develop a series of organic quinone compounds as cathode materials for AZIBs [85]. Of the five carbonyl-containing quinone compounds they tested – 1,4-naphthoquinone (1,4-NQ), 9,10-phenanthrenequinone (9,10-PQ), cavitand [4] quinone (C4Q), 1,2-naphthoquinone (1,2-NQ), and 9,10-anthraquinone (9,10-AQ) – C4Q showed the highest specific capacity, reaching 335 mAh g^{-1}. However, the strong interaction between zinc ions and the two adjacent carbonyl groups in C_4Q resulted in significant dissolution of the discharge products in aqueous electrolytes, limiting the battery's cycle life. To address this issue, Wang et al. introduced a ZIB with a pyrene-4,5,9,10-tetrone (PTO) cathode [86], which achieved a high specific capacity of 336 mAh g^{-1} for Zn^{2+} ion storage and demonstrated stable cycling performance for 1000 cycles at 3 A g^{-1}.

In contrast to small organic molecules, polymer materials can offer longer conjugated frameworks and more active sites, effectively alleviating solubility issues in aqueous electrolytes. Zhang et al. synthesized a poly(2,5-dihydroxy-1,4-benzoquinone sulfide) (PDBS) material via a simple polymerization reaction [87]. Compared to the monomer, the sulfur and hydroxyl groups in PDBS enhance electrochemical activity and better facilitate Zn^{2+} ion intercalation and deintercalation. The PDBS-based aqueous zinc-organic battery demonstrated impressive rate performance, retaining 79% of its capacity after 2000 cycles at a current density of 2 A g^{-1}. In another study, Wang and coworkers coated carbon nanotube fibers (CNF) with polydopamine (PDA) to create a flexible fiber battery [88], which exhibited both high specific capacity and outstanding rate performance, maintaining 80% of its initial capacity after 1700 cycles at 1 A g^{-1}.

Although high-activity polymers provide more active sites, their low electrical conductivity remains a key factor limiting further applications in AZIBs. Currently, one of the most effective strategies to enhance conductivity is the hybridization of carbon materials. Xu et al. introduced a novel polymer material known as polycatechol (PC) and investigated the discharge performance of PC/graphene (PC/G) composites [75]. Their findings indicated that incorporating graphene significantly improved the conductivity of the PC cathode materials, thereby enhancing the reaction kinetics of the electrode. The PC/G-2 battery achieved a high specific capacity of 355 mAh g^{-1} at 0.1 C and maintained 74.4% of its initial capacity after

3000 cycles at 2.0 C. However, while covalent bonding between carbon frameworks and small molecules can enhance the solubility of cathode materials, an increase in molecular weight often results in reduced specific capacity and diminished rate performance. Thus, modifying carbonyl compound cathode materials for AZIBs remains an ongoing challenge.

6.3.5.3 Imine Compounds

Imine compounds with C-N functional groups demonstrate high redox activity and significant structural tunability, making them a focal point of research in the realm of AZIBs. Their multi-ion insertion mechanism allows for rapid kinetics and substantial capacity. However, one notable drawback is their typically low average discharge voltage, which poses challenges for practical applications in energy storage devices. Wang et al. conducted a comparative study on the electrochemical behaviors of phenazine (PNZ) and 2,3-diaminophenazine (DAP) in aqueous electrolytes [89]. The hydrophilic groups ($-NH_2$) in DAP form hydrogen bonds with water, increasing its solubility in the electrolyte. Conversely, PNZ allows for the reversible insertion and extraction of zinc ions due to the spatial hindrance caused by adjacent hydrogen atoms. Zhang et al. designed an aqueous organic zinc battery system using 2,3-a:2′,3′-c]phenazine (DQP) [90], which features a large π-conjugated system that enhances rapid charge transfer and collection, thereby improving structural stability throughout charge and discharge cycles. This battery exhibits an impressive specific capacity of 413 mAh g^{-1} at a current density of 50 mA g^{-1}, retaining 86% of its initial capacity after 1000 cycles at a higher current density of 5 A g^{-1}. Furthermore, incorporating electron-withdrawing groups (such as –F, –Cl, and –CN) can lower the lowest unoccupied molecular orbital (LUMO) energy level in organic materials, which is advantageous for boosting discharge voltage. Ye et al. introduced –CN groups into the HATN edges, successfully reducing the energy gap (ΔE) between the highest occupied molecular orbital (HUMO) and LUMO [91]. This modification not only enhances the battery's operating voltage but also improves charge distribution among nitrogen atoms. As a result, this battery system maintains 90.7% of its initial capacity after 5800 cycles at a current density of 5 A g^{-1}. Additionally, Xu et al. prepared poly(o-phenylenediamine) (PoPD) organic cathodes under varying conditions, [92]. Among these, the PoPDH variant produced under acidic conditions achieved a specific capacity of 318 mAh g^{-1} at a current density of 0.05 A g^{-1}, while still delivering a capacity of 95 mAh g^{-1} at a higher current density of 5 A g^{-1}, indicating good cycling stability.

6.3.5.4 Nitroxide Radicals Compounds

Nitroxide radical compounds, consisting of carbon, nitrogen, oxygen, and hydrogen elements alongside spin-polarized electrons, are widely utilized across various fields due to their distinctive properties. When employed as cathode active materials, nitroxide radicals offer notable advantages such as substantial reversible capacity, commendable cycling performance, and excellent rate capability. These benefits stem from their high redox activity and rapid electron transfer processes. Koshika et al. reported a ZIB that utilized poly(2,2,6,6-tetramethylpiperidinyl-4-yl)oxyethyl ether (PTVE) as the cathode material [93]. The Zn||PTVE battery demonstrated good rate performance and achieved a high discharge voltage plateau of 1.7 V. However, some ions tended to decompose at elevated voltages, resulting in a limited cycle life within a relatively wide potential window for the Zn||PTVE battery. Consequently, designing aqueous electrolytes with a high stability voltage window is essential to complement high-voltage organic cathodes. Luo et al. compared the effects of three anions – SO_4^{2-}, $CF_3SO_3^-$, and ClO_4^- – on the electrochemical performance of the battery [94]. The batteries utilizing these different electrolytes exhibited discharge voltage plateaus of 1.77, 1.58, and 1.53 V, respectively. DFT calculations further supported that the stronger the binding energy between the anions and PTVE, the higher the operating voltage of the battery.

6.3.5.5 COFs and MOFs

COFs and metal-organic frameworks (MOFs) are characterized by their highly ordered structures, intrinsic high porosity, and excellent physical and chemical properties. The ordered porous frameworks of COFs and MOFs provide efficient pathways for rapid ion transport, while their large dimensions and nanoscale thickness result in high surface areas and active sites, making them attractive for various applications, including energy storage. COFs are assembled using predesigned symmetrical organic building blocks through π–π stacking, resulting in abundant redox-active sites, improved stability, long-range ordered open channels, and reduced solubility. Their robust conductive networks and open pores facilitate the penetration of electrolytic ions, enabling Faradaic processes for Zn^{2+} electrochemical storage. Khayum et al. developed a hydrogen quinone-based COF (HqTp) to serve as a polymer cathode material for AZIBs [95]. The C=O and N—H functional groups present in the polymer play a crucial role in effectively anchoring a significant number of zinc ions during electrochemical reactions. This material demonstrates a notable specific capacity of 276 mAh g^{-1} at a current density of 125 mA g^{-1} and retains 94% of its initial capacity even after 1000 cycles at a high current density of 3750 mA g^{-1}. Moreover, a novel ortho-phenanthroline-based COF (PA-COF) has been utilized as a cathode material for ZIBs [96]. Theoretical simulations indicate that the ortho-phenanthroline units act as active sites for zinc-ion binding. Through inductively coupled plasma atomic emission spectroscopy (ICP-OES) and solid-state nuclear magnetic resonance (NMR), the capacity contributions from zinc and hydrogen ions have been quantified. PA-COF delivers a high capacity of 247 mAh g^{-1} at a current density of 0.1 A g^{-1} and demonstrates excellent cycling stability, withstanding 10 000 cycles at 1.0 A g^{-1} and showing an average capacity decay of just 0.38% per cycle.

MOFs are generally formed by coordinating metal nodes (such as Fe, Ni, or Co) with conjugated planar organic ligands that contain multifunctional groups (–OH, $-NH_2$, –SH, etc.). These structures possess abundant micropores or mesopores, enabling the reversible insertion and extraction of ions. Additionally, the redox activity of the metal ion clusters within MOFs facilitates efficient electron transport pathways.

For example, Nam et al. synthesized a 2D conductive MOF material, $Cu_3(HHTP)_2$, using 2,3,6,7,10,11-hexahydroxytriphenylene (HHTP) [97]. The unique 2D framework and 1D channel structure create an ideal host for the insertion of hydrated zinc ions. Non-*in situ* XPS analysis revealed that both C=O groups and Cu centers in $Cu_3(HHTP)_2$ act as redox-active sites during the charge and discharge process. Furthermore, the $Cu_3(HHTP)_2$ cathode operates through an intercalation pseudocapacitive mechanism, where its kinetics are not hindered by cation diffusion. As a result, this material achieves a high reversible capacity of 228 mAh g^{-1} at a current density of 50 mA g^{-1} and retains 75% of its capacity after 500 cycles at a high current density of 4000 mA g^{-1}. The development of MOFs offers an effective solution to the solubility challenges commonly faced by organic cathode materials.

6.4 Zinc Metal Anode Materials

Due to the high hydrogen evolution overpotential of zinc metal in neutral or weakly acidic aqueous electrolytes, thermodynamically, zinc anodes can undergo reversible stripping and deposition in such environments. Additionally, zinc metal offers advantages such as low cost, non-flammability, and environmental friendliness. Furthermore, its exceptionally high theoretical specific capacity (mass specific capacity: 820 mAh g^{-1}; volumetric capacity: 5855 mAh cm^{-3}) and low redox potential (−0.76 V vs. SHE) make zinc metal one of the most promising and widely used anode materials for AZIBs.

Despite these benefits, the application of zinc anodes still faces significant challenges, particularly dendritic growth and parasitic side reactions, which considerably reduce their lifespan and Coulombic efficiency. Excessive dendritic growth can penetrate the separator, leading to battery short circuits and shortening the overall battery life. Moreover, compared to alkaline electrolytes, the proton-rich environment of weakly acidic electrolytes enhances the thermodynamic drive for HER. The excessive release of hydrogen gas can cause separation between the cathode, anode, and separator, resulting in an open-circuit condition. Additionally, the irreversible consumption of the electrolyte during hydrogen evolution and corrosion processes can diminish the long-term cycling performance of ZIBs, increasing the risks of battery swelling, explosion, and electrolyte leakage. These issues exacerbate one another, further deteriorating anode performance. Therefore, implementing economically viable strategies to mitigate or even resolve the array of challenges associated with zinc anode materials is crucial for the large-scale application of AZIBs.

To develop zinc metal anodes that exhibit good reversibility and high utilization in weakly acidic or neutral electrolytes, researchers have conducted extensive studies, particularly focusing on interface modification of zinc anodes. These strategies are typically categorized into: anode substrate design, electrolyte optimization, interface modification, and separator design. This chapter targets to provide a comprehensive summary of the latest advancements in zinc anodes for AZIBs. First, it systematically analyzes the major challenges related to dendrite growth, hydrogen evolution, chemical/electrochemical corrosion, and their interactions to understand zinc deposition behavior. Then, it categorizes and discusses recent strategies for improving anode performance, including interface modification, anode structure design, current collector design, separator design, and electrolyte optimization.

6.4.1 Challenges and Issues of Zinc Anodes in Aqueous Zinc-Ion Batteries

6.4.1.1 Zinc Deposition and Dendrite Growth

The anode/electrolyte interfacial reactions play a crucial role in the energy storage and output processes of zinc-ion batteries. Unlike battery systems with alkaline electrolyte using zinc salts ($Zn(OH)_4{}^{2-}$) as charge carriers, the mild AZIBs with neutral/weakly acidic electrolyte involves the reversible deposition/dissolution of zinc on the anode surface during charge and discharge cycles. The zinc deposition process is similar to that of alkali metal anodes and primarily follows a nucleation-growth mechanism. Initially, zinc ions migrate to nucleation sites under the influence of an electric field, overcoming the energy barrier for the reduction reaction (i.e., forming a pristine solid phase) to nucleate. Following nucleation, Zn continues to plate onto the initial deposition layer. Zinc growth is a diffusion-controlled process influenced by both the electric field and concentration gradient, as illustrated in Figure 6.28. The quality of zinc plating can be assessed using two key parameters: initial nucleation overpotential and subsequent platform overpotential. Nucleation overpotential refers to the difference between the tip potential and the stable potential that follows, indicating the thermodynamic cost for forming critical atomic clusters. The platform overpotential is associated with zinc growth following initial nucleation. Reducing both types of overpotentials is important to decrease resistance during nucleation and growth, leading to improved plating quality and lower energy consumption.

It is well-known that zinc dendrites result from uneven deposition during the charging process. Specifically, at the onset of zinc deposition, Zn^{2+} ions are

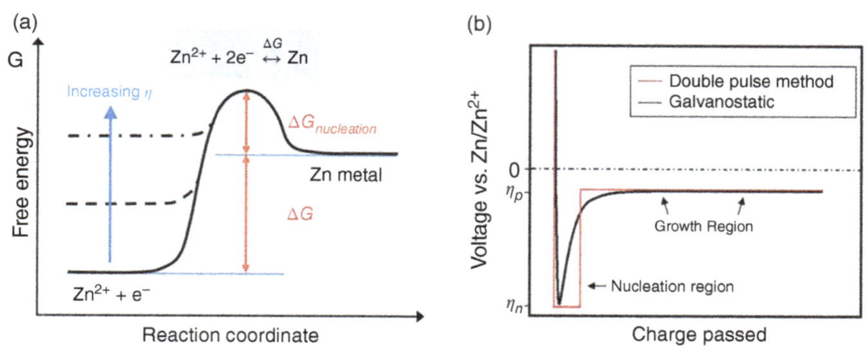

Figure 6.28 (a) Free energy schematic showing the effects of increasing overpotential on the nucleation energy barrier. (b) Schematic plot comparing the typical voltage profiles of galvanostatic Zn deposition (black) and double pulse potentiostatic Zn deposition (red). Source: Reproduced from Pei et al. [98]/with permission of American Chemical Society.

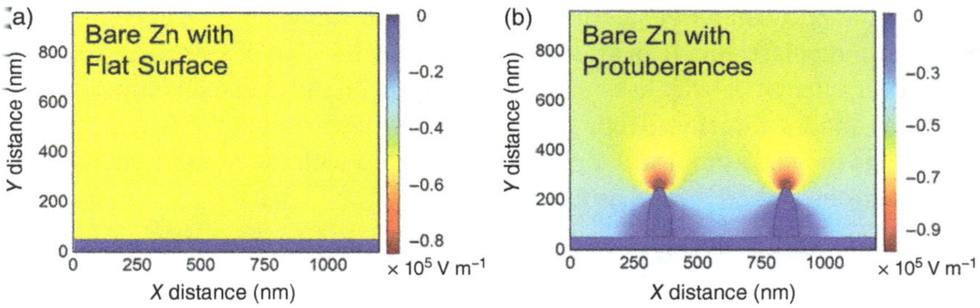

Figure 6.29 Simulated electric field distributions on bare Zn electrode (a) without and (b) with protuberances. Source: Reproduced from Yang et al. [99]/with permission of John Wiley & Sons.

transferred to the anode/electrolyte interface under the dual influence of the ions concentration gradient and electric field, and then gain electrons to nucleate. Based on the metal nucleation theory, nucleation sites should be randomly distributed on the surface of the metal anode. However, due to the inherent roughness of the metal anode surfaces, which are not perfectly smooth, localized protrusions can occur. As illustrated in Figure 6.29, when an electric field is applied to the electrodes, zinc ions preferentially accumulate at these protruding areas, which exhibit higher curvature. This leads to a greater surface charge density and consequently a stronger local electric field, a phenomenon known as the "tip effect". Furthermore, even on an ideally smooth plane, the initial deposition of Zn nuclei can form "protrusions" that influence the subsequent localized accumulation and preferential deposition of Zn^{2+} ions. Thus, the "tip effect," driven by the electric field and ion concentration gradient, is a crucial factor in the dendritic growth of zinc anodes. Implementing strategies to homogenize the surface electric field and induce preferential or synchronous deposition of zinc ions away from the tips can effectively enhance the reversibility of zinc anodes.

In addition to the "tip effect", several other factors contribute to the uneven deposition of zinc metal electrodes and dendrite growth, including polarization issues such as electrochemical polarization and concentration polarization. Under specific electrolyte concentrations and current densities, Zn^{2+} ions migrate toward the anode interface due to the electric field. At the inner Helmholtz layer, these ions receive electrons from the external circuit and are reduced to metallic zinc. However, the migration rate of Zn^{2+} ions is much slower than that of electrons, resulting in a situation where the consumption of zinc ions leads to a depletion of Zn^{2+} ions at the reaction interface. This creates an ion-depleted layer and a significant concentration gradient normal to the zinc metal anode interface, referred to as concentration polarization. Greater concentration polarization increases the overpotential at the anode surface. According to the double-layer model of electrode surfaces, excessive overpotential results in uneven electric field distribution, leading to localized preferential deposition of zinc and ultimately dendrite formation. Additionally, this uneven surface electric field can promote parasitic side reactions, negatively impacting the utilization and reversibility of the zinc metal anode. Therefore, employing appropriate strategies to reduce interface concentration polarization can also help prolong the lifespan of zinc metal anodes.

Moreover, current density has a significant impact on the electrochemical properties of zinc anodes, directly affecting the rate of zinc deposition. In diffusion models, the "τ" time in Sand's empirical formula is associated with the transfer characteristics of electrons and Zn^{2+} ions, as follows:

$$\tau = \pi D \frac{eC_0\left(\mu_a + \mu_{Zn^{2+}}\right)^2}{2J\mu_a} \tag{6.19}$$

where τ represents the time before zinc dendrites initiate growth, while D denotes the diffusion coefficient. The electronic charge is represented by e, and C_0 refers to the initial concentration of zinc salt. The migration rates of the anion and Zn^{2+} are indicated by μ_a and ${\mu_{Zn}}^{2+}$, respectively, and J signifies the effective electrode current density. Lower effective current densities (J) and higher Zn^{2+} ion migration rates (${\mu_{Zn}}^{2+}$) contribute to an extended Sand's time (τ), signifying a prolonged period before the onset of dendrite formation. This phenomenon can be attributed to the fact that reduced current densities promote a more uniform distribution of local current density and minimize distortion of the surface electric field. Therefore, in neutral or mildly acidic electrolytes, an increase in current density results in the gradual emergence and accelerated growth of zinc dendrites. It has been noted that elevated current densities heighten electrode polarization, leading to irregular zinc deposition, while lower current densities may alleviate dendrite formation issues.

6.4.1.2 Water-Induced Interface Side Reactions

Apart from dendrite growth, various side reactions at the interface of zinc metal anodes in aqueous electrolytes, such as hydrogen evolution, corrosion, and the accumulation of by-products, significantly impact the electrochemical performance and cycling life of zinc batteries. In weakly acidic or neutral electrolytes, the hydrogen evolution reaction (HER) is primarily driven by protons and water molecules. Although zinc metal possesses a relatively high overpotential for hydrogen evolution, the presence of dendrite formation, uneven electric field distribution, and localized pH changes during electrode operation makes the zinc anode susceptible to severe hydrogen evolution issues in practice. The HER occurring on the zinc surface can be described by the following reaction equation:

$$2H_2O + 2e^- \leftrightarrow 2OH^- + H_2 \uparrow \tag{6.20}$$

The OH^- produced from these reactions increases the localized pH at the interface, resulting in the formation of basic zinc salts on the surface of the zinc anode. Taking zinc sulfate as an example, the formation of basic zinc sulfate can be summarized by the following reaction equation:

$$4Zn^{2+} + 6OH^- + SO_4^{2-} + xH_2O \leftrightarrow Zn_4SO_4(OH)_6 \cdot xH_2O \tag{6.21}$$

In addition to the basic zinc sulfate formed during the deposition process, the dissolution of the zinc metal anode can also lead to the formation of basic zinc sulfate. This is primarily due to the high concentration of zinc ions released, which results in a local accumulation of positive charges at the interface. This encourages the diffusion of SO_4^{2-} and OH^- from the electrolyte to the interface, where they complex with zinc ions to form by-products.

Due to the poor reversibility associated with the formation and dissolution of basic zinc sulfate, it continues to accumulate on the surface of the zinc anode, leading to the loss of active zinc. Furthermore, the ionic and electronic conductivity of these by-products is generally inadequate, and their growth often occurs in a direction perpendicular to the anode plane. This results in uneven electric field distribution, establishing basic zinc sulfate as a significant interfacial side product influencing the reversibility of the zinc anode. Under different testing conditions, particularly in high-temperature environments, these by-products may also manifest in the form of ZnO [100]. The HER and side product accumulation occurring during the electrochemical plating/stripping of the zinc metal anode can also be characterized as electrochemical corrosion.

In addition to the aforementioned issues, the non-uniformity of the zinc anode surface (such as roughness and edge effects) can lead to higher current densities in localized regions, thus increasing the local overpotential and potentially causing localized corrosion. This situation may result in rapid loss of zinc and non-uniform deposition. Due to its high metal activity, zinc metal can also experience corrosion from saline solutions and dissolved oxygen in aqueous electrolytes under steady-state conditions or as the battery ages, stating this type of corrosion as chemical corrosion. The primary mechanism involves the oxidation of zinc in the presence of water or protons, as well as dissolved oxygen in the electrolyte, producing oxide or hydroxide passivation layers that lower the rate of electrochemical reactions and decrease electric field uniformity (which can trigger dendrite growth), while also consuming active zinc.

6.4.2 Zinc Anode Protection Strategies

Given the previous discussion, controlling nucleation and suppressing side reactions are critical for achieving stable, dendrite-free plating zinc and, consequently, high battery performance. To accomplish these goals, it is essential to manage interface electron transfer kinetics, ion diffusion patterns, and the water/oxygen content at the interface. These parameters can be controlled by optimizing various components of the battery, such as the zinc electrode, electrolyte, and separator, as these assembly parts are closely related to battery performance. This section will delve into high-performance zinc anode strategies based on material selection and structural design. Four main aspects will be analyzed and discussed: (i) surface coating modification; (ii) optimization of internal electrode structure; (iii) electrolyte optimization; and (iv) multifunctional separator modification.

6.4.2.1 Surface-Coating Modification of Zinc Anodes

Planar zinc anodes, such as zinc foils/plates, are widely used due to their availability and industrial advantages. However, pure zinc foils/plates have some drawbacks. For instance, the limited surface area of planar zinc anodes impedes uniform electric field distribution, which hinders uniform nucleation. Additionally, a simple planar structure is prone to becoming electrically insulating once a passivation layer forms on its surface. These issues can be addressed by designing surface coatings, which is an effective way to enhance zinc anode performance while maintaining practicality. An artificial interfacial layer, similar to the solid electrolyte interface of lithium anodes, positively impacts the anode by not only regulating the electrochemical environment of the anode surface but also providing physical protection. Thus, surface modification can act as a multifunctional layer, protecting or regulating interfacial chemical/electrochemical properties and processes. Through material and structural design-based surface engineering, zinc anode performance can be significantly improved.

A variety of functional materials have been proposed as candidates for modifying the surface of zinc anodes. These include carbon-based materials (such as GO, rGO, carbon quantum dots, carbon fiber, MOF-derived carbon and graphyne), transition metal oxides (such as Sn-InO_3, TiO_2, CuO_2, CeO_2, ZrO_2, and $ZnWO_4$), hydroxides (such as ZnNiCo-LH), chalcogenide (such as ZnS and ZnSe), phosphates (such as $Zn_3(PO_4)_2$ and ZnP_2O_6, $ZrPO_4$, amorphous zinc phosphate), PBA, precious metals (Au, Ag), liquid metals (such as Ga, In), MOF materials (such as ZIF-67, MOF-5W, MOF-808), COF materials, polymers (such as PPy, PVDF, PAM, PAA, cyclodextrin polymer), and multicomponent composite materials (such as MOF/rGO, PPy/rGO). Typically, coatings can serve multiple functions depending on the material and specific structure.

Uniform charge distribution on the zinc anode surface: Various materials, such as metal oxides with high dielectric constants, can facilitate uniform charge distribution through surface space charge polarization effects, like Maxwell–Wagner polarization.

Regulation of zinc-ion diffusion at the interface: The random, 2D diffusion of zinc ions at the interface is crucial for achieving uniform electrodeposition. Such random diffusion can result in preferential deposition, leading to uneven ion flux distribution and, consequently, significant dendrite growth. Certain materials can enhance zinc-ion diffusion at the interface through electrostatic interactions and the induction of ion channels.

Inhibition of water/oxygen diffusion at the interface: Some coatings feature dense structures or strong molecular interactions, serving as physical barriers that prevent water and oxygen from reaching the electrode interface. This helps to minimize the water and oxide content at the interface. As noted, these coatings typically provide multiple functions, thereby enhancing the performance of the zinc anode.

6.4.2.2 3D Structure Design of Zinc Metal Anode Materials

Although planar zinc foil anodes exhibit relatively high economic benefits in AZIBs due to lower electrode design costs and simplified battery assembly processes, their electrochemical performance remains challenging to meet demands under high-rate operating conditions. This is primarily due to localized charge accumulation on the surface, which exacerbates dendrite growth and parasitic side reactions. Furthermore, when increasing the active zinc utilization in zinc foil anodes – by enhancing the discharge depth of the zinc anode – unbacked zinc anodes are prone to local depletion, leading to irreversible capacity loss and significantly reduced Coulombic efficiency.

To address these issues, constructing a 3D structure for zinc anodes or designing a 3D current collector to support the active zinc is an effective strategy. The design of a 3D structure can provide a larger specific surface area and abundant nucleation sites, resulting in a more uniform surface electric field that enhances the rate performance of zinc metal anodes. For instance, zinc metal anodes can be synthesized by *in situ* growing carbon layers on the surface of zinc powder, followed by sintering via powder metallurgy to achieve a continuous 3D carbon network within the bulk [101]. This interpenetrating 3D carbon network can mitigate the substantial electronic transport resistance caused by point contacts between zinc particles, thereby facilitating electron transport and enhancing the conductivity of the zinc anode. Additionally, the dispersed 3D continuous carbon network can offer more zincophilic sites, inducing uniform deposition of zinc ions on the electrode surface and effectively preventing the "tip effect". Moreover, designing a graphene-based matrix with high porosity and radial carbon channels can effectively reduce localized current density, manipulate the Zn^{2+} ion concentration gradient, and homogenize the electric field distribution, allowing the zinc metal anode to stably operate at an ultra-high rate of 80 mA cm^{-2} for over 2000 hours.

While the design of 3D electrode structures can effectively reduce localized current density, the longitudinal distribution of nucleation active sites may lead to preferential deposition of zinc at the top of the electrode, thereby promoting dendrite growth. Implementing a zinc-affinity gradient in the 3D structure can alleviate this issue. For example, a nickel foam substrate can be utilized to fabricate a zinc metal negative electrode with three gradients – zincophilic gradient, conductivity gradient, and porosity gradient – using a simple rolling technique [102]. This tri-gradient design optimizes the electric field distribution, modulates the Zn^{2+} ion flux and deposition pathway, and induces bottom-up zinc metal deposition, effectively preventing dendritic growth at the electrode top that could lead to short-circuiting. Furthermore, this multi-gradient design strategy for 3D-structured electrodes demonstrates significant versatility and can guide the multidimensional construction of other metal electrodes.

6.4.2.3 Zinc Alloy Anodes

Zinc alloys typically consist on zinc anodes in the form of protective layers or nanoparticles. They can be categorized into two types: the first type consists of zinc metal as the solvent with other metallic elements dissolved as solutes in the zinc lattice, forming a solid solution where neither the lattice type nor the properties of the individual elements are altered. The second type involves the formation of intermetallic compounds through interactions between metallic elements, which possess different lattice types and properties compared to any single element. Various intermetallic compounds offer distinct advantages, such as inducing uniform deposition of zinc ions, enhancing corrosion resistance, and suppressing hydrogen evolution reactions. Mechanisms by which zinc alloys optimize zinc anode performance can be divided into four aspects: enhancing the corrosion resistance of the zinc anode, delaying hydrogen evolution reactions, reducing by-products, and inhibiting dendritic growth. Each zinc alloy possesses one or more mechanisms that enhance the performance of the zinc anode.

Zinc alloys exhibit excellent electrochemical properties, and the controllability and variability of their constituent elements endow them with the ability to suppress dendritic growth, reduce corrosion, promote passivation, and inhibit hydrogen evolution reactions in battery applications. Different solute elements within zinc alloys have varying effects on improving anode performance. Significant progress has been made in the study of binary alloys. For instance, due to the inherent electrochemical inertness of copper and its corrosion potential being higher than that of zinc, the addition of copper to zinc anodes effectively improves their corrosion resistance. Additionally, directly depositing zincophilic nanomaterials onto 3D frameworks is regarded as an effective method for spatially regulating zinc deposition. In this process, incorporating silver as a zincophilic element in the zinc anode allows for the formation of $AgZn_3$ alloys, which serve as a zinc source and mitigate the irreversible loss of active zinc during cycling. Silver nanoparticles can also act as heterogeneous seeds, promoting uniform nucleation and deposition of zinc ions, thereby significantly suppressing the formation of zinc dendrites and the associated hydrogen evolution reactions.

Moreover, incorporating aluminum into zinc alloys has shown positive effects. Previous studies indicate that aluminum ions can create positively charged electrostatic shielding, inhibiting the growth of metallic dendrites while preferentially consuming dissolved oxygen in the electrolyte to prevent the generation of irreversible by-products such as zinc oxide or zinc hydroxide. The surface of zinc-aluminum alloys can also form an Al_2O_3 electrostatic shielding layer *in situ*, guiding the deposition of zinc ions at designated zinc sites and further suppressing dendrite formation.

6.4.2.4 Electrolyte Design to Optimize Zinc Anode

In AZIBs, the side reactions encountered by zinc metal anodes are largely attributed to the high free water content and the weakly acidic pH of the aqueous electrolyte. Therefore, in addition to optimizing the structure and surface composition of the zinc anode itself, designing the electrolyte for better optimization of the zinc anode can lead to improved performance and higher economic feasibility. For example, to address the issue of localized charge accumulation at the zinc metal anode surface caused by the tip effect, introducing zincophilic cations with lower reduction potentials (which can be metal ions or nonmetal ions) into the electrolyte can form an electrostatic shielding layer on the tips or initial nucleation sites of the zinc surface during the deposition of zinc ions. This mechanism drives subsequent zinc ions diffusing to the surface to deposit at lower areas, thereby preventing dendrite formation.

Achieving regulation of the internal components of the electric double-layer (EDL) structure through electrolyte design is also regarded as an effective strategy to improve the electrochemical behavior of the zinc anode interface. According to classical EDL theory, the internal structure of the EDL at the surface of the zinc metal anode is defined by the inner Helmholtz plane (IHP), which is the region

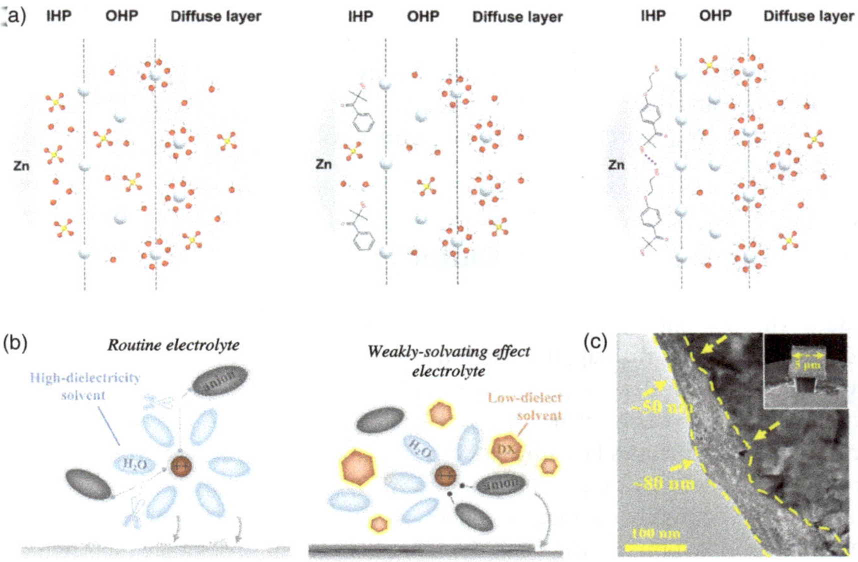

Figure 6.30 (a) Schematic illustration of the characteristic EDL structure and interfacial behaviors on the zinc anode surface. Source: Reproduced with permission of [103]/American Chemical Society. (b) Schematic illustrations of the solution structures for the routine electrolyte and the weakly solvating effect electrolyte. (c) TEM image of SEI derived from weakly solvating effect electrolyte. Source: Reproduced with permission from [104]/John Wiley & Sons.

where interfacial reactions occur, while the outer Helmholtz plane (OHP) is related to the desolvation process. For instance, in conventional $ZnSO_4$ electrolytes, due to the lower solubility of hydrophilic sulfate anions and free water solvent molecules compared to the solubility of the cation Zn, their diffusion kinetics are much faster. As a result, a large amount of hydrophilic sulfate anions and free water solvent molecules preferentially occupy the IHP (Figure 6.30a), leading to instability at the zinc anode interface. Introducing highly polar and zincophilic components into the aqueous electrolyte competes with the existing water molecules and sulfate ions in the IHP, thereby reducing the tendency for interfacial side reactions. Moreover, these adsorbents' electronegative groups can modulate the distribution of interfacial charges, promoting uniform deposition of zinc ions.

Additionally, the *in situ* construction of a SEI on the zinc anode surface by adjusting the electrolyte composition has also been widely proven to be an effective strategy for protecting the zinc anode. The *in situ* formed SEI not only effectively shields dendrite growth and suppresses interfacial side reactions but also exhibits better adaptability to environmental changes (including local variations in the electrolyte environment and changes in electrochemical behavior during battery operation). The optimization strategies for constructing *in situ* SEI include introducing low-reduction-potential additive components and inducing anion decomposition. The working mechanism of the additives is that they reduce and decompose before zinc deposition, with the decomposition products complexing with zinc ions and depositing on the anode surface to form the SEI. Furthermore, as shown in Figure 6.30b,c, designing a weakly solvated environment in the electrolyte encourages some lower LUMO anions (such as OTF^-, $TFSI^-$) in the electrolyte itself to participate in the first solvation sheath of zinc ions, which can induce anion decomposition, and the decomposition products combine with zinc ions to form the SEI [104].

6.5 Summary and Future Outlook

Aquatic zinc-ion batteries (AZIBs) stand as a promising new type of green secondary battery with broad application prospects, having demonstrated tremendous potential in large-scale energy storage fields such as green grids. This chapter summarizes the working principles and key materials of AZIBs. It particularly focuses on the challenges faced by cathode and anode materials and the strategies for their modification. However, as it stands, AZIBs still face difficulties in securing a position in the highly competitive energy storage market. The further development and optimization of AZIBs based on material design to enhance their market competitiveness are of great significance for their future large-scale application. Herein, several challenges and potential directions for the future development of AZIBs are proposed, aiming to achieve the industrial production of stable and high-energy-density AZIBs.

1) The energy storage mechanism of cathode materials for AZIBs remains a subject of considerable debate, especially the proton contribution requires further investigation.
2) The service life and utilization rate of the zinc anode need to be further improved. The reported modification strategies require scalability verification.
3) The majority of reported AZIBs systems are assembled using glass fiber separators, which are costly and require a large amount of electrolyte. The development of new separators holds profound significance for the large-scale application of AZIBs.
4) Considering the potential application scenarios and desired performance metrics of AZIBs, which differ significantly from organic electrolyte-based lithium-ion batteries, the current electrochemical performance evaluation protocols used for lithium-ion batteries are not applicable to AZIBs. Therefore, establishing an efficient and reliable electrochemical performance evaluation testing protocol is crucial.
5) The integration of advanced machine learning and other artificial intelligence methods should be employed to accelerate the development, screening, and optimization of electrode materials for AZIBs.
6) The differences between laboratory battery systems and industrial battery systems in terms of industrial stability and production requirements are emphasized. Establishing systematic guidance from laboratory-scale batteries to scalable soft-pack or pouch cells is necessary, including aspects such as cathode loading, N/P ratio, and E/C ratio.

References

1 Williams, J.B. (2018). *The Electric Century: How the Taming of Lightning Shaped the Modern World* (ed. J.B. Williams). Cham: Springer International Publishing.

2 Xu, C., Li, B., Du, H., and Kang, F. (2012). Energetic zinc ion chemistry: the rechargeable zinc ion battery. *Angewandte Chemie International Edition* 51: 933.

3 Kundu, D., Adams, B.D., Duffort, V. et al. (2016). A high-capacity and long-life aqueous rechargeable zinc battery using a metal oxide intercalation cathode. *Nature Energy* 1: 16119.

4 Yang, Y., Tang, Y., Fang, G. et al. (2018). Li^+ intercalated $V_2O_5{\cdot}nH_2O$ with enlarged layer spacing and fast ion diffusion as an aqueous zinc-ion battery cathode. *Energy & Environmental Science* 11: 3157.

5 Sun, W., Wang, F., Hou, S. et al. (2017). Zn/MnO_2 battery chemistry with H^+ and Zn^{2+} coinsertion. *Journal of the American Chemical Society* 139: 9775.

6 Gao, X., Wu, H., Li, W. et al. (2020). H^+-insertion boosted α-MnO_2 for an aqueous Zn-ion battery. *Small* 16: 1905842.

7 Wan, F., Zhang, L., Dai, X. et al. (2018). Aqueous rechargeable zinc/sodium vanadate batteries with enhanced performance from simultaneous insertion of dual carriers. *Nature Communications* 9: 1656.

8 Yan, M., He, P., Chen, Y. et al. (2018). Water-lubricated intercalation in $V_2O_5{\cdot}nH_2O$ for high-capacity and high-rate aqueous rechargeable zinc batteries. *Advanced Materials* 30: 1703725.

9 Chen, X., Wang, L., Li, H. et al. (2019). Porous V_2O_5 nanofibers as cathode materials for rechargeable aqueous zinc-ion batteries. *Journal of Energy Chemistry* 38: 20.

10 Zhu, K., Wu, T., and Huang, K. (2019). A high capacity bilayer cathode for aqueous Zn-ion batteries. *ACS Nano* 13: 14447.

11 Sun, T., Nian, Q., Zheng, S. et al. (2020). Layered $Ca_{0.28}MnO_2{\cdot}0.5H_2O$ as a high performance cathode for aqueous zinc-ion battery. *Small* 16: 2000597.

12 Zhai, X.-Z., Qu, J., Hao, S.-M. et al. (2020). Layered birnessite cathode with a displacement/intercalation mechanism for high-performance aqueous zinc-ion batteries. *Nano-Micro Letters* 12: 56.

13 Pan, H., Shao, Y., Yan, P. et al. (2016). Reversible aqueous zinc/manganese oxide energy storage from conversion reactions. *Nature Energy* 1: 16039.

14 Chao, D., Zhou, W., Ye, C. et al. (2019). An electrolytic Zn–MnO_2 battery for high-voltage and scalable energy storage. *Angewandte Chemie* 131: 7905.

15 Liu, W., Zhang, X., Huang, Y. et al. (2021). β-MnO_2 with proton conversion mechanism in rechargeable zinc ion battery. *Journal of Energy Chemistry* 56: 365.

16 Soundharrajan, V., Sambandam, B., Kim, S. et al. (2020). The dominant role of Mn^{2+} additive on the electrochemical reaction in $ZnMn_2O_4$ cathode for aqueous zinc-ion batteries. *Energy Storage Materials* 28: 407.

17 Liang, G., Mo, F., Li, H. et al. (2019). A universal principle to design reversible aqueous batteries based on deposition–dissolution mechanism. *Advanced Energy Materials* 9: 1901838.

18 Guo, C., Yi, S., Si, R. et al. (2022). Advances on defect engineering of vanadium-based compounds for high-energy aqueous zinc–ion batteries. *Advanced Energy Materials* 12: 2202039.

19 Liao, M., Wang, J., Ye, L. et al. (2020). A deep-cycle aqueous zinc-ion battery containing an oxygen-deficient vanadium oxide cathode. *Angewandte Chemie* 132: 2293.

20 Deniz, A.B.D.C., Valt, R.B.G., Ponte, M.J.J.D.S., and Ponte, H.D.A. (2020). Evaluation of transport phenomena for removing vanadium from petrochemical industry solid waste. *Chemical Engineering & Technology* 43: 1393.

21 Dai, Y., Zhang, C., Li, J. et al. (2024). Inhibition of vanadium cathodes dissolution in aqueous Zn-ion batteries. *Advanced Materials* 36: 2310645.

22 He, P., Yan, M., Zhang, G. et al. (2017). Layered VS_2 nanosheet-based aqueous Zn ion battery cathode. *Advanced Energy Materials* 7: 1601920.

23 Ding, J., Du, Z., Gu, L. et al. (2018). Ultrafast Zn^{2+} intercalation and deintercalation in vanadium dioxide. *Advanced Materials* 30: 1800762.

24 Cui, F., Zhao, J., Zhang, D. et al. (2020). VO_2 (B) nanobelts and reduced graphene oxides composites as cathode materials for low-cost rechargeable aqueous zinc ion batteries. *Chemical Engineering Journal* 390: 124118.

25 J. Wu, Q. Kuang, K. Zhang, J. Feng, C. Huang, J. Li, Q. Fan, Y. Dong, Y. Zhao, Spinel $Zn_3V_3O_8$: a high-capacity zinc supplied cathode for aqueous Zn-ion batteries, *Energy Storage Materials* 2021, 41, 297.

26 T. Wang, S. Li, X. Weng, L. Gao, Y. Yan, N. Zhang, X. Qu, L. Jiao, Y. Liu, Ultrafast 3D hybrid-ion transport in porous V_2O_5 cathodes for superior-rate rechargeable aqueous zinc batteries, *Advanced Energy Materials* 2023, 13, 2204358.

27 P. Hu, Z. Zou, X. Sun, D. Wang, J. Ma, Q. Kong, D. Xiao, L. Gu, X. Zhou, J. Zhao, S. Dong, B. He, M. Avdeev, S. Shi, G. Cui, L. Chen, Uncovering the potential of M1-site-activated NASICON cathodes for Zn-ion batteries, *Advanced Materials* 2020, 32, 1907526.

28 McNulty, R.C., Penston, K., Amin, S.S. et al. (2023). Self-assembled surfactant-polyoxovanadate soft materials as tuneable vanadium oxide cathode precursors for lithium-ion batteries. *Angewandte Chemie* 135: e202216066.

29 Chen, D., Rui, X., Zhang, Q. et al. (2019). Persistent zinc-ion storage in mass-produced V_2O_5 architectures. *Nano Energy* 60: 171.

30 He, Q., Hu, T., Wu, Q. et al. (2024). Tunnel-oriented VO_2 (B) cathode for high-rate aqueous zinc-ion batteries. *Advanced Materials* 36: 2400888.

31 Soundharrajan, V., Sambandam, B., Kim, S. et al. (2018). $Na_2V_6O_{16}{\cdot}3H_2O$ barnesite nanorod: an open door to display a stable and high energy for aqueous rechargeable Zn-ion batteries as cathodes. *Nano Letters* 18: 2402.

32 Duan, W., Zhao, M., Li, Y. et al. (2020). Excellent rate capability and cycling stability of novel $H_2V_3O_8$ doped with graphene materials used in new aqueous zinc-ion batteries. *Energy & Fuels* 34: 3877.

33 Chen, J., Liu, Y., Xiao, B. et al. (2024). Using MXene as a chemically induced initiator to construct high-performance cathodes for aqueous zinc-ion batteries. *Angewandte Chemie International Edition* 63: e202408667.

34 Xia, C., Guo, J., Li, P. et al. (2018). Highly stable aqueous zinc-ion storage using a layered calcium vanadium oxide bronze cathode. *Angewandte Chemie* 130: 4007.

35 Wang, Y., Wei, S., Qi, Z.-H. et al. (2023). Intercalant-induced V t_2g orbital occupation in vanadium oxide cathode toward fast-charging aqueous zinc-ion batteries. *Proceedings of the National Academy of Sciences* 120: e2217208120.

36 Zhu, K., Wu, T., and Huang, K. (2019). $NaCa_{0.6}V_6O_{16}{\cdot}3H_2O$ as an ultra-stable cathode for Zn-ion batteries: the roles of pre-inserted dual-cations and structural water in V_3O_8 layer. *Advanced Energy Materials* 9: 1901968.

37 Liu, S., Zhu, H., Zhang, B. et al. (2020). Tuning the kinetics of zinc-ion insertion/extraction in V_2O_5 by in situ polyaniline intercalation enables improved aqueous zinc-ion storage performance. *Advanced Materials* 32: 2001113.

38 Bin, D., Huo, W., Yuan, Y. et al. (2020). Organic-inorganic-induced polymer intercalation into layered composites for aqueous zinc-ion battery. *Chem* 6: 968.

39 Ye, J.-J., Li, P.-H., Zhang, H.-R. et al. (2023). Manipulating oxygen vacancies to spur ion kinetics in V_2O_5 structures for superior aqueous zinc-ion batteries. *Advanced Functional Materials* 33: 2305659.

40 Xiong, T., Zhang, Y., Lee, W.S.V., and Xue, J. (2020). Defect engineering in manganese-based oxides for aqueous rechargeable zinc-ion batteries: a review. *Advanced Energy Materials* 10: 2001769.

41 Xu, C., Du, H., Li, B. et al. (2009). Reversible insertion properties of zinc ion into manganese dioxide and its application for energy storage. *Electrochemical and Solid-State Letters* 12: A61.

42 Islam, S., Alfaruqi, M.H., Mathew, V. et al. (2017). Facile synthesis and the exploration of the zinc storage mechanism of β-MnO_2 nanorods with exposed (101) planes as a novel cathode material for high performance eco-friendly zinc-ion batteries. *Journal of Materials Chemistry A* 5: 23299.

43 T. Yamamoto, T. Shoji, Rechargeable Zn| $ZnSO_4$| MnO_2-type cells, *Inorganica Chim. Acta* 1986, 117, L27.

44 Alfaruqi, M.H., Mathew, V., Gim, J. et al. (2015). Electrochemically induced structural transformation in a γ-MnO_2 cathode of a high capacity zinc-ion battery system. *Chemistry of Materials* 27: 3609.

45 Alfaruqi, M.H., Gim, J., Kim, S. et al. (2015). A layered δ-MnO_2 nanoflake cathode with high zinc-storage capacities for eco-friendly battery applications. *Electrochemistry Communications* 60: 121.

46 Jin, Y., Zou, L., Liu, L. et al. (2019). Joint charge storage for high-rate aqueous zinc–manganese dioxide batteries. *Advanced Materials* 31: 1900567.

47 Li, Y., Wang, S., Salvador, J.R. et al. (2019). Reaction mechanisms for long-life rechargeable Zn/MnO_2 batteries. *Chemistry of Materials* 31: 2036.

48 Yuan, C., Zhang, Y., Pan, Y. et al. (2014). Investigation of the intercalation of polyvalent cations (Mg^{2+}, Zn^{2+}) into λ-MnO_2 for rechargeable aqueous battery. *Electrochimica Acta* 116: 404.

49 Hao, J., Mou, J., Zhang, J. et al. (2018). Electrochemically induced spinel-layered phase transition of Mn_3O_4 in high performance neutral aqueous rechargeable zinc battery. *Electrochimica Acta* 259: 170.

50 Kim, J.M., Huh, Y.S., Han, Y.-K. et al. (2012). Facile synthesis route to highly crystalline mesoporous γ-MnO_2 nanospheres. *Electrochemistry Communications* 14: 32.

51 C. Guo, H. Liu, J. Li, Z. Hou, J. Liang, J. Zhou, Y. Zhu, Y. Qian, Ultrathin δ-MnO_2 nanosheets as cathode for aqueous rechargeable zinc ion battery, *Electrochimica Acta* 2019, 304, 370.

52 Liu, D.-S., Mai, Y., Chen, S. et al. (2021). A 1D–3D interconnected δ-MnO_2 nanowires network as high-performance and high energy efficiency cathode material for aqueous zinc-ion batteries. *Electrochimica Acta* 370: 137740.

53 Alfaruqi, M.H., Gim, J., Kim, S. et al. (2015). Enhanced reversible divalent zinc storage in a structurally stable α-MnO_2 nanorod electrode. *Journal of Power Sources* 288: 320.

54 Wang, Y., Wu, Z., Jiang, L. et al. (2019). A long-lifespan, flexible zinc-ion secondary battery using a paper-like cathode from single-atomic layer MnO_2 nanosheets. *Nanoscale Advances* 1: 4365.

55 Liu, Y., Chi, X., Han, Q. et al. (2019). α-MnO_2 nanofibers/carbon nanotubes hierarchically assembled microspheres: approaching practical applications of high-performance aqueous Zn-ion batteries. *Journal of Power Sources* 443: 227244.

56 Chen, X., Li, W., Zeng, Z. et al. (2021). Engineering stable Zn-MnO_2 batteries by synergistic stabilization between the carbon nanofiber core and birnessite-MnO_2 nanosheets shell. *Chemical Engineering Journal* 405: 126969.

57 Li, X., Zhou, Q., Yang, Z. et al. (2023). Unraveling the role of nitrogen-doped carbon nanowires incorporated with MnO_2 nanosheets as high performance cathode for zinc-ion batteries. *Energy & Environmental Materials* 6: e12378.

58 Wu, B., Zhang, G., Yan, M. et al. (2018). Graphene scroll-coated α-MnO_2 nanowires as high-performance cathode materials for aqueous Zn-ion battery. *Small* 14: 1703850.

59 Bi, S., Wu, Y., Cao, A. et al. (2020). Free-standing three-dimensional carbon nanotubes/amorphous MnO_2 cathodes for aqueous zinc-ion batteries with superior rate performance. *Materials Today Energy* 18: 100548.

60 Zeng, Y., Zhang, X., Meng, Y. et al. (2017). Achieving ultrahigh energy density and long durability in a flexible rechargeable quasi-solid-state Zn–MnO_2 battery. *Advanced Materials* 29: 1700274.

61 Zhao, Y., Zhou, R., Song, Z. et al. (2022). Interfacial designing of MnO_2 half-wrapped by aromatic polymers for high-performance aqueous zinc-ion batteries. *Angewandte Chemie* 134: e202212231.

62 Shi, M., Wang, B., Chen, C. et al. (2020). 3D high-density MXene@ MnO_2 microflowers for advanced aqueous zinc-ion batteries. *Journal of Materials Chemistry A* 8: 24635.

63 Lian, S., Sun, C., Xu, W. et al. (2019). Built-in oriented electric field facilitating durable ZnMnO_2 battery. *Nano Energy* 62: 79.

64 Fang, G., Zhu, C., Chen, M. et al. (2019). Suppressing manganese dissolution in potassium manganate with rich oxygen defects engaged high-energy-density and durable aqueous zinc-ion battery. *Advanced Functional Materials* 29: 1808375.

65 Kim, S., Koo, B.-R., Jo, Y.-R. et al. (2021). Defect engineering via the F-doping of β-MnO_2 cathode to design hierarchical spheres of interlaced nanosheets for superior high-rate aqueous zinc ion batteries. *Journal of Materials Chemistry A* 9: 17211.

66 Song, J., Wang, L., Lu, Y. et al. (2015). Removal of interstitial H_2O in hexacyanometallates for a superior cathode of a sodium-ion battery. *Journal of the American Chemical Society* 137: 2658.

67 Yang, D., Xu, J., Liao, X.-Z. et al. (2023). Prussian blue/RGO with less coordinated water as superior cathode material for sodium-ion batteries. *Chemical Communications* 59: 211.

68 Ren, W., Qin, M., Zhu, Z. et al. (2017). Activation of sodium storage sites in Prussian blue analogues via surface etching. *Nano Letters* 17: 4713.

69 Wang, J., Mi, C., Nie, P. et al. (2018). Sodium-rich iron hexacyanoferrate with nickel doping as a high performance cathode for aqueous sodium ion batteries. *Journal of Electroanalytical Chemistry* 818: 10.

70 Shi, H.Y., Ye, Y.J., Liu, K. et al. (2018). A long-cycle-life self-doped polyaniline cathode for rechargeable aqueous zinc batteries. *Angewandte Chemie* 130: 16597.

71 Du, W., Xiao, J., Geng, H. et al. (2020). Rational-design of polyaniline cathode using proton doping strategy by graphene oxide for enhanced aqueous zinc-ion batteries. *Journal of Power Sources* 450: 227716.

72 Yue, X., Liu, H., and Liu, P. (2019). Polymer grafted on carbon nanotubes as a flexible cathode for aqueous zinc ion batteries. *Chemical Communications* 55: 1647.

73 Ma, L., Ying, Y., Chen, S. et al. (2021). Electrocatalytic iodine reduction reaction enabled by aqueous zinc-iodine battery with improved power and energy densities. *Angewandte Chemie International Edition* 60: 3791.

74 Wang, C., Ji, X., Liang, J. et al. (2024). Activating and stabilizing a reversible four electron redox reaction of I^-/I^+ for aqueous Zn-iodine battery. *Angewandte Chemie* e202403187.

75 Zhang, S., Zhao, W., Li, H., and Xu, Q. (2020). Cross-conjugated polycatechol organic cathode for aqueous zinc-ion storage. *ChemSusChem* 13: 188.

76 Zou, Y., Liu, T., Du, Q. et al. (2021). A four-electron Zn-I_2 aqueous battery enabled by reversible $I^-/I_2/I^+$ conversion. *Nature Communications* 12: 170.

77 Lee, J.-H., Byun, Y., Jeong, G.H. et al. (2019). High-energy efficiency membraneless flowless Zn–Br battery: utilizing the electrochemical–chemical growth of polybromides. *Advanced Materials* 31: 1904524.

78 Zhang, Y., Wei, C., Wu, M.-X. et al. (2023). A high-performance COF-based aqueous zinc-bromine battery. *Chemical Engineering Journal* 451: 138915.

79 Gao, L., Li, Z., Zou, Y. et al. (2020). A high-performance aqueous zinc-bromine static battery. *iScience* 23: 101348.

80 Li, X., Wang, S., Wang, T. et al. (2022). Bis-ammonium salts with strong chemisorption to halide ions for fast and durable aqueous redox Zn ion batteries. *Nano Energy* 98: 107278.

81 Wei, H., Qu, G., Zhang, X. et al. (2023). Boosting aqueous non-flow zinc–bromine batteries with a two-dimensional metal–organic framework host: an adsorption-catalysis approach. *Energy & Environmental Science* 16: 4073.

82 Tie, Z. and Niu, Z. (2020). Design strategies for high-performance aqueous Zn/organic batteries. *Angewandte Chemie International Edition* 59: 21293.

83 Wan, F., Zhang, L., Wang, X. et al. (2018). An aqueous rechargeable zinc-organic battery with hybrid mechanism. *Advanced Functional Materials* 28: 1804975.

84 Wang, J., Liu, J., Hu, M. et al. (2018). A flexible, electrochromic, rechargeable Zn//PPy battery with a short circuit chromatic warning function. *Journal of Materials Chemistry A* 6: 11113.

85 Zhao, Q., Huang, W., Luo, Z. et al. High-capacity aqueous zinc batteries using sustainable quinone electrodes. *Science Advances* 4: eaao1761.

86 Guo, Z., Ma, Y., Dong, X. et al. (2018). An environmentally friendly and flexible aqueous zinc battery using an organic cathode. *Angewandte Chemie* 130: 11911.

87 Sun, T., Li, Z.-J., Zhi, Y.-F. et al. (2021). Poly (2, 5-Dihydroxy-1, 4-Benzoquinonyl sulfide) as an efficient cathode for high-performance aqueous zinc–organic batteries. *Advanced Functional Materials* 31: 2010049.

88 Wang, C., He, T., Cheng, J. et al. (2020). Bioinspired interface design of sewable, weavable, and washable fiber zinc batteries for wearable power textiles. *Advanced Functional Materials* 30: 2004430.

89 Wang, Q., Liu, Y., and Chen, P. (2020). Phenazine-based organic cathode for aqueous zinc secondary batteries. *Journal of Power Sources* 468: 228401.

90 Zhang, H., Xie, S., Cao, Z. et al. (2021). Extended π-conjugated system in organic cathode with active C= N bonds for driving aqueous zinc-ion batteries. *ACS Applied Energy Materials* 4: 655.

91 Ye, Z., Xie, S., Cao, Z. et al. (2021). High-rate aqueous zinc-organic battery achieved by lowering HOMO/LUMO of organic cathode. *Energy Storage Materials* 37: 378.

92 Zhang, S., Long, S., Li, H., and Xu, Q. (2020). A high-capacity organic cathode based on active N atoms for aqueous zinc-ion batteries. *Chemical Engineering Journal* 400: 125898.

93 Koshika, K., Sano, N., Oyaizu, K., and Nishide, H. (2009). An ultrafast chargeable polymer electrode based on the combination of nitroxide radical and aqueous electrolyte. *Chemical Communications* 836.

94 Luo, Y., Zheng, F., Liu, L. et al. (2020). A high-power aqueous zinc–organic radical battery with tunable operating voltage triggered by selected anions. *ChemSusChem* 13: 2239.

95 A. Khayum, M. Ghosh, V. Vijayakumar, A. Halder, M. Nurhuda, S. Kumar, M. Addicoat, S. Kurungot and R. Banerjee, Zinc ion interactions in a two-dimensional covalent organic framework based aqueous zinc ion battery, *Chemical Science* 2019, 10, 8889.

96 Wang, W., Kale, V.S., Cao, Z. et al. (2020). Phenanthroline covalent organic framework electrodes for high-performance zinc-ion supercapattery. *ACS Energy Letters* 5: 2256.

97 Nam, K.W., Park, S.S., dos Reis, R. et al. (2019). Conductive 2D metal-organic framework for high-performance cathodes in aqueous rechargeable zinc batteries. *Nature Communications* 10: 4948.

98 Pei, A., Zheng, G., Shi, F. et al. (2017). Nanoscale nucleation and growth of electrodeposited lithium metal. *Nano Letters* 17: 1132.

99 Yang, X., Li, C., Sun, Z. et al. (2021). Interfacial manipulation via in situ grown ZnSe cultivator toward highly reversible Zn metal anodes. *Advanced Materials* 33: 2105951.

100 Qu, G., Wei, H., Zhao, S. et al. (2024). A temperature self-adaptive electrolyte for wide-temperature aqueous zinc-ion batteries. *Advanced Materials* 36: 2400370.

101 Wang, J., Zhang, H., Yang, L. et al. (2024). In situ implanting 3D carbon network reinforced zinc composite by powder metallurgy for highly reversible Zn-based battery anodes. *Angewandte Chemie International Edition* 63: e202318149.

102 Gao, Y., Cao, Q., Pu, J. et al. (2023). Stable Zn anodes with triple gradients. *Advanced Materials* 35: 2207573.

103 Li, T.C., Lin, C., Luo, M. et al. (2023). Interfacial molecule engineering for reversible Zn electrochemistry. *ACS Energy Letters* 8: 3258.

104 Yang, Y., Qu, G., Wei, H. et al. (2023). Weakly solvating effect spawning reliable interfacial chemistry for aqueous Zn/Na hybrid batteries. *Advanced Energy Materials* 13: 2203729.

Index

Electrode Materials in Energy Storage Technologies: Applications in Lithium-, Sodium-, Potassium-, Sulfur- and Zinc-Based Rechargeable Batteries, First Edition. Edited by Liqiang Xu.

r

Z

www.ingramcontent.com/pod-product-compliance
Lightning Source LLC
Chambersburg PA
CBHW080932240926
51849CB00003BA/434